Students' Manual

Biology A Functional Approach

M. B. V. Roberts MA PhD
Biology Department, Marlborough College

Nelson

Equipment and Materials

Most of the laboratory equipment and materials specified in this manual are obtainable from Philip Harris Biological Ltd. (Oldmixon, Weston-super-Mare, Somerset BS24 9BJ) with whom the author and publisher have been in consultation during the preparation of the manual. However, many of the items can be obtained from alternative sources and the teacher is advised to shop around. Useful pieces of apparatus can sometimes be scrounged from research laboratories.

General biological suppliers in the United Kingdom, apart from Philip Harris Biological Ltd., include Griffin and George Ltd. (Head Office: 285 Ealing Road, Wembley HA0 1HJ), and Gerrard and Haig Ltd. (Gerrard House, Worthing Road, East Preston, Sussex BN16 1AS). Gerrard and Haig Ltd. are sole agents in the United Kingdom for the Carolina Biological Supply Company, U.S.A.

Media for chick tissue culture can be obtained from Flow Laboratories, Victoria Park, Heatherhouse Road, Irvine, Scotland.

Marine specimens are generally available from Marine Biological Laboratories such as those at Plymouth and Oban.

© M. B. V. Roberts 1974
First published 1974
Reprinted 1975, 1976, 1977

ISBN 0 17 448002 4

Thomas Nelson and Sons Ltd.
Lincoln Way Windmill Road Sunbury-on-Thames Middlesex TW16 7HP
P.O. Box 73146 Nairobi Kenya

Thomas Nelson (Australia) Ltd.
19–39 Jeffcott Street West Melbourne Victoria 3003

Thomas Nelson and Sons (Canada) Ltd.
81 Curlew Drive Don Mills Ontario

Thomas Nelson (Nigeria) Ltd.
8 Ilupeju Bypass PMB 1303 Ikeja Lagos

Designer Phil Kay
Diagrams Colin Rattray and Associates
Diagrams © Thomas Nelson and Sons Ltd. 1974

Printed photolitho in Great Britain by
Ebenezer Baylis and Son Limited
The Trinity Press, Worcester, and London

Note to Teachers and Students

The purpose of this manual is to provide a repertoire of practical laboratory investigations and homework questions, and a source book of further information suitable for Advanced Level biology students in schools and colleges. It is a sequel to, and is designed to be used in conjunction with, my textbook *Biology: A Functional Approach* and its complementary series of slides (*see* Appendix). However, the manual is self-contained and can be used independently of its parent book.

In compiling this manual I lay little claim to originality. Most of the laboratory investigations have been attempted before and many of the questions and problems have featured in past examination papers. In selecting material I have been greatly influenced by what we, at Marlborough College, have found useful, since we adopted a functional approach some years ago.

Broadly speaking, laboratory work in biology falls into four categories: direct observation, dissection, microscopic study, and experimental work. There are, of course, no hard and fast distinctions between these disciplines, and a full investigation of a biological problem often involves all four. I believe that in a practical biology course all have an important part to play and should be given the right emphasis. I hope that in this manual I have achieved the correct balance. I make no apology for including a fair amount of morphological work, for a functional approach is only meaningful if it is based on sound structural principles, and nowhere can these principles be better acquired than in the laboratory.

Whatever the approach and whatever the techniques, the ultimate purpose of laboratory work is to explore and investigate the world of living things. Wherever possible, I have tried to present each unit of laboratory work in such a way that it can be seen as a genuine scientific investigation and not merely as a practical exercise to be done for the purpose of passing an examination.

All the investigations are designed to fit into clearly defined periods of time. Most of the experimental investigations can be carried out in $1\frac{1}{2}$ hours. The times for microscopic work and dissection will vary according to the aptitude and ability of the student. I have deliberately reduced experiments that are unduly time-consuming, expensive or capricious to a minimum.

It would, of course, be impossible for all the laboratory work suggested here to be completed in a two-year course, but I have included more than can be covered in the time so as to give teachers and students some degree of choice. The same applies to the questions and problems.

A common criticism of laboratory exercises, particularly experimental investigations that are tailor-made to fit into a prescribed period of time, is that they are contrived. This is bound to be so, but I hope that some of the investigations in this manual may stimulate the student to carry out project work on his own.

I strongly urge that the student should make his own drawings and notes and record his own observations and write up the experiments which he, himself, carries out. Advice on this is given in the appendix.

Many of the homework questions involve interpretation of biological phenomena and analysis of data. In the belief that learning to express ideas clearly and succinctly is an essential part of a biologist's training, I have also included a number of questions requiring the writing of a short essay. On the other hand, I have not included many questions demanding purely descriptive answers, on the grounds that such questions can easily be set by individual teachers.

In preparing this manual I have not had any one syllabus in mind, but the range of laboratory investigations and homework questions should make it suitable for students following the Advanced Level syllabuses of the various universities and examining boards in Great Britain and the Commonwealth, and for students pursuing introductory biology courses in universities and colleges of further education. It is envisaged that in the United States and Canada the manual would be suitable for students doing introductory biology courses in the first and second years at colleges and universities.

Acknowledgments

In preparing a book of this kind one is more than ever dependent on the constructive advice of one's friends. For their continual help and encouragement I must thank, first and foremost, my colleagues at Marlborough College: Mr Jack Halliday, Mr John Emmerson, Mr Malcolm Hardstaff and Mr Roderick Putman. Our present course is as much a result of their endeavours as mine, and without their help this manual would have been impossible. Mr Peter Holway of St Dunstan's College kindly commented on some of the problems, for which I wish to record my thanks; and for their helpful comments on the text I am grateful to Mr Beverley Heath and Mrs Morag Putman. I am also indebted to many of my students for testing the investigations, questions and problems, and for their valuable comments on the text. They include Mr David Howard, Mr Robert Wright, Mr Graham Fergusson, Mr Nigel Bruce, Mr Robin Bertaut, Mr Paul Goldsmith, Mr Mark Davies, and Mr Tony Brown, to all of whom I owe warm thanks. I am grateful to our laboratory staff, Mr Charles Hughes, Mr Ivor Radford, Mr Michael Ward, and Miss Lynn Thompson, for their technical assistance, and to Mr John Haller and Mr Stephen Wood of Philip Harris Biological Ltd. for their useful suggestions and cooperation over the provision of laboratory materials. For bearing the burden of typing the manuscript I am indebted to Mrs Jeanette Radford and also to my wife for her support and assistance. Finally I wish to thank my publishers, Thomas Nelson and Sons Ltd., particularly Miss Elizabeth Johnston and Mr David Worlock for their encouragement and courtesy at all times during the preparation of this manual.

In thanking all these helpers, I must make it clear that I alone am responsible for the many imperfections which doubtless remain. I hope that teachers and students will not hesitate to point these out to me as they use the manual at home and in the laboratory.

M. B. V. Roberts
June, 1974.

Examination questions

The author and publisher wish to thank the following examining bodies for permission to reproduce, either direct or in modified form, certain questions from their past examination papers. The sources are acknowledged in the manual by the abbreviations shown in parentheses. The questions are all A level or the equivalent.

Associated Examining Board (*AEB*); Biological Sciences Curriculum Study of the American Institute of Biological Sciences (*BSCS*); Cambridge Colleges Joint Examination (*CCJE*); Cambridge Local Examinations Syndicate (*CL*); Joint Matriculation Board (*JMB*); New South Wales Department of Education (*NSW*); Oxford and Cambridge Schools Examination Board (*O and C*); Oxford Colleges Joint Examination (*OCJE*); Oxford Delegacy of Local Examinations (*OL*); Public Examinations Board of South Australia (*SA*); Schools Board of Tasmania (*SBT*); University of London (*UL*); Victorian Universities and Schools Examinations Board (*VUS*); Joint Matriculation Board— Nuffield (*JMB—N*); Hong Kong University (*HK*).

Contents

1 Introducing Biology

Background Summary

1 **Biology**, the study of life and living organisms, is divided into numerous subjects which include **zoology, botany, microbiology** (bacteriology and virology), **anatomy, physiology, biochemistry, cytology** (cell biology), **heredity** (genetics), **molecular biology, behaviour** and **ecology**.

2 Properties shared by all living organisms are: **movement** (which may be internal), **responsiveness, growth** by internal assimilation, **reproduction** (involving replication of nucleic acid), **release of energy** by breakdown of adenosine triphosphate, and **excretion.**

3 **Scientific method** starts with an observation which leads to the formulation of an **hypothesis**. Predictions made from the hypothesis are tested by **experiment.** Every experiment must be accompanied by the appropriate **controls**.

4 Animal and plant species run into millions, so a system of classification (**taxonomy, systematics**) is essential.

5 Organisms as a whole are divided into the **animal** and **plant kingdoms**, within each of which they are further divided into **phyla, classes, orders, families, genera** and **species**. Bacteria, viruses and certain other microorganisms do not fit readily into either the animal or plant kingdoms and may be classified separately.

6 Basic biological concepts include survival, adaptation, a close relationship between structure and function, and evolution.

Investigation 1.1

Construction of an identification key

The diversity of organisms is prodigious and it is therefore necessary for each organism to be classified and named. They are classified according to their similarities with one another. Once a system of classification has been constructed, a method must be devised whereby other people can quickly determine the name of a particular organism. This is done by constructing an **identification key** based on the classification.

Principles involved

To illustrate how a classification and identification key can be constructed, consider the following example.

Nine students, all belonging to the same class in a school, have the following features. They are listed in alphabetical order:

Alan   — dark hair, blue eyes
Ann   — auburn hair, brown eyes
David   — dark hair, brown eyes
Elizabeth   — auburn hair, blue eyes
Jane   — fair hair, hazel eyes
John   — fair hair, brown eyes
Pamela   — auburn hair, green eyes
Philip   — fair hair, blue eyes
Susan   — fair hair, blue eyes

It would be possible to classify these students in various ways. A **dichotomous classification** is one which splits them into successive **pairs** of sub-groups of approximately equal size. Such a classification is given below and from it a dichotomous key can be constructed.

This enables a stranger, unfamiliar with the class, to quickly determine the name of any student.

1a male . . . . go to 2
1b female . . . . go to 5
2a fair hair . . . . go to 3
2b dark hair . . . . go to 4
3a blue eyes . . . . Philip
3b brown eyes . . . . John
4a blue eyes . . . . Alan
4b brown eyes . . . . David
5a fair hair . . . . go to 6
5b auburn hair . . . . go to 7
6a blue eyes . . . . Susan
6b hazel eyes . . . . Jane
7a blue eyes . . . . Elizabeth
7b eyes not blue . . . . go to 8
8a green eyes . . . . Pamela
8b brown eyes . . . . Ann

Using the key, determine the name of the brown-eyed boy with dark hair, and the blue-eyed girl with fair hair.

Exactly the same principles apply to the construction and use of an animal or plant key, as you will see if you consult a **flora**, that is a key to the plant kingdom.

Construction of classification

With these principles in mind, try classifying one of the groups of objects laid out before you. Each group contains between 15 and 20 objects, each identified by a **letter**. Using *visible* characteristics, arrange the objects into two sub-groups of approximately equal size. Then divide each of these sub-groups into two further groups. Continue to split the groups until each individual object is in a group of its own.

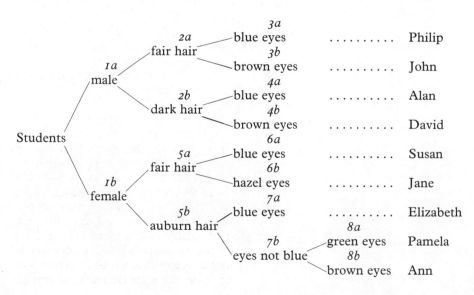

Construction of key

Now construct a dichotomous key based on your classification. As other people must be able to use your key, it should be based on visible characteristics which can be readily observed by anyone looking at the specimens. Resist the temptation to place specimens in a given group because you happen to know they belong there: i.e. do not use ready-made taxonomic groups such as 'plant', 'insect', 'bird', 'reptile', etc.

If there is any ambiguity about the meaning of a term used in describing a specimen, write out a definition and incorporate it into your key.

When completed, invite someone to test your key: choose a specimen at random, cover its identifying letter, and see if the specimen can be identified correctly, using only the key you have constructed.

For Consideration

Animals and plants in existence today are believed to have evolved from pre-existing ancestors by a process of gradual change. A natural classification is designed to show how closely, or distantly, organisms are related to each other in evolution. With this in mind, try to answer these questions:
(1) To what extent do you think your classification reflects evolutionary relationships?
(2) What explanation can you offer of the fact that some organisms are strikingly similar, others different?
(3) How would you explain the fact that in a natural classification some organisms that appear to be similar are placed in different groups?

Requirements

For each group of students: approximately 16 objects, animate or inanimate, each labelled with a letter. The range of objects should be such that they can be readily classified dichotomously.

Investigation 1.2

Animal and plant classification

In the previous investigation you based your classification on easily observable features. Although this is satisfactory for purposes of 'pigeon-holing' and identification, it can be scientifically misleading for it may result in some organisms being grouped together on the basis of some superficial similarity when, in fact, they have little in common.

Animals and plants are believed to have arisen by a process of **evolution**. Those that are closely related in this evolutionary process would be expected to share certain fundamental features in common, whereas those that are distantly related would not be expected to do so.

In classifying organisms a biologist aims to group them according to their evolutionary closeness to one another. The science of classification is known as **taxonomy**.

THE TAXONOMIC HIERARCHY
Organisms which share detailed features in common, but which do not normally interbreed, are grouped together as a **species**. A **genus** is a somewhat larger group which includes additional organisms, similar to one another in many respects, but not sufficiently close to merit putting them in the same species. It is customary to name organisms by their genus and species: the **generic name** is written first, followed by the **specific name**. Both are normally printed in italics, or underlined. The generic name begins with a capital letter, the specific name with a lower case letter. Thus, the proper name for man in *Homo sapiens*, the earthworm is *Lumbricus terrestris*, and the common buttercup is *Ranunculus acris*.

Organisms are grouped together into progressively larger groups, creating a kind of hierarchy. Thus genera are grouped together into **families**, families into **orders**, orders into **classes**, classes into **phyla**, and phyla into **kingdoms**. Intermediate divisions are sometimes used, for example, between a phylum and class, but we will not concern ourselves with those here.

Two kingdoms are generally recognized: the **animal kingdom** and **plant kingdom**.

It follows that, as one progresses *down* the hierarchy, the smaller the number of organisms belonging to each group and the more they have in common. Thus a phylum may contain a large number of organisms, held together by certain fundamental features but at the same time displaying a wide range of variety. On the other hand, the different members of a genus may be so similar as to be virtually indistinguishable except by an expert.

Taxonomic group	Plant example	Animal examples	
Kingdom	Plant	Animal	Animal
Phylum	Angiospermae/Tracheophyta	Annelida	Chordata
Class	Dicotyledon/Angiospermae	Oligochaeta	Mammalia
Order	Ranales	Terricolae	Primates
Family	Ranunculaceae	Lumbricidae	Hominidae
Genus	Ranunculus	Lumbricus	Homo
Species	bulbosus	terrestris	sapiens
Common name	bulbous buttercup	earthworm	man

Table 1.1 Three examples of classification from the plant and animal kingdoms.

EXAMPLES OF CLASSIFICATION
Three examples of classification are given in Table 1.1. The features on which the main phyla and classes of the animal and plant kingdoms are based are summarized in the classification of the animal and plant kingdoms (*see* pp. 400–417).

There is sometimes dispute as to what features a group should be based on. Thus some botanists consider that the phylum to which the buttercup belongs should be the **Angiospermae**, characterized by the possession of **flowers**; others consider that such plants should be placed in a phylum called the **Tracheophyta**, characterized by the possession of **vascular tissues** (*see* p. 406). Although, in the last analysis, this is a matter of opinion, it does make a difference because the former would embrace *only* the flowering plants, whereas the latter would also include the conifers and ferns (*see* p. 416).

CLASSIFICATION OF A PLANT GROUP
Examine different species of a common plant growing in your part of the world. As an example we can take the genus *Ranunculus* which includes the buttercup and its relatives.

Twelve species of *Ranunculus*, together with their common names, times of flowering, and occurrence, are listed in Table 1.2. Examine specimens of some or all of these species, noting their similarities and differences. From your observations can you say what features cause them

to be placed by botanists in the same genus? What are the differences which cause them to be placed in separate species?

Ranunculus belongs to the family **Ranunculaceae** (*see* Table 1.1). This contains a number of other genera besides *Ranunculus*. These include:

Anemone, e.g. *A. nemorosa*: wood anemone

Caltha, e.g. *C. palustris*: marsh marigold★

Clematis, e.g. *C. vitalba*: traveller's joy, old man's beard

Delphinium, e.g. *D. ambiguum*: larkspur

Helleborus, e.g. *H. viridis*: green hellebore

Paeonia, e.g. *P. mascula*: paeony

Examine specimens of some or all of the above genera. What features do they have in common with *Ranunculus*, and how do they differ from it? The various members of the Ranunculaceae are in fact held together by the structure of their flowers. What features of the flowers do they all have in common?

★ *C. palustris* is also known as king-cup, golden cup, brave celandine, horse-blob, mare-blob, may-blob, Mary-bud, soldier's button, and publicans and sinners. In parts of the U.S.A. it is called cowslip, a name which elsewhere is applied to *Primula veris*, a member of the Primulaceae. This is a splendid example of how misleading the use of common names can be.

Proper name	Common names	Time of flowering	Occurrence
R. acris	Meadow buttercup Common meadow buttercup Tall buttercup Common buttercup	April–September	sloping meadows
R. aquatilis	Water crowfoot	May–August	rivers
R. arvensis	Corn buttercup	May–July	cornfields in calcareous soil
R. auricomus	Goldilocks Wood crowfoot	April–July	thickets and woods
R. bulbosus	Bulbous buttercup	May–August	dry fields and meadows
R. flammula	Lesser spearwort	May–September	wet areas
R. ficaria	Lesser celandine	March–May	shade
R. fluitans	Water crowfoot River crowfoot	June–August	fast-flowing streams
R. lingua	Great(er) spearwort	June–September	marshes, fens, ditches
R. parviflorus	Small-flowered buttercup	May–July	short grass, arable land
R. repens	Creeping buttercup	May–August	wet meadows in valleys
R. sceleratus	Celery-leaved buttercup	May–September	sides of ponds and ditches

Table 1.2 Names, times of flowering, and occurrence of 12 species of *Ranunculus*.

The Ranunculaceae is a family within the phylum **Angiospermae**. This contains many other families besides the Ranunculaceae. These include the following:

Compositae (daisy, etc.)
Convolvulaceae (bindweed, etc.)
Cruciferae (wallflower, etc.)
Geraniaceae (geranium, etc.)
Labiatae (deadnettle, etc.)
Leguminosae (pea, bean, etc.)
Primulaceae (primrose, etc.)
Rosaceae (rose, etc.)
Scrophulariaceae (snapdragon, etc.)
Violaceae (violet, etc.)

Examine representatives of some or all of the above families. In what respects do they differ from the Ranunculaceae, and from each other? What features hold them together and cause them to be placed in the same phylum?

CLASSIFICATION OF OTHER ORGANISMS

What you have done in this exercise is to look at a **genus**, then at the **family** to which that genus belongs, and finally at the **phylum**.

This can be done with any organism, though the procedure is easier for some organisms than for others. By way of contrast, try it with the earthworm (*see* Table 1.1). In this case the genus is *Lumbricus*. First look at different species of *Lumbricus*; then look at other worms belonging to the same

Requirements

Ranunculus species (*see* Table 1.2)
Selection of genera of Ranunculaceae
Selection of angiosperms
Other organisms and groups as required.

class (**Oligochaeta**). Then broaden your survey further to include the whole phylum (**Annelida**). This includes leeches, fanworms, ragworms and lugworms, as well as the earthworm and its relatives (*see* p. 404). What features do the various groups have in common? How do they differ?

For Consideration

(1) It is a basic biological principle that organisms are adapted to their environments. In what respects are the various species of *Ranunculus*, which you have examined, adapted to their respective environments?

(2) Construct a dichotomous key enabling the different species of *Ranunculus* to be identified (*see* Investigation 1.1).

(3) What problems might arise if animals and plants were only called by their common names? (Consult Table 1.2 and the footnote on p. 4).

Investigation 1.3

Who's who in the animal and plant kingdoms

The animal and plant kingdoms are subdivided into **phyla**, which in turn are subdivided into smaller groups. The members of each phylum, though often displaying considerable diversity of form, are held together by certain features which they all possess.

The purpose of this investigation is to examine representatives of each major phylum of the animal and plant kingdoms. In doing this, you are urged to notice the variety within each phylum, but also the more obvious features uniting its various members.

Classifications of the animal and plant kingdoms are given on pp. 400–417. Compare your observations with the information given in these classifications.

Procedure

ANIMAL KINGDOM
Examine several representatives of each of the following phyla: **Protozoa** (single-celled animals), **Porifera** (sponges), **Coelenterata** (*Hydra*, etc.), **Platyhelminthes** (flatworms), **Nematodes** (thread worms, round worms), **Rotifera** ('wheel animalcules'), **Annelida** (ringed worms, earthworms, etc.), **Mollusca** (snails, etc.), **Arthropoda** (insects, etc.), **Echinodermata** (starfish, etc.), **Chordata** (vertebrates, etc.).

From your own observations try to determine the criteria upon which each group of animals are combined into one phylum. Then check against the classification in Appendix 1.

PLANT KINGDOM
Examine several representatives of each of the following phyla: **Thallophyta** (including representative **Algae** and **Fungi**), **Bryophyta** (mosses, etc), **Pteridophyta** (ferns, etc.), **Spermatophyta** (including representative **gymnosperms** and **angiosperms,** coniferous and flowering plants respectively).

As with the animal kingdom, try to determine, *from your own observations*, the criteria on which the different plants are grouped together into phyla.

For Consideration

(1) How would you explain the similarities that exist between the members of a phylum?

(2) How would you explain the fact that some phyla appear to have more in common than others?

(3) From the organisms which you have looked at, give examples of cases where the animal or plant is clearly adapted to its particular habitat and way of life.

(4) Give examples of cases where there is a clear correlation between a specific structure possessed by an organism, and its function.

Requirements

Live and/or preserved animals and plants belonging to the phyla listed above. (*See also* Classification of the Animal Kingdom, pp. 400–412, and Classification of the Plant Kingdom, pp. 413–417).

Questions and Problems

1 What are the basic properties of living things? What arguments would you present to show that (a) a crystal, and (b) a candle flame are not alive, and that an oak tree is alive?

2 What characteristics of life are exploited in:
(a) the use of flashing signs;
(b) giving a green plant extra light;
(c) manuring soil;
(d) the use of a carrier pigeon;
(e) making wine?

3 'The body was found in the neighbouring water meadows. Beside it was a rusty and bloodstained iron bar. At the inquest it was stated that the victim was Fortescue-Watson, a member of Merlberry College, and that death resulted from a series of blows on the back of the head with a blunt instrument.

'The police searched the area of the crime and the only clue they found was two sets of footprints in the mud; one of these fitted the victim's shoes, while the other had been made by shoes with steel quarter heels and a hole in the left sole. A search was made, and a pair of shoes belonging to Snooks, a member of the same school, was found to fit the footprints.

'An examination of Snook's clothes was then made and showed recent mud stains on the trousers, seeds of a weed currently in flower in the water meadows in the turn-ups, and on the jacket bloodstains which had been treated, but not obliterated, by the application of ammonia.'

Discuss how far the police investigations in this story illustrate the method of science, and point out any differences in method between these investigations and those carried out by a scientist studying natural phenomena in a laboratory.

Do you consider that, on the evidence given, Snooks' guilt is proved? If not, explain why you consider the evidence to be insufficient. What other investigations should the police carry out?

4 Put forward hypotheses to explain the following observations, and suggest how you might test your hypotheses experimentally:
(a) Plants are sometimes seen to droop.
(b) Woodlice are generally found under logs or stones, rarely in the open.
(c) The cut shoot of a water weed often exudes a continuous stream of bubbles.
(d) Mosses occur in greater abundance on north-facing than south-facing walls.
(e) In a certain type of malaria fever occurs at regular 48-hour intervals.

5 Explain fully how you would test the hypothesis that:
(a) the apical bud is essential for vertical growth of the main stem of a flowering plant;
(b) vitamin B_1 (thiamine) is required for the growth of hens.

6 In answering the following question refer, if necessary, to the classification of the animal and plant kingdoms on pp. 400–417.

Place each of the following organisms into its correct kingdom, phylum and sub-group within the phylum. In each case state what you consider to be the most easily observed characteristic which shows what phylum and sub-group the organism belongs to.

Clam, man, bread mold, crab, pine, fanworm, buttercup, *Amoeba*, moss, bee.

7 From your own observations give one example of (a) an animal, and (b) a plant that is strikingly adapted to a particular way of life.

2 Structure and Function in Cells

Background Summary

1 Cells were first described in 1665 by Robert Hooke and are now known to be of almost universal occurrence in organisms. The **cell theory** states that the cell is the basic unit of an organism, the whole organism being little more than a collection of independent cells; the rival **organismal theory** states that the whole organism is the basic unit, the cells being incidental sub-units with no independent life of their own.

2 Cells may be observed with various kinds of **microscope**, e.g. optical (light); phase-contrast, polarizing and electron microscopes. A typical light microscope magnifies about 800 times and has a resolving power of approximately 0.2 μm. The electron microscope can magnify objects 300,000 times and has a resolving power of approximately 1.0 nm. (Compare the naked eye whose resolving power is about 0.1 mm.)

3 Structures characteristic of animal cells: **nucleus** with nuclear membrane perforated by pores; nucleolus and chromatin granules; **cytoplasm** with endoplasmic reticulum, food granules (glycogen, lipids), ribosomes and/or polyribosomes (polysomes), Golgi body, secretory vesicles and granules, mitochondria, lysosomes, centrioles and microtubules; **cell membrane (plasma membrane)** from which may project microvilli, cilia, or flagella with basal bodies.

4 Typical plant cells differ from animal cells in lacking cilia, flagella and centrioles; and in possessing chloroplasts, starch grains (instead of glycogen), sap vacuole and cellulose wall. The cellulose is laid down on the inside of a primary wall consisting largely of calcium pectate, the latter being represented in mature cells by the middle lamella. The secondary wall may sometimes be absent locally, giving rise to a pit, and adjacent cells are linked by plasmodesmata.

5 The **plasma membrane**, approximately 7.5 nm thick, is thought to consist of a layer of lipid sandwiched between two layers of protein. There is evidence that it is perforated by pores of less than 1.0 nm diameter.

6 Though containing much in common, cells show considerable diversity in their contents, shape and functions. In all cases there is a close relationship between cell structure and function.

Investigation 2.1
Structure of cells

The various structures that often are seen crammed into a theoretical diagram of a generalized cell are not all visible in any one cell. So to piece together the structure of a 'typical' cell—animal or plant—it is necessary to look at more than one type of cell.

To see a particular structure it may be necessary to stain the cell. The choice of stain is important because certain stains are specific to certain structures; thus aceto-carmine stains the nucleus and its contents, iodine stains starch grains, and so on.

Procedure

(1) Gently scrape the inside of your cheek with a spatula and mount the scrapings in a drop of water on a microscope slide. Cover with a coverslip. Observe under the microscope. (If you have not used a microscope before, consult Appendix 7). Locate a **single cell** and examine it under high power. Many of the cells will be crumpled and irregular in outline because the cell membrane is extremely thin and delicate.

Make out as much as you can of the **nucleus** and **cytoplasm**. You will find this fairly easy provided you don't let too much light through the microscope. Try dark ground illumination and, if available, phase contrast. How do these different techniques affect what you observe?

To get a better picture of the nucleus, stain with methylene blue. The stain may be introduced by a technique called **irrigation** (Fig. 2.1). A drop of stain is placed on the slide so it just touches the edge of the coverslip. Fluid is then withdrawn from the opposite side of the coverslip by a piece of absorptive paper and the stain flows in, replacing the fluid removed by the paper.

Sketch the cell, putting in as many structures as you have been able to observe. Label: **nucleus, nuclear membrane, chromatin granules, nucleolus, granular cytoplasm, cell membrane**.

(2) Strip off a piece of epidermis from one of the inner 'fleshy scales' of an onion, mount in iodine and observe one cell under low and high powers. In addition to the nucleus, observe the distribution of granular cytoplasm surrounding the **vacuole**. Also notice: **cellulose cell wall, nucleoli** (how many?), **chromatin granules**. The onion is a plant organ but it does not contain chloroplasts: explain. What can you say about the 3-dimensional shape of the cells?

(3) Mount a small leaf of Canadian pondweed, *Elodea*, in water and examine its cells under high power. The cells are so packed with **chloroplasts** that little else can be seen. Look for **cytoplasmic streaming** resulting in circulation of the chloroplasts. What do you suppose the function of this is?

(4) Streaming of the cytoplasm can be seen more clearly in cells of the staminal hairs of *Tradescantia*. Open up one of the flowers and remove a stamen. Mount the stamen in water and examine a hair under high power. Adjust the illumination carefully or, better still, use phase contrast. What might be the function of streaming in this case? Make sketches to illustrate.

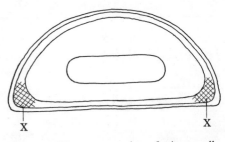

Fig. 2.2 Transverse section of pine needle. The cells at the corner (marked X in the diagram) have thick cellulose walls perforated by channels connecting adjacent cells.

(5) Examine the large cells at the corners of a transverse section of a pine needle (Fig. 2.2). Notice the thick **cellulose walls** which have been laid down in layers. The thin line separating the cellulose walls of adjacent cells, the **middle lamella**, is clearly visible. What does it represent? Fine channels in the cellulose walls connecting adjacent cells may be seen. What do they represent?

These observations can be performed either on prepared slides or on sections which you have cut yourself (*see* p. 96). If you are cutting your own sections mount them in either fresh Schultz' solution or FABIL. Both stain cellulose purple.

(6) Many plant cells store starch in the form of **starch grains**. Nowhere can this be better seen than in a potato, a swollen stem (stem tuber) specially adapted as a plant storage organ. Scrape some tissue from the cut surface of a potato and mount in water. Observe starch grains under high power.

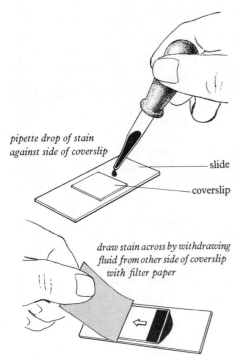

pipette drop of stain against side of coverslip

— slide

— coverslip

draw stain across by withdrawing fluid from other side of coverslip with filter paper

Fig. 2.1 The technique of irrigation.

Requirements

Slides and coverlips
Spatula
Optical microscope if possible
  with phase contrast
Filter paper

Iodine
Methylene blue solution
Schultz' solution or FABIL
Noland's solution

Onion
Elodea (Canadian pondweed)
Tradescantia flowers
Potato
Euglena
Pine needles and material for
  section-cutting, or prepared
  transverse sections of pine
  needles.

Now irrigate with iodine and watch the starch grains turn blue. Notice that the starch grains are located inside tightly packed hexagonal cells.

(7) Examine *Euglena* or some other comparable unicellular flagellate, under high power, using low illumination or phase contrast, and watch the **flagellum** in action. Then irrigate with Noland's solution which fixes flagella and stains them blue.

For Consideration

(1) From your observations made in this investigation, draw up a table comparing the structure of animal and plant cells.

(2) Make a list of the structures visible in the electron microscope but which you have been *unable* to see in this investigation.

Investigation 2.2

Examination of some specialized cells

In the previous investigation we were concerned with those features which are common to cells in general. In this investigation we will examine seven cells which all depart from the common pattern in some way and show various degrees of specialization. Four of the cells are free-living unicells (protists), which are specialized in the sense that all their life processes have to be carried out within the one cell. The other three cells belong to multicellular organisms.

Procedure

EUGLENA

Examine the unicellular flagellate *Euglena* by pipetting a very small drop of culture onto a slide and putting on a coverslip. Watch the organism's method of locomotion and record your observations. Observe a stationary specimen under high power. How many of the structures shown in Fig. 2.3A can you see? Be honest! Try to make out: **canal, pigment spot, contractile vacuole, flagellum, nucleus, chloroplasts, cytoplasmic granules**. How would you account for the difference in detail between your observations and the diagram in Fig. 2.3A?

How do you think *Euglena* feeds? In some older texts the canal is referred to as a gullet. However, there is no evidence that *Euglena* takes in particulate food either through the canal

or any part of the body surface. But in the absence of light *Euglena* feeds saprophytically on *soluble* organic food which it absorbs through the pellicle.

Euglena swims towards light. The mechanism guiding it towards light depends on the opaque **pigment spot** working in conjunction with the **light-sensitive swelling** at the base of the flagellum. From your observations on the way *Euglena* swims, can you offer an explanation as to how the light-directing mechanism works?

With filter paper carefully withdraw water from beneath the coverslip. Unable to swim freely by flagellate locomotion, the organisms may resort to a worm-like wriggling movement (**euglenoid locomotion**). What sort of structures do you imagine might be responsible for this type of movement?

If you want to examine the **flagellum** in detail, irrigate with Noland's solution. This fixes flagella and stains them blue.

CHLAMYDOMONAS

Examine *Chlamydomonas*, another unicellular flagellate, under high power and note: two **flagella**, cellulose **cell wall**, two small **contractile vacuoles**, **cytoplasm**, cup-shaped **chloroplast**, **pigment spot**, **pyrenoid** (Fig. 2.3B).

More detailed information may be obtained by staining: irrigate with methyl green in acetic acid and/or with acetocarmine to see the **nucleus**;

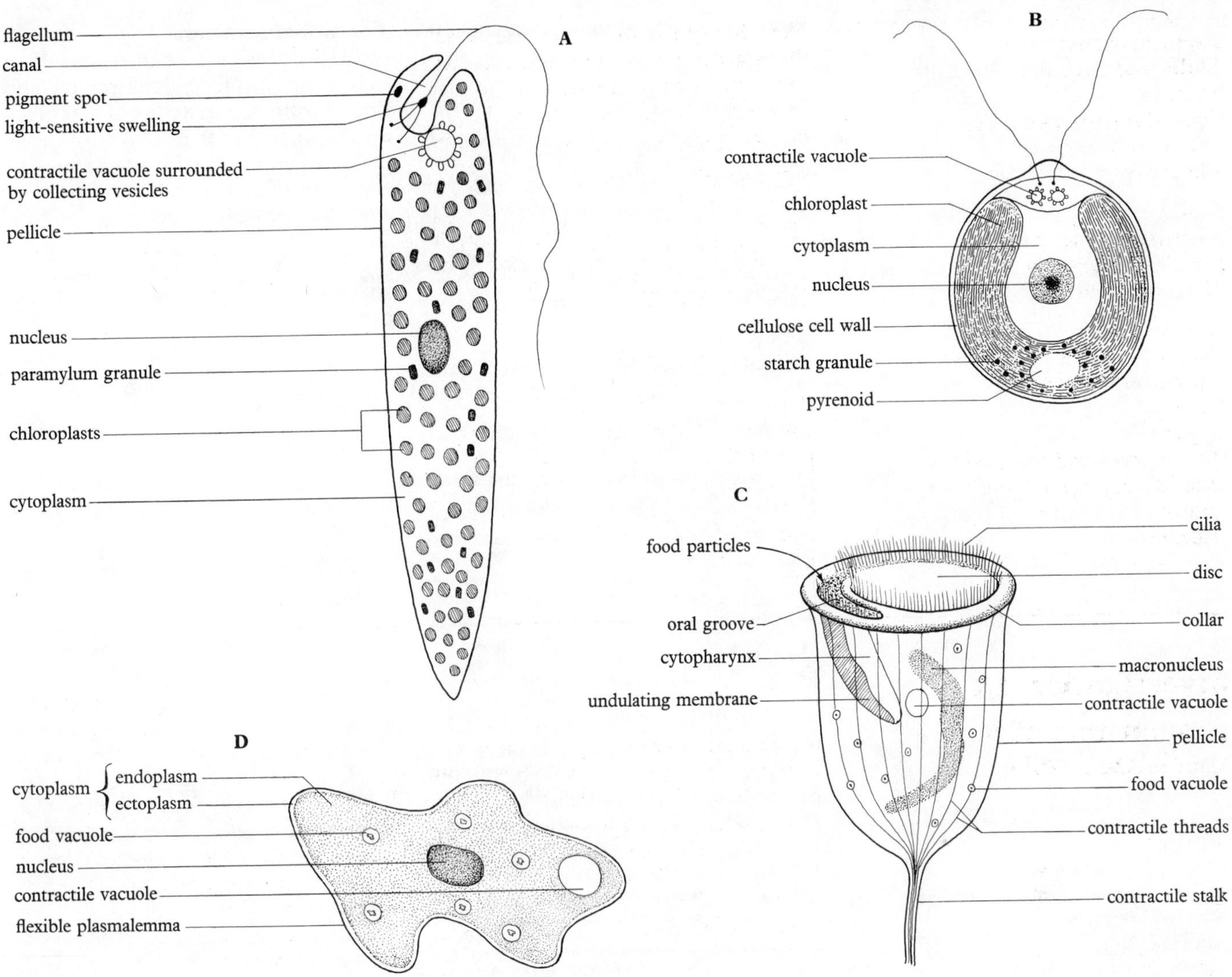

A

flagellum
canal
pigment spot
light-sensitive swelling
contractile vacuole surrounded
by collecting vesicles
pellicle

nucleus
paramylum granule

chloroplasts

cytoplasm

B

contractile vacuole
chloroplast
cytoplasm
nucleus
cellulose cell wall
starch granule
pyrenoid

C

food particles
oral groove
cytopharynx
undulating membrane

cilia
disc
collar
macronucleus
contractile vacuole
pellicle
food vacuole
contractile threads
contractile stalk

D

cytoplasm { endoplasm
            ectoplasm
food vacuole
nucleus
contractile vacuole
flexible plasmalemma

Fig. 2.3 Four representative free-living uni-cells (protists). **A** *Euglena*, based on electron micrographs by Leedale; **B** *Chlamydomonas*, based on electron micrographs by Vickerman; **C** *Vorticella*, (*based on* Bullough); **D** *Amoeba*.

with Noland's solution to see the **flagella**; with iodine to see **starch grains** in the vicinity of the pyrenoid.

VORTICELLA
Examine the stalked ciliate *Vorticella* under both low and high powers (Fig. 2.3C). This unicellular organism is a common sessile inhabitant of dirty ditches, where it is found attached to weeds and stones; it is also commonly found attached to the head region of the fresh water shrimp *Gammarus*, and other animals. Observe: spiral rows of **cilia** whose beating draws small food particles into the **oral groove, body**, elongated **macronucleus**; **undulating membrane** of fused cilia in **cytopharynx**; **food vacuoles**, contractile **stalk**. Stain as necessary.

AMOEBA
Mount an *Amoeba* in a drop of water on a slide under a coverslip supported by a strip of paper. Ex-

amine under low and high powers (Fig. 2.3D). Notice how this uni-cellular animal changes in shape as it moves. Notice: flowing of the granular **endoplasm**, the clear **ectoplasm**, **pseudopodia, food vacuoles** (how are they formed?), **nucleus, contractile vacuole** (can you see it collapsing periodically?). Irrigate with aceto-carmine which fixes the animal and stains the nucleus.

BLOOD CELLS
Amoeba is a free-living unicell, but amoeboid cells are found in the bodies of multicellular animals. Examine **white blood corpuscles** in a prepared smear of human blood. Observe the **granulocytes** (*see* p. 163), which in life move by amoeboid loco-motion and engulf bacteria by phago-cytosis. What other unusual feature is apparent in these cells?

While looking at the blood smear notice the **red blood corpuscles**.

How do they differ from normal animal cells?

PIGMENT CELLS

Look carefully at a prepared slide of **frog's skin** under high power. Immediately beneath the epidermis you will see black **pigment cells** (**chromatophores**). What special features do these cells possess? What is their function?

SPIROGYRA CELL

Mount a filament of the fresh water alga *Spirogyra* in water and observe under low and high powers.

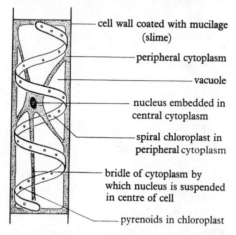

cell wall coated with mucilage (slime)

peripheral cytoplasm

vacuole

nucleus embedded in central cytoplasm

spiral chloroplast in peripheral cytoplasm

bridle of cytoplasm by which nucleus is suspended in centre of cell

pyrenoids in chloroplast

Fig. 2.4 Diagram of a cell of the filamentous green alga *Spirogyra*. To what extent does your own specimen conform to this diagram?

To what extent does the structure of its cells agree with the diagram in Fig. 2.4? Observe one of its elongated cells and note: **cell wall**, spiral **chloroplast(s)** with **pyrenoids**; **cytoplasm**; **nucleus** suspended in centre of **vacuole** by **cytoplasmic bridles**. What is the three-dimensional shape of the cell?

Investigate the detailed structure of the cells by staining with acetocarmine for the **nucleus**, with methylene blue for the **mucilaginous cell wall**, with iodine for **starch grains** formed in the vicinity of the pyrenoids.

Other animal and plant cells are dealt with elsewhere.

For Consideration
(1) How does each of the cells which you have examined conform to, and depart from, the structure of a 'typical' animal or plant cell?
(2) Do you agree with the statement that unicellular organisms such as *Euglena* provide examples of cell specialization?
(3) From your observations can you suggest a possible function of the pyrenoids in *Chlamydomonas* and *Spirogyra*?

Requirements
Microscope
Slides and coverslips
Filter paper

Iodine
Methylene blue solution
Noland's solution
Methyl green in acetic acid
Acetocarmine

Euglena
Chlamydomonas
Vorticella
Amoeba
Spirogyra
Human blood smear stained with
 Leishman's or Wright's stain
WM or VS frog skin

Questions and Problems

1 Explain the difference between the cell theory and organismal theory. List the evidence for and against each.

2 What is the difference between magnifying power and resolving power of a microscope? How do the optical and electron microscopes compare as regards magnification and resolution, and how would you explain the difference?

3 Distinguish between (a) chromatin granules and chromosomes; (b) rough and smooth endoplasmic reticulum; (c) ribosomes and polyribosomes (polysomes); (d) cell membrane and cell wall; (e) leucoplast and chloroplast.

4 A drawing of the fine structure of part of a certain animal cell is shown in Fig. 2.5. Comment.

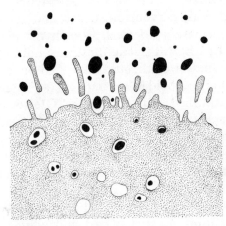

Fig. 2.5 Drawing of an electron micrograph of part of a certain animal cell.

5 Draw a diagram of a generalized *plant* cell and label it fully. Include structures that can only be seen with an electron microscope, as well as those that can be seen with the optical microscope. Write *A* against the labels corresponding to structures that are found in animal as well as plant cells.

6 The Davson–Danielli hypothesis explaining the structure of the cell membrane postulates that the lipid (fat) 'core' of the membrane consists of two layers of lipid molecules whose polar (water-soluble) ends point outwards and hydro-carbon chains inwards. What physico-chemical evidence supports this idea? What criticisms can you level against the Davson–Danielli hypothesis?

7 What features do the following cells possess over and above those that are common to cells generally: (a) chromatophore; (b) spermatozoon; (c) smooth muscle cell; (d) *Euglena*; (e) musculo-epithelial cell of *Hydra*? In each case explain how the cell is adapted to perform its particular functions.

8 Certain cells have arm-like processes projecting from them. Give examples of such cells and in each case relate the presence of their arm-like extensions to the functions which these cells perform.

3 Tissues, Organs and Organization

Background Summary

1 Cells are united to form **tissues**, and tissues to form **organs**.

2 **Animal tissues** may be classified into epithelial, connective, skeletal, blood, nervous, muscular, and reproductive; **plant tissues** into meristematic, epidermal, parenchyma, photosynthetic, mechanical and vascular.

3 **Epithelial tissue** illustrates how a tissue can be built up. The different types of epithelium, classified on the basis of their cellular composition, are: cuboidal, pavement (squamous), columnar, ciliated, glandular, stratified and transitional. **Glandular epithelium** may be invaginated and folded to form different kinds of glands.

4 **Connective tissue** consists of cells and fibres embedded in an organic ground substance or matrix. It is classified according to its inclusions into areolar, collagen (white fibrous), elastic (yellow elastic) and adipose (fatty) tissue. **Skeletal tissue** (cartilage and bone) is also composed of cells embedded in a matrix but in this case the matrix is harder.

5 **Blood** consists of red and white cells (corpuscles) and platelets; and **muscular tissue** is divided into visceral (smooth), cardiac (heart) and skeletal (striated) muscle.

6 Most animal tissues have their counterparts in plants. Thus plant **epidermal tissue** is equivalent to epithelium, **mechanical tissues** (collenchyma and sclerenchyma) to skeletal tissue. Plant **vascular tissue** is equivalent to the circulatory system of animals. **Photosynthetic tissue** is unique to plants.

7 The chief differences between animal and plant tissues (and between animals and plants generally) can be related to their different methods of nutrition.

8 Organs are generally interrelated to form **organ systems** in which several different organs co-operate and work together to fulfil a single overall function.

9 In organisms three different **levels of organization** are recognized: the unicellular, tissue and organ levels. Most animals are organized on an organ basis; some animals (e.g. *Hydra*) and most plants are constructed on the tissue level; and a wide range of organisms are unicellular.

10 **Unicellular organisms**, though consisting of only one cell, are by no means simple as is shown by an examination of *Paramecium*. *Paramecium* displays an astonishingly high degree of intracellular complexity.

11 Some organisms, e.g. sponges, can be regarded as **colonies of cells** showing little or no co-operation and integration between cells.

12 The **multicellular state** carries with it several advantages but it also carries certain disadvantages.

Investigation 3.1

Examination of some basic animal tissues

The purpose of this laboratory work is to introduce you to fundamental animal tissues which you will encounter repeatedly in later microscopic studies.

Animal tissues can be investigated by mounting a small piece of living tissue on a slide and then observing it unstained or stained; by making a permanent preparation of it; or by examining a stained section.

Procedure

CUBOIDAL EPITHELIUM

Examine a prepared section of thyroid gland and observe the **cuboidal epithelial cells** lining the follicles (*see* p.189). Alternatively examine a prepared section of the kidney and notice cuboidal cells lining the tubules and collecting ducts (*see* p.140). In common with other types of epithelium, the cells rest on a non-living **basement membrane**.

PAVEMENT (SQUAMOUS) EPITHELIUM

Your slide of thyroid or kidney should also show **pavement epithelial cells in section**. These will be seen lining small blood vessels and, in the case of kidney, Bowman's capsules. The cells are generally so flat that except in the region of the nuclei they appear as no more than a thin line (Fig. 3.1).

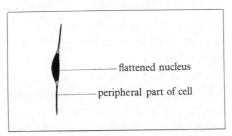

Fig. 3.1 A typical pavement epithelical cell as it appears in section under the optical microscope.

Mount a smallish piece of frog peritoneum or pericardium in water on a slide, or examine a stained preparation of mesothelium, and observe **pavement epithelial cells in surface view**. The peritoneum is the thin membrane lining the body cavity, the pericardium is the thin membrane surrounding the heart. Both are known as **mesothelium** because they line internal cavities and are derived from mesoderm. Note how the cells fit together.

COLUMNAR EPITHELIUM

Examine a prepared section of mammalian small intestine (*see* p. 80) and observe the inner lining under high power. Notice **columnar epithelial cells** interspersed with goblet mucus-secreting cells (*see* below). Examine a representative columnar cell in detail, noting its shape and **brush border** on its free surface (Fig. 3.2a). What does the brush border represent?

Columnar epithelium can also be seen lining the gall bladder: in this case no goblet cells are present.

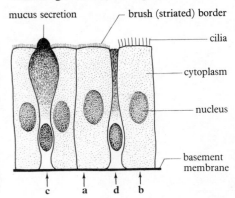

Fig. 3.2 Three types of cell commonly found in columnar epithelium. **a**, normal columnar cell with brush (striated) border on its free surface; **b**, ciliated columnar cell; **c, d**, mucus-secreting goblet cells before and after discharge of mucus.

CILIATED EPITHELIUM

Examine macerated **ciliated epithelium** under high power and observe **cilia** projecting from one of the surfaces of each cell.

Examine a prepared section of trachea and observe its inner lining under high power. Notice cilia projecting from the free surface of columnar epithelial cells (Fig. 3.2b). As in the small intestine, mucus-secreting goblet cells are present (Fig. 3.2c,d).

Ciliated epithelium can also be seen lining the roof of the mouth cavity of the frog.

GLANDULAR EPITHELIUM

If secretory cells are present in an epithelial tissue, the latter qualifies as **glandular epithelium**. The inner lining of the small intestine is, therefore, glandular. Examine, using oil-immersion if possible, one of the **secretory cells** lining the bottom of a crypt of Lieberkühn, which is seen in the wall of the ileum (*see* p. 80). Secretory cells usually have prominent granules in the secretory area of the cell.

The epithelial lining of the small intestine also contains mucus-secreting **goblet cells**. Examine one in detail, if necessary using oil-immersion. The mucus droplets accumulate at the free end of the cell, giving rise to a clear cup-shaped region—which is why they are called goblet cells. The goblet appearance is further enhanced by the fact that the end of the cell containing the mucus may be swollen, the remainder of the cell being constricted. The exact shape of the cells depends on their age and whether or not active secretion has taken place (Fig. 3.2c,d).

Glandular epithelium is even better demonstrated by the large intestine (colon or rectum) where goblet cells greatly outnumber the supporting columnar epithelial cells.

Glands are formed by invagination of glandular epithelium. Examine a crypt of Lieberkühn in the mammalian small intestine as a representative **simple tubular gland** (*see* p. 80), and the flask-shaped glands in the skin of the frog as representative **simple saccular glands** (Fig. 3.3). What kind of epithelial cells line each of these glands and what are their functions?

STRATIFIED EPITHELIUM

Examine a prepared vertical section of frog skin and observe the **epidermis** (Fig. 3.3). This is composed of **stratified epithelium**. Notice the **formative (Malpighian) layer** at the base, above which are successive layers of increasingly flattened cells. You will find the nuclei easy to detect but the cell membranes are more difficult to see.

Examine a vertical section through the wall of the mammalian bladder. How does its inner lining differ from the epidermis of the frog? It is known as **transitional epithelium**: the cells are approximately the same size and can change their shape according to circumstances. Transitional epithelium lines tubes and cavities that are liable to stretching.

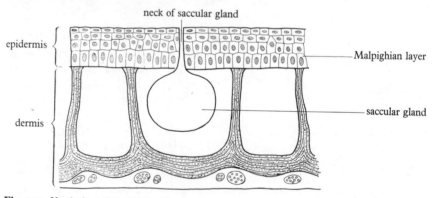

Fig. 3.3 Vertical section of frog skin showing location of saccular gland and epidermis.

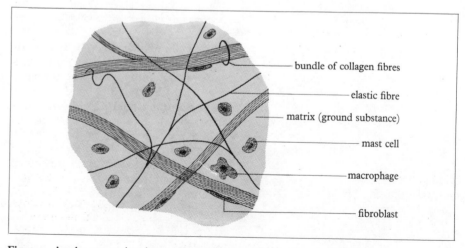

Fig. 3.4 Areolar connective tissue as it appears in a typical microscopical preparation.

Requirements

Microscope (if possible with oil-
 immersion and phase contrast)
Slides and coverslips

0.75 per cent salt solution (NaCl)
1 per cent acetic acid
90 per cent ethanol
Eosin
Methylene blue

Frog peritoneum or
 pericardium
Frog's leg   } in Ringer's
Xiphisternum of frog   solution
Frog's bladder

Section of thyroid gland
Section of kidney
Pavement epithelium (e.g. meso-
 thelium)
VS frog skin (with epidermis intact)
TS mammalian small intestine
 (section of gall bladder)
Macerated ciliated epithelial tissue
TS trachea
TS colon or rectum
Areolar tissue
LS tendon
LS ligament
Adipose tissue
Hyaline cartilage
Fibro-cartilage
Elastic cartilage
TS and LS compact bone
Visceral (smooth) muscle

CONNECTIVE TISSUE

Remove a small piece of **connective
tissue** from between the muscles in
the thigh of a frog. Spread it out on a
dry slide, cover with a dry coverslip
and allow a drop of 0.75 per cent salt
solution to flow under the coverslip.
Observe connective tissue cells and
fibres, using phase contrast if available.
Irrigate with 1 per cent acetic acid to
show nuclei of cells.

Examine a prepared slide of **areolar
tissue** (Fig. 3.4). Notice the **matrix
(ground substance)**, bundles of **collagen (white) fibres**, and irregular
network of **elastic (yellow) fibres**.
How many cells are discernible depends on the quality of the slide. Try
to distinguish elongated **fibroblasts**,
amoeboid **mast cells** and **macrophages**. Spherical **fat cells** may also
be seen.

Tease out a tendon (e.g. the Achilles
tendon of the frog, *see* p. 219) on a slide
and mount in methylene blue. What
can you observe?

Compare prepared longitudinal sections of tendon and ligament (Fig. 3.5).

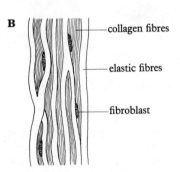

Fig. 3.5 Longitudinal sections of **A** tendon,
and **B** ligament.

A — collagen fibres — fibroblast

B — collagen fibres — elastic fibres — fibroblast

The tendon consists of densely packed
parallel bundles of collagen fibres
between which are rows of fibroblasts
(**collagen tissue**). The ligament is
composed mainly of elastic fibres
(**elastic tissue**).

How does the composition of these
two structures relate to their respective
locations and functions in the body?

Examine a section of **adipose
tissue**, e.g. in the dermis of mammalian skin, (*see* p. 150). Notice clear
fat-filled cells, little cytoplasm discernible, flattened nucleus lying
against edge of cell (Fig. 3.6).

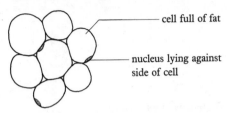

cell full of fat

nucleus lying against
side of cell

Fig. 3.6 Fat cells as they appear in a typical
section of adipose tissue.

CARTILAGE

Strip off the free edge of any part of the
thin **cartilage** of the xiphisternum of
the frog. Put it in 0.75 per cent salt
solution and scrape it clean; then
mount it and examine under the
microscope. Observe **cartilage cells
(chondroblasts)** lying in the **matrix
(chondrin)**. Irrigate with 1 per cent
acetic acid: this should make the cells
more distinct and show up their
nuclei.

Examine a prepared section of
hyaline cartilage, e.g. in the wall of
the trachea, and note the chondroblasts in the matrix (Fig. 3.7). Does
the grouping of the cells give any clues
as to how they have been formed?

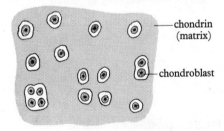

chondrin
(matrix)

chondroblast

Fig. 3.7 Hyaline cartilage as it appears in a
typical microscopical section.

Examine sections of **fibro-cartilage** (cartilage containing collagen
fibres) and **elastic cartilage** (cartilage containing elastic fibres). Compare these structurally and functionally with hyaline cartilage.

A   Four adjacent osteons in transverse section

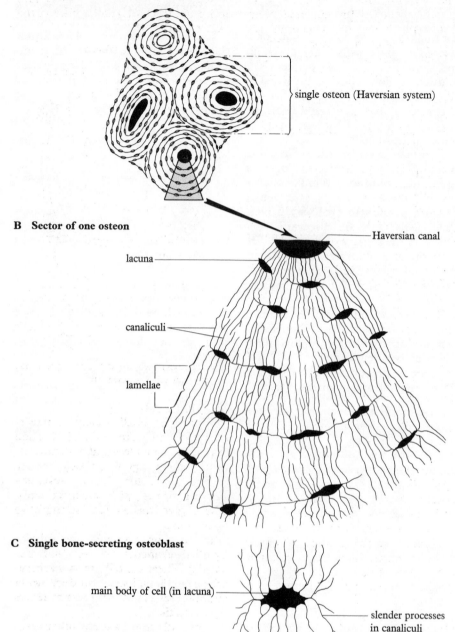

single osteon (Haversian system)

B   Sector of one osteon

Haversian canal

lacuna

canaliculi

lamellae

C   Single bone-secreting osteoblast

main body of cell (in lacuna)

slender processes
in canaliculi

Fig. 3.8   Structure of compact bone as seen in transverse sections.

BONE

Examine a prepared transverse section of **compact bone** and observe the **Haversian pattern** (Fig. 3.8). In each unit (known as an osteon or Haversian system) observe the central **Haversian canal** surrounded by concentric **lamellae**. Also note **lacunae** and **canaliculi** which house the bone-forming osteoblasts and their fine processes. To gain a three dimensional picture of bone examine longitudinal as well as transverse sections.

VISCERAL (UNSTRIATED, SMOOTH) MUSCLE

Spread a piece of frog's bladder, inner surface uppermost, on a dry slide. Allow the edges to dry and stick to the glass. Rub off the inner epithelium with your finger. Put on a drop of 90 per cent ethanol, followed by a drop of eosin and methylene blue. Examine under high power noting long thin **muscle fibres**, each one a single cell with an elongated nucleus (Fig. 3.9).

Compare your temporary mount with a permanent preparation of visceral muscle.

If you wish to make permanent preparations of any of these tissues you are advised to stain with borax carmine or haematoxylin and eosin (*see* p. 436).

A   Sheet of smooth muscle fibres

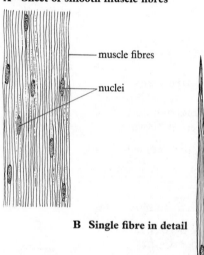

muscle fibres

nuclei

B   Single fibre in detail

nucleus

membrane surrounding muscle fibre

sarcoplasm containing fine myofibrils

Fig. 3.9   Structure of visceral (smooth) muscle as seen in longitudinal section.

For Consideration

How is the structure of each of the animal tissues which you have investigated related to the functions which it performs?

Investigation 3.2

Examination of some basic plant tissues

The object of this investigation is to introduce you to fundamental plant tissues which you will encounter repeatedly in later microscopic studies.

Plant tissues can be investigated by mounting a piece of the living tissue or a thin section of it, on a slide and observing it stained or unstained; by making a permanent preparation of it; or by examining a prepared section or whole mount.

Procedure

EPIDERMIS
Strip off a piece of **epidermis** from the lower side of a leaf (e.g. privet, laurel, holly, iris, geranium) and mount it in glycerine or dilute iodine. Examine the shape and form of the **epidermal cells** and compare with the inner epidermis of an onion scale and a piece of epidermis removed from the side of a smooth stem.

How does plant epidermal tissue compare with the epithelia of animals?

PARENCHYMA
Remove a small amount of pulpy tissue from just beneath the skin of a tomato. Place it on a slide in a drop of water and spread it out with needles. Put on a coverslip and examine the **parenchyma cells**. Irrigate with fresh Schultz' solution and notice closely packed cells with cellulose walls, nuclei, and vacuoles. How would you describe the shape of the cells?

Compare with sections of potato tuber mounted in Schultz' solution. What does this tell us about the functions of parenchyma tissue?

PHOTOSYNTHETIC TISSUE
Mount a small leaf of moss in a drop of water on a slide and examine the chloroplast-packed **photosynthetic tissue** making up the leaf's structure. (*see* also p. 97).

COLLENCHYMA
Examine transverse and longitudinal sections of **collenchyma tissue**. This is particularly abundant at the four thickened corners of deadnettle stem (Fig. 3.10). If you cut your own sections mount them in FABIL. (Instructions for cutting sections are given on pp. 96 and 121.) FABIL clears the material and stains the cellulose light purple. From your examination of transverse and longitudinal sections, reconstruct the shape of an individual collenchyma cell, ascertain how they fit together to form the tissue, and notice the thick **cellulose ribs** at the corners of the cells.

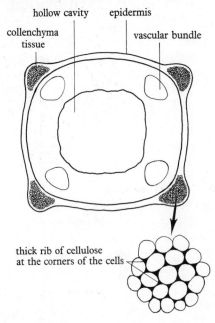

Fig. 3.10 Collenchyma tissue in a transverse section of the stem of deadnettle.

SCLERENCHYMA
Mount a very small amount of macerated woody tissue in FABIL solution and examine it under low and high powers. FABIL stains lignin (wood) red to brown. Look for slender **sclerenchyma fibres** with tapering ends. These give mechanical strength to plant stems.

In addition you will see tubular lignified elements (vessels and tracheids). These contribute to mechanical strength and also conduct water and mineral salts from roots to leaves (vascular tissue).

You could now examine other plant tissues: if you want to make permanent preparations you should stain with safranin and light green (*see* p. 437).

For Consideration
Which tissue (if any) in animals is functionally equivalent to each of the plant tissues which you have investigated? What are the structural similarities between them?

Requirements
Microscope
Slides and coverslips
Razor blade
Petri dishes

Dilute glycerine
Iodine solution
Schultz' solution
FABIL

Leaf and stem with easily removable epidermis
Onion
Tomato
Potato tuber
Leafy moss plant
Deadnettle stem (fresh or preserved)
Macerated woody twig (*see* below)

Woody twig can be macerated as follows (Franklin's method):

Cut into strips approximately 1 × 10 mm and immerse in a mixture of equal parts of glacial acetic acid and 20 volumes of hydrogen peroxide. Either boil under reflux for an hour or maintain at 60°C for 24 hours (times depend on material). Shake vigorously to disintegrate the tissues. Decant off the fluid and wash with several changes of water. Neutralize with a little ammonium hydroxide and store in 70 per cent ethanol.

Investigation 3.3

Unicellular level of organization: examination of *Paramecium*

Paramecium is a single-celled animal that has to carry out within one cell all the functions that in a multicellular organism are performed by numerous different cells. Accordingly it possesses a high degree of internal organization which will become apparent as you examine it (Fig. 3.11).

Procedure

(1) Mount a drop of water containing *Paramecium* on a slide; put on a coverslip and examine under the low power of the microscope. Can you make a rough estimate of *Paramecium*'s speed of swimming? Can you see **cilia** beating? What happens when the animal hits an obstacle?

(2) Slow them down by mixing a drop of *Paramecium* culture with an equal amount of methyl cellulose or some other equally viscous medium. Alternatively they can be trapped in the fine fibrous meshwork obtained by pulling lens paper to pieces. Now make a thorough examination of a single specimen under high power. Observe: **macronucleus**, granular **cytoplasm**, **pellicle**, **cilia** beating rhythmically (note metachronal rhythm), **trichocysts** beneath pellicle, **oral groove**, **oral vestibule**, **cytopharynx**, **food vacuoles** in cytoplasm, two **contractile vacuoles** with collecting channels. Can you see the contractile vacuoles discharging?

(3) Irrigate your slide with either methyl green in acetic acid, or acetocarmine. Both fix the animals and stain the **nuclei**, green in the case of the first stain, red in the second. The micronucleus will be difficult to see because it is usually covered by the macronucleus.

(4) Feed *Paramecium* on yeast suspension stained with Congo Red. Make a vaseline enclosure on a slide just large enough to fit under a coverslip. Place within the enclosure one drop of *Paramecium* culture. Shake the stained yeast suspension and dip in a needle to a depth of about 1 cm and then stir the drop of culture on the slide with it. Put on a coverslip and press gently.

Watch the feeding process and note

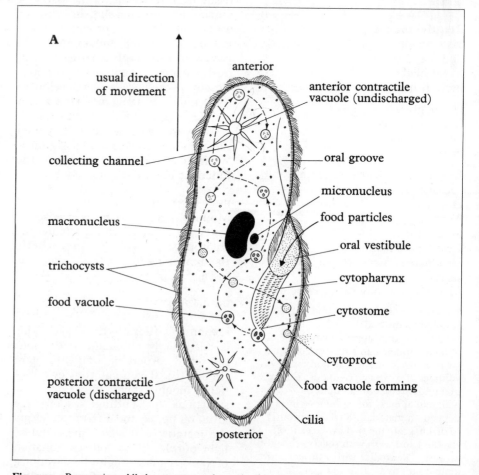

A

usual direction of movement

anterior

anterior contractile vacuole (undischarged)

collecting channel

oral groove

micronucleus

macronucleus

food particles

oral vestibule

trichocysts

cytopharynx

food vacuole

cytostome

cytoproct

posterior contractile vacuole (discharged)

food vacuole forming

cilia

posterior

Fig. 3.11 *Paramecium.* All the structures shown in this diagram can be detected under the light microscope if the correct procedures are adopted (*see* text).

Requirements

Microscope
Slides and coverslips
Lens paper
Filter paper
Mounted needle

Vaseline
Acid ethanol for cleaning slides
Methyl cellulose
1 per cent solution of methyl green in 1 per cent acetic acid
Acetocarmine
2 per cent aqueous solution of silver nitrate
Canada balsam

Stained yeast suspension made by mixing thoroughly:
  Yeast (fresh baker's yeast or dried yeast) (3 g)
  Congo Red (30 mg)
  Distilled water (10 cm³)
  and boiling the mixture gently for 10 minutes.
Paramecium

the contribution of the cilia. Stained yeast cells should be visible in **food vacuoles** within 5 minutes.

. Trace the movement of the food vacuoles and notice any change in colour of the stain. Congo Red is red in an alkaline medium and blue in an acid medium: it changes from red to blue at between pH 5 and 3.

Record any changes in the behaviour of the animals while you are watching them. Can you observe them feeding on anything else besides the yeast cells? Record and interpret any observations you make.

(5) Beneath the sculptured pellicle the basal bodies of the cilia are interconnected by fine threads (kinetodesmata). Something of this **sub-pellicular apparatus** can be seen by staining with silver nitrate as follows:

Clean a slide with acid ethanol and dry it thoroughly. Spread a drop of the culture on the slide and dry it at room temperature. Cover the smear with a 2 per cent aqueous solution of silver nitrate and leave for 6–8 minutes.

Then rinse the smear thoroughly with distilled water and place the slide on a white background in bright daylight. This reduces the silver nitrate with the result that the preparation turns brown to the naked eye. When reduction is complete dry the slide and then put on a drop of Canada balsam and a coverslip.

Find a specimen which has stained satisfactorily and view it under high power. The large dots along the lines of the kinetodesmata are **basal bodies**, the smaller dots in between are **trichocysts**. Small dots between the kinetodesmata are probably junctions in the pellicular lattice.

For Consideration

(1) What structures in man are equivalent to those which you have observed in *Paramecium*?

(2) Does the behaviour of *Paramecium* show any obvious similarities to that of higher animals?

Investigation 3.4

Tissue level of organization: observations on *Hydra*

Hydra shows some degree of cellular specialization. Seven different types of cell are structurally integrated into two distinct tissue layers, the ectoderm and endoderm, and the activities of some of the cells are coordinated by a nerve net. However, behaviour tends to be very simple and slow and, apart from the ovary and testis, there are no organs.

Procedure

(1) Examine living green *Hydra* either in a watchglass under a binocular microscope; or in a watchglass (or on a slide beneath a supported coverslip) under the low power of an ordinary microscope.

Identify: **mouth, tentacles, foot, body wall** surrounding **enteron**. The **ectoderm** is colourless; the **endoderm** is green because it contains symbiotic green algae (zoochlorella). Depending on the time of year, you may see **buds** developing into new individuals; or, alternatively, **testes** near the tentacles and/or **ovaries** towards the foot.

(2) Separate the cells of a specimen whose intercellular material has been partially dissolved in boric acid, by teasing it into pieces with fine needles. Put on a coverslip and gently press it so as to cause further separation of the cells. Examine under high power. How many types of cell can you discern? In particular look out for **musculo-epithelial cells**.

(3) **Nematoblasts (stinging cells)** can be examined as follows: Mount a live specimen in water under a coverslip and irrigate with 1.0 per cent acetic acid. This should cause discharge of some of the nematoblasts which can then be examined under high power. Staining with methylene blue may help you to see them more clearly.

(4) To see how the different types of cell are structurally integrated into tissue layers, examine a prepared transverse or longitudinal section of *Hydra*. Observe the different cells of the **ectoderm** and **endoderm** under high power and notice the non-cellular **mesogloea** in between (Fig. 3.12).

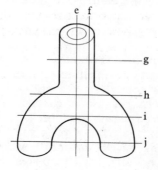

ectoderm   endoderm
↓ mesogloea ↓

musculo-epithelial cell

digestive cell
with flagellum

discharged nematoblast

interstitial cell

symbiotic algae
(zoochlorella)

undischarged nematoblast

muscle tail

vacuole

Fig. 3.12   Body wall of the green hydra *Chlorohydra viridissima* as it appears in a typical longitudinal section.

Requirements
Microscope
Binocular microscope
Watchglass
Slides and coverslips
Mounted needles

1 per cent acetic acid
Methylene blue

Hydra macerated with boric acid
TS or LS *Hydra*
Live green hydra
Daphnia

Can you make out the direction in which the muscle tails of the musculo-epithelial cells are orientated in the ectoderm and endoderm?

(5) Investigate the feeding behaviour of a live specimen of *Hydra* by feeding it with live water fleas (e.g. *Daphnia*). With your knowledge of *Hydra*'s structure in mind, try to interpret the sequence of events that takes place.

For Consideration
In what respects would you consider *Hydra* to be more advanced than *Paramecium* and less advanced than man? In what sense is the word advanced being used in this context?

Questions and Problems

1   Fig. 3.13 shows various planes in which a tubular structure, such as a blood vessel, with a wall of uniform thickness might be cut in a microscopical section. In each case the plane of the section depends on the orientation of the structure relative to the cutting blade. Draw simple sketches showing the appearance of the tube when sectioned in each of the planes **a–j**.

Imagine that the structure instead of being tubular is a hollow ovoid. What will be its appearance when sectioned in planes **k–o**?

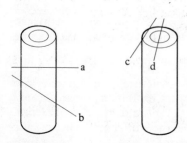

Fig. 3.13

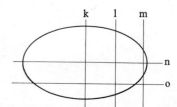

Fig. 3.14

2   What criticisms would you level against classifying epithelial tissue into cuboidal, pavement, columnar, ciliated, glandular and stratified epithelia? Can you suggest any alternative classifications?

3   In some classifications of tissues, cartilage, bone and blood are included with connective tissue; and in plants parenchyma is included with mechanical tissue. Do you think this is justified?

4   What general principle emerges from your answers to questions (2) and (3)?

5   In the course of its development simple hyaline cartilage may become impregnated with calcium salts to form calcified cartilage. Under these circumstances the chondroblasts die and the cartilage becomes dead tissue. Bone, however, is living tissue: when developing bone becomes impregnated with calcium salts the osteoblasts remain alive. Explain this difference between cartilage and bone.

6   The organic matrix of bone, secreted by the osteoblasts, consists of collagen and a polysaccharide impregnated with the inorganic salts, calcium phosphate and calcium carbonate. What raw materials must be supplied to the osteoblasts to enable them to manufacture the matrix and by what route would you suppose that they get there?

7   The osteons (Haversian systems) in a limb bone, and the sclerenchyma fibres in a plant stem, are both tightly packed and both are orientated parallel with the long axis. Is there a functional explanation for this similarity?

8   Bone marrow is soft tissue found in the centre of bones. Surrounded by compact bone, it is the site of formation of red blood cells. What are the advantages and disadvantages of red blood cell formation taking place here?

9   By means of a table compare a typical animal, as exemplified by a mammal, with a typical plant, as exemplified by an angiosperm (flowering plant).

10   The chief differences between animals and plants can be related to their different methods of nutrition. Explain.

11   To what extent can meaningful comparisons be made between the tissues of animals and plants?

12   Discuss the evidence for considering a sponge to be (a) a colony of single cells, and (b) a simple multicellular animal.

13   Which structures in (a) *Hydra* and (b) *Paramecium* are functionally equivalent to the following structures in man:
(i) skin (ii) mouth (iii) anus (iv) kidney (v) limbs.

14   It used to be thought that the kinetosomes and kinetodesmata beneath the pellicle of *Paramecium* were responsible for coordinating the beating of the cilia. This idea is now considered unlikely. What kind of evidence do you suppose might have caused a change of opinion on this?

15   Make a list of the advantages and disadvantages of being (a) unicellular and (b) multicellular.

16   Some biologists prefer to regard protists like *Paramecium* as acellular (non-cellular) rather than unicellular (single celled). What do you think?

4 Movement in and out of cells

Background Summary

1 Materials move in and out of cells by diffusion, osmosis, active transport, phagocytosis, and pinocytosis.

2 **Diffusion** is the net movement of molecules (or ions) from a region of high concentration to a region of lower concentration.

3 Diffusion satisfies the metabolic needs of small organisms whose **surface: volume ratio** is large and where the diffusion distance is small. Larger organisms, whose surface:volume ratio is small, and where the diffusion distance is great, require special adaptations for acquiring essential materials and transporting them within the body.

4 **Osmosis**, the one-way net diffusion of solvent (particularly water) molecules across a semipermeable membrane, is particularly important to cells because the plasma membrane is semipermeable.

5 An animal cell immersed in water or a hypotonic solution swells and may burst unless it has a means of disposing of the water that enters by osmosis. In a hypertonic solution the cell shrinks and the plasma membrane crinkles.

6 A plant cell immersed in water or a hypotonic solution becomes **turgid**. If immersed in a hypertonic solution the protoplast shrinks and **plasmolysis** occurs. (The protoplast membranes are semipermeable but the cellulose wall is fully permeable.)

7 The equation summarizing the water relations of a plant cell is as follows:

$$DPD = OP - WP$$

where DPD is the **diffusion pressure deficit**, the net tendency for water to enter the cell; OP is the **osmotic pressure** (**osmotic potential**) exerted by the cell sap; and WP is the **inward pressure** exerted by the cellulose wall against the protoplast. (Diffusion pressure deficit used to be known as suction pressure).

8 The suction pressure of a plant tissue can be determined by balancing it with an external solution which produces no weight or volume change in the tissue. The osmotic potential can be determined by balancing it with an external solution that produces incipient plasmolysis.

9 **Active transport**, the movement of molecules or ions against a concentration gradient, is an energy-requiring process that is thought to involve the movement of **carriers** in the plasma membrane.

10 **Phagocytosis** is the intake of solid particles at the cell surface by invagination of the plasma membrane. The contents are then digested and absorbed.

11 **Pinocytosis** and **micropinocytosis**, also involving invagination of the plasma membrane, but on a smaller scale than phagocytosis, provide a means by which liquids and macromolecules may be brought into cells.

Investigation 4.1

Cells and osmosis

Requirements

4 slides and coverslips
Lancet for drawing blood
5 pipettes
Filter paper

Distilled water
Salt (NaCl) solutions: 0.75, 1.0 and
   3 per cent
Sucrose solution 1.0 M

Onion bulb or rhubarb stem

The cell membrane is semipermeable and if a cell is placed in a solution whose osmotic pressure differs from that of the cell's contents, water either enters or leaves the cell. It enters if the external solution is hypotonic, it leaves if the external solution is hypertonic. In plant cells the cellulose wall is fully permeable to both water and solutes.

Procedure

(1) Place a drop of blood on each of four slides, labelled A–D. To A add a drop of distilled water; to B, C and D add a drop of 0.75 per cent, 1.0 per cent and 3.0 per cent salt solution. Observe the **red blood cells** under high power for some period of time and interpret your observations as fully as you can.

(2) Strip off two small pieces of epidermis from a plant stem or leaf. (The epidermis from the inner surface of an onion scale, or the red pigmented epidermis from a rhubarb stem are recommended). Mount one piece in water and the other in a strong solution of sucrose (1.0 M is satisfactory). Compare the two slides at intervals and interpret your observations. Look for **plasmolysis** and compare with

Fig. 4.1. If you are using rhubarb epidermis can you say where the red pigment is located within the cell?

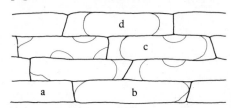

Fig. 4.1 Appearance of onion epidermal cells in a hypertonic external solution. Cell **a** is unplasmolysed; **b**, **c**, and **d** show progressive plasmolysis with the protoplast shrinking away from the cell wall.

(3) Irrigate the piece of epidermis, which you mounted in sucrose, with distilled water. Use filter paper to draw the water beneath the coverslip, and look down the microscope as you do so. Observe what happens and interpret your results.

For Consideration

What is the relevance of your observations on the osmotic properties of cells to the animals and plants concerned?

Investigation 4.2

The osmotic relations of red blood corpuscles

Like any other cell, red blood corpuscles are enclosed in a semipermeable membrane. If a corpuscle is placed in a hypotonic solution, water enters by osmosis and the corpuscle swells. In extreme cases, i.e. when the external solution is markedly hypotonic to the cell's contents, the corpuscle bursts and its haemoglobin is released (**haemolysis**). Even within the same individual, red blood corpuscles vary in their resistance to the haemolysing effect of a hypotonic external solution, and there are also differences between species. The purpose of this investigation is to examine the effect of hypotonic solutions on human red blood corpuscles and to investigate the mechanical properties of the corpuscular membrane.

Procedure

(1) Place 10 cm³ of sodium chloride (NaCl) solution into a series of five test tubes, one tube each of the following strengths: 0.1, 0.3, 0.5, 0.7, 0.9 per cent.

   The 0.9 per cent NaCl has approximately the same osmotic concentration

as mammalian blood; the other solutions are increasingly hypotonic.

(2) To each test tube add 0.1 cm³ of **defibrinated human blood** with a micropipette. It is most important to stir the blood thoroughly before pipetting it so as to ensure uniform distribution of the corpuscles. Leave the tubes standing for a while and examine them at intervals.

   In some of the test tubes the contents will be cloudy: what does this indicate? In others the contents will be clear: what does this indicate?

   In certain of the test tubes a clearing may develop in a restricted region at the top of the tube if you leave it long enough. What is this caused by? What is the colour of this clear liquid? Explain fully.

   Plainly, haemolysis has occurred to a greater or lesser extent in some of the tubes. From looking at the tubes what can you conclude about the extent of haemolysis in each case?

(3) Using a **haemocytometer**, estimate the number of red blood corpuscles per mm³ in each tube. Start

Requirements
Microscope
Haemocytometer with accessories
(Slide and coverslip if no
 haemocytometer is available)
10 test tubes (to fit centrifuge)
Test-tube rack
Pipette for 1.0 cm^3 of blood
Micropipette for 0.1 cm^3 of blood
Rod for stirring blood
Labels or wax pencil
Centrifuge

NaCl solutions (0.1, 0.3, 0.5, 0.7,
 0.9 per cent) (each student
 requires 10 cm^3 of each solution)
Defibrinated blood (1.0 cm^3 per
 student)

with a sample of blood from the tube containing 0.9 per cent NaCl, which you know has not undergone haemolysis, and then go on to the other tubes.

The technique for counting red blood cells is described on p.102. Mix the blood thoroughly before you suck it into the haemocytometer pipette. There is no need to dilute the blood any more than it has been already.

If you have not got a haemocytometer, a rough estimate of the red cell concentration can be obtained by pipetting a small volume of blood onto a slide, putting on a coverslip and counting the number of corpuscles visible in the high-power field of view.

(4) Centrifuge the remaining blood in the tubes until the contents are clear. (About three minutes should be sufficient.) In each case examine the amount of sediment at the bottom of the tube, and the colour of the supernatant. Conclusions?

Results
Assuming that no haemolysis has occurred in the 0.9 per cent NaCl and that the blood in this tube gives a normal blood count, calculate the number of haemolysed corpuscles per mm^3 of blood in the other tubes. In each case this figure can be obtained

by subtracting the number of intact corpuscles present in the sample from the number in the 0.9 per cent NaCl.

Now calculate the percentage of corpuscles haemolysed in each tube:

$$\text{percentage haemolysis} = \frac{a}{b} \times 100$$

a = no. of haemolysed corpuscles per mm^3

b = no. of intact corpuscles per mm^3 in 0.9 per cent NaCl

Plot percentage haemolysis (vertical axis) against the strength of the NaCl solutions.

For Consideration
(1) Can you account for the shape of your curve?
(2) Why don't all the red blood corpuscles haemolyse at the same NaCl concentration?
(3) Do you think the results of this experiment might have any medical significance?
(4) What light does this investigation throw on the properties of blood plasma?
(5) How are the properties mentioned in answer to (3) maintained in the human body?

Investigation 4.3
Determination of the diffusion pressure deficit of potato tuber cells

The diffusion pressure deficit (DPD)[1] of a plant cell is the net tendency for water to pass into the sap vacuole from outside. It is equal to the osmotic potential of the cell sap (OP) minus the inward pressure exerted by the cellulose cell wall (WP).

$$DPD = OP - WP$$

It follows that if a plant cell is in equilibrium with an external solution of solute concentration such that there is no net loss or gain of water, the osmotic potential of the external solution will be equal to the diffusion pressure deficit of the cell.

Use of this fact can be made in estimating the diffusion pressure deficit of a plant tissue. Samples of tissue are immersed in a range of external solutions of different strengths. The solution that induces neither an increase nor a decrease in the volume or weight of the tissue has the same osmotic potential as the diffusion pressure deficit of the tissue.

The cells to be investigated in this experiment are those of the potato tuber. Changes in weight will be used as an indication of whether the cells are taking up or losing water.

Procedure
(1) Label six specimen tubes: distilled water, 0.5 M, 0.4 M, 0.3 M, 0.2 M, 0.1 M. Place approximately one-third of a tube of **distilled water** in the first, and an equal volume of each of a series of **sucrose solutions** of different strengths (molarities) in the remainder. Each tube should be firmly stoppered.

(2) Using a cork borer and razor blade, prepare six solid cylinders of **potato**. Each cylinder should be approximately 1 cm diameter and 12 mm long. Slice up each cylinder into six discs of approximately equal thickness. Place each group of discs on a separate piece of filter paper.

[1] The former term for diffusion pressure deficit was suction pressure. The term osmotic potential is synonymous with osmotic pressure; the former is becoming more widely used nowadays.

(3) Weigh each group of discs. (In each case weigh them on the piece of filter paper, then weigh the filter paper alone, and subtract the one from the other to get the weight of the discs). Record the weight of each group of discs.

(4) Put the groups of discs in each of the labelled tubes. Stopper the latter firmly and leave for not less than 24 hours.

(5) After about 24 hours remove the discs from each tube, remove any surplus fluid from them quickly and gently with filter paper and re-weigh them. Record the new weight of each group of discs.

(6) Graph your results by plotting the percentage change in weight (change in weight multiplied by 100 divided by original weight) against the molarity of the sucrose solutions. The latter, being the independent variable, should be on the horizontal axis; the former on the vertical axis.

(7) Calculate the **diffusion pressure deficit** of the potato cells as follows. Find the point on your graph corresponding to a percentage weight change of zero. The molarity of sucrose corresponding to this zero weight change can now be read from the horizontal axis. From Table 4.1 find the osmotic potential of a sucrose solution of that molarity. That is the diffusion pressure deficit of your sample of potato cells. Express your result in N/mm^2.

For Consideration

(1) Criticize this method of finding the diffusion pressure deficit of plant cells. How might it be improved?

(2) What was the reason for dividing each cylinder into six discs?

(3) With what kind of plant tissue might it be possible to use a change in volume rather than weight for estimating the diffusion pressure deficit?

(4) How does the value of the diffusion pressure deficit differ from the osmotic potential of the solution in the vacuole.

(5) In constructing your graph did you join up the points with straight lines or a smooth curve? Justify whichever technique you used.

Requirements
6 specimen tubes with stoppers
Cork borer of 1 cm diameter
Labels or wax pencil
Razor blade
Filter papers
Balance (preferably rapid-weighing)

Distilled water
Sucrose solutions of the following molarities:
0.5 M, 0.4 M, 0.3 M, 0.2 M, 0.1 M

Potato tuber

Investigation 4.4
Determination of the osmotic potential of cell sap of plant epidermal cells

Bearing in mind that $DPD = OP - WP$, it follows that if $WP = O$, then $OP = DPD$. Use can be made of this in determining the osmotic potential of plant tissue. The cells are placed in a range of external solutions of different concentrations. The one which plasmolyses the cells to the extent that the protoplasts just lose contact with the cellulose walls (**incipient plasmolysis**) can be regarded as having the same osmotic potential as the cell sap. This is because, under these circumstances, no inward pressure is exerted by the cellulose wall, i.e. $WP = O$ and $OP = DPD$.

In practice the cells of a piece of plant tissue, such as the epidermal tissue to be used in this experiment, plasmolyse at different rates. For practical purposes we can regard incipient plasmolysis as the condition in which half the cells are visibly plasmolysed.

Procedure
(1) Remove one of the fleshy scale leaves of an onion. With a razor blade cut the **inner epidermis** into six squares of approximately 5 mm side. With fine forceps peel off each square of epidermal cells and place it in distilled water in a petri dish.

(2) Label five stoppered tubes and into each one place approximately 10 cm^3 of sucrose solution of the following concentration: 0.3 M, 0.35 M, 0.4 M, 0.5 M, 1.0 M. Stopper the tubes, and gently shake the contents so as to bring all parts of the epidermis into full contact with the solution. Leave the tissue in the solutions for about 20 minutes.

(3) After 20 minutes remove each strip, one at a time, and mount it in a drop of the solution in which it has been immersed. Observe under low power.

(4) Count *all* the cells visible within the low power field of view. Now count all those that are plasmolysed: include all cells which show a visible separation of the protoplast from the cell wall, however slight this may be.

(5) Plot a graph of percentage plasmolysis (vertical axis) against molarity of sucrose solution (horizontal axis).

Requirements
Microscope
Razor blade
Petri dish
5 specimen tubes with stoppers
Graph paper

Distilled water
10 cm^3 of each of the following
  sucrose solutions:
  0.3 M, 0.35 M, 0.4 M,
  0.5 M, 1.0 M.

Onion scale

(6) From your graph read off the molarity of sucrose that corresponds to 50 per cent plasmolysis. This solution may be regarded as having the same osmotic potential as the cell sap of the tissue. From Table 4.1 you can work out the osmotic potential in N/mm^2 corresponding to this molarity.

For Consideration
(1) Why should the pieces of epidermis be placed in distilled water before being put into the experimental solutions?
(2) What are the major sources of inaccuracy in this method of determining the osmotic potential of plant cells?
(3) Suggest a better method which overcomes the objections raised in your answer to (2).

Molarity	Osmotic potential (in N/mm^2)
0.05	0.13
0.10	0.26
0.15	0.41
0.20	0.54
0.25	0.68
0.30	0.86
0.35	0.97
0.40	1.12
0.45	1.28
0.50	1.45
0.55	1.62
0.60	1.80
0.65	1.98
0.70	2.18
0.75	2.37
0.80	2.58
0.85	2.79
0.90	3.00
0.95	3.25
1.00	3.50

Table 4.1   Relationship between molarity and osmotic potential of sucrose solutions.

Questions and Problems

1   It is said that as an object increases in size its surface:volume ratio decreases. Prove this relationship mathematically for (a) a cube, (b) a sphere, and (c) a cylinder. What is the significance of this principle to organisms?

2   Consider a cylindrical-shaped organism filled with densely packed cells. What problems are attendant on getting oxygen from the surrounding atmosphere to its innermost cells? By what means might the delivery of oxygen to these cells be facilitated?

3   Justify the statement that osmosis is a special case of diffusion. What is the relevance of osmosis to (a) a human liver cell, and (b) a parenchyma cell in the stem of a flowering plant?

4   Pieces of well washed epidermis of the stem of *Lamium album* (deadnettle) were placed in each of five sucrose solutions and the percentage of plasmolysed cells in each solution determined at intervals of time until no further change took place. The results were as follows:

Molar concentration of sucrose solution	0.55	0.6	0.65	0.7	0.75
Percentage of cells plasmolysed	0	5	20	80	100

Explain these results and state, with reasons, what conclusions you draw from them.
                                                                            (*O and C*)

5   When placed in distilled water mammalian red blood cells burst but a plant parenchyma cell remains intact. Why the difference?

6   The hollow scape (flower stalk) of a dandelion is split longitudinally into six portions of length 3 cm. The strips immediately bend outwards as shown in Fig. 4.2 A. The curvature is the same for all six strips.

   The strips are now placed in six sucrose solutions of different concentrations. The curvatures adopted by the strips after 15 minutes are shown in Fig. 4.2 B–G, together with the concentration of each sucrose solution.
   (a) Why do the strips bend outwards as soon as they have been removed from the scape?
   (b) Why do the strips bend further outwards in B and C?
   (c) Why do the strips bend inwards relative to A in E–G?
   (d) Why does strip D remain unchanged relative to A?
   (e) What are the limitations of this experiment as a method for measuring the suction pressure of dandelion scape cells?

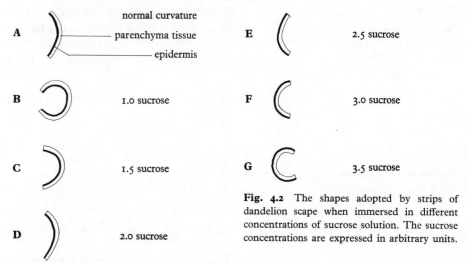

Fig. 4.2 The shapes adopted by strips of dandelion scape when immersed in different concentrations of sucrose solution. The sucrose concentrations are expressed in arbitrary units.

7   Mammalian red blood cells are sensitive to a change in salt concentration of the external solution. If they are transferred from plasma to a less concentrated solution they swell and, if they swell sufficiently, they burst; in which case they are said to have haemolysed. In an experiment to find the percentage of human red cells haemolysed at different concentrations of salt solution, the following results were obtained:

percentage salt concentration (g/100 cm³)	0.33	0.36	0.38	0.39	0.42	0.44	0.48
percentage red cells haemolysed	100	90	80	68	30	16	0

(a) Plot the results on graph paper, using the horizontal axis for the varying percentages of salt concentration.
(b) Explain why red cells swell and burst when placed in a less concentrated salt solution.
(c) At what percentage salt solution is the proportion of haemolysed to non-haemolysed cells equal?
(d) Suggest a hypothesis to account for the red cells haemolysing over a range of salt concentrations rather than at one particular salt concentration.
(e) Of what significance are these observations as far as the working of the human body is concerned?   (*AEB modified*)

8 What are the similarities and differences between phagocytosis, pinocytosis and micropinocytosis? What are the functions of each?

In what sense can the contents of a phagocytic vesicle be said to be outside the cell?

9 The following data illustrate the relative absorption by carrot discs of potassium ions at different concentrations of oxygen from an external solution at 23°C.

Concentration of O_2 (per cent)	2.7	12.2	20.8
Relative absorption of K^+	22	96	100

What explanations can you suggest for the form of the relationship, and what experiments would you carry out to test your suggestions? *(O and C)*

10 A scientist collected a number of sandworms from a beach. He selected 50 worms of equal weight and placed equal numbers in five different concentrations of sea water. After twelve hours he re-weighed the worms and determined the average weight for each group. From this he calculated the percentage changes in weight. His results are shown in Fig. 4.3. Explain them as fully as possible.

(VUS)

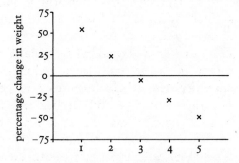

Fig. 4.3 Results of experiment on weight change of sandworms placed in different concentrations of sea water.

5 The Chemicals of Life

Background Summary

1. The principal organic constituents of organisms are carbohydrates, lipids, proteins, nucleic acids and vitamins. Inorganic constituents include acids, bases, salts and water.

2. The importance of **water** as a medium for life derives from its solvent properties, heat capacity, surface tension and freezing properties.

3. **Acids** and **bases** are important in that they determine the **pH** of an organism's body fluids. The correct functioning of cells depends on the pH being kept constant.

4. **Salts**, or their **ions**, perform a wide range of functions in different animals and plants. Many are essential metabolites without which disabilities and death may occur.

5. **Carbohydrates** (general formula $(CH_2O)_n$), contain energy and are also important structurally (e.g. cellulose). They are classified into **monosaccharides**, **disaccharides** and **polysaccharides**. Monosaccharides like glucose can be built up into polysaccharides like starch and cellulose by condensation, and the latter can be broken down by hydrolysis. In both cases the reactions are catalysed by specific enzymes.

6. The chemical composition of **starch** and **cellulose** explain their different physical properties, which are related to their functions in cells. Cellulose, a material of great commercial importance, may become lignified to form wood.

7. **Lipids** (fats and oils) are compounds of **glycerol** ($C_3H_8O_3$) and **fatty acids** ($R(CH_2)_nCOOH$) which can be united by condensation or split by hydrolysis. Lipids are important sources of energy and also carry out structural roles (e.g. in the plasma membrane).

8. **Proteins** are composed of **amino acids** ($RCHNH_2COOH$). Amino acids can be combined by condensation to form **polypeptide chains** and ultimately proteins, and the latter can be hydrolysed into free amino acids. About 20 naturally occurring amino acids are known.

9. Proteins are classified as **globular** or **fibrous**. The former perform mainly regulatory functions in the body (e.g. **enzymes**), whilst the latter fulfil structural roles (e.g. collagen). Proteins are often combined with another compound, e.g. a porphyrin, to form a **conjugated protein** (e.g. haemoglobin).

10. Proteins may be analysed by chromatography, isotope-labelling, and X-ray crystallography. **Chromatography** enables their constituent amino acids to be identified, and when combined with **labelling techniques** it enables the amino-acid sequence to be determined (e.g. insulin). **X-ray crystallography** enables the three-dimensional shape of the protein molecule to be established (e.g. haemoglobin and myoglobin).

11. Proteins are constructed on a helical plan, the α **helix**. The helix is commonly single (e.g. globular proteins and keratin), or triple (e.g. collagen).

12. **Nucleic acids**, like proteins, are very large molecules. Formed from

building blocks called **nucleotides**, they are concerned with the formation of genetic material (*see* chapter 30).

13 **Vitamins**, a mixed assortment of chemical compounds, are required by organisms for various metabolic purposes. Some of them function as coenzymes.

Investigation 5.1

Identification of some biologically important chemical compounds

The chemical composition of cells can be investigated in two main ways. One is to grind up the tissue, or extract the juices from it, and perform chemical analyses on the material so obtained. Alternatively the tissue can be sectioned, stained with a reagent specific to a certain chemical constituent, and examined under the microscope. This approach has the advantage that the distribution of individual compounds within the cells can be studied in detail.

Procedure

STARCH

To approximately 4.0 cm^3 (quarter of a test tube) of starch suspension add two drops of **dilute iodine**. A blue-black colour indicates the presence of starch.

To examine the distribution of starch in a plant organ or tissue, cut a thin section of the material and mount in dilute iodine. Starch grains stain blue and can be located in the cells. Try this on, for example, potato tuber. If you irrigate a section of potato tuber on a slide with dilute iodine, you can see the starch grains turning blue as the iodine reaches them.

GLUCOSE AND FRUCTOSE

All monosaccharide sugars and certain disaccharides will reduce copper sulphate, producing a precipitate of cuprous oxide on heating. To approximately 4.0 cm^3 of glucose add an equal quantity of **Benedict's reagent**. Shake and bring to the boil. A precipitate indicates reducing sugar.

The colour and density of the precipitate gives a rough indication of the amount of sugar present. A green precipitate means relatively little sugar; a brown or red precipitate means progressively that more sugar is present.

Cut a cube of apple of side 1.0 cm. Grind it up in a mortar with a small quantity of water. Transfer the material to a test tube and add water to make up a total volume of approximately quarter of a test tube. Add an equal volume of Benedict's reagent and heat to boiling. Record your results and conclusions.

The reducing sugar in apple tissue is fructose. Investigate its distribution as follows. Cut a thin section with a razor blade and mount it on a slide in a few drops of Benedict's reagent. Examine the cells under the microscope. Now heat the slide *gently* over a flame until the tissue turns brown. Add water if the tissue looks like drying up. Allow the slide to cool and then re-examine the cells. What changes have occurred in them? Is sugar present in all the cells or only certain ones?

SUCROSE

Sucrose (cane sugar) is a disaccharide that does not reduce copper sulphate. However, it can be detected by first hydrolysing it into its constituent monsaccharides and then testing with Benedict's reagent. Boil a little sucrose solution with dilute hydrochloric acid, neutralize with sodium bicarbonate, then test with Benedict's reagent.

CELLULOSE

Cellulose stains purple with **Schultz' solution** (*see* p. 438). Put a drop of the stain onto a small piece of teased-out cotton wool and notice the reaction.

Mount a section of plant tissue in Schultz' solution and examine the cells. Note the colour reaction shown by some of the cell walls.

*Monosac - either fehlings
sugar or benedict's*

LIGNIN

Lignin (wood) stains red with **acidified phloroglucinol** (*see* p. 438). Put some phloroglucinol into a watch glass and add a few drops of concentrated hydrochloric acid. Dip a match stick into this mixture and note the colour reaction.

With a razor blade cut thin transverse and/or longitudinal sections of a woody stem. Stain your sections in acidified phloroglucinol in a watch glass for five minutes and then mount in water, or (if you wish to keep them for long) in dilute glycerine. Examine under the low power and observe the distribution of lignin in your section.

LIPIDS

Lipids produce a red stain with **Sudan III.** An oil can be detected by adding it to a small amount of water in a test tube and shaking it up with a few drops of Sudan III. On standing, the oil separates from the water and will be seen to have taken up the red stain. Try this test on olive oil.

Alternative test for lipid: shake with 5 cm³ **ethanol (absolute)** until the fat dissolves; add an equal volume of water. A cloudy white precipitate indicates lipid.

The presence of lipid in a tissue can be detected as follows. Grind the tissue in a mortar. Transfer it to a test tube containing water, and boil. If lipids are present oil droplets will escape from the tissue and rise to the surface of the water. Add Sudan III and shake. Allow the oil to settle and notice that, as before, it is stained red. Try this on a fat-containing seed such as castor oil or linseed.

The distribution of fat in a tissue can be explored by staining the tissue in Sudan III and then washing with several changes of water and/or 70 per cent ethanol. Any fat present will retain the red colour of the stain. Section a castor oil seed longitudinally and investigate the distribution of fat this way.

PROTEIN *eggwhite*

To about 2 cm³ of a solution or suspension of the protein add about six drops of **Millon's reagent**, and boil. A brick-red colour indicates protein. Try this test on some egg albumen. What happens to the albumen when you boil it? Explain your observations.

Split open a soaked pea and slice off a thin section. Mount it in Millon's reagent on a slide, and heat gently to boiling. Add water if necessary. Cool and examine under the microscope for the distribution of protein.

The following alternative tests are particularly suitable for soluble proteins:

Xanthoproteic test: add a little concentrated nitric acid, heat till lighter colour: cool to cold; add 880 ammonia. A deep-yellow partial precipitate, which disappears on shaking, indicates protein. *eggwhite*

Biuret test: add a little potassium hydroxide till the solution clears; add a drop of copper sulphate down the side of the test tube. Do not heat. A blue ring at the surface of the solution indicates protein; on shaking, the blue ring disappears and the solution turns purple.

INVESTIGATING UNKNOWNS

Investigate the occurrence and distribution of different sugars, starch, cellulose, lignin, fat and protein in a variety of plant tissues and organs, e.g. apple, orange, grape, carrot, parsnip, etc. In the case of a complex structure like an orange you should test the different components (e.g. skin, matrix, juice, pips) separately. Where appropriate you will need to consider regional differences within each component.

In some cases it is necessary to carry out your analysis on a solution prepared by grinding up the tissue in a mortar, in other cases it is more appropriate to analyse the solid tissue. Where possible, carry out the tests on microscope slides and examine the distribution of the substance in question under low power.

For Consideration

(1) Summarize the functions performed by the various chemical substances which you have investigated.
(2) What is the value of obtaining information on the occurrence and distribution of different substances in a plant or animal?
(3) Select one of the plant organs in which you have investigated the distribution of substances. Explain, as far as you can, how the various substances came to be located in their specific situations.

Requirements

Microscope
Test tubes
Slides and coverslips
Bunsen burner or spirit lamp
Pestle and mortar
Razor blade
Watch glass

Dilute iodine
Benedict's reagent
Schultz' solution
Phloroglucinol
Concentrated hydrochloric acid
Dilute hydrochloric acid
Sodium bicarbonate
Sudan III
Millon's reagent
Concentrated nitric acid
880 ammonia
Potassium hydroxide
Copper sulphate (one per cent)

Starch solution
Potato tuber
Glucose (dextrose)
Sucrose (cane sugar)
Apple
Cotton wool
Woody stem
Match sticks
Olive oil
Castor oil seed
Egg albumen
Soaked peas
Variety of plant tissues and organs

Investigation 5.2

Analysis of amino acids in a protein by paper chromatography

The building blocks of protein are amino acids. This experiment provides an opportunity to determine which amino acids are present in the protein **albumen**. The protein is first hydrolysed by treating it with the digestive enzyme **trypsin**; the amino acids are then separated and identified by **paper chromatography**.

The principle behind paper chromatography is as follows. A small amount of solvent is put at the bottom of a jar. A strip of absorptive paper, with a concentrated spot of the mixed amino acids near the bottom, is suspended in the jar so that its end dips into the solvent. The latter moves slowly up the strip of paper, carrying the amino acids with it. As the amino acids travel at different speeds they separate from one another. The paper is then treated with a reagent which stains the amino acids so they can be detected and identified.

Procedure

This experiment takes several days so it is advisable to draw up a time schedule before you start.

BREAKING DOWN THE PROTEIN

Place 0.5 g of powdered **trypsin** in a large test tube. Half fill the test tube with **distilled water** and shake until the trypsin has dissolved. Add about 5 cm^3 of **sodium bicarbonate solution** to make the mixture alkaline.

Now add the **protein**: either 2 g of egg white, or 5.0 cm^3 of albumen.

Finally drop in a crystal of **thymol** to kill any bacteria. Leave the mixture to incubate at 37°C for 48–72 hours.

SETTING UP THE CHROMATOGRAPHY APPARATUS

Separation of the amino acids can be carried out in a glass jar of height not less than 30 cm. There should be a cap for covering the open end of the jar to which the strip of chromatography paper can be attached (see below).

Pour the **solvent** into the jar to a depth of about 3 cm. Cover the open end of the jar so the atmosphere inside becomes saturated with vapour.

Next cut a strip of **chromatography paper** of sufficient length for one end to dip into the solvent to a depth of about 5 mm, and the other end to stick out of the top of the jar by about 2 cm. Handle the paper as lightly as possible and make sure your hands are clean and dry, otherwise amino acids on your skin may get onto the paper.

About 4 cm from one end of the strip of paper draw a line in *pencil* across the strip. Using a fine pipette place a small drop of the **amino-acid mixture** in the centre of the pencil line. Let this dry, and then place another drop on top of the first, and dry again. Repeat this about six times keeping the spot as small as possible. Now the chromatography can be started.

Lower the strip into the glass jar, pencilled end first, taking care not to let it come into contact with the sides of the jar. Allow the bottom end of the strip to dip about 5 mm into the solvent, then bend the top end over and attach it to the cap (Fig. 5.1).

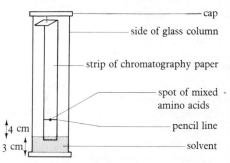

Fig. 5.1 Diagram of chromatography apparatus for the separation of amino acids.

The experiment should now be left for not less than eight hours, and not longer than 16.

DEVELOPING THE CHROMATOGRAM
(8–16 HOURS LATER)

Remove the strip from the jar. The solvent should have risen some 20–25 cm from the pencil line. Draw another line across the strip at the highest point reached by the solvent. Then hang the strip to dry in a warm place.

The chromatogram is developed (i.e. the amino acids are stained) by means of a dilute solution of **ninhydrin in butan-1-ol.**

Caution: ninhydrin is poisonous: do not inhale its fumes, and wash it off immediately if it gets onto your skin.

Pour a small amount of the ninhydrin reagent into a petri dish, and slowly draw the chromatography strip through the liquid. Ensure that the whole of the area between the two pencil lines is thoroughly soaked.

Now dry the strip rapidly by holding it close to a source of heat. If you continue heating it after it is dry, purple spots should appear at points along its length. Continue heating until the spots are as dense as possible.

INTERPRETATION OF THE CHROMATOGRAM

Each purple spot corresponds to one or more amino acids. Certain spots are clear and easily identified; others tend to merge together and are less distinct.

To identify the amino acids we make use of a measurement called the R_f **value**. R_f stand for relative front. This is the ratio of the distance moved by the spot to the distance moved by the solvent.

$$R_f = \frac{\text{distance moved by spot}}{\text{distance moved by solvent}}$$

Draw a line through the centre of each spot and calculate the R_f values. By comparing your R_f values with the list given in Table 5.1, try to identify the amino acid responsible for each spot in your chromatogram.

PRESENTATION OF RESULTS

Attach your chromatogram to a sheet of paper and write alongside each spot the name of the amino acid, together with your estimated R_f value.

For Consideration

(1) According to your analysis how many different amino acids are present in albumen?

(2) It is probable that you will have identified between six and eight amino acids with reasonable certainty. But there are known to be 15 (*see* Table 5.1). Why don't they all appear in your chromatogram? (There are several possible reasons.)

(3) One possible reason for (2) is that all 15 amino acids *are* present in your chromatogram but some of the spots may be so close to each other that it is impossible to distinguish between them. How could you extend the chromatographic technique in order to show whether or not this explanation is correct?

Requirements

Large test tube (24 × 150 mm)
Glass jar approximately
  40 cm high × 7 cm diameter
  (gas jar recommended)
Lid for glass jar
Strip of chromatography paper
  (approximately 40 cm × 2 cm)
Dropping pipette with fine point
Measuring cylinder (10 cm^3)
Sellotape
Crystallizing dish (125 mm
  diameter)
Incubator at 37°C

0.5 g trypsin
5.0 cm^3 sodium bicarbonate
2.0 g egg white or 5.0 cm^3 fluid
  albumen
Thymol crystals
200 cm^3 solvent (see below)

Distilled water
Ninhydrin reagent (sufficient to
  fill crystallizing dish to depth of
  10 mm)

Make up solvent as follows: 4 parts of butan-1-ol, 1 part glacial acetic acid, and 1 part distilled water.

Make up ninhydrin reagent as follows: 1.0 per cent solution of ninhydrin in butan-1-ol.

Amino Acid Concentration	(%)	R_f value
Glutamic acid	16.5	0.30
Aspartic acid	9.3	0.24
Leucine	9.2	0.73
Serine	8.2	0.27
Phenylalanine	7.7	0.68
Valine	7.1	0.60
Isoleucine	7.0	0.72
Alanine	6.7	0.38
Lysine	6.3	0.14
Arginine	5.7	0.20
Methionine	5.2	0.55
Threonine	4.0	0.35
Tyrosine	3.7	0.45
Proline	3.6	0.43 (yellow)
Glycine	3.1	0.26

Table 5.1 The amino acids present in albumen listed in order of decreasing concentrations.

Questions and Problems

1   Using Table 5.2 (pp. 38–41) to help you, write a short essay on the importance of mineral elements to organisms. In doing this, don't reproduce all the data in the table, but *select* that information which, in your view, illustrates the basic principles involved.

2   Describe the experiments you would perform to determine which mineral elements are required for the growth of barley plants, and in what form?

3   Carbon is quadrivalent. Why does this make it ideal as a constituent of biological molecules? Illustrate your answer by referring to the structure of carbohydrates, fats and proteins.

4   What are the physical properties of starch and cellulose, and how are the properties related to the functions of these two compounds in plant cells? How does a knowledge of the chemical structure of starch and cellulose help to explain their physical properties?

5   Compare and contrast carbohydrates and proteins from a structural and functional point of view.

6   The amino acids present in a protein can be identified by hydrolysing the protein with an enzyme and then separating the amino acids by chromatography. However, such an analysis gives no information on the sequence of amino acids in the protein molecule. What kind of techniques can be used to investigate this aspect of protein structure? To what extent have successful experiments been carried out along these lines in recent years?

7   It was discovered accidentally in the East Indies towards the end of the nineteenth century that fowls fed on cooked polished rice developed paralysis, but when fed on cheaper unhusked rice they recovered. Without being prejudiced by modern knowledge, suggest hypotheses to explain this. How would you test your hypotheses?

   Now refer to Table 5.3 and explain as fully as possible what the fowls were actually suffering from.

8   Using the information in Tables 5.2 and 5.3, and your general biological knowledge, make an abbreviated list of the constituents of a complete human diet. Give the common source of each constituent, state its functions in the body, and summarize the consequences that result from its absence.

9   What is meant by (a) a vitamin, and (b) an essential amino acid? Make clear the physiological and nutritional differences between them.

   Separate samples of young rats were fed on different diets as shown below. After a period of 28 days the rats were killed and sections of their limb bones were made and examined. The results are shown below.

Diet given	Bone section showed
(i) Kitchen scraps	Normal development
(ii) Maize, calcium carbonate, sodium chloride, yeast, wheat protein, water	Severe rickets
(iii) As in (ii) plus olive oil	Severe rickets
(iv) As in (ii) plus ergosterol	Severe rickets
(v) As in (ii) plus olive oil and ergosterol	Normal development

What can you deduce from these experiments? Explain carefully how you arrive at your conclusions.

(*AEB*)

10 How can the shape of a protein molecule be determined? Why is it that the fibres of wool are easily stretched but collagen fibres are virtually unstretchable?

11 On the side of a packet of a certain well-known brand of corn flakes is printed the following information.

'One ounce of our Corn Flakes provides at least one quarter of the recommended daily intake of these vitamins for the average adult.

Vitamin	Quantity per ounce	Recommended daily intake
Niacin	4.50 mg	16.50 mg
Riboflavin B$_2$	0.40 mg	1.50 mg
Thiamine B$_1$	0.30 mg	1.10 mg

Niacin is the anti-pellagra vitamin
Riboflavin is essential to growth, normal digestion, normal vision, healthy skin
Thiamin aids nerve function and the utilization of carbohydrate foods.'

Discuss this information from a scientific point of view.

The elements are listed in alphabetical order. 'General' functions are those that apply to both animals and plants. The data relate to the flowering plant and man unless otherwise stated.

Table 5.2   Mineral elements required by organisms

Element	Obtained as	Functions	
		general	*plant*
Boron (trace)	$BO_3{}^{3-}$ or B_4O^{2-}	—	Required for uptake and utilization of Ca^{2+}, pollen germination and cell differentiation
Calcium	Ca^{2+}	—	Required for development of stem and root apex; as calcium pectate in plant cell wall
Chlorine	Cl^- by plants NaCl by animals	With Na^+ and K^+, helps determine osmotic pressure and anion-cation balance in inter- and intracellular fluid	May be involved in light stage of photosynthesis
Cobalt (trace)	Co^{2+}	—	Involved in nitrogen-fixation by certain plants
Copper (trace)	Cu^{2+}	—	Activates certain plant enzymes
Fluorine (trace)	F^-	—	May be required by certain plants but function unknown
Iodine (trace)	I^-	—	Not required
Iron	Fe^{2+}	Metallic radical in prosthetic group of cytochromes; also constituent of catalase	Required for synthesis of chlorophyll
Magnesium	Mg^{2+}	Activates enzymes in phosphate metabolism	Metallic radical in prosthetic group of chlorophyll
Manganese (trace)	Mn^{2+}	Activates certain enzymes	Activates carboxylases

imal	Deficiency Symptoms		Main sources for man
	plants	animals	
ot required	Brown heart disease	—	—
equired for skeleton and eeth formation (more than) per cent of body calcium in skeleton); muscular ontraction, blood clotting	Stunted growth of root and stem	Poor growth of skeleton, soft bones, muscular spasms, delayed clotting times	Milk, cheese, fish, drinking water if hard
equired for activity of xcitable tissue (nerve, uscle and receptors)	Negligible	Shortage of NaCl causes muscular cramps	See sodium
onstituent of vitamin B_{12}	—	Same as vitamin B_{12} deficiency	Most foods
equired for formation of aemoglobin; metallic adical in prosthetic group of aemocyanin in blood of ertain invertebrates. onstituent of certain ertebrate enzymes and eruloplasmin, a globular lasma protein	Dieback of shoots	Certain metabolic disorders	Most foods
ound in bones and teeth; revents dental caries	—	Weak teeth, especially in children	Drinking water
onstituent of thyroxine	—	—	Sea fish, shellfish, drinking water and vegetables if soil contains iodine
Metallic radical in rosthetic group of aemoglobin and nyoglobin in vertebrate lood and certain other lood pigments in nvertebrates	Chlorosis (Yellowing of leaves)	Anaemia	Liver, kidneys, beef, eggs, cocoa powder, apricots, drinking water if soil contains iron
Enters into composition of ones and teeth	Chlorosis	—	Nearly all foods
Activates oxidases, ntestinal aminopeptidase nd bone phosphatase particularly important in irds)	Chlorosis, grey spots on leaves	Malformation of skeleton	Most foods

Element	Obtained as	Functions	
		general	*plant*
Molybdenum (trace)	Mo^{3+} or Mo^{4+}	—	Activates certain enzymes in nitrogen metabolism
Nickel (trace)	Ni^{2+}	—	May be required by certain plants but function unknown
Nitrogen	NO_3^- or NH_4^+ by plants; protein by animals	Constituent of proteins, nucleic acids and porphyrins	—
Phosphorus	$H_2PO_4^-$ (ortho-phosphate) by plants; combined, with, e.g. protein by animals	Constituent of unit membranes (as phospholipid), certain proteins, all nucleotides (e.g. ATP) and nucleic acids. Required for phosphorylation of sugar in carbohydrate metabolism	—
Potassium	K^+	Helps determine anion-cation balance particularly in intracellular fluid. Involved in protein synthesis	Involved in formation of cell membrane. Increases hardiness
Selenium (trace)	Se^{2+}	Involved in formation of coenzyme Q	—
Silicon	$H_2SiO_4^{2-}$	—	Constituent of certain grasses
Sodium (trace in plants)	Na^+	Helps determine osmotic pressure and anion-cation balance	Not essential
Strontium (trace)	St^{2+}	—	—
Sulphur	SO_4^{2-} by plants; protein by animals	Constituent of certain proteins	—
Zinc (trace)	Zn^{2+}	Required in various enzyme systems	Activates carboxylases

…mal	Deficiency symptoms		Main sources for man
	plants	*animals*	
…und in prosthetic group …certain enzymes involved …uric acid formation	Slight retardation of growth	—	Most foods
…t required	—	—	—
	Stunted growth; chlorosis	Stunted growth and weakness (kwashiorkor) due to lack of protein	Protein foods
…quired for formation of …eleton	Poor growth, leaves dull green	—	Most foods
…quired for activity of …xcitable tissue	Yellow edges to leaves premature death	—	Prunes, potatoes, Brussels sprouts, mushrooms, cauliflower, beef, liver, fish
	—	—	Most foods
…ot required	Slight decrease in weight	—	—
…equired for activity of …citable tissue	Negligible	Muscular cramp	As NaCl in table and cooking salt, bacon, salty fish, cheese
…resent in bones and teeth …t no known …hysiological function	—	—	Water, vegetables
…-	Chlorosis	—	Sulphur-containing protein foods
…resent in carbonic …hydrase	Malformed leaves	—	Most foods

Table 5.3 Vitamins required by man

Letter designation	Synonym	Deficiency disease	Functions	Principal sources
Fat soluble				
A (group) (long chain alcohol)	**Retinol** Anti-xeroph-thalmic vitamin	Poor dark adaptation. Xerophthalmia (drying and degeneration of cornea). Drying of conjunctiva	Essential in light perception by rods by uniting with protein (opsin) to form the photochemical pigment rhodopsin. Light breaks down this pigment	The carotenoid pigments (yellow and orange pigments in vegetables, particularly carrots) contain precursors of vitamin A
D* (group) (sterols)	**Calciferol** anti-rachitic vitamin	Rickets in children. Osteomalacia in adults. (Rickets involves enlarge-ment of ends of long bones and bending of shaft)	Increases absorption of calcium and phosphorus from intestine, and increases uptake of Ca and P by bone: hence influences calcification and hardening of bone	Liver of fish; small amounts in eggs and cheese. Ultraviolet light on skin causes its formation from ergosterol, a deriva-tive of cholesterol
E (an alcohol)	**Tocopherol** anti-sterility vitamin	Degeneration of testes and ovary. Muscular dystrophy and nervous defects (animals)	Poorly understood. May be important in cell respiration by exerting action on ubiquinone which is an electron carrier between FAD and cytochromes	Many plants (lettuce, grasses, pea-nuts) milk, egg yolk
K (Ring compounds attached to hydrocarbon chain)	**Phyllo-quinone** etc. anti-haemorrhage vitamin	Haemorrhage in mucous membranes and organs. Prolonged clotting times	Required for synthesis by the liver of certain coagulation (clotting) factors in blood plasma, notably prothrombin, proconvertin (Factor VII) and Christmas Factor (Factor IX). Mode of action unknown	Cabbage, spinach, etc. Also tomatoes and pig's liver
Water soluble				
B$_1$ (pyrimidine ring and thiazole)	**Thiamine** antineuritic vitamin	'Beri-beri': wasting of muscles, gastric upsets, circulatory failure and paralysis	Thiamine, as thiamine pyrophosphate, is required as a coenzyme for the decarboxylation of pyruvic acid to Acetyl CoA. (NAD and CoA are also required for this reaction which probably involves many steps)	Yeast, rice, cereals and most plant and animal tissues
B$_2$ (isoalloxa-zine derivative)	**Ribo-flavine**	Soreness and lesions of mouth, ulceration, eye irritation, facial dermatitis, eye lesions etc.	Forms two important coenzymes: FAD (flavine adenine dinucleotide) and FMN (flavine mononucleotide). These are concerned in hydrogen transport	Leafy vegetables, fish, eggs
PP (pyridine carboxylic acid)	**Nicotinic acid** (niacin) pellagra-preventive factor	Pellagra: diarrhoea, dermatitis and mental disorder. Characteristic pigmentation of neck (Casal's necklace)	Forms two important coenzymes: NAD and NADP, both concerned in hydrogen transport	Meat, fish, wheat
B$_5$	**Panto-thenic acid**	Headache, fatigue, poor motor coordination, muscle cramps, gastro-intestinal disturbances	Forms coenzyme A which 'activates' certain carboxylic acids, notably acetate derived from pyruvic acid, thereby rendering them capable of further chemical interactions. Acetate, as acetyl CoA can undergo all manner of conversions	All animal and plant tissues. Yeast and eggs particularly rich

Letter Designation	Synonym	Deficiency disease	Functions	Principal sources
B_6 (group) (alcohol with its aldehyde and amine)	**Pyridoxine**	Anaemia, vomiting, diarrhoea, convulsions (animals)	Pyridoxal phosphate, a coenzyme derived from pyridoxine, forms the prosthetic group of a number of important enzymes including certain transaminases which transfer amino groups to certain keto acids, thus making it possible for amino acids to be synthesized from carbohydrate intermediates	In low concentration in all animal and plant tissues
B_{12} (group) (porphyrin with central cobalt atom)	**Cobalamin** anti-pernicious anaemia vitamin	Pernicious anaemia: variation in size of RBCs due to abnormal RBC-formation in the bone marrow	Required by liver for production of haematinic principle needed for proper RBC-formation. Controversial but thought to be necessary for the reduction of single carbon fragments in the synthesis of labile methyl groups which are in turn required for the synthesis of purine and pyrimidine bases, choline, creatine etc. Absorption of B_{12} from gut (ileum) requires presence of an 'intrinsic factor' which is present in gastric juice. Intrinsic factor probably interacts with B_{12}, facilitating its absorption; it may also facilitate action of B_{12} at the cellular level	Beef (lean), ox kidney and liver, pigs' hearts, herrings. Original source probably exclusively microbial synthesis
M or Bc (group) (ring compound attached to glutamic acid)	**Folic acid**	Macrocytic anaemia	Concerned in transfer of single C atoms. Its general functions parallel those of vitamin B_{12}. Its precise relationship with B_{12} is not fully understood	All green-leaved vegetables; liver and kidney
C (enolic form of keto-gulofurano lactone)	**Ascorbic acid** anti-scorbutic vitamin	Scurvy: bleeding gums, loose teeth, weakness, muscle pains, weight loss. Delay in wound-healing	Fundamental biochemical function unknown. Known to be associated with iron metabolism and formation of collagen. Also required for formation of intercellular substance which binds cells together	Citrus fruits and green vegetables etc.
H (Cyclic derivative of urea)	**Biotin**	Natural deficiency rare. Experimental deficiency: dermatitis	Important coenzyme in carboxylation reactions in carbohydrate, fat and protein metabolism, e.g. carboxylation of pyruvic acid to oxaloacetic acid; conversion of acetyl CoA into long chain fatty acids; formation of citrulline	All animal and plant tissues, particularly yeasts, liver and kidney

* Since sufficient quantities are formed in the skin, vitamin D should more correctly be regarded as a hormone. Only in climates where there is very little sunlight or where children are kept under covers is an external source of this chemical substance required.

6 Chemical Reactions in Cells

Background Summary

1 Chemical reactions occurring in cells constitute **metabolism**. Metabolic processes proceed in small steps which constitute a metabolic pathway.

2 **Metabolic pathways** can be analysed and reconstructed by various techniques, for example the use of enzyme inhibitors, isotope labelling and chromatography.

3 Metabolic reactions may be synthetic (**anabolism**) or breakdown (**catabolism**). The former absorb energy (**endergonic**), the latter release energy (**exergonic**).

4 The energy released by catabolic reactions is required for driving anabolic reactions, for work, e.g. muscular contraction, and for maintenance purposes.

5 Oxidative breakdown of sugar yields carbon dioxide, water, and energy (**cell respiration**). The chemical products of this catabolic process can be re-synthesized into sugar by green plants (**photosynthesis**). A proportion of the energy from respiration can be used for the establishment of chemical bonds in organic molecules, the rest being lost as heat.

6 To initiate chemical reactions such as the oxidative breakdown of sugar, **activation energy** must be supplied.

7 Chemical reactions in organisms are catalysed by **enzymes**. Enzymes effectively lower the amount of activation energy required to initiate the reaction.

8 **Enzymes are proteins** and their properties are as follows:
(a) They generally work very rapidly
(b) They are not destroyed by the reactions they catalyse
(c) They are inactivated by excessive heat
(d) They are sensitive to pH
(e) They are usually specific.

9 Enzymes work by temporarily combining with substrate molecule(s) to form an **enzyme-substrate complex**. The enzyme molecule has an **active site** to which specific substrate molecules become attached. This **lock-and-key hypothesis** explains many properties of enzymes.

10 Enzymes are inhibited by poisons which either compete with the normal substrate molecules for the active site (**competitive inhibition**), or block the active site permanently (**non-competitive inhibition**).

11 Sometimes the end-product of a metabolic reaction itself acts as an enzyme inhibitor, thereby cutting down its own production and preventing itself from accumulating.

12 **Coenzymes** and **prosthetic groups** work in conjunction with enzymes and play an important part in transferring chemical groups from one enzyme to another.

Investigation 6.1

Action of the enzyme catalase

Nearly every chemical reaction which takes place in a living organism is catalysed by an **enzyme** which greatly speeds up the rate of the reaction. The enzymes are not consumed during these reactions and are considered to be organic catalysts. We now know that enzymes are always proteins, and their properties are therefore the properties of proteins.

The enzyme **catalase** breaks down **hydrogen peroxide** into water and oxygen. Hydrogen peroxide is a highly active chemical, often used for bleaching or for cleansing minor wounds. It is also formed continually as a by-product of chemical reactions in living cells. It is toxic, and if it were not immediately broken down by the cells it would destroy them. Hence the importance of the enzyme. It is, in fact, the fastest enzyme known. In this investigation you will be able to watch the action of catalase in a test tube and compare it with an inorganic catalyst that catalyses the same reaction.

Procedure

(1) Pour hydrogen peroxide solution into each of two test tubes to a depth of about two centimetres. Into one of the test tubes sprinkle about 0.1 g of **fine sand**. Note the result, if any. Into the second test tube sprinkle the same amount of either **manganese dioxide powder** or **iron filings**. Observe what happens and test for oxygen with a splint. Make notes on your observations.

(2) Pour fresh hydrogen peroxide into a clean test tube to the same depth as before. Now cut a **cube** of **liver** 1-cm square and drop it into the test tube of hydrogen peroxide. Observe carefully and record what happens.

(3) Take a piece of liver the same size as before and place it in a mortar along with a little fine sand. Grind, and then transfer the **ground-up liver** (along with the sand which will not interfere with the reaction) to a test tube containing fresh hydrogen peroxide. Note how the activity of the ground-up liver compares with the activity observed for the whole piece of liver. Explain fully.

(4) Take another piece of liver and put it in a beaker of **boiling water** for three minutes. Then drop the piece of liver into fresh hydrogen peroxide and find out if the enzyme is still capable of breaking down hydrogen peroxide. Explain your results as fully as possible.

(5) Do experiments to determine if catalase occurs in other organs and tissues besides liver. Mammalian kidney, muscle, and blood might be tried; also potato and apple. Test ground-up material as well as whole pieces (why?).

(6) Do an experiment to find out as accurately as you can the temperature at which the enzyme is destroyed. For most enzymes this critical temperature is about 40°C. Does the same apply to catalase?

For Consideration

(1) These experiments on catalase are qualitative, or only very roughly quantitative. How could you *measure* the rate at which hydrogen peroxide is broken down?

(2) What conclusions can you draw from these experiments as to how enzymes work, and what further experiments could you do to test your ideas?

Requirements

Test tubes
Pestle and mortar
Beaker (240 cm³ approx.)
Thermometer
Bunsen burner, tripod, and gauze
Splint

Fine sand
Hydrogen peroxide
Manganese dioxide powder
Iron filings

Mammalian liver, kidney, blood, and muscle
Potato and apple

Investigation 6.2

Effect of temperature on the action of an enzyme

Enzymes, being proteins, are readily inactivated by excessive heat. One way of investigating the influence of heat on the action of an enzyme is to expose the enzyme to a given temperature for a known period of time, and then estimate how long it takes to catalyse its particular reaction. In this experiment samples of the enzyme **diastase** are exposed to different temperatures for exactly five minutes. The time required for each sample to hydrolyse a given volume of starch is then estimated. We shall use iodine as an indicator: in the presence of iodine starch turns blue.

Procedure

(1) Label five test tubes: room temperature (control), 25°C, 40°C, 60°C, and 100°C. To each add 5 cm³ of **diastase solution**.

(2) Place each tube in the appropriate water bath for exactly five minutes. The first tube should be kept at room temperature.

Requirements

Five test tubes with labels or marked with wax pencil[1]

Test tube rack

Pipette (5 cm³)

White tile

Six glass rods

Water baths maintained at 25°C, 40°C, 60°C and 100°C.

Diastase solution (enzyme) (25 cm³ of 0.2 per cent)

Starch solution (substrate) (25 cm³ of 0.5 per cent)

Dilute iodine

(3) At the end of the five-minute period remove the tubes from the water baths, and cool them rapidly to room temperature.

(4) Now add to each tube 5 cm³ of **starch solution** and mix with a clean glass rod.

(5) At intervals of five minutes test each tube for the presence of starch: withdraw one drop of the starch-diastase mixture, place it on a white tile, and add one drop of **iodine solution**. Avoid contaminating the mixtures by using one glass rod for each tube and a separate one for the iodine. Continue doing this for at least one hour. Make a complete record of your observations, noting in particular how long it takes in each case before a blue colour ceases to be given when iodine is added to the mixture.

For Consideration

(1) As a method of investigating the effect of temperature on enzyme action, what are the shortcomings of this experiment? What could be done to improve the experiment?

(2) Suggest further experiments you might carry out to examine the effect of temperature on the inactivation of the enzyme.

(3) Interpret your results as fully as you can in terms of what you know about how enzymes work.

Investigation 6.3

Specificity of enzymes: investigation into which sugars can be metabolized by yeast

In anaerobic conditions yeast cells break down sugar into ethanol and carbon dioxide. This process is called **alcoholic fermentation**:

$$C_6H_{12}O_6 \rightarrow 2C_2H_5OH + 2CO_2\uparrow + E$$

The process consists of a series of reactions catalysed by specific enzymes. The enzymes present in yeast cells are specific, i.e. they will catalyse the fermentation of certain sugars but not others.

In this experiment you will investigate which of four different 6-carbon sugars can be broken down by yeast. We shall use the evolution of carbon dioxide as an indication that fermentation has taken place. After finding out which sugars can, and cannot, be fermented, the structural formulae of the four sugars can be compared so as to determine the particular configuration of atoms required for the yeast enzymes to work.

Procedure

(1) Label five fermentation tubes 1–5, writing the numbers with a wax pencil upside down near the bottom of each tube.

(2) Pipette the following into each tube:

No. 1: 10 drops of distilled water
No. 2: 10 drops of glucose
No. 3: 10 drops of fructose
No. 4: 10 drops of galactose
No. 5: 10 drops of sorbose

To each tube add 10 drops of **yeast** and top up with **distilled water**. N.B. Mix the yeast suspension thoroughly by swishing the flask before adding the 10 drops to each tube.

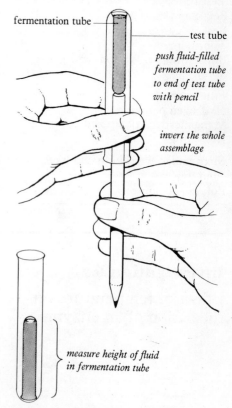

Fig. 6.1 Assembling a fermentation tube.

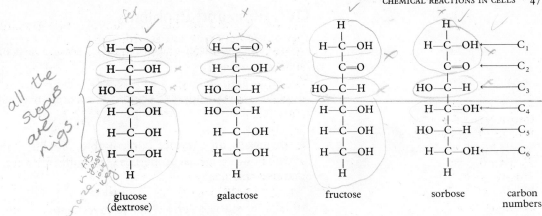

Fig. 6.2 The structural formulae of the four sugars studied in Investigation 6.3.

glucose (dextrose)   galactose   fructose   sorbose   carbon numbers

H—C=O
HO—C—H
HO—C—H
H—C—OH
H—C—OH
H—C—OH
H

mannose

H—C=O
H—C—OH
HO—C—H
HO—C—H
H—C—OH
H

arabinose

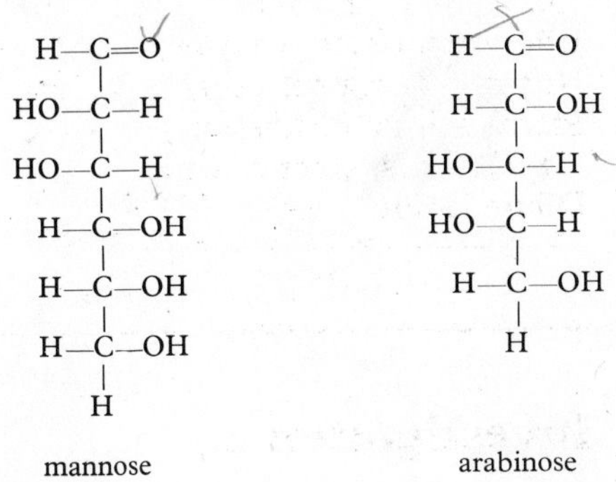

Fig. 6.3 The structural formulae of two other sugars.

(3) Set up the fermentation tubes as follows. Hold the filled tube in your hand and insert it into an inverted test tube. With a pencil, or some other suitable instrument, push the small tube into the inverted test tube as far as it will go and then invert the whole assemblage (Fig. 6.1). Tap the outer tube firmly to release any bubbles trapped at the mouth of the inner tube.

(4) When all the fermentation tubes have been set up, record the height of the fluid in mm in each one. Do this by holding the test tube vertically on a table and placing a millimetre ruler alongside it.

(5) Now place all the test tubes in a constant-temperature water bath at 37°C for at least one hour.

(6) Remove the test tubes from the constant-temperature bath, and re-measure the height of the fluid in each fermentation tube. If enough gas has collected in the fermentation tube to make it buoyant, you must push it down when taking your reading.

(7) Record the difference in height between the initial and final reading for each of the tubes.

For Consideration

(1) Fig. 6.2 shows the structural formulae of four sugars investigated in this experiment. All four have the general formula $C_6H_{12}O_6$, but they differ in the positions of their hydrogen and oxygen atoms, and OH groups, in relation to the carbon atoms. The carbon atoms are numbered 1–6.

From your results, which positions are important, and which ones unimportant, in determining whether or not the yeast enzymes will break down the sugar?

(2) Fig. 6.3 shows the structural formulae of two other sugars, mannose and arabinose. Compare these formulae with those of the other sugars, and predict which ones can be fermented by yeast. If time and facilities permit, test your predictions.

(3) Interpret the results of this experiment in terms of what you know about how enzymes work?

Requirements
Constant-temperature water bath at 37°C
Polypropylene test tube rack
Five fermentation tubes (Durham: 50 × 7.5 mm)
Five test tubes
Wax pencil
Ruler with mm scale
Flask of brewer's yeast
Distilled water
20 per cent solutions of the following sugars:
  Glucose (dextrose)
  Fructose
  Galactose
  Sorbose
  (Mannose)
  (Arabinose)

* Since wax writing tends to melt at high temperatures a non-water soluble felt pen may be preferred.

Questions and Problems

1 Discuss the significance of the first and second laws of thermodynamics in cell biology.

2 More than 1,000 different chemical reactions may take place in a single cell. How are confusion and chaos prevented?

3 In what ways do enzymes (a) resemble, and (b) differ from other catalysts? Explain how the special properties of enzymes are a consequence of the structure of enzyme molecules. *(JMB modified)*

4 An experiment was carried out to investigate the effect of temperature on the rate of an enzyme-controlled reaction. The concentrations of enzyme and substrate were kept constant at all the temperatures investigated. The results were as follows:

Temperature (°C)	Rate of reaction (mg of products per unit time)
5	0.3
10	0.5
15	0.9
20	1.4
25	2.0
30	2.7
35	3.3
40	3.6
45	3.6
50	2.3
55	0.9
60	0

Plot the results on graph paper. Interpret and explain them as fully as you can.

5 The table below shows the activity of an enzyme (in arbitrary units) in relation to pH:

pH	4.5	5.5	6.5	7.5
Units of enzyme activity	3.1	9.6	14.5	10.1

(a) Have you any criticisms of the data?
(b) What general features of enzyme action are illustrated by these figures?
(c) How are the results related to the chemical structure of enzymes?
(d) Give an example of a particular enzyme that might give such results.
(e) Give examples of enzymes which would *not* be expected to give these results.
(JMB modified)

6 Two types of enzyme inhibitor are recognized. In the first, the extent to which the enzyme-controlled reaction is inhibited depends on the relative concentration of the substrate and inhibitor. In the second, the extent of the inhibition depends only on the concentration of the inhibitor, and cannot be varied by changing the amount of substrate present. From this information, can you suggest a hypothesis explaining the action of each type of inhibitor?

7 An experiment was carried out to investigate the effect of varying the concentration of substrate on the enzymatic hydrolysis of adenosine triphosphate (ATP). The concentration of enzyme (ATPase) was kept constant at each substrate concentration investigated. The results were as follows:

Concentration of substrate (ATP) (millimoles/dm^3)	Rate of hydrolysis (micromoles/ dm^3/s)
0.01	0.06
0.02	0.09
0.04	0.15
0.08	0.18
0.17	0.19
0.25	0.19

Plot these results on a graph and interpret them as fully as you can. Would you expect the shape of the graph to be the same whatever enzyme-controlled reaction was investigated?

8 Suggest ways in which enzyme inhibitors might be made use of in (a) agriculture, (b) medical practice, and (c) biochemical research.

catalysts eg iron, platinum, sulphuric acid etc

enzymes can be distinguished from non biological catalysts - whilst such catalysts as Raney nickel will readily catalyse a whole class of reactions of the same general type - that of reduction of hydrogen - enzymes are very highly selective about the reactions they choose to speed up & those they will not catalyse. The cell requires a different enzyme for nearly every reaction it carries out.

(the majority of enzymes - protein

13 6 87 9
24 8 25 1   P 89

catalysts inorganic
enzymes organic

enzymes - more quick & efficient

7 The Release of Energy

Background Summary

1 Energy is released by **cell respiration** (**tissue** or **internal respiration**) which generally involves the oxidation of sugar:

$$C_6H_{12}O_6 + 6O_2 \rightarrow 6H_2O + 6CO_2 + Energy$$

2 Simple experiments can be carried out to show that oxygen is used and carbon dioxide produced in respiration.

3 Spirometry and/or analysis of inspired and expired air enables oxygen consumption and carbon dioxide production to be determined quantitatively. This provides a measure of the **metabolic rate**. Oxygen consumption and carbon dioxide production depend on the degree of muscular activity occurring in the body.

4 Dividing the amount of CO_2 produced by the amount of O_2 consumed gives the **respiratory quotient** (RQ). Knowledge of the RQ gives information on the type of food being respired, and the kind of metabolism that is taking place.

5 Another method of determining the metabolic rate, apart from measuring O_2 consumption and CO_2 production, is to estimate the energy released per unit time. This can be done by measuring heat production using a calorimeter.

6 The minimum amount of energy on which the body can survive is the **basal metabolic rate** (BMR). Actual metabolic rates usually exceed the BMR, and depend on the activity of the individual.

7 The energy value of different foods can be determined by means of a bomb calorimeter which measures the amount of heat produced when a given quantity of food is burned.

8 Sufficient energy-containing food must be consumed to maintain the BMR and to satisfy such additional energy needs as the individual's activities and circumstances demand.

9 The immediate source of energy for biological functions is **adenosine triphosphate** (ATP) which, when hydrolysed in the presence of the appropriate enzyme, liberates free energy.

10 Energy for synthesis of ATP comes from the breakdown of **sugar**, stored in cells as **glycogen** (animals) or **starch** (plants). Fat and protein are also potential sources of energy for ATP synthesis.

11 The step-by-step breakdown of sugar yields energy for ATP synthesis in two ways, of which the second is by far the most productive:
(a) Several of the reactions are exergonic, each releasing sufficient energy for synthesis of an ATP molecule.
(b) Many of the reactions involve the removal of two hydrogen atoms (dehydrogenation) which are taken up by an acceptor. The hydrogen atoms, or their electrons, are then passed along a series of carriers, the energy released at each transfer being used for synthesis of a molecule of ATP.

12 Aerobic breakdown of sugar occurs in two main stages: **glycolysis** (sugar → pyruvic acid) followed by **Krebs' citric acid cycle.** Most of the energy for ATP synthesis is derived from the Krebs cycle.

13 Evidence suggests that glycolysis takes place in the cytoplasm, Krebs cycle in the matrix of the mitochondria, and electron transfer on the cristae of the mitochondria.

14 In anaerobic breakdown of sugar, pyruvic acid is converted into **lactic acid** (animals) or **ethanol** (plants), Krebs cycle being omitted. Considerably less energy is released this way.

15 Carbohydrate metabolism is connected with fat and protein metabolism for which reason these two groups of compounds may also be used as sources of energy for ATP synthesis.

Investigation 7.1

Multiplication of yeast cells in aerobic and anaerobic conditions

Requirements
Microscope
Haemocytometer
Conical flasks (3) of capacity
    50 cm³, 150 cm³, and 500 cm³
Measuring cylinder (50 cm³)
Muslin (cheese cloth)
Pipette
Elastic band

Rich culture of brewer's yeast
Uncontaminated dry cider

Cell division, like any other vital activity, requires energy, and the rate at which cells divide depends on the amount of energy available. Because of the difference in energy yield of aerobic and anaerobic respiration, we would expect cell division to proceed faster in aerobic than in anaerobic conditions. The following experiment is designed to test this hypothesis.

Procedure
(1) Obtain three conical flasks of the following sizes: 50 cm³, 150 cm³, and 500 cm³ (or close equivalents).

(2) Pour 50 cm³ of **cider** into each flask, and to each one add *one drop* of **yeast suspension** (make sure that the yeast suspension is mixed thoroughly first).

(3) Cover the opening of each flask with four layers of muslin (cheese cloth) held in place by a rubber band. Leave your flasks in a warm cupboard for about a week.

(4) **One week later:** examine the distribution of yeast cells in your three flasks. Now distribute the yeast cells uniformly. For the two larger flasks this can be done by swirling the flasks; in the case of the small flask it can be done by sucking and expelling the contents back and forth with a pipette.

(5) After the contents of each flask are thoroughly mixed, remove a sample with a clean pipette, and estimate the concentration of cells using a **haemocytometer** (*see* p. 102). For counting the cells select a type of square that contains between 8 and 12 cells per square. Count in as many squares as you consider necessary to give a reasonably accurate result, and carry out at least two separate counts for each flask. Express the concentration of yeast cells in each flask as number of cells per cm³.

For Consideration
(1) How would you summarize conditions in the three flasks with reference to the availability of oxygen?

(2) Is our original hypothesis that cell division occurs more rapidly in aerobic than anaerobic conditions confirmed or refuted?

(3) Are there any other possible explanations for the different cell concentrations in the three flasks apart from availability of oxygen?

(4) Which do you consider is the critical factor determining the availability of oxygen: the depth of medium through which the oxygen has to diffuse to the yeast cells, or the surface area of the medium exposed to oxygen —or both?

(5) Can you, from the results of this investigation, make predictions concerning the shape and complexity of organisms in relation to their size? To answer this question think of the *medium* in each flask as an *organism*.

Investigation 7.2

Oxygen consumption in man

For the cells of the body to maintain their vital activities, a continual expenditure of energy is required. This energy comes from the oxidative breakdown of food. Estimating oxygen consumption gives a measure of the **metabolic rate**. In this investigation changes in the metabolic rate will be related to different kinds of activity.

Method

The technique involves the use of a **spirometer**. The model recommended for this experiment consists of an inverted perspex lid floating in a tank of water (Fig. 7.1). The chamber is filled with oxygen and connected to the subject by a rubber mouthpiece at the end of a length of concertina tubing. A cannister of soda lime is inserted between the mouthpiece and the spirometer chamber so all the carbon dioxide expired by the subject is absorbed. At the side of the spirometer is a two-way tap enabling the subject to be quickly connected to, or disconnected from, the spirometer

chamber. The subject's nose must be clipped so that, when connected, his lungs and respiratory tract form a closed system with the chamber and tubing of the spirometer. As he breathes in and out, the lid goes up and down in time with his respiration, and slowly sinks as oxygen is used up.

The lid is counter-balanced by an adjustable weight which should be set so that the lid falls very slowly when the spirometer chamber is open to the atmosphere. When the chamber is closed off, the lid should remain stationary: if it falls there is a leak in the system.

Changes in the volume of oxygen in the spirometer chamber can be read off the scale attached to the side of the lid. Alternatively, movements of the lid can be recorded by a pen writing on a slowly revolving **kymograph** drum fitted with paper calibrated horizontally for volume. It is necessary to know the speed at which the drum rotates so the rate of oxygen consumption can be measured. The kymo-

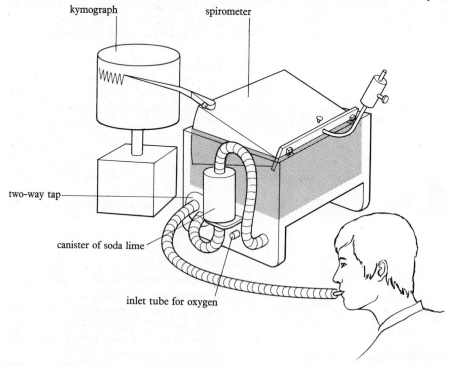

kymograph

spirometer

two-way tap

canister of soda lime

inlet tube for oxygen

Fig. 7.1 Diagram of recording spirometer. *See* text for explanation.

graph should be set at a speed of approximately 20 mm/min.

Procedure for doing an experiment

With the two-way tap closed, fill the spirometer chamber with oxygen from a cylinder. Make a note of the volume, or if you are using a kymograph, put ink in the pen and bring it into contact with the top of the drum surface.

Rinse the rubber mouthpiece in antiseptic and insert it into the subject's mouth. Clip the subject's nose. The two-way tap being closed, the subject is connected to the outside atmosphere and should remain so until he has got accustomed to breathing through the mouthpiece.

When you are ready to proceed, turn the two-way tap so as to connect the subject with the spirometer chamber. This must be done *at the end of a normal expiration*, so the first movement of the spirometer lid will be downwards.

When you have finished, turn the tap at the end of a normal expiration so as to disconnect the subject from the spirometer and re-connect him with the atmosphere. If you are not using a kymograph, make a note of the new volume.

The subject's oxygen consumption can be determined by subtracting the new volume from the initial volume. This can be done using the scale on the lid or, if you are using a kymograph, from your recordings. It is most important that both the initial and final volume are taken at the end of a normal expiration.

Experiments

OXYGEN CONSUMPTION AT REST

With the subject sitting as relaxed as possible, connect him to the spirometer chamber for five minutes. Estimate his **total oxygen consumption** in cm^3 during the five-minute period. What is his oxygen consumption as measured in cm^3/min?

OXYGEN CONSUMPTION DURING VIGOROUS MUSCULAR EXERTION

Refill the spirometer with oxygen. The same subject, with mouthpiece in position and two-way tap open to the atmosphere, should now engage in vigorous muscular activity (e.g. running on the spot) for 12 minutes. Measure his oxygen consumption during the 9th and 10th minutes.

During the 11th and 12th minutes, while the subject continues to exert himself, the partner should top up the spirometer with oxygen.

After 12 minutes' exercise the subject should rest. As soon as he does so, his partner should estimate his oxygen consumption over a further five minutes, i.e. during the subject's recovery period.

Estimate the oxygen consumption in cm^3/min for the 9th and 10th minutes of exercise, and for the five-minute recovery period. Compare with the results obtained in the first experiment.

For Consideration

(1) What general conclusions can be drawn from your results?

(2) Oxygen consumption gives a measure of metabolic rate. The subject should convert his weight into grammes and calculate his oxygen consumption in $cm^3 O_2/g/h$ (a) in resting conditions, (b) during muscular activity. What significance would you attach to the difference?

(3) How does the five-minute recovery oxygen consumption compare with the resting oxygen consumption? Why should a high metabolic rate continue after muscular exertion has ceased?

Requirements

Recording spirometer with concertina tubing, mouthpiece, nose clip, CO_2-absorber, and recording pen

Eosin or non-clogging ink

Kymograph set at speed of about 20 mm/min

Kymograph paper calibrated horizontally for volume (500 cm^3 divisions)

Oxygen cylinder

Questions and Problems

1 Without going into experimental details, discuss the principles underlying the different methods by which an organism's metabolic rate may be determined. Assess the advantages and disadvantages of each method.

2 Measurements of gas exchange between a living tissue and its surroundings may be made using an apparatus of the kind shown in Fig. 7.2. The manometer

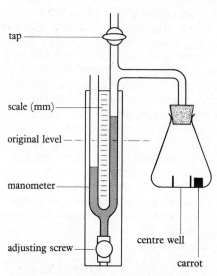

Fig. 7.2 Apparatus for measuring gas exchange between living tissue (in this case carrot root) and its surroundings.

registers changes in the volume of gases in the flask (and its connecting tube) that result from gases being absorbed or produced by living tissues placed in the flask. Changes in the level of the manometer fluid (measured in mm) are converted to changes in gas volume (measured in mm^3), at standard temperature and pressure, by multiplying by a constant factor.

If, given this apparatus, you were required to determine the rate of oxygen uptake in fresh carrot root tissue, answer the following questions to show how you would proceed:

(a) What will you require to know about the carrot root before it is placed in the flask?

(b) What substance will you add to the centre well of the flask?

(c) Describe your procedure for taking manometer readings during the course of an experiment.

(d) Do you consider that any additional measurements are necessary in carrying out an experiment of this kind?

(e) How could the apparatus be modified to enable you to ascertain the steps in a metabolic pathway? *(CL modified)*

3 What is meant by respiratory quotient (*RQ*) and how would you measure it for a small animal like a mouse?

The *RQ* for carbohydrate is 1.0, for fat 0.7, and for protein 0.9. What would you expect the *RQ* to be for most animals most of the time?

Under what circumstances would you expect the *RQ* of an organism to be (a) greater than 1.0, (b) lower than 0.7, and (c) somewhere between 1.0 and 0.7?

The *RQ* of a germinating cereal seed may vary from 0.7 to 1.0 and then rise to infinity before settling back to 1.0 again. Account for this. *(AEB modified)*

4   Refer to Appendix 3 in order to complete the following assignments:
(a) Make a rough estimate of your own daily energy intake.
(b) How adequate, or otherwise, is your own diet from the energy point of view?
(c) Find out the current cost per unit weight of six types of food listed in Appendix 3. Convert to cost per kilojoule. Assuming a dietary requirement of 10,500 kJ/day, calculate the cost per day of (i) the cheapest and (ii) most expensive meal that satisfies this energy requirement.
(d) Briefly discuss the social implication of your answer to (c) above.

5   The table gives approximate figures for the daily requirement of kilojoules and protein for human males of different age groups. The average body weight for each age group is included for comparison:

Age (years)	Weight (kg)	Kilojoules per day	Protein per day (g)
10–12	35	10 500	70
13–15	49	13 400	85
16–20	63	16 000	100
25	77	13 400	65
45	77	12 200	65
65	77	10 900	65

Explain the variation in energy and protein requirements at different ages.

(JMB modified)

6   Give a concise account of how energy is made available in cells. (A detailed account of the Krebs cycle is not required).

7   By reference to Table 5.3 discuss the importance of vitamins in cell respiration.

8   Aerobic breakdown of sugar yields 2,880 kJ of energy per mole, whereas anaerobic breakdown of sugar yields not more than 210 kJ/mole. Explain this difference in terms of the chemistry of cell respiration.

9   By what experimental techniques could it be ascertained whereabouts inside a cell the reactions of the Krebs cycle take place?

10   Discuss the circumstances in which (a) carbohydrate, (b) fat, and (c) protein are the substrates for cell respiration; and the circumstances in which (a) carbon dioxide, (b) ethanol, and (c) lactic acid are the end products.

11   Glutamate dehydrogenase catalyses the reaction in which the amino acid glutamine is converted into α-ketoglutaric acid, the latter being one of the intermediates in the Krebs cycle. Comment on the following data which show the effect of salicylic acid on the yield of the reaction. The salicylate concentration was kept constant at 40mM. (*Data after Lehninger.*)

Glutamine concentration (mM)	1.5	2.0	3.0	4.0	8.0	16.0
mg product/min without salicylate	0.21	0.25	0.28	0.33	0.44	0.40
mg product/min with salicylate	0.08	0.10	0.12	0.13	0.16	0.18

What physiological significance do you attach to the fact that glutamine can be converted into α-ketoglutaric acid in the body?

12 The following metabolic pathway represents the first four steps in the Krebs citric acid cycle. Examine it carefully and then answer the questions below:

(a) The five compounds in the pathway are all acids. To what do they owe their acid properties?

(b) The citric acid cycle is also known as the tricarboxylic acid cycle. Which of the above compounds are tricarboxylic acids and which are dicarboxylic acids?

(c) What kind of enzymes are responsible for catalysing each of the steps depicted?

(d) Citric acid is the first acid of the cycle. What do you know about the reaction by which citric acid is formed?

(e) How would you explain the fact that water is removed at step **1** and then added again at step **2**?

(f) Why is isocitric acid so-called?

(g) What happens to the two hydrogen atoms removed from the isocitric acid at stage **3**, and what is the biological usefulness of this?

(h) Why do you think it is necessary for carbon dioxide to be removed from oxalosuccinic acid at stage **4**?

(i) In general terms what happens to the α-ketoglutaric acid in the completion of the cycle?

(j) What sort of techniques would be used to elucidate the steps in a metabolic pathway such as the one outlined above?

13 What do you understand by the term glycolysis?

At the beginning of glycolysis two hydrogen atoms are removed from each of two 3-carbon sugars. The latter are formed from the initial splitting of glucose. In aerobic conditions the two hydrogen atoms are either used for various reduction processes in the cytoplasm, or in certain situations they may be shunted into a mitochondrion where they go through the hydrogen carrier system.

(a) In each case what will be the total number of ATP molecules formed every time a molecule of glucose goes through glycolysis and the Krebs citric acid cycle?

(b) In what tissues and/or circumstances would you expect the hydrogen atoms to be shunted into a mitochondrion?

(c) What happens to the two hydrogen atoms in anaerobic conditions, and what will be the consequences of this in terms of energy released?

14 Suggest experiments that might be carried out to test the hypothesis that glycolysis takes place in the cytoplasm, the reactions of the citric acid cycle in the matrix of the mitochondria, and the hydrogen carrier system in the cristae and membranes of the mitochondrion.

8 Gaseous Exchange in Animals

Background Summary

1 In small or flattened animals the **surface:volume ratio** is large enough for diffusion across the body surface to satisfy their respiratory needs. Larger animals, with a smaller S:V ratio, possess special **respiratory surfaces**, e.g. gills, lungs, tracheal tubes.

2 In man air is drawn by expansion of the thorax into the **lungs** where gaseous exchange occurs by diffusion across a greatly folded and highly vascularized respiratory surface.

3 This may be compared with fishes in which water is pumped over much folded and vascularized **gills**.

4 For efficient gaseous exchange in aquatic animals a **counterflow system** is better than parallel flow. The gills of bony fishes (teleosts) are such that a more direct counterflow is achieved than in the gills of cartilaginous fishes, e.g. dogfish.

5 In teleosts, and probably also in the dogfish, the ventilation mechanism, involving the combined actions of a force pump and suction pump, ensures a continuous stream of water over the gills at all stages of the respiratory cycle.

6 In insects gaseous exchange occurs in the **tracheal system**. Air reaches the tissues by diffusion, aided in some species by rhythmical movements of the thorax or abdomen.

7 In general rhythmical respiratory movements are coordinated by the nervous system, and can be influenced by changes in the level of carbon dioxide, etc.

Investigation 8.1

Dissection of the respiratory apparatus of the rat

In all mammals the respiratory and alimentary tracts cross each other at the back of the pharynx. Air in the nasal cavity is drawn via the nasopharynx, through the glottis and larynx into the trachea whence it reaches the lungs via the bronchi.

In the lungs oxygen from the inspired air comes into close association with an extensive and highly vascularized epithelium.

Procedure

(1) Pin out the animal, ventral side uppermost. Open up the thorax by cutting along the dotted lines shown in Fig. 8.1A. Tie a thread round the xiphoid cartilage and pull it back so as to stretch the diaphragm (Fig. 8.1B). Note the **muscles of the diaphragm** and the **intercostal muscles** between the ribs.

(2) Remove the thymus gland from the surface of the heart. Be careful not to damage the underlying blood

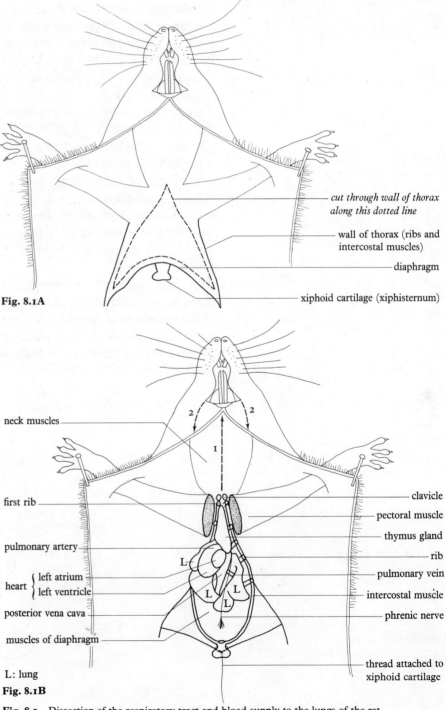

Fig. 8.1A

L: lung

Fig. 8.1B

Fig. 8.1 Dissection of the respiratory tract and blood supply to the lungs of the rat.

vessels. Identify **heart** and **lungs**. Remove superfluous fat sufficiently to see the **pulmonary arteries** and **veins** (Fig. 8.1B). Also note the **phrenic nerve** which innervates the diaphragm.

(3) With a scalpel cut along the centre of the neck muscles and remove them. (Arrow 1 in Fig. 8.1B). Be careful not to damage the trachea underneath.

(4) Dislocate the lower jaw (mandible) by cutting along the angle of the jaw on both sides of the head. (Arrows 2 in Fig. 8.1B). *Do not remove the tongue.*

(5) Grasp the tongue with forceps and cut along the walls of the pharynx on either side as far back as the **glottis**. This is the point where the alimentary and respiratory tracts cross.

Pull back the tongue and floor of the pharynx and notice the **epiglottis** guarding the glottis (Fig. 8.1C). Insert a probe into the glottis and ascertain that it enters the **larynx** and **trachea**.

Just dorsal to the glottis (i.e. on the far side of the glottis as you look at your dissection) is the opening into the **oesophagus**. Insert your probe into this opening, directing it posteriorly. Note that if you push the point of the probe to one side it distends the wall of the oesophagus.

(6) Now identify the **soft palate**. Immediately above this is the **nasal cavity**. Push a bristle into the nasal cavity from the rear and go on pushing until it comes out of one of the **nostrils**.

(7) Insert the tip of a rubber pipette into the glottis and inflate the lungs. What happens to the respiratory tract when the animal swallows?

(8) Follow the trachea down to the thorax. It disappears on the dorsal side of the heart where it divides into a pair of **bronchi**, one to each lung. To see the connection between the bronchi and the lungs necessitates removing the heart and the great vessels, which you may not want to do at this stage.

For Consideration

(1) Trace the path taken by a molecule of oxygen from the atmosphere just outside a person's nose to the blood in his pulmonary vein. What structures play a part in the process by which the air is moved?

(2) What are the possible advantages and disadvantages of the fact that the respiratory and alimentary tracts cross each other in the throat?

Requirements
Dissecting instruments
Bristle
Rubber pipette

Fresh-killed or deep-frozen rat

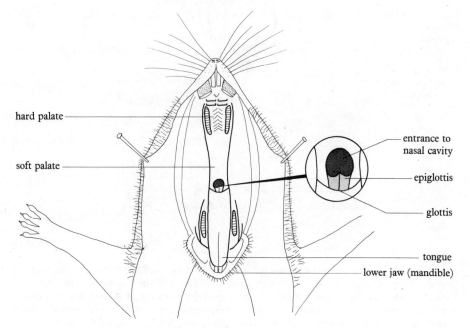

hard palate

soft palate

entrance to nasal cavity

epiglottis

glottis

tongue

lower jaw (mandible)

Fig. 8.1C Dissection of the respiratory tract and lungs of the rat. For explanation see text.

Investigation 8.2
Microscopic structure of the mammalian respiratory apparatus

Functionally the respiratory apparatus of the mammal can be distinguished into two major components, the respiratory surface itself where gaseous exchange takes place, and the respiratory tract through which air is moved to and from the respiratory surface by the animal's ventilation mechanism.

It follows that in its microscopic structure the respiratory surface would be expected to show an intimate relationship between the inspired air and the bloodstream; and the respiratory tract would be expected to show adaptations for keeping the tubes permanently open and preventing anything other than air reaching the respiratory surface.

Structure of the Trachea
Examine a transverse section of trachea under low and high powers. Use Fig. 8.2 to help you identify the various structures in the tracheal wall. In particular notice how the inner lining composed of **ciliated columnar epithelium** is interspersed with **goblet cells**; in some sections mucus glands are seen in the tissue immediately beneath the epithelium (submucosa) opening to the surface by ducts; and the incomplete **ring of cartilage** embedded in the centre of the wall. Why is the ring incomplete?

What are the functions of the cilia, glands and cartilage? Is there anything odd about the epithelial lining? Explain.

Can you make out any other tissues in the wall of the trachea? Explain their functions.

Structure of the Lung
Examine a section of mammalian lung, first under low power and then under high power. Using Fig. 8.3 to guide you, identify alveoli, atria, bronchioles, bronchi, pulmonary arteries, veins, and capillaries.

The following notes may help you:
Alveoli and **atria** have very thin walls of pavement epithelium.

Bronchi can be distinguished from **bronchioles** as follows: bronchi are larger, and contain glands and cartilage in their walls. Bronchioles are smaller and have no glands or cartilage.

The bronchi and larger bronchioles are lined with ciliated epithelium.

Blood vessels can be distinguished from other cavities by the fact that they contain numerous red blood corpuscles. **Arteries** can be distinguished from **veins** by their thicker muscular walls. **Capillaries** are lined with a single layer of pavement epithelium.

Notice how numerous are the alveoli in your section. Using high power (with oil immersion if available) explore the intimate association between a capillary and adjacent alveolus (**X** in Fig. 8.3). Can you see that the thin barrier between them is only two cells in thickness? What sort of cells are they? If you have a micrometer eye piece, estimate the thickness of the barrier in micrometers (μm). Is the barrier of uniform thickness or uneven? Explain.

For Consideration
(1) What functions are performed by (a) the ciliated cells, (b) the mucus-secreting cells, and (c) the muscle fibres in the wall of the trachea and bronchi?

(2) Review the ways in which the mammalian lung is adapted to perform its function of ensuring rapid gaseous exchange.

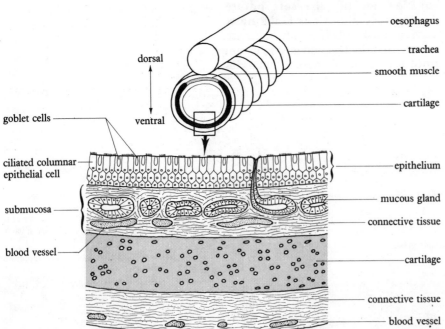

Fig. 8.2 Diagram of a transverse section through the wall of the mammalian trachea. The epithelial lining consists of several layers of cells only some of which reach the free surface. Because of its superficial resemblance to stratified epithelium (*see* p. 16) it is known as pseudo-stratified epithelium. The connective tissue on either side of the cartilaginous ring consists of a mixture of collagen and elastic fibres. Some smooth muscle may also be seen, particularly in the vicinity of the free ends of the cartilage.

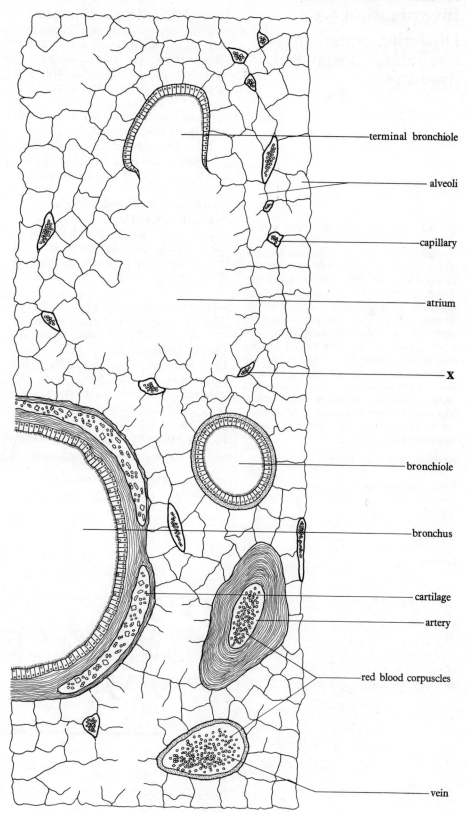

terminal bronchiole

alveoli

capillary

atrium

X

bronchiole

bronchus

cartilage

artery

red blood corpuscles

vein

Fig. 8.3 Semi-diagrammatic section of mammalian lung.

Requirements
Microscope
Oil immersion if available

TS trachea
Section of mammalian lung

Investigation 8.3
Dissection of the respiratory apparatus of the dogfish

Fig. 8.4 Diagram to show the spatial relationship between the afferent and efferent branchial arteries, the skeletal arches, and hypobranchial muscles of the dogfish. I–V, gill pouches; a_1–a_5, afferent branchial arteries; e_1–e_4, epibranchial (efferent) arteries. The gills lie between successive gill pouches. Within each gill the afferent and efferent arteries are linked by capillaries (not shown in the diagram). The vessels shown between the gill pouches in the diagram are all located in the base of each gill, along with the branchial arch. Successive efferent loops are interconnected by shunt vessels which ensure an equal distribution of blood pressure. The heart lies in a depression in the coracoid, the pericardial depression. Direction of blood flow indicated by arrows.

In the dogfish water is drawn into the pharynx whence it flows through five pairs of gill pouches to the exterior, thus coming into close association with the gills. The gills are well vascularized: deoxygenated blood is pumped to them from the heart; after oxygenation the blood flows to the various parts of the body.

The gills contain an extensive capillary system and present a large surface area to the water that flows over them.

Dissection of Blood Vessels Serving the Gills

Before dissecting the blood supply to the gills it is necessary to understand the relationship between the various structures involved.

THE CIRCULATORY SYSTEM

The relevant parts of the circulatory system and related structures are shown in Fig. 8.4 in side view. Deoxygenated blood from the veins is pumped by the **heart** into the **ventral aorta** beneath the floor of the pharynx. From the ventral aorta blood is conveyed into the base of each gill by an **afferent branchial artery**

(a_1–a_5 in Fig. 8.4). There are five pairs of gills and therefore five afferent branchial arteries. In the dogfish the first two afferent branchial arteries on each side of the body arise from a common branch of the ventral aorta; the remaining afferents arise separately. After oxygenation in the gills, the blood is collected into **efferent branchial arteries** which form a series of interconnected loops round the bases of the gill pouches. The last loop is incomplete and is confined to the anterior side of the gill pouch. It follows that the base of each gill will contain one afferent branchial artery and two efferents. It is supported by a cartilaginous **branchial arch**.

From the efferent branchial arteries blood is conveyed via four pairs of **epibranchial arteries** (e_1–e_4 in Fig. 8.4) to the dorsal aorta in the roof of the pharynx. The dorsal aorta takes blood to the various parts of the body.

THE HYPOBRANCHIAL MUSCLES

The afferent branchial arteries lie between a series of longitudinal muscles which lie beneath the gill region. Collectively known as the

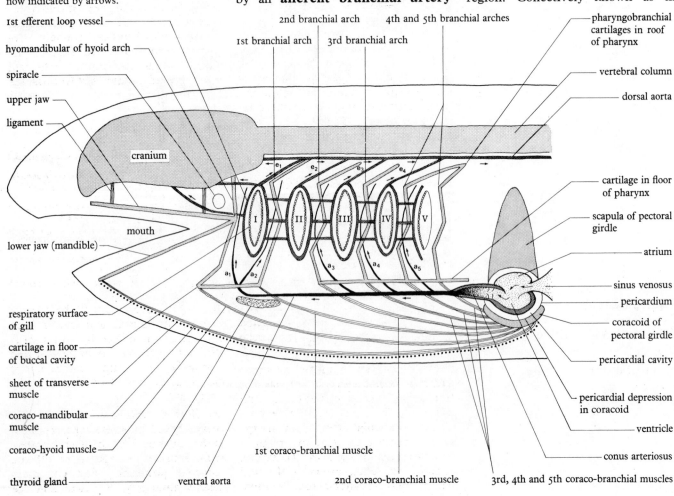

1st efferent loop vessel

hyomandibular of hyoid arch

spiracle

upper jaw

ligament

cranium

mouth

lower jaw (mandible)

respiratory surface of gill

cartilage in floor of buccal cavity

sheet of transverse muscle

coraco-mandibular muscle

coraco-hyoid muscle

thyroid gland

ventral aorta

2nd branchial arch   4th and 5th branchial arches
1st branchial arch   3rd branchial arch

1st coraco-branchial muscle

2nd coraco-branchial muscle

pharyngobranchial cartilages in roof of pharynx

vertebral column

dorsal aorta

cartilage in floor of pharynx

scapula of pectoral girdle

atrium

sinus venosus

pericardium

coracoid of pectoral girdle

pericardial cavity

pericardial depression in coracoid

ventricle

conus arteriosus

3rd, 4th and 5th coraco-branchial muscles

hypobranchial muscles, they run from the cartilaginous lower jaw and the floor of the buccal cavity and pharynx to the ventral part of the pectoral girdle (coracoid). These muscles, which are responsible for inspiration, are named according to their origins (anterior attachments) as follows:

(1) The **coraco-mandibular muscle** (single) runs from the lower jaw (mandible) to the coracoid.
(2) The **coraco-hyoid muscles** (one pair) run from the ventral components of the hyoid (jaw supporting) arch to the coracoid.
(3) The **coraco-branchial muscles** (five pairs) run from the ventral components of the branchial arches to the coracoid.

On contraction the coraco-mandibular muscle pulls the lower jaw downwards, opening the mouth.

The ventral component of the hyoid and 1st branchial arch form the floor of the buccal cavity, so contraction of the coraco-hyoid and first pair of coraco-branchial muscles pulls the floor of the buccal cavity downwards, creating a suction which draws water in through the mouth and spiracles.

The ventral components of branchial arches 2–5 form the floor of the pharynx, so contraction of the 2nd to 5th coraco-branchial muscles pulls the floor of the pharynx downwards, thus expanding the pharynx and drawing water in from the buccal cavity.

The above muscles lie beneath a superficial sheet of **transverse muscle** whose contraction raises the floor of the buccal cavity and pharynx. This, together with contraction of various other muscles, forces water through the gill pouches (expiration).

RELATIONSHIP BETWEEN THE AFFERENT BRANCHIAL ARTERIES AND HYPOBRANCHIAL MUSCLES

The positions of the afferent branchial arteries relative to the hypobranchial muscles is shown in Fig. 8.4. Plainly, to expose afferent branchial arteries 1 and 2 it is necessary to remove the transverse, coraco-mandibular and coraco-hyoid muscles. To expose afferent branchial artery 3 it is necessary to remove the first two pairs of coraco-branchial muscles. To expose afferent branchial artery 4 the 3rd pair of coraco-branchial muscles must

be removed. And to expose afferent 5 necessitates removing the 4th pair of coraco-branchial muscles.

DISSECTION OF THE AFFERENT SYSTEM

(1) Skin the throat region from the lower jaw to the pectoral girdle and outwards on each side as far as the gills. Identify the **transverse sheet of muscle** immediately beneath the skin (Fig. 8.5A).

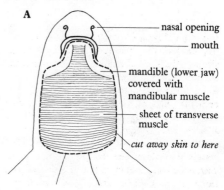

Fig. 8.5 Dissection of the afferent branchial arteries of the dogfish: **A, B** this page, **C, D, E** p. 64.

(2) Fillet away the sheet of transverse muscle and note the longitudinal **coraco-mandibular muscle** beneath. Immediately underneath the coraco-mandibular muscle you will see the paired **coraco-hyoid muscles** (Fig. 8.5B).

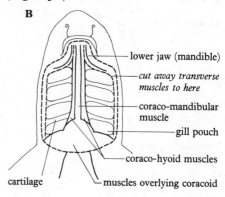

(3) Insert a seeker under the coraco-mandibular muscle and free it from the underlying coraco-hyoid muscles. With a scalpel scrape it off the coracoid posteriorly and with scissors cut it cleanly at its point of attachment to the lower jaw.

(4) Now do the same with the two coraco-hyoid muscles: separate them in the mid-line. Free them from underlying muscles, scrape them off the coracoid posteriorly and cut them cleanly anteriorly. You will now see

the **thyroid gland**. This lies immediately above the common origin from the ventral aorta of the **first two afferent branchial arteries** (Fig. 8.5C). Identify these arteries and remove the thyroid gland. Remove the connective tissue covering the arteries.

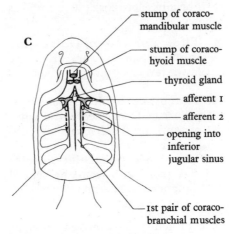

(5) Immediately behind the 2nd afferent branchial artery on either side is the opening of a vein, the **inferior jugular sinus**. Insert a seeker into the sinus in order to establish its course. Then, cutting upwards towards you with a scalpel, slice vertically through the tissue ventral to the sinus. This cut should be orientated antero-posteriorly immediately between the gill pouches and the coraco-branchial muscles and should follow the course indicated by the arrows in Fig. 8.5C.

Identify the **3rd afferent branchial artery** crossing the roof of the inferior jugular sinus.

(6) Now separate the **first pair** of **coraco-branchial muscles** in the mid-line and notice the ventral aorta plunging downwards beneath them. Remove these muscles in exactly the same way as you removed the coraco-hyoid muscles, scraping them off the coracoid posteriorly and cutting them clearly anteriorly.

Now remove the **2nd pair** of **coraco-branchial muscles**. This will expose the origin of the **3rd pair** of **afferent branchial arteries** from the ventral aorta (Fig. 8.5D). In cutting the second pair of coraco-branchial muscles be careful that you don't damage either the afferent branchial arteries beneath them or the ventral aorta between them.

(7) Cut into the inferior jugular sinus on either side of the 3rd pair of coraco-branchial muscles (arrows Fig. 8.5D), thereby exposing the **4th pair** of

afferent branchial arteries. Remove the muscles to show the origin of the arteries from the ventral aorta.

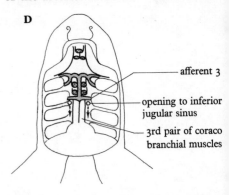

(8) In just the same way remove the **4th pair** of **coraco-branchial muscles**, thus exposing the **5th pair** of **afferent branchial arteries**.

(9) Cut out the central portion of the coracoid so as to expose the **heart**. Identify the **sinus venosus** posteriorly, the thin-walled **atrium**, and thick-walled **ventricle** leading via the **conus arteriosus** to the ventral aorta (Fig. 8.5E).

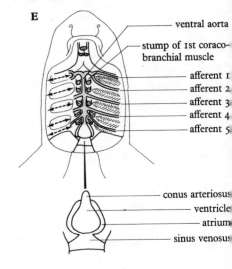

(10) Open the **gill pouches** by cutting along the dotted lines shown on the left side of Fig. 8.5E. Trim the **gills** to the level of the arteries as shown on the right side of Fig. 8.5E and trace each afferent branchial artery into its gill as far as you can. In the case of the first two afferent branchial arteries (which are easier to trace into the gills than the others), notice the small holes where the artery gives off branches to the gill lamellae.

Apart from removing superfluous tissue the dissection is now complete and should look like the right-hand side of Fig. 8.5E.

DISSECTION OF THE EFFERENT SYSTEM

(1) With scissors cut along the angle of the jaws on one side of the body. Cut through each branchial arch and down the side of the oesophagus as indicated by the dotted line in Fig. 8.6A. Deflect the floor of the buccal cavity and pharynx to one side as shown in Fig. 8.6B. Don't force it back too much, particularly at the posterior end or you may break some of the arteries.

(2) Fillet away the skin and mucous membrane from the roof of the mouth, pharynx and oesophagus. Identify the structures shown in Fig. 8.6B, noting the topographical relationship between the **pharyngo-branchial cartilages** (which form the skeleton in the roof of the pharynx) and the **epibranchial arteries**.

(3) Carefully remove the pharyngo-branchial cartilages (including their heads) on both sides of the body. Do this by lifting the cartilage from beneath with a sharp scalpel, keeping the blade as close as possible to the cartilage so as not to damage any blood vessels underneath. Be especially careful when removing the head of each cartilage for the epibranchial artery splits into its two **efferent branches** immediately underneath.

Note that the first pair of pharyngo-branchial cartilages are joined in the mid-line. The origin of the first pair of epibranchial arteries from the dorsal aorta is immediately *beneath* this cartilage, and as these epibranchial arteries pass outwards they adhere closely to the front of the cartilage. So be careful as you remove this cartilage.

(4) When all the cartilages have been disposed of, remove the connective tissue from the dorsal aorta and epibranchial arteries so as to expose them fully (Fig. 8.6C).

A   Cutting along the angle of the jaw

B   Before removal of pharyngo-branchial cartilages

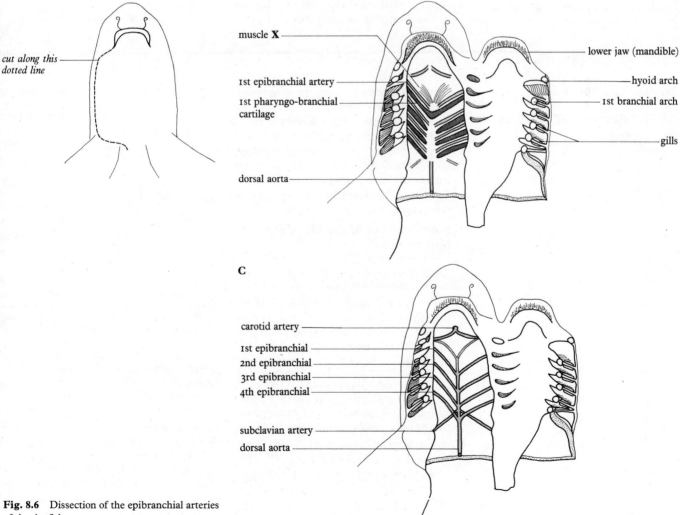

cut along this dotted line

muscle **X**

1st epibranchial artery

1st pharyngo-branchial cartilage

dorsal aorta

lower jaw (mandible)

hyoid arch

1st branchial arch

gills

C

carotid artery

1st epibranchial

2nd epibranchial

3rd epibranchial

4th epibranchial

subclavian artery

dorsal aorta

Fig. 8.6 Dissection of the epibranchial arteries of the dogfish.

Identify the **subclavian arteries** arising from the dorsal aorta just in front of the point of origin of the 4th epibranchials. Remove the muscles marked X in Fig. 8.6B. Notice that the **dorsal aorta** splits just in front of the origin of the 1st pair of epibranchials to join the **carotid** arteries anteriorly. Your dissection is now sufficiently complete and should look like Fig. 8.6C.

For Consideration

(1) How would you explain the fact that in the dogfish there are five afferent branchial arteries but only four epibranchials? How many efferent loop vessels are there?

(2) How does blood get from the efferent loop belonging to the last gill pouch to the dorsal aorta?

(3) Review the structures responsible for moving water (a) into the pharynx, and (b) through the gill pouches in the dogfish.

(4) Compare the mechanism by which the respiratory medium is brought into contact with the respiratory surface in the dogfish and mammal.

Requirements
Dissecting instruments

Fresh-killed or preserved dogfish

Investigation 8.4
Structure of the gills of the dogfish

In the mammalian lung the numerous alveoli create a very large surface area for gaseous exchange. A comparable situation is seen in the gills of fishes except that here the large surface area is achieved not by alveoli but by numerous flattened epithelial surfaces. Like alveoli, these have a very good blood supply.

Examination of the Gills and Gill Pouches

(1) Examine a thick slice (hand section) of dogfish cut through the gill region. What structures can you identify? Can you judge the level at which the cut has been made? Notice in particular the relationship between the **pharyngeal cavity**, **gill pouches**, and **gills**.

(2) Examine a single gill that has been removed from a dogfish (Fig. 8.7). You will be able to feel the cartilaginous **branchial arch** supporting the base of the gill. Projecting from the branchial arch are flattened **lamellae**. Can you detect the delicate **gill plates** on each lamella? (Running a needle along the surface of the lamellae may help.) Notice the **branchial valve**

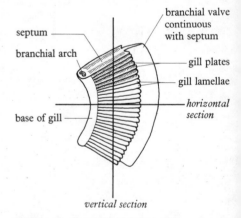

Fig. 8.7 Isolated gill of dogfish. The entire gill is known as a holobranch, the portion on each side of the septum being called a hemibranch ('half-gill'). The base of the gill, stiffened by the cartilaginous branchial arch, contains the afferent and efferent branchial arteries which are interconnected by an extensive capillary system in the gill lamellae and plates.

projecting beyond the end of the gill. What is its function?

Examine the gill under a hand lens or binocular microscope and see the plates in more detail. Approximately how many lamellae are there per gill? How many plates per lamella?

Fig. 8.8 Structure of the gill of dogfish. **A**, horizontal section of whole gill (holobranch) showing a pair of lamellae and associated structures. **B**, vertical section of gill cut in the plane X–Y; this is drawn on a larger scale and shows six pairs of lamellae together with the gill plates which project from them: this is how the gills appear in a transverse section of a dogfish embryo. **a** and **b** are small blood vessels: **a** carries deoxygenated blood from the afferent branchial artery in the base of the gill to the lamella, **b** carries oxygenated blood from the lamella to the efferent branchial artery in the base of the gill. Within each lamella **a** and **b** are interconnected by tiny blood spaces which run through the gill plates. Gaseous exchange takes place during passage of the blood through the lamellae and plates.

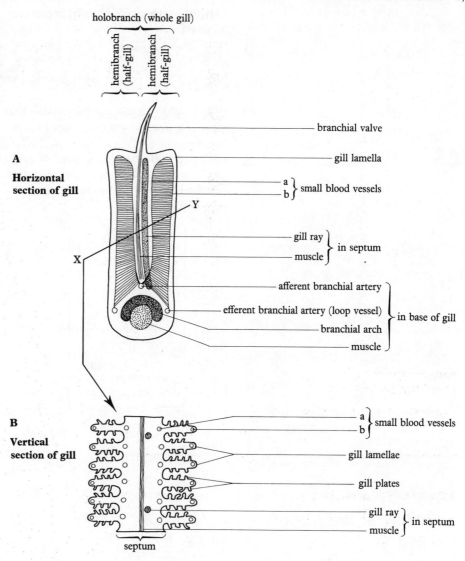

Requirements

Microscope
Hand lens or binocular microscope

Hand section through gill region of dogfish
HS dogfish gill
VS dogfish gill

(3) Having studied the structure of an individual gill, you should be able to interpret microscopic sections of the gill. First examine a horizontal section of a gill, i.e. one which passes parallel to the lamellae (Fig. 8.8A), and then look at a vertical section, i.e. one which passes at right angles to the lamellae (Fig. 8.8B). Gills cut in the latter plane can be seen in a transverse section of a dogfish embryo.

In the horizontal section observe: **afferent** and **efferent branchial arteries** and the cartilaginous **branchial arch** in the gill base; a cartilaginous **gill ray** may be seen in the **septum**; **branchial valve** continuous with septum; **blood vessels**.

In the vertical section notice: **lamellae** and **gill plates** at right angles to each other; vertical **septum** running down centre of gill; cartilaginous **gill rays** in septum; **blood vessels**.

Examine the blood vessels in a single gill plate. How many layers of cells are there between the lumen of the vessels and the exterior?

For Consideration

(1) How do Figs. 8.7 and 8.8 relate to Fig. 8.4? Show precisely how the gills illustrated in the first two figures would fit into the diagram in Fig. 8.4.
(2) Compare the respiratory surface of the dogfish with that of the mammal. What particular structures are directly comparable in these two animals?

Questions and Problems

1 How is the structure of an animal's respiratory surface, and its ventilation mechanism, related to (a) its size, (b) its activities, and (c) its environment? Illustrate your answer with specific examples.

2 How would you test the hypothesis that feathery structures projecting from the sides of a marine worm are used for gaseous exchange?   (OCJE)

3 Evidence indicates that groups of nerve cells, collectively known as the respiratory centre, in the hindbrain are involved in the control of respiration in the mammal. Fig. 8.9 shows how the respiratory centre is connected to the breathing apparatus by nerves:

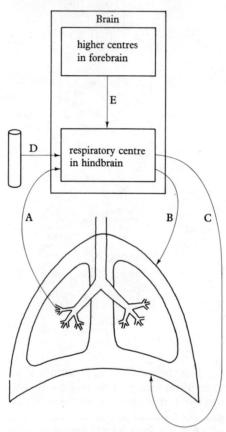

Fig. 8.9 Diagram showing neural connections between the respiratory apparatus of a mammal and the respiratory centre in the hindbrain.

A represents the vagus nerves which transmit impulses (messages) from the walls of the bronchial tubes in the lungs to the medulla.
B represents the intercostal nerves which transmit impulses from the respiratory centre to the intercostal muscles.
C represents the phrenic nerves which transmit impulses from the respiratory centre to the diaphragm.
In addition two other nervous pathways are shown:
D represents nerves which transmit impulses from the walls of certain blood vessels to the respiratory centre.
E represents nerve tracts within the brain which connect the higher centres (cerebral cortex) with the respiratory centre.
Consider the following experiments and their results:
(i) A, B and C are cut. Result: breathing ceases.

(ii) A, D and E are cut, but B and C are left intact. Result: rhythmical breathing continues but slower and deeper than normal.

(iii) B, C, D and E are cut. The lungs are inflated by means of a pump and nervous impulses are recorded from A with an oscilloscope. Result: as the lung is inflated the frequency of impulses increases.

(iv) A and E are cut. The concentration of carbon dioxide in inspired air is increased to three per cent. Result: ventilation rate increases.

(v) Experiment 4 is repeated. After the ventilation rate has increased, D is cut. Result: ventilation rate decreases.

(vi) A and E are cut. The concentration of carbon dioxide in inspired air is increased to five per cent. Result: ventilation rate increases.

(vii) Experiment 6 is repeated. After the ventilation rate has increased, D is cut. Result: ventilation rate remains at the increased rate.

What conclusions can you draw about the control of breathing from these results?

Predict the effect of cutting E but leaving all the other nerves intact, and of cutting A and D but leaving B, C and E intact. If your predictions turned out to be correct, what conclusions would you draw concerning the role of the higher centres in the control of breathing?

4 Fig. 8.10 shows the respiratory apparatus of the frog. The lungs, a pair of sacs with folded lining, lie in the general body cavity and are not enclosed by ribs and diaphragm. They are ventilated by movements of the bucco-pharynx.

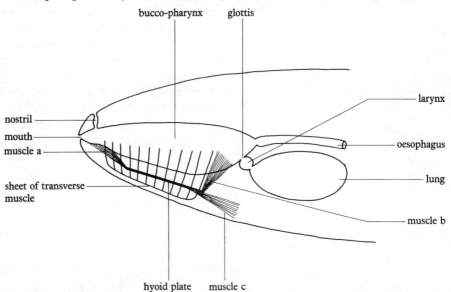

Fig. 8.10 Diagram of the respiratory apparatus of the frog in side view. Only one of the two (paired) nostrils and lungs are shown. The hyoid plate forms the floor of the bucco-pharynx and has three sets of muscles attached to it: muscle **a** has its insertion on the mandible (lower jaw), muscles **b** and **c** on the pectoral girdle. The nostrils, mouth and glottis can be opened or closed according to circumstances.

From the information given, put forward a hypothesis explaining how air is conveyed from the exterior to the lungs and from the lungs to the exterior.

What does the frog's respiratory apparatus suggest to us about the evolutionary position of frogs relative to fishes and mammals? What else do you know about amphibians that might have a bearing on this question?

In the frog gaseous exchange also occurs across the skin and the lining of the bucco-pharynx.

(a) why should this be necessary?

(b) what structural features must the skin possess for it to be suitable as a respiratory surface?

(c) to what extent does the skin of the frog fulfil these structural requirements?

5 It has been suggested that in many insects, including the locust, air is drawn into the tracheal system through the thoracic spiracles and leaves via the abdominal spiracles (Fig. 8.12). How would you verify this experimentally and what might be the mechanism ensuring such a unidirectional flow of air?

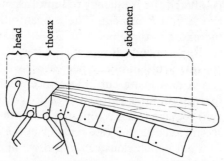

Fig. 8.12 Simplified side view of locust. The spiracles are indicated by dots.

6 What features of the structure of a fish's gills and ventilation mechanism are similar to mammalian lungs and their ventilation?

7 Fig. 8.11 shows part of the tracheal system of an insect.
(a) suggest possible functions for the hairs round the spiracles?
(b) what is the function of the rings of chitin round the tracheae?
(c) why do the tracheoles have no such rings?
(d) by what mechanism might oxygen be conveyed through the tracheal system?
(e) the tracheoles are made up of rows of 'drainpipe' cells—i.e. each individual cell is a hollow cylinder with the nucleus to one side. What is the possible functional significance of this?
(f) what would you expect to happen to the watery fluid in the tracheoles when the muscle is actively contracting?
(g) of what respiratory significance is your answer to (f)?
(h) what would you expect to happen if the tracheal system was blocked at X?

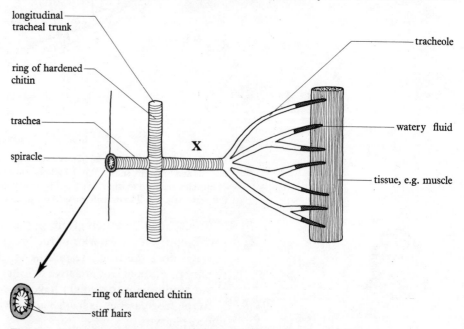

Fig. 8.11 Diagrammatic representation of part of the tracheal system of an insect.

8 The following table shows a comparison between a sample of fresh water and air. Oxygen content is given in cm^3 per dm^3, other parameters in arbitrary units:

	Air	Water
oxygen content	210	8
diffusion rate	1	10^{-5}
density	1	1,000
viscosity	1	100

With reference to these figures, compare air and water as respiratory media.

9 In what respects can the respiratory apparatus of a cod be considered more efficient than that of a shark?

9 Heterotrophic Nutrition

Background Summary

1   Heterotrophic nutrition, feeding on organic food, can be classified into three types: **holozoic**, **saprophytic** and **parasitic**. Saprophytic nutrition, carried out by many bacteria and fungi, is important in bringing about decay.

2   The problem facing any heterotroph is how to acquire and take in organic compounds so they can be **assimilated**. In most heterotrophs this function is performed by a **gut** (**alimentary canal**) in conjunction with a circulatory system. The food is ingested, digested, absorbed, transported and assimilated. Indigestible remains are egested.

3   **Digestion**, the breaking down of the food into an absorbable form, takes place by physical and chemical means. The former is achieved by teeth, or equivalent structures, the gut muscles, and the action of solutions such as bile. Chemical breakdown is achieved by **digestive enzymes**.

4   In general digestive enzymes fall into three groups: **carbohydrases**, **lipases** and **proteases**—which attack carbohydrates, fats and proteins respectively. The enzymes work by splitting chemical bonds.

5   **Absorption** of the products of digestion is generally aided by the surface area of the gut wall being increased by **villi** and **microvilli**.

6   Digestion is either entirely **extracellular** (e.g. man), entirely **intracellular** (e.g. *Amoeba* and sponges), or extracellular and intracellular (e.g. *Hydra*).

7   The human gut is differentiated into a series of specialized regions, each showing, in both its coarse and microscopic anatomy, a close relationship between structure and function.

8   In man, physical digestion is achieved by the teeth, stomach contractions, and bile. Movement of food along the gut is achieved by **peristalsis**.

9   Chemical digestion is achieved by enzymes contained in **saliva**, **gastric juice**, **pancreatic juice** and **intestinal juice**, secreted by the salivary glands, stomach wall (gastric glands), pancreas and wall of the small intestine respectively. Secretion is initiated by expectation, reflex stimulation, hormones or direct mechanical stimulation, depending on the gland in question.

10   **Heterotrophs** can be described in terms of the food they eat as herbivores, carnivores, omnivores, liquid-feeders, and microphagous feeders.

11   **Herbivores** have special adaptations for digesting plants, e.g. serrated molar teeth of horse and elephant, mandibles of locust and grasshopper, radula of snail. A number of herbivores harbour cellulase-secreting micro-organisms in the gut.

12   **Carnivores** have adaptations for catching and killing prey (e.g. canine teeth, tentacles, stinging cells) and for chewing it (e.g. carnassial teeth). Certain specialized plants are carnivorous with adaptations for trapping and digesting small animals, e.g. butterwort, sundew, Venus fly-trap, pitcher plants.

13 **Liquid feeders** include 'wallowers' (e.g. tapeworm) and suckers (e.g. certain insects). The mouthparts of sucking insects are adapted in different ways to form various types of proboscis.

14 **Microphagous feeders** devour small particles suspended in water which are collected and filtered. They are hence also known as **filter feeders,** e.g. bivalve molluscs.

Investigation 9.1

Dissection of the mammalian alimentary canal

As food passes along the alimentary canal it is digested physically and chemically and the soluble products of digestion are absorbed into the bloodstream. Solid indigestible matter remains in the gut and is voided. The overall function of the alimentary canal is therefore to process and absorb food. In both its coarse and microscopical anatomy the gut shows a close relationship between structure and function.

Procedure

(1) Pin the rat to a dissecting board, ventral surface upwards and head pointing away from you. Make a mid-ventral incision through the skin and cut forward as far as the lower jaw, and backwards to the anus. Cut either side of the urino-genital openings as shown in Fig. 9.1. Free the skin from the underlying body wall.

(2) Pin back the skin and cut through the body wall as shown in Fig. 9.2.

Male

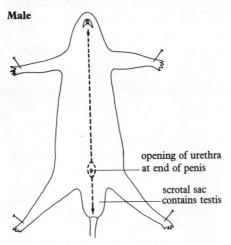

opening of urethra at end of penis

scrotal sac contains testis

Female

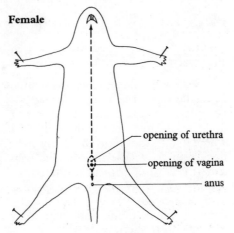

opening of urethra

opening of vagina

anus

Fig. 9.1 Opening up the rat. Cut through the skin as indicated by the dotted line.

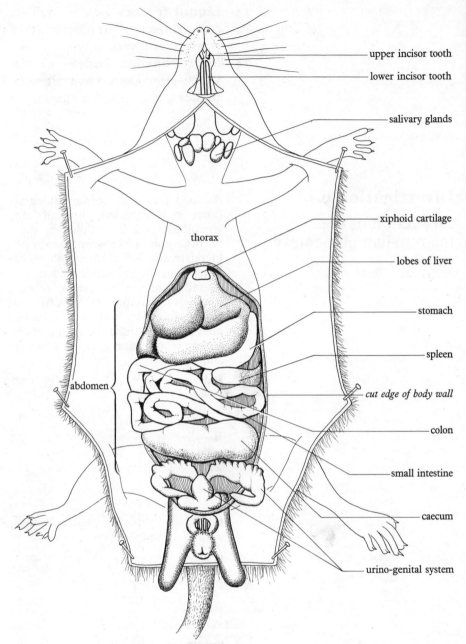

Fig. 9.2 Contents of abdominal cavity of rat *in situ*.

Identify all the structures shown in Fig. 9.2.

(3) Wet your fingers so they slide easily between the organs. Look between the stomach and the liver and find the lower end of the **oesophagus** where it joins the stomach. Using your fingers to move the organs this way and that, and *without cutting the mesentery by which the gut is suspended in the abdominal cavity*, follow the alimentary canal in the abdomen from oesophagus to rectum.

Identify: **cardiac** and **pyloric** regions of **stomach** (note **spleen** clinging to stomach), **pyloric sphincter**, **duodenum**, **ileum**, **caecum** with short appendix, **colon and rectum**. The duodenum and ileum together make up the **small intestine**; the colon and rectum make up the **large intestine**.

(4) Push the liver upwards, the stomach to your right and the ileum to your left. Deflect the duodenum downwards towards you as shown in Fig. 9.3. This will enable you to see: the **pancreas** within the loop of the duodenum, and the **bile duct** (white tube) running from liver to duodenum (there is no gall bladder in the rat).

Numerous small pancreatic ducts open into the bile duct as it runs towards the duodenum. Running in

A *deflect duodenum as indicated by arrow*

B *and note pancreas enclosed within the loop*

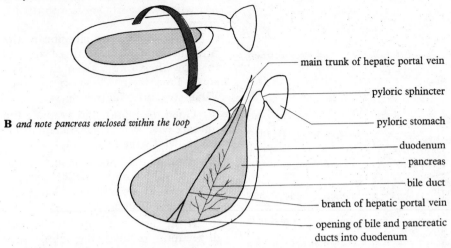

main trunk of hepatic portal vein

pyloric sphincter

pyloric stomach

duodenum

pancreas

bile duct

branch of hepatic portal vein

opening of bile and pancreatic ducts into duodenum

Fig. 9.3 Technique for revealing the pancreas, bile duct, and hepatic portal vein of the rat.

the same mesentery as the bile duct is the main trunk of the **hepatic portal vein** which conveys blood from the gut to the liver.

(5) Without breaking the mesentery, spread out the ileum to your left and notice numerous branches of the hepatic portal vein in the mesentery (Fig. 9.4).

Running alongside the hepatic portals are branches of the anterior mesenteric artery but these are generally obscured by fat.

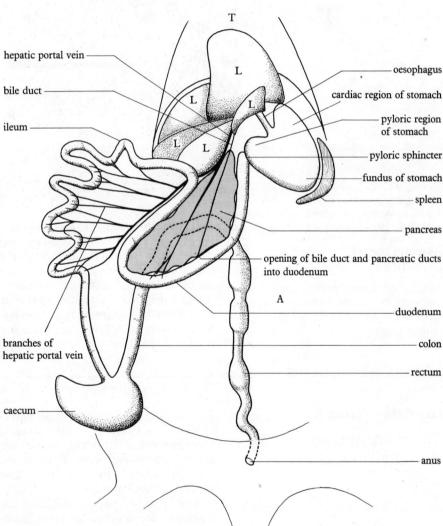

hepatic portal vein

bile duct

ileum

oesophagus

cardiac region of stomach

pyloric region of stomach

pyloric sphincter

fundus of stomach

spleen

pancreas

opening of bile duct and pancreatic ducts into duodenum

duodenum

colon

rectum

branches of hepatic portal vein

caecum

anus

Fig. 9.4 Alimentary canal of the rat as seen with the duodenum deflected downwards (*see* Fig. 9.3), the stomach pushed to the right, and the ileum spread out to the left. **L**, lobes of liver; **T**, thorax; **A**, abdominal cavity.

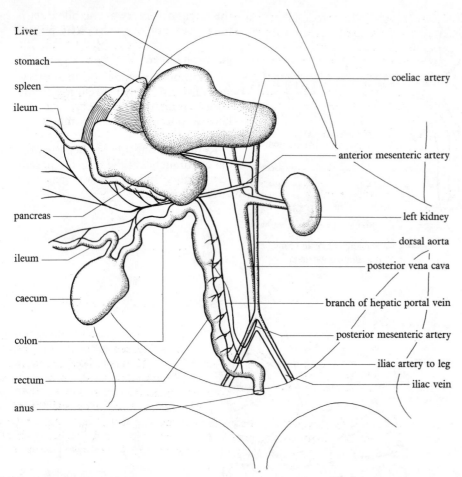

Liver
stomach
spleen
ileum
pancreas
ileum
caecum
colon
rectum
anus

coeliac artery
anterior mesenteric artery
left kidney
dorsal aorta
posterior vena cava
branch of hepatic portal vein
posterior mesenteric artery
iliac artery to leg
iliac vein

Fig. 9.5 Arteries supplying the alimentary canal of the rat, and associated structures.

Requirements
Dissecting instruments

Rat, killed immediately before the laboratory session (deep-frozen rats are unsatisfactory for this dissection).

(6) With the liver pushed forward, deflect the whole of the stomach and intestines to your left and stretch the mesentery. Running in the mesentery are three median (unpaired) arteries to the gut (Fig. 9.5): **coeliac, anterior mesenteric**, and **posterior mesenteric arteries**. All three are branches of the **dorsal aorta**, the first two

arising at the level of the kidneys, the last where the aorta splits into the leg arteries.

Pluck away the fat clinging to the proximal end of these three arteries to show their origin from the dorsal aorta. Trace them to their destinations. What structures do they serve?

(7) Open the mouth and examine the buccal cavity. What special adaptations are shown by the teeth?

(8) Arrange the viscera in such a way as to display the whole of the alimentary canal and its blood supply to maximum advantage.

(9) If the rat is to be used for further dissection, remove the alimentary canal as follows. Ligature the main trunk of the hepatic portal vein, cut through the oesophagus where it enters the stomach, and cut through the rectum where it disappears under the urino-genital organs. Cut through the suspensory mesenteries so as to remove the whole of the gut, spleen and pancreas. Leave the liver.

For Consideration
The veins serving most organs of the body take blood straight to the heart. However those serving the alimentary canal take blood to the liver, after which it flows on to the heart. Why should there be this difference between the venous supply of the alimentary canal and other organs?

Investigation 9.2
Structure and action of mammalian teeth

In mammals, unlike most other vertebrates, the teeth are differentiated (**heterodont**). Typically there are cutting **incisors**, piercing **canines**, and grinding **pre-molars** and **molars** (**cheek teeth**).

The teeth of lower vertebrates are numerous and are constantly replaced. In contrast, mammals have relatively few teeth at any given time, and they only have two sets in the course of post-natal life. First there are the **milk teeth**, consisting of incisors, canines and pre-molars, but no molars; these are later replaced by the **permanent teeth** characteristic of the adult.

It is a convenient convention to

summarize a mammal's complement of teeth in the form of a dental formula. In this the number of teeth of each type belonging to the upper jaw is written above the number of teeth of each type belonging to the lower jaw. Since the jaws of all mammals are symmetrical about the mid-line, the number of teeth of each type represented is scored on one side only. The tooth types are designated by their initial letters. Thus the dental formula of the human adult is:

$$i\frac{2}{2}, c\frac{1}{1}, p\frac{2}{2}, m\frac{3}{3} = \frac{8 \times 2}{8 \times 2} = 32$$

All these teeth have the same basic structure (Fig. 9.6): hard **enamel** on the outside, then a layer of somewhat softer **dentine**, and in the centre a **pulp cavity** containing blood vessels and nerve fibres. The tooth is attached to the jaw bone by **cement**. Examine a vertical section of a tooth and verify its structure. Note distinction between **crown** and **root**.

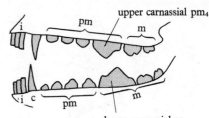

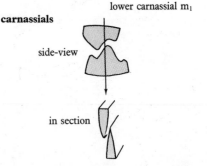

carnassials

Fig. 9.7 Teeth of dog, a carnivore.

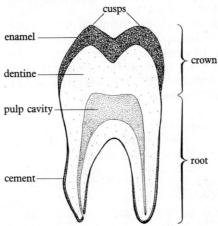

Fig. 9.6 Vertical section of a human molar tooth.

In their microscopic structure teeth bear a striking similarity to the **placoid scales** of the rough hound dogfish to which they are no doubt related in evolution. The teeth of dogfish, sharks, etc., are, in fact, nothing more than greatly enlarged placoid scales which project from the skin covering the jaws and lining of the mouth cavity. Examine the teeth of, for example, dogfish or shark.

Despite this basic histological similarity between the different types of tooth, there are certain characteristic differences between them. For example, incisors are always more or less chisel-shaped with single roots, canines are pointed and single-rooted, whilst the premolars and molars have two or more cusps on the surface of the crown and are double-rooted.

In this exercise we will look at the teeth of a selection of mammals, noting in particular the way the teeth are adapted both in the relative numbers of each type and in their structure, to deal with particular kinds of diet.

MEAT-EATERS

Examine the skull of a dog and note the teeth (Fig. 9.7). Verify that the dental formula is:

$$i\frac{3}{3}, \ c\frac{1}{1}, \ pm\frac{4}{4}, \ m\frac{2}{3} = \frac{10 \times 2}{11 \times 2} = 42$$

Whatever you may feed your own dog on, it is supposed to be a meat-eater (carnivore). How are the teeth adapted to deal with such food? What is the exact function of each type of tooth?

Notice the large **carnassial teeth**. These are the last pair of pre-molars in the upper jaw and the first pair of molars in the lower jaw. What are they for?

Close the upper and lower jaws together. Note how the upper and lower teeth fit (**occlusion**). What is the importance of this? What sort of jaw movements will make best use of the teeth. Do dogs display such jaw movements?

Look at an individual molar tooth. Notice that the surface of the crown bears a series of triangular **cusps**. Such cusps are typical of the cheek teeth of primitive mammals. They have been modified for different purposes in the course of mammalian evolution.

If available, compare the teeth of the dog with those of other carnivores such as cats.

INSECT-EATERS

Moles, hedgehogs, shrews—these are insect-eating mammals (insectivores). A diet of insects does not demand very special adaptations and the insectivore probably possesses the most primitive type of mammalian dentition.

Examine the skull of a representative insectivore, e.g. hedgehog. Note the **small teeth with pointed cusps**. How do the teeth compare with the dog's?

FISH-EATERS

Dolphins and certain types of whales eat squid and fish which are taken alive and swallowed whole. If available, examine the teeth of, e.g. dolphin. Note the large number of **sharp peg-like teeth**. Their main function is to prevent the fish escaping from the jaws.

MOLLUSC-EATERS

An example of a mammal which feeds on molluscs is the walrus. Its **large upper canines** ('tusks') are used, not only for fighting, but also for digging for clams. The shells are crushed with its **flattened back teeth**.

OMNIVORES

An omnivore lives on a mixed diet of animal and plant material.

Examine the skull of an adult man and pig, both omnivores, and compare their teeth with those of the dog. Note that the cusps are lower and more rounded than the dog's. The dental formula of man is given on p. 76.

The dental formula of the pig is:

$$i\frac{3}{3}, \ c\frac{1}{1}, \ pm\frac{4}{4}, \ m\frac{3}{3} = \frac{11 \times 2}{11 \times 2} = 44$$

FRUIT-EATERS

Examine the skull of a sloth. This tropical, tree-living mammal lives on soft fruit. Its **peg-like teeth**, which lack enamel, are quite adequate for this kind of diet.

LEAF, STEM AND BARK-EATERS

These include the rodents, e.g. rats, mice and beavers. The most striking feature of their teeth is seen in the large **chisel-like incisors** which are adapted for gnawing. Their effectiveness is admirably demonstrated by the beaver, which uses them to cut down trees for building dams.

Examine the skull of a rodent such as rat. Note reduction in the total number of teeth. The dental formula of the rat is:

$$i\frac{1}{1}, \ c\frac{0}{0}, \ pm\frac{3}{2}, \ m\frac{3}{3} = \frac{7 \times 2}{6 \times 2} = 26$$

The teeth of most mammals cease to grow at a certain stage. This is because the opening into the pulp cavity constricts, so isolating the tooth from its blood supply. The incisors of rodents, however, have a persistent pulp cavity, which enables them to continue growing throughout life. This enables the teeth to maintain their length despite constant wearing.

The absence of canines, creates a space behind the incisors. This is called the **diastema**; here food is rolled into a ball before being swallowed. What do you think the cheek teeth are used for?

Examine the skull of a rabbit. This is not a rodent, but it has a similar dentition. Compare its teeth with those of the rat. In what respects are they similar, and different? How do their diets differ?

GRASS-EATERS

Feeding on grass presents special problems particularly the necessity for grinding the food. In a grass-eater like the horse, the incisors are used for cutting and the cheek teeth for grinding.

The cheek teeth of herbivores show varying degrees of elongation of the crown to allow for wear. This is achieved by growth of the cusps as shown in Fig. 9.8A. The cusps are covered with cement. With continued wear, the tops of the cusps are worn away. Since the cement and dentine are softer than the enamel, they wear away more quickly, thus leaving a series of **enamel ridges** on the surface of the crown. This makes the cheek teeth ideally suited to grinding tough plant food.

The cheek teeth of herbivorous mammals are thought to have evolved from a primitive type of tooth which bore six cusps. The cusps have fused, and undergone other modifications to give two types of teeth (Fig. 9.8B): (1) **Lophodont teeth**. The cusps are fused to give two ridges running at *right angles* to the antero-posterior axis. Examples are tapir, rhinoceros, elephant and horse.

Examine the skull of tapir or rhinoceros, which show the lophodont condition in a comparatively unmodified form. What sort of jaw movements would be expected to make best use of these teeth? The horse shows a rather modified version of the lophodont condition. In what respects is it modified?

The lophodont pattern can be seen particularly clearly in the elephant. Apart from the tusks (which are elongated upper incisors), the elephant's teeth are reduced to a few enormous molars, each bearing numerous enamel ridges.

A Molars in longitudinal section

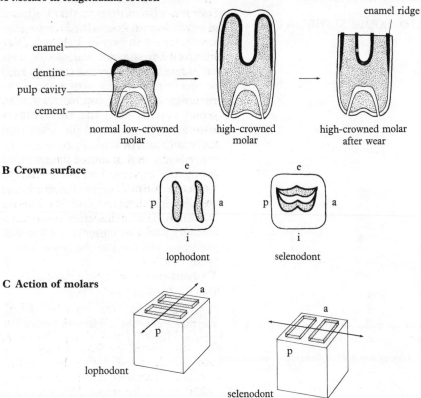

B Crown surface

C Action of molars

Fig. 9.8   Molar teeth of herbivore. In **B** and **C**: **a**, anterior; **p**, posterior; **e**, external; **i**, internal. In **C** the arrows indicate the direction of movement of the upper and lower jaws relative to each other.

Requirements
Microscope

VS tooth
Teeth of dogfish or shark
Skin of dogfish
VS dogfish skin (placoid scales)
Skulls or photographs of the
   following:

dog	sloth
cat, tiger	rat, beaver
hedgehog, mole	rabbit
dolphin	rhinoceros, tapir
walrus	horse
pig	elephant
man	ox, sheep, deer

(2) **Selenodont**. The cusps are modified into crescent-shaped ridges running parallel to the antero-posterior axis. Examples are ox, sheep and deer.

Examine the lower jaw of, e.g. sheep. The pre-molars and molars display the selenodont condition very clearly. What kind of jaw movements would make best use of these teeth? Do the jaws of sheep and cows display such movements?

Note that the upper incisors and canines are absent. They are replaced by a hardened upper lip which the lower incisors bite against. The pre-molars show the selenodont pattern very clearly.

For Consideration

(1) In this investigation we have, for convenience, classified mammals according to their diets. Which of the mammals featuring in this investigation would you describe as carnivores, and which herbivores?

(2) Which do you consider to be the most primitive kind of mammalian cheek teeth from which the others may have been derived in the course of evolution?

(3) The teeth of fishes, amphibians and reptiles are, in the main, numerous, all alike, and pointed. There are no cusped teeth of the molar type. What general effects do you think such a dentition has on the feeding habits of these lower vertebrates? Give examples.

(4) What functions are performed by chewing (mastication) and what are the consequences of its not taking place?

Investigation 9.3

Microscopic structure of the gut wall

Although the mammalian gut shows regional differentiation, its wall can always be distinguished into the seven layers shown in Fig. 9.9. The **epithelium** showing various degrees of folding is protective, secretory, and/or absorptive; the **mucosa** and **sub-mucosa** consist of connective tissue, blood vessels, etc.; the **muscularis mucosa** is composed of two thin layers of smooth muscle, an inner circular layer, and an outer longitudinal layer; the **external muscle coat** is made up of thick layers of circular and longitudinal muscle; and the **serosa** is composed of connective tissue continuous with the mesentary by which the gut is attached to the body wall.

Procedure

SMALL INTESTINE
(1) Examine a transverse or longitudinal section of the **ileum** under low power. Identify the various layers of

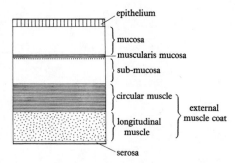

Fig. 9.9 The principal layers in the wall of the alimentary canal.

the wall using Fig. 9.9. to help you. Then examine under high power and answer these questions:
(a) What sort of cells occur in the epithelium lining the **villi** and **crypts of Lieberkühn** and what are their functions?
(b) In what ways is the epithelium adapted for absorption of the products of digestion?

Fig. 9.10 Microscopic structure of the wall of the mammalian small intestine.

A Sectors of the wall of ileum and duodenum

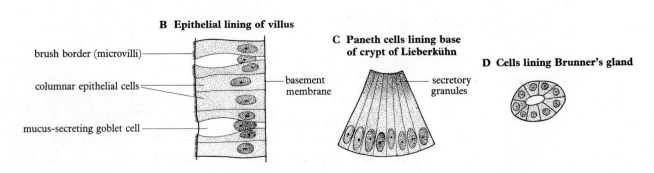

(c) What functions are performed by the various muscles which are visible in your section?

(d) In what other respects is the wall of the ileum adapted to perform its various functions?

(2) Examine a section of the wall of the **duodenum** (Fig. 9.9). How does this differ from the ileum?

STOMACH

Examine a vertical section of the wall of the stomach. Identify the various layers of the wall (Fig. 9.11C) and compare with the ileum.

Examine a **gastric gland** in detail. Towards the surface where it opens into the gastric pit it will be cut in longitudinal section, but deeper down where the glands twist and turn they will have been cut transversely and obliquely (Fig. 9.11A).

Identify the epithelial cells lining the gastric glands, noting: **mucus-secreting cells** lining the neck and gastric pit, pear-shaped **parietal (oxyntic) cells** at intervals along most of the length of the gland, and closely-packed **chief (peptic) cells** lining the deeper part of the gland (Fig. 9.11A and B).

(a) What are the functions of these cells?

(b) In what ways is the wall of the stomach adapted to perform its functions?

A Gastric glands in mucosa of stomach wall

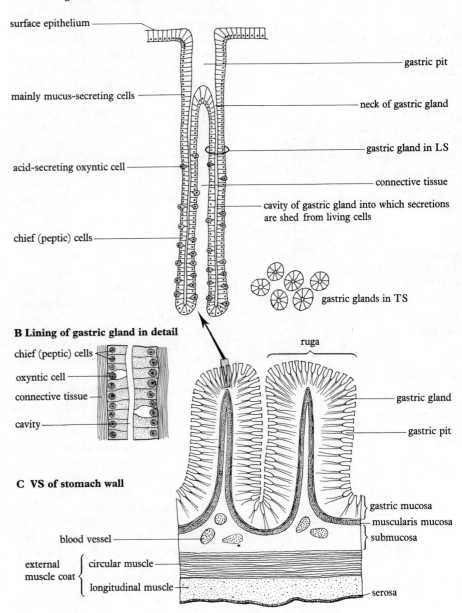

B Lining of gastric gland in detail

C VS of stomach wall

Fig. 9.11 Microscopic structure of the wall of the mammalian stomach.

OTHER REGIONS OF THE GUT

If available, examine sections or photomicrographs of the tongue, oesophagus, colon and rectum and compare them with the stomach and small intestine with respect to their epithelial lining and muscles. In particular notice:

Tongue: taste buds (flask-shaped bundles of sensory and supporting cells) embedded in stratified epithelium; striated muscles are seen deeper down (*see* p. 212).

Oesophagus: inner lining of stratified epithelium; outer layer of striated circular muscles.

Colon and **rectum**: wall contains simple tubular glands lined by numerous goblet cells.

While examining the above sections relate structure to function where possible.

Requirements

TS or LS ileum
TS or LS duodenum
VS wall of stomach
Sections or photomicrographs of tongue, oesophagus, colon, rectum.

For Consideration

(1) From your observations of different regions of the alimentary canal, construct a *generalized diagram* of a section through the wall of the gut, indicating the component tissues of each layer.

(2) In what respects is the structure of the wall of different parts of the gut adapted to perform its particular functions?

Investigation 9.4
Digestion of starch in man

Starch is a polysaccharide which has to be broken down into free monosaccharide molecules before absorption can take place.

The purpose of this experiment is firstly to test the hypothesis that saliva contains an enzyme which breaks down starch to sugar; and secondly to investigate the properties of the enzyme and the conditions under which it works most effectively.

Starch is mixed with saliva under various conditions and after a given period of time the mixture is tested for starch with iodine and for sugar with Benedict's reagent. (For details of these tests see p. 436).

It is of course important to set up the necessary controls and to carry out standard tests on starch and sugar with which to compare the experimental results. The sugar used should be a reducing sugar, e.g. glucose.

Procedure

Collect a test tube full of uncontaminated **saliva**: the flow can be increased by chewing paraffin wax. Transfer one quarter of the saliva to another test tube and place this in a boiling bath for 15 minutes. Transfer a further quarter to another test tube: add four drops of M hydrochloric acid, mix well and leave for at least 15 minutes. Keep the rest of the saliva at room temperature.

Now set up six pairs of test tubes and label each with a wax pencil as follows:

Pair A (1) ¼ test tube of **sugar** only
(2) ¼ test tube of **sugar** only

Pair B (3) ¼ test tube of **starch** only
(4) ¼ test tube of **starch** only

Pair C (5) ¼ test tube of **saliva** only
(6) ¼ test tube of **saliva** only

Pair D (7) ¼ test tube of **starch** plus ⅛ test tube of **saliva**

(8) ¼ test tube of **starch** plus ⅛ test tube of **saliva**

Pair E (9) ¼ test tube of **starch** plus ⅛ test tube of **pre-heated saliva**

(10) ¼ test tube of **starch** plus ⅛ test tube of **pre-heated saliva**

Pair F (11) ¼ test tube of **starch** plus ⅛ test tube of **acidified saliva**

(12) ¼ test tube of **starch** plus ⅛ test tube of **acidified saliva**

The contents of all test tubes must be left for at least 10 minutes (why?). The contents of each pair of test tubes should now be tested for (a) starch by adding two drops of **iodine** and (b) sugar by heating with one-eighth test tube of **Benedict's reagent**. Do the starch test on the first of each pair of test tubes, and the sugar test on the second of each pair.

It is important to be able to compare the results: this is why separate test tubes are used and it is also why the same quantities of substrate, saliva and test-reagent must be used in each case. To facilitate comparison place the test tubes in a rack in numerical order.

Record your results in a table, indicating which test tubes give a positive and which ones a negative result for starch and sugar. It is possible that some may give a result in between: if so, say so.

Requirements

Test tubes (12) with rack
Water bath (250-cm³ beaker)
Bunsen burner with tripod and gauze
Wax pencil

Starch suspension
Dextrose (glucose)
Iodine
Benedict's reagent
M HCl
Paraffin wax

For Consideration

(1) What is the purpose of testing test tubes 1 to 6?

(2) In which test tubes have you actually tested the hypothesis that saliva contains an enzyme which breaks down starch to sugar? Does the hypothesis turn out to be correct? Would you say it was proved?

(3) Which test tubes provide information on the properties of the enzyme?

(4) What tentative conclusions can be drawn as to the chemical nature of the enzymes and what further experiments could be done to confirm these conclusions?

(5) What is the name of the enzyme in saliva and what compound does it produce from starch?

Investigation 9.5

The mouthparts of insects

Insects employ a wide variety of feeding techniques involving the use of mouthparts which are adapted to suit the type of food eaten. It is thought that the biting and chewing mouthparts of the cockroach, locust, etc. are the basic type, and that the more specialized mouthparts of other insects have arisen by modification of the different components found in this primitive pattern.

Procedure

COCKROACH

(1) Remove the head from a cockroach (or locust) and boil it gently in a test tube containing a little caustic potash (KOH solution) until it is soft and semi-transparent. (Put a few fragments of broken pottery in the test tube to prevent 'bumping'). It is normally necessary to boil for approximately five minutes. This facilitates the removal of the mouthparts later.

A Head showing positions of mouthparts

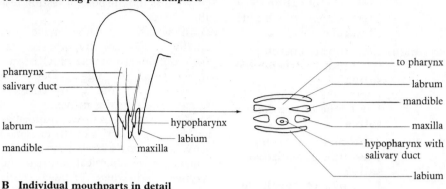

B Individual mouthparts in detail
mandible

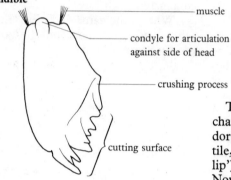

maxilla

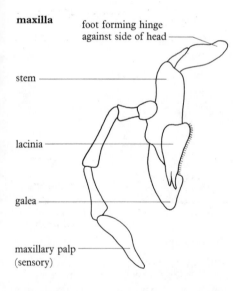

labium

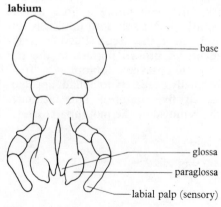

Fig. 9.12 Mouthparts of cockroach.

Then wash the head in several changes of water. Hold the head down, dorsal surface uppermost, on a white tile, and cut off the **labrum** ('upper lip') which shields the mouthparts. Now remove each of the mouthparts by grasping them *at their base* with small forceps and pulling them gently but firmly from their attachment to the head. Remove the **mandibles** first, then the two **maxillae**, and finally the **labium**. Notice the **hypopharynx** adhering to the inner surface of the labium. This forms the floor of the pharynx and is pierced by the opening of the **salivary duct**. Separate the hypopharynx from the labium.

Mount the mouthparts in dilute glycerine and examine under low power. Compare with Fig. 9.12.

(2) To make a permanent preparation proceed as follows. Dehydrate the mouthparts by placing them in a covered watchglass of 70 per cent ethanol for five minutes, then 90 per cent (five minutes), then two changes of absolute ethanol (five minutes each). Clear in xylene or clove oil (at least five minutes) and mount in Canada balsam under a supported coverslip.

(3) What is the function of the different components · of the cockroach's mouthparts? Observe the mouthparts of a live specimen whose labrum has been removed. Present it with a small speck of bread on the end of a needle and observe the mouthparts working. Notice that the bread becomes moistened with saliva.

OTHER INSECTS

Either make your own slides or examine prepared slides of the heads of some or all of the following insects (Fig. 9.13):

Mosquito: in life the proboscis sheath (labium) envelops six slender processes: suction food tube (formed from labrum), hypopharynx with salivary tube, and two pairs of sharp stylets for piercing skin (mandibles and maxillae). In your preparation some or all of these will probably have come out of the sheath. Note also the pair of short maxillary palps (sensory).

Aphid: here the proboscis sheath (labium again) encloses three slender processes: suction food tube which also contains the salivary tube (maxillae) and one pair of mandibular stylets for piercing plant tissues.

Butterfly: mouthparts reduced or absent except for labial palps and long, flexible proboscis formed by galeae (part of maxilla).

Honey bee: in this case the suction food tube, formed by the glossa (part of labium) is surrounded by four supports which together form a flexible, probe-like proboscis; the supports are formed from the galeae and labial palps; the mandibles are spatulate and used for moulding wax in building the comb.

House fly: numerous pseudotracheal tubes in the pair of expanded labellar lobes at the end of the labium form a filtering device which leads to the food tube formed by the labrum.

For Consideration
Review the use to which (a) the mandibles, (b) maxillae, and (c) labium are put in different insects.

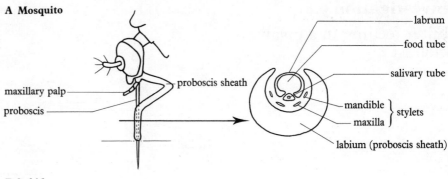

A Mosquito

maxillary palp
proboscis
proboscis sheath
labrum
food tube
salivary tube
mandible } stylets
maxilla
labium (proboscis sheath)

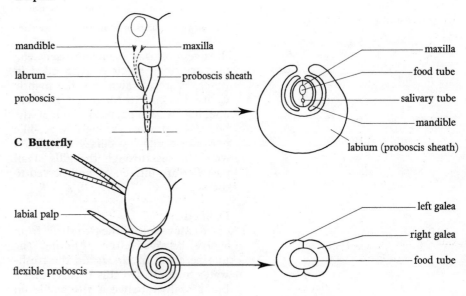

B Aphid

mandible
labrum
proboscis
maxilla
proboscis sheath
maxilla
food tube
salivary tube
mandible
labium (proboscis sheath)

C Butterfly

labial palp
flexible proboscis
left galea
right galea
food tube

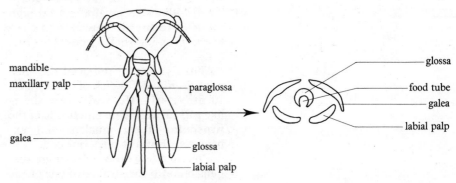

D Honey bee (worker)

mandible
maxillary palp
galea
paraglossa
glossa
labial palp
glossa
food tube
galea
labial palp

Requirements
Broken pottery fragments
Test tube
White tile
Dissecting instruments
Slides and coverslips
Watchglasses
Microscope
Binocular microscope

KOH solution (5 per cent)
Dilute glycerine
70 per cent ethanol
90 per cent ethanol
Absolute ethanol
Xylene or clove oil
Canada balsam

Fresh-killed or preserved
  cockroach or locust
Live cockroach or locust, dorsal
  side set in wax and labrum
  removed
Prepared slides of mosquito (head),
  aphid, butterfly (head), honey
  bee (head) and house fly (head)

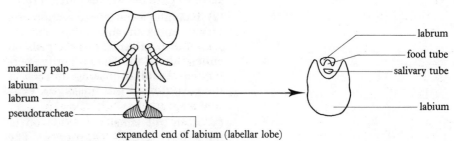

E House fly

maxillary palp
labium
labrum
pseudotracheae
expanded end of labium (labellar lobe)
labrum
food tube
salivary tube
labium

Fig. 9.13 Mouthparts of various insects. Notice how the same structures have been adapted for a variety of functions in different insects.

Investigation 9.6

Filter feeding in a bivalve mollusc

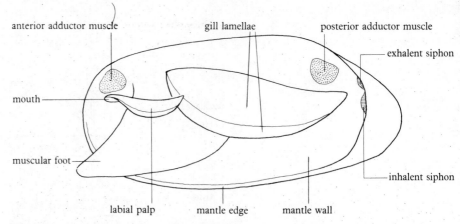

anterior adductor muscle · gill lamellae · posterior adductor muscle

exhalent siphon

mouth

inhalent siphon

muscular foot

labial palp · mantle edge · mantle wall

Fig. 9.14 Contents of swan mussel after removal of left shell.

Bivalve molluscs (clams, mussels, etc. feed on small particles suspended in water. Water, drawn into the mantle cavity by cilia through an inhalent opening at the posterior end of the body, is drawn through sheet-like perforated gills by ciliary action. As water passes through the gills small particles are filtered off and passed to the mouth.

Procedure

(1) Remove one of the two shells from a live bivalve (either *Mytilus*, the marine clam, or *Anodonta*, the freshwater mussel). Do this by inserting a blunt scalpel between the shells on the ventral side and turning it sideways so as to force the shells apart. Then with a sharp scalpel cut through the two **adductor muscles** as close to the nearside shell as possible (Fig. 9.14).

When you have removed the shell put the animal in a dish of clean water. Identify: **mantle wall**, two sheet-like **gill lamellae** on either side of the **muscular foot**, inhalant and exhalant **siphons** at posterior end.

Pipette a few drops of carmine suspension onto the surface of one of the gills. Record what happens to the carmine particles in the course of time.

(2) Examine a transverse section of a bivalve under the microscope and observe the arrangement of the **gills** on either side of the foot. Identify: perforated **lamellae** making up each of the gills, **dorsal gill passages**, **ventral food grooves** (Fig. 9.15A).

(3) Each gill lamella is made up of numerous vertical **filaments**. The perforations through which water passes are located between successive filaments.

Individual filaments can best be seen by examining a horizontal section of a gill. Observe the cilia projecting from each filament (Fig. 9.15B,C). The **lateral cilia** draw water through the gills, whence the water is drawn upwards to the dorsal gill passage. The **latero-frontal** cilia prevent particles passing through the gill, directing them onto the outer surface of the filament. Here the particles, caught up in mucus, are drawn downwards by the **frontal cilia** to the ventral food groove. Cilia lining the ventral food groove draw them towards the mouth.

(4) When you have finished watching the movement of particles in your live bivalve, cut out a very small piece of one of the gills and squash it under a coverslip. Watch for evidence of ciliary action.

For Consideration

(1) How would you ascertain the precise nature of the small particles on which the mussel feeds?

(2) The filtering apparatus of the mussel has here been described as a gill. This implies that it is also used for respiration. What might be the evidence for such an assumption?

A Transverse section of whole animal

- shell
- dorsal gill passage
- foot
- edge of mantle
- gill lamella
- ventral food groove

B Horizontal section of gill lamellae

- filaments

perforation

- inter-filamentary junction

inter-lamellar junction

inter-lamellar space

- lamellae

- blood vessel

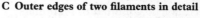

C Outer edges of two filaments in detail

food particles

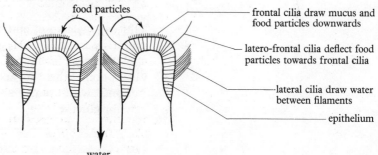

- frontal cilia draw mucus and food particles downwards
- latero-frontal cilia deflect food particles towards frontal cilia
- lateral cilia draw water between filaments
- epithelium

water

Requirements

Clean dissecting dish
Blunt and sharp scalpel
Rubber pipette
Slide and coverslip

Carmine suspension

Live *Anodonta* or *Mytilus*
TS bivalve
HS gill of bivalve

Fig. 9.15 Structure of the gill of *Anodonta*.

Questions and Problems

1 Make a list of the organ systems involved in the assimilation of food in (a) a mammal, and (b) a planarian worm.

2 Explain each of the following statements:
(a) If you stand on your head it is still possible to convey food upwards into your stomach.
(b) When food touches the back of the pharynx it is difficult not to swallow.
(c) When food touches the back of the pharynx swallowing normally occurs, but if the back of the pharynx is touched with a feather vomiting is initiated.
(d) Secretion of gastric juice may start before the food reaches the stomach.
(e) If the bile duct is blocked (for example by gall-stones in obstructive jaundice), digestion of fats is impaired.

3 Using Table 9.1 to help you, trace what happens to a ham sandwich from the moment it is taken into the mouth to the absorption of its chemical constituents into the bloodstream.

4 Explain how it is that a man whose stomach has been removed in the surgical operation of gastrectomy may continue to eat and digest a fairly normal diet. What special precautions do you think a gastrectomized man should observe in his eating habits?

5 Junket is prepared by warming milk with a commercial substance called rennet to a temperature of about 36°C. What is happening chemically and what is the biological significance of this process?

6 To what extent does a study of the anatomical and microscopic structure of the alimentary canal throw light on the processes which go on within it? Discussion need not be confined to mammals. *(O and C)*

7 Glucose, galactose, and fructose are examples of hexose sugars, having the empirical formula $C_6H_{12}O_6$. Xylose and arabinose are pentose sugars, having the empirical formula $C_5H_{10}O_5$. Results from experiments on the rate of absorption of these sugars by pieces of living intestine and by pieces of intestine poisoned with cyanide are given below. The rates are shown as relative to the rate for glucose.

	Rate of absorption	
	By living intestine	By poisoned intestine
glucose	1.00	0.33
galactose	1.10	0.53
fructose	0.43	0.37
xylose	0.30	0.31
arabinose	0.29	0.29

Comment fully on these results. *(JMB Nuffield)*

8 How are the teeth of mammals related to their diets?

9 What is the chemical structure of cellulose and how must this structure be altered in order for cellulose to be assimilated by an animal?
It is known that snails can digest cellulose. How would you investigate the source of the enzyme that achieves this?

Table 9.1 Digestive enzymes and other secretions produced by the mammalian alimentary canal and associated organs

Secretion	Source	Site of action	Induced by	Enzymes	Substrate	Products
Saliva (slightly alkaline)	Salivary gland	Mouth cavity	Expectation and reflex	Amylase (ptyalin)	Starch	Maltose
Gastric juice (acidic)	Stomach wall (gastric glands)	Stomach	Reflex and hormone (gastrin)	HCl (not an enzyme)		
				Pepsin	Protein	Polypeptides
				Rennin	Caseinogen (soluble)	Casein (insoluble)
				Lipase (small amount)	Fats	*Fatty acids and glycerol*
Bile (alkaline)	Liver (via gall bladder)	Duodenum	Reflex and hormone (cholecystokinin)	Bile salts (not enzymes)	Fats	Fat droplets (*chylomicrons*)
Pancreatic juice (alkaline)	Pancreas	Duodenum	Hormone (secretin)	Amylase	Starch	Maltose
				Trypsin	Protein	Polypeptides and *amino acids*
				Chymotrypsin	Casein	Polypeptides
				Carboxy-peptidase	Polypeptides	*Amino acids*
				Lipase	Fats	*Fatty acids and glycerol*
				Nucleases	Nucleic acid (DNA and RNA)	Nucleotides
Intestinal juice (alkaline)	Wall of small intestine (crypts of Lieberkühn, particularly Paneth cells)	Small intestine	Mechanical stimulation	Enterokinase	Trypsinogen (inactive)	Trypsin (active)
				Amylase (small amount)	Starch	Maltose
				Maltase	Maltose	*Glucose*
				Sucrase (invertase)	Sucrose	*Glucose and fructose*
				Lactase	Lactose (milk sugar)	*Glucose and galactose*
				Lipase (small amount)	Fats	*Fatty acids and glycerol*
				Peptidases (erepsin)	Polypeptides	*Amino acids*
				Nucleotidase	Nucleotides	*Phosphoric acid, pentose sugars, and organic bases*

The products in italics are soluble and capable of being absorbed through the lining of the gut into the blood stream.

D

10   What is the composition of human saliva and what are its functions? How do its functions in man compare with the mosquito and aphid?

11   *Anodonta*, the freshwater mussel, has large gills which, when the shells are prized apart, are clearly exposed. If small pieces of silver paper are now placed on the top edge of the gill they can be seen to move downwards by ciliary action.
     Such mussel preparations were made and the rate of movement of the silver paper down the gill was followed when the mussel was immersed in solutions of different pH at constant temperature. The following results were obtained:

pH	3	5	7	9	11
mm/s	0	0.1	1.7	3.5	0.12

A further set of experiments were made at pH 9, varying the temperature. The following results were obtained:

Temperature °C	6	10	15	20	25
mm/s	0.3	1.2	2.1	3.5	2.0

How can the above results be interpreted?   (*AEB*)

10 Autotrophic Nutrition

Background Summary

1 Autotrophic nutrition, the synthesis of organic compounds from inorganic sources, takes place by **photosynthesis** in plants and **chemosynthesis** in certain bacteria.

2 Photosynthesis and chemosynthesis play an important part in the **carbon cycle**: carbon dioxide is built up into complex carbon compounds by photosynthesis and chemosynthesis; it is subsequently released back into the atmosphere by respiration.

3 The raw materials of photosynthesis are carbon dioxide and water; the products are carbohydrate and oxygen. Sunlight is the source of energy and **chlorophyll** traps the light energy.

4 Experiments can be performed to show that the conditions required for photosynthesis are carbon dioxide, water, light, a suitable temperature, and chlorophyll.

5 Comparison of the **absorption** and **action spectra** of chlorophyll indicates that red and blue light are the most effective wavelengths in photosynthesis.

6 Experiments show that photosynthesis is subject to the **law of limiting factors**, i.e. when a chemical process depends on more than one condition being favourable, its rate is limited by that factor which is nearest its minimum value.

7 Probably the most important single factor controlling the rate of photosynthesis is light, and plants exhibit a wide variety of adaptations for securing adequate illumination.

8 Experiments indicate that photosynthesis occurs in the **chloroplasts** where chlorophyll is located.

9 Chlorophyll consists of five pigments which can be separated and identified by chromatography. The pigments are: **carotene**, **phaeophytin**, **xanthophyll, chlorophyll a,** and **chlorophyll b**.

10 Studies on the fine structure of the chloroplast show that the chlorophyll molecules are laid out on a series of parallel membranes (**lamellae**), which are organized into **grana**, thereby providing maximum surface in a minimum volume.

11 The chloroplasts are mainly in the **leaves** whose anatomy shows a close relationship between structure and function. Loosely packed **photosynthetic cells** (**palisade** and **spongy mesophyll**) carry out photosynthesis and the lower epidermis is pierced by numerous **stomata** for gaseous exchange. Other tissues are responsible for strengthening and transport.

12 Evidence indicates that photosynthesis is a two-stage process, the first requiring light and the second capable of occurring in the dark. The function of the **light stage** is to produce ATP, and to split water, thus providing hydrogen atoms for the subsequent reduction of carbon dioxide. In the **dark stage** the carbon dioxide is reduced and carbohydrate synthesized.

13   In the light stage electrons are removed from chlorophyll and either passed back to chlorophyll via a series of carriers with the production of ATP, or combined with hydrogen ions (from the splitting of water) to form hydrogen atoms for the dark reactions.

14   In the dark stage carbon dioxide is fixed, reduced by the hydrogen formed in the light stage, and built up, using energy from the ATP formed in the light stage, into carbohydrate (**Calvin cycle**).

15   Within the chloroplast the light reactions take place on the lamellae where the chlorophyll is located, the dark reactions in the stroma.

16   Some autotrophic bacteria undergo photosynthesis but most are chemosynthetic. In chemosynthesis organic compounds are synthesized from inorganic raw materials, the necessary energy coming from the oxidation of, e.g. ferrous salts, nitrates and nitrites.

17   Chemosynthetic bacteria are important in the **nitrogen cycle**, the process in which nitrogen compounds circulate in nature. **Nitrifying bacteria** obtain energy for chemosynthesis by converting ammonium compounds, released from animal and plant protein during decay, into nitrites and nitrates. Nitrates are then assimilated by plants.

18   Working against the nitrogen cycle are **denitrifying bacteria** which, in order to obtain oxygen for aerobic respiration, convert nitrates into nitrites, ammonia or free nitrogen.

19   **Nitrogen-fixing bacteria**, either free in soil or symbiotic in the roots of certain plants, can convert atmospheric nitrogen into protein.

Investigation 10.1
Effect of light intensity on the rate of photosynthesis

Reference
Nuffield A-level Biological Science, *Maintenance of the Organism: a Laboratory Guide* (Longman and Penguin).

A convenient way of investigating the effect of light intensity on the rate of photosynthesis is to measure the evolution of oxygen from an aquatic plant at different levels of illumination. In this experiment we will use Canadian pondweed, *Elodea*, from whose cut stem gas may be seen emerging as a stream of bubbles.

A simple way of estimating the oxygen evolved by *Elodea* is to count the number of bubbles given off per unit time. If this method is used it is important to select a piece of *Elodea* which is exuding bubbles of approximately equal size at a constant rate.

A more accurate method is to estimate the volume of gas given off by the plant in a given time.

Apparatus
The apparatus that is recommended for collecting and measuring the gas consists of a capillary tube, flared at one end for collecting the gas, and connected at the other end via a plastic tube to a 2-cm^3 syringe (Fig. 10.1). Gas evolved by the plant over a known period of time is collected in the flared end of the capillary tube. It is then drawn into the capillary tube by the syringe, and its volume recorded in terms of the length of the capillary tube that it occupies. This is repeated at different light intensities. The plastic tube serves as a depository for the gas that has been collected and pulled through the capillary tube.

Procedure
(1) Cut a piece of well-illuminated *Elodea* about 10 cm long and make sure bubbles are emerging from the cut end of the stem. Place it, bubbling end upwards, in a test tube containing the same water that the pondweed has been kept in. Stand the test tube in a beaker containing water at room temperature. Take the temperature of the water now, and also at intervals during the experiment: try to ensure that it remains constant.

(2) Fill the apparatus with tap water as follows: Remove the plunger of the syringe, and direct a gentle stream of water from a tap into the barrel until the whole of the syringe and plastic tube are full of water. Then replace the plunger and gently expel water from the flared end of the capillary

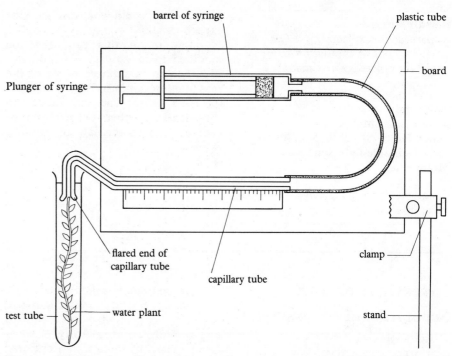

barrel of syringe

plastic tube

board

Plunger of syringe

flared end of
capillary tube

capillary tube

clamp

test tube

water plant

stand

Fig. 10.1 Apparatus for collecting and measuring the amount of gas given off by a water plant.

tube until the plunger is almost at the end of the syringe. Make sure the whole apparatus is full of water and that there are no air bubbles in the capillary tube.

(3) Place the end of the capillary tube in the test tube in such a way that the bubbles emerging from the piece of *Elodea* accumulate in the flared end. Stand the test tube in a beaker of water to prevent the temperature of the plant from increasing during the experiment.

(4) With the room darkened, position a light source 100 mm from the plant. A bench lamp with 40W bulb will do, though a lantern slide projector is better. (Why?).

(5) Wait for the plant to come into equilibrium with this light intensity. Then collect the gas given off during a given period of time (five minutes is generally sufficient). Then draw the bubble of gas into the capillary tube by gently pulling the plunger of the syringe. Measure its length and record this together with the distance of the plant from the light source.

(6) Now repeat the process with the light source at increasing distances from the plant, e.g. 5, 10, 15, 20, 25, 30, 40, 80 cm. In each case measure the amount of gas evolved in a standard period of time. Record your readings.

Results

The intensity of light falling on a given object from a constant source is inversely proportional to the square of the distance between them. In other words:

$$\text{Intensity} = \frac{1}{d^2}$$

where d is the distance between the light source and the object.

Work out the light intensity as $\frac{1}{d^2}$

(or, more conveniently, $\frac{100}{d^2}$) for each distance used in your experiment and record this together with the amount of gas given off.

Plot your results on a graph, putting light intensity on the horizontal axis and the amount of oxygen evolved (as length of bubble in mm) on the vertical axis.

For Consideration

(1) What relationship between gas production and light intensity is demonstrated by your results?

(2) What difficulties, if any, did you encounter in the course of this experiment and how did you overcome them?

(3) Consider any sources of inaccuracy in the technique used in this experiment.

(4) What do you suspect to be the composition of the gas evolved by the

Requirements

Apparatus for collecting gas (*see* Fig. 10.1) (The capillary tube bore should be approximately 0.5 mm).

Test tube
Beaker (400 cm³)
Bench lamp with 40W bulb (or other suitable light source)
Metre rule

Canadian pondweed, *Elodea* (well illuminated)

plant? How could you confirm this experimentally?

(5) How might this procedure be modified to investigate the influence of carbon dioxide concentration on the rate of photosynthesis?

(6) How would you use *Elodea* to investigate whether or not photosynthesis obeys the law of limiting factors?

Investigation 10.2

Separation of chlorophyll pigments by paper chromatography

What is commonly called chlorophyll consists of five different pigments. These can be separated and identified by **paper chromatography**. Absorptive paper containing a concentrated spot of chlorophyll extract is dipped in a suitable solvent. The various pigments have different solubilities in the solvent with the result that as the solvent ascends the absorptive paper it carries the pigments with it at different rates. In this way they become separated from one another and can be identified by their different colours and positions.

Procedure

Cut a strip of chromatography paper (filter paper will do) of sufficient length to almost reach the bottom of a large test tube and of such width that the edges do not touch the sides of the tube.

Rule a pencil line across the strip of paper 30 mm from one end. Fold the other end through 90° and by means of a pin attach it to the stopper as shown in Fig. 10.2. Make sure that the lower end of the strip almost reaches the bottom of the tube and that the edges do not touch the sides.

Remove the paper from the boiling tube and, using the head of a small pin as a dropper, place a drop of **chlorophyll solution** at the centre of the pencil line. Let the drop dry, then place a second small drop on the first. Repeat this process for about 15 minutes, building up a small area of concentrated pigment: the smaller and more concentrated, the better. Drying the spot can be hastened each time by gently warming with the heat from a lamp.

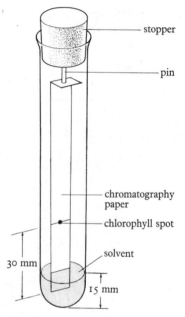

Fig. 10.2 Set-up for the chromatographic separation of chlorophyll pigments.

While preparing your chlorophyll spot, pour some **solvent** (a mixture of acetone and petroleum ether) into the boiling tube to a depth of not more than 15 mm. Seal the tube with a stopper for about 10 minutes so the atmosphere inside becomes saturated with vapour.

Now suspend the strip of paper in the boiling tube. The bottom edge of the paper should dip into the solvent, but make sure the chlorophyll spot is not immersed.

The solvent will ascend rapidly and the pigments will separate in about 10 minutes. When the solvent is approximately 20 mm from the top of the paper, remove the strip, rule a pencil line to mark the solvent front, and dry the paper.

Requirements
Large test tube (24 × 150 mm)
Stopper to fit test tube
Pin
Chromatography or filter paper
Scissors
Rack for test tube

Chlorophyll pigment solution
Solvent (5 cm³)

Prepare chlorophyll solution as follows: Grind up fresh leaves of, e.g. nettle, killed by rapid immersion in boiling water, with pure acetone.

Make up solvent as follows: Add one part 90 per cent acetone to 9 parts petroleum ether (boiling point 80–100°C).

Your chromatogram is now complete and you can proceed to identify the pigments.

Results
If you are lucky you should be able to detect the five pigments listed in Table 10.1. The pigments can be identified by their colours and R_f values.

$$R_f = \frac{a}{b}$$

where a = distance moved by substance from its original position.

b = distance moved by solvent from same position.

Measure the distance from the pencil line to the leading edge of each clearly detectable pigment, and work out the R_f values for each one. Compare your results with the information given in Table 10.1.

Name	Colour	R_f
Carotene	Yellow	0.95
Phaeophytin	Yellow-grey	0.83
Xanthophyll	Yellow-brown	0.71
Chlorophyll a	Blue-green	0.65
Chlorophyll b	Green	0.45

Table 10.1 Colours and R_f values of the pigments found in chlorophyll.

For Consideration
(1) How could you ascertain if there are more pigments present in chlorophyll than the ones you have identified? (2) What might be the functional significance of the fact that there are several different pigments, not just one?

Investigation 10.3
Structure of leaves

The leaf is, first and foremost, the plant's organ of photosynthesis. As such it would be expected to contain photosynthetic tissue held in such a position as to secure optimum illumination of the chloroplasts. The leaf must obviously be in functional connection with the rest of the plant, transport tissues bringing water and mineral salts to the photosynthetic tissues and taking the products of photosynthesis away from them.

Procedure

EXTERNAL FEATURES
Study the external features of a leaf of a representative dicotyledon (e.g. privet, laurel, holly, etc.) and monocotyledon (e.g. lily, grasses, etc.). Dicotyledons generally display a network of veins, monocotyledons parallel veins.

Note the terminology in Fig. 10.3A. Why is the upper surface of the leaf a darker green and more shiny than the bottom surface? What are the functions of the midrib and veins?

With small forceps strip off a piece of epidermis from the upper and lower surfaces of the leaf. Mount in water, outer side of the epidermis uppermost, and put on a coverslip. Examine under the microscope noting epidermal cells in surface view, stomata, and guard cells (Fig. 10.3B).

A **Whole leaf**

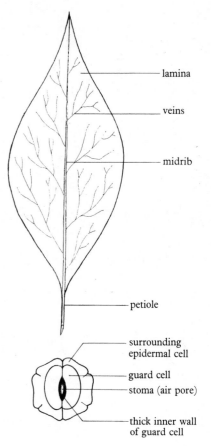

B **Stoma in surface view**

Fig. 10.3 External features of a generalized dicotyledonous leaf.

INTERNAL STRUCTURE

Cut transverse sections of a dicotyledonous leaf. Holly is recommended since it is available throughout the year and is comparatively easy to cut. As the leaf is thin it must be mounted in a firm position while the sections are cut: do this by inserting the piece of leaf into a vertical slit made down the centre of a piece of moistened elder pith or carrot tuber (Fig. 10.4). Hold the pith or carrot in one hand and cut your sections rapidly and smoothly with a sharp safety razor blade or 'cut-throat' razor held in the other hand. Place the sections in a dish of water.

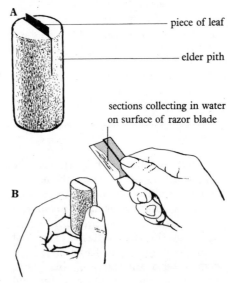

A — piece of leaf

— elder pith

sections collecting in water on surface of razor blade

B

Fig. 10.4 Technique for cutting sections of a leaf. **A** shows how the piece of leaf is mounted in elder pith prior to sectioning. **B** shows the actual sectioning procedure.

Select one or two thin sections, including at least one that is cut through the midrib, and mount them in water on a slide. Transfer the sections with a fine brush, not with forceps which may damage them.

(1) Examine the **lamina** of the leaf on either side of the midrib and,—using Fig. 10.5A as a guide—identify the following tissues:

Upper epidermis: one or more layers of rectangular cells, generally with thick cuticle.

Photosynthetic tissue: Cells with chloroplasts, subdivided into:
Palisade layer :
One to three cells thick, dense green due to presence of numerous chloroplasts; cells elongated at right angles to leaf surface; narrow intercellular air spaces appear as dark lines flanking many of the cells.
Spongy mesophyll :
Cells rounded or sausage-shaped; fewer chloroplasts; extensive intercellular air spaces.

Lower epidermis: single layer of cells with cuticle. (Is the cuticle as thick as that on the upper epidermis? There are generally many stomata perforating the lower surface of the leaf and your section may be cut through several pairs of guard cells. These are sunk below the general surface of the epidermis (Fig. 10.5B)).

Vascular bundles (veins): these may have been cut through transversely or obliquely. The upper part of the vascular bundle consists of water-conducting xylem tissue; the lower part of food-conducting phloem tissue. Thick-walled sclerenchyma fibres may be seen immediately beneath the phloem. If the bundle is cut obliquely, spirally thickened xylem elements may be seen (Fig. 10.5C).

(2) Examine the **midrib** and observe how this differs from the rest of the leaf. Note:

Vascular tissue: large vascular bundle(s) in centre; phloem is generally confined to the underside of the xylem but in some plant species it surrounds it as shown in Fig. 10.6.

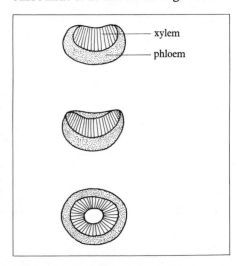

— xylem
— phloem

Fig. 10.6 Arrangement of vascular tissues in petiole and midrib of different species of plant. In all three examples the xylem is cross-hatched and the phloem stippled.

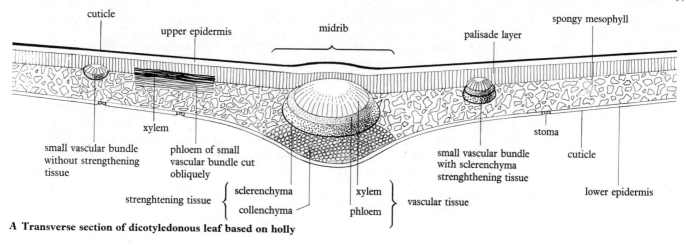

cuticle
upper epidermis
midrib
palisade layer
spongy mesophyll
xylem
small vascular bundle without strengthening tissue
phloem of small vascular bundle cut obliquely
small vascular bundle with sclerenchyma strengthtening tissue
stoma
cuticle
lower epidermis

strenghtening tissue { sclerenchyma / collenchyma
xylem / phloem } vascular tissue

A Transverse section of dicotyledonous leaf based on holly

Fig. 10.5 Structure of dicotyledonous leaf based on holly.

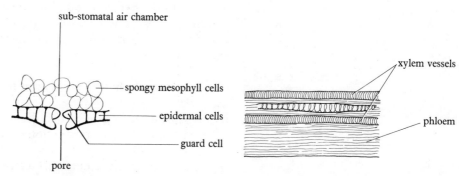

sub-stomatal air chamber
spongy mesophyll cells
epidermal cells
guard cell
pore

B Generalized stoma

xylem vessels
phloem

C Detail of vascular bundle cut obliquely

Requirements
Slides and coverslips
Dish for sections
Fine brush
Safety razor or 'cut-throat' razor

Iodine solution
Sudan III
FABIL

Fresh leaves of dicotyledon and monocotyledon
Elder pith and/or carrot tuber

Strengthening tissue: variable amount of parenchyma tissue surrounds vascular area; collenchyma immediately beneath lower epidermis; sclerenchyma close to phloem.

Irrigate your sections with iodine solution. All lignified walls will be stained yellow and any starch present will be stained dark blue. If no starch is present what is the explanation?

Mount another section in Sudan III. This stains fat and is a good way of showing up the cuticle.

Examination of the vascular and strengthening tissues is greatly aided by staining sections in FABIL (see p. 437). The staining reaction varies but generally xylem elements stain brown, sclerenchyma pink, cellulose pale blue, cytoplasm and nuclei dark blue, starch black.

COMPARISON WITH LEAF
OF MONOCOTYLEDON
Cut transverse sections of a monocotyledon leaf, e.g. lily, iris, etc. How does its internal structure differ from that of the dicotyledon leaf? Why the differences?

For Consideration
(1) If you are an artist attempt to draw a stereoscopic diagram illustrating the three-dimensional structure of the dicotyledonous leaf which you have studied in this investigation. Include the vascular and strengthening tissues as well as the photosynthetic tissue.

(2) Summarize the adaptive features of the leaf, i.e. the ways in which it is adapted to perform its particular functions.

(3) Although the leaves of all dicotyledons are basically similar, they differ in various details. Investigate the internal structure of other species of dicotyledons of your own choice. How do they differ from one another and how would you explain the differences?

Questions and Problems

1 The highest yield of wheat ever recorded on an agricultural scale is stated to have been one of 8925 kg of grain per hectare obtained near Doncaster in 1962 (*Guinness Book of Records*). In northern England, the total energy of sunlight and infra-red radiation received per unit area of ground surface during March to July inclusive is, in an average year, approximately 200 J/cm². Make a rough estimate of the percentage of this energy converted into chemical energy in the grain. Outline briefly what would have happened to the rest of the energy.

[Assume: that the heat given out by the combustion of dry carbohydrate = 16.8 J/g; 1 hectare = 1.0 × 10⁴ m²]   (*CJE modified*)

2 How would you demonstrate:
(a) that carbon dioxide is necessary for photosynthesis,
(b) that red and blue light are more effective than yellow light in photosynthesis,
(c) that chlorophyll is required for photosynthesis,
(d) that photosynthesis occurs in the chloroplasts,
(e) that the oxygen evolved in photosynthesis is derived not from carbon dioxide but from water?

3 A plant's rate of photosynthesis can be estimated by the following alternative methods:
(a) measuring the rate of carbon dioxide uptake,
(b) measuring the rate of evolution of oxygen gas,
(c) measuring increase in the dry-weight of leaves during a given period,
(d) measuring the amount of carbohydrate formed in the leaves during a given period.

Compare the relative merits and disadvantages of each method.

4 Results of experiments to measure the rate of photosynthesis under different conditions are shown in the graphs in Fig. 10.7.

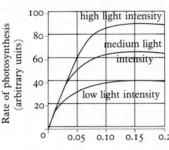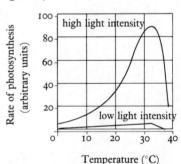

Fig. 10.7 Results of experiments in which the rate of photosynthesis was measured under different conditions. The rate of photosynthesis is expressed in arbitrary units.

(a) Explain the effect of variation in the carbon dioxide concentration on the rate of photosynthesis.
(b) In what way is the rate of photosynthesis dependent on temperature?
(c) What general principle is illustrated by the results of changing the light intensity?
(d) What conclusions can you draw from these experiments concerning the nature of the process of photosynthesis?   (*AEB modified*)

5 Write an essay on the interdependence of animals and plants.

6 Account for the following:
(a) In dropping its leaves a deciduous tree loses its photosynthetic equipment, and yet it survives the winter.
(b) The CO_2 content of the air in early spring has been found to be slightly higher than in the later summer.
(c) Crops planted in the vicinity of certain factories have been observed to show more rapid and prolific growth than those planted in identical soil elsewhere.
(d) It is customary for the windows of greenhouses to be whitewashed in summer.

7 The effect of temperature on the rates of (a) apparent photosynthesis (net CO_2 uptake in the light) and (b) respiration (CO_2 produced in the dark) was determined. The results, expressed as mg CO_2, taken up or released, per gramme dry weight of leaf per hour, are given in the following table:

mg CO_2/g/h	Temperature (°C)						
	7	10	15	19	22	28	31
Uptake (apparent photosynthesis)	1.3	2.3	2.8	3.1	2.8	2.5	1.8
Release (respiration)	0.3	0.6	0.7	1.2	1.8	2.1	2.7

(a) Calculate the rates of true photosynthesis at each temperature, assuming that the rate of respiration in the light is equal to the rate of respiration in the dark.
(b) Plot on a graph the results for apparent photosynthesis, true photosynthesis and respiration.
(c) Comment briefly on these observations. (*O and C*)

8 The relative rates of photosynthesis at three temperatures were determined at five-minute intervals over a period of 20 minutes. Comment on the results obtained which are given below. What experiment would you perform in order to obtain results like these?

Temp. °C	Relative rates of photosynthesis			
	5 mins	10 mins	15 mins	20 mins
30	38	39	39	36
35	61	60	58	47
40	85	80	57	30

(*O and C*)

9 Discuss how the structure of (a) a leaf of a flowering plant, and (b) an individual chloroplast are adapted for the performance of photosynthesis.

10 Photosynthesis involves the removal of electrons from chlorophyll, the splitting of water, the production of ATP, and the reduction of carbon dioxide. Explain concisely how these four processes are linked. What kind of experiments have been carried out to investigate the chemistry of photosynthesis?

11 The effect of the length of the dark intervals between flashes of light on the amount of synthesis per flash was measured at two different temperatures with the following results:

Amount of photosynthesis per light flash (arbitrary units)

Temp. °C	Length of dark intervals (seconds)					
	0.02	0.06	0.1	0.2	0.3	0.4
1.0	2.1	3.5	4.0	4.6	4.8	5.0
24.0	5.0	5.1	5.0	4.9	5.0	5.0

The amount of photosynthesis is given in arbitrary units and in all cases the light flash was 1.0×10^{-5} seconds.

Comment on these results in relation to your knowledge of the effects of other factors on photosynthesis and consider their significance in relation to the mechanism of the process.

(*O and C*)

11 Transport in Animals

Background Summary

1 The volume of an animal is generally too large for diffusion to supply the needs of the tissues, and as a result transport systems have been developed. These range from water-filled canals to blood-filled circulatory (vascular) systems.

2 Mammalian blood is composed of **red** and **white blood cells** (**erythrocytes** and **leucocytes** respectively) and **platelets** suspended in **plasma**. Dissolved food substances are transported in the plasma, respiratory gases by the red blood cells.

3 Biconcave, disc-shaped to increase their surface:volume ratio, red blood cells contain **haemoglobin** which, experiments show, has a high affinity for oxygen.

4 Haemoglobin's affinity for oxygen is lowered by the presence of carbon dioxide. Thus loading of haemoglobin with oxygen is favoured in the lungs, whereas unloading is favoured in the tissues.

5 **Myoglobin** (**muscle haemoglobin**) and **foetal haemoglobin** have a higher affinity for oxygen than adult haemoglobin. The invertebrate blood pigments **chlorocruorin**, **haemoerythrin** and **haemocyanin** have comparable affinities to haemoglobin though their total oxygen capacities are generally lower.

6 The lowering of haemoglobin's affinity for oxygen in the presence of carbon dioxide is explained by the mechanism of carbon dioxide carriage in the blood. Due to the presence of the enzyme **carbonic anhydrase**, most of the carbon dioxide is carried in solution in the red blood cells. Hydrogen ions resulting from the dissociation of carbonic acid are readily taken up by haemoglobin which, therefore, acts as a buffer.

7 The mammalian circulation with its heart, arteries, capillaries, and veins is well adapted for delivering oxygen at high speed to the tissues.

8 The heart is divided into four chambers: two **atria** and two **ventricles**. Blood is propelled through the heart by a series of electrical and mechanical events which constitute the **cardiac cycle**. The ventricles have thick muscular walls, and **valves** prevent blood flowing in the wrong direction.

9 The heart beat is initiated by the **sino-atrial node** (pacemaker) which, though it has an innate rhythm, is influenced by its nerve supply. The sympathetic nerve accelerates the heart, the vagus slows it.

10 The **capillaries** are narrow and thin-walled and, coming into intimate association with the tissue cells, are well adapted for exchanges between the blood and tissues.

11 The mammal has a **double circulation**, blood flowing through the heart twice for every complete circuit of the body. The heart is completely divided into right and left sides. In contrast, fishes have a **single circulation** with an undivided heart. Amphibians have a double circulation with a partially divided heart. Octopuses and squids have a single circulation with two sets of hearts.

12 The circulation of most animals, with tubular blood vessels, is described as a **closed circulation**. This contrasts with the **open circulation** of insects where blood flows through large cavities and sinuses.

Investigation 11.1
The number of red corpuscles in human blood

Since it is obviously impossible to count all the blood cells present in a circulatory system, one resorts to sampling—that is, one counts the cells in a representative volume.

A device for sampling cells in this way is the **haemocytometer**, a modified microscope slide. Although in this investigation the haemocytometer is used for counting blood corpuscles, it can be used for counting any cells that are uniformly distributed (*see* p. 51, for example).

In the present experiment a measured volume of blood is diluted a known number of times. The corpuscles are counted in a known volume of the diluted blood, from which the number of corpuscles per mm³ of undiluted blood may be calculated.

Apparatus
The haemocytometer consists of a special slide with a ruled area in the centre, a coverslip and two graduated pipettes with rubber tubing.

The slide is a delicate piece of equipment: be careful not to scratch it; wipe it only with lens paper. Ensure that both the slide and coverslip are clean: if necessary wash them in distilled water followed by acetone; when the acetone has evaporated rub them with lens paper.

Examine the slide under the low power of the microscope and locate the ruled area in the centre. The middle of the ruled area consists of a grid 1 mm² in area. We will call this the type-A square (Fig. 11.1). If you use the × 10 objective and × 10 eyepiece, the type-A square should just about fill your field of view.

A   Surface of slide

B   Section of slide showing coverslip in position

C   Central part of grid

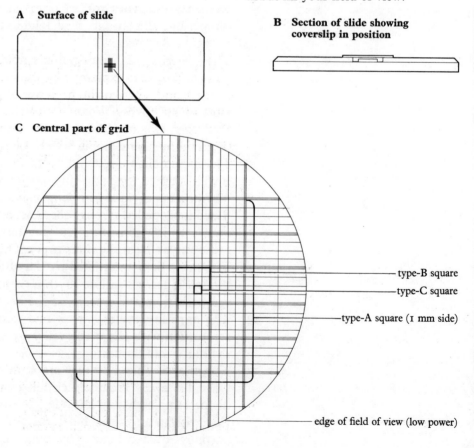

type-B square
type-C square
type-A square (1 mm side)
edge of field of view (low power)

Fig. 11.1   Neubauer haemocytometer slide.

Notice that the type-A square is subdivided by triple lines into 25 type-B squares, each of which has an area of $\frac{1}{25}$ mm². Each type-B square is further subdivided by single lines into 16 type-C squares which have an area of $\frac{1}{400}$ mm².

The surface of the slide between the two deep grooves is 0.1 mm lower than the rest of the slide on either side of the grooves. So when the coverslip is put on, its lower surface clears the ruled surface of the slide by 0.1 mm. The volume subtended by the type-A square is therefore 0.1 mm³; the volume subtended by a type-B square is 0.004 mm³; and the volume subtended by a type-C square is 0.00025 mm³.

The pipette for red blood corpuscles is the one with the mark 101 above the bulb. The other pipette, with the mark 11 above the bulb, is used for white corpuscles.

The pipettes must be cleaned and dried immediately after use. Wash by sucking distilled water (or, if necessary, acetic acid) into the pipette; dry with acetone.

Procedure for Counting Red Blood Corpuscles

(1) Obtain a large drop of blood by pricking the end of your thumb with a lancet or sterilized needle. First sterilize your skin by swabbing it with ethanol. Do not squeeze the blood out too violently for this will force out a lot of tissue fluid which will dilute the corpuscles.

(2) With the rubber tubing attached to the pipette, suck the blood up to the '1' mark. Quickly dry the tip of the pipette. If the 1 mark has been passed, touch the tip of the pipette with filter paper until the blood drops back to the 1 mark.

(3) Now suck up three per cent sodium chloride (salt solution, NaCl) until the contents of the pipette reach the 101 mark. If the 101 mark is passed, the pipette must be emptied and the procedure started again.

(4) Remove the rubber tubing from the pipette. Close the two ends of the pipette with thumb and finger and rock the pipette for at least a minute. When the blood and NaCl are thoroughly mixed, blow out six drops so as to expel excess NaCl solution.

Your blood sample is now diluted 1 in 100.

(5) Place the coverslip in the centre of the slide. Put one drop of diluted blood from the pipette onto the slide alongside the coverslip in the area between the two deep grooves. The blood should be drawn under the coverslip by capillary action. Wait five minutes to allow the corpuscles to settle. If the blood flows into the grooves, clean the slide and put on another drop.

(6) Place the slide under the microscope and adjust the illumination so the grid and the corpuscles can be clearly seen. If the corpuscles are unevenly distributed, clean the slide and start again.

(7) Count the red blood corpuscles in 100 type-C squares. Record your results by ruling out 100 squares and writing the number of corpuscles in each square.

Note: In each square count all the corpuscles which lie entirely within it, plus those which are touching or overlapping the top and left-hand sides. Do not include those touching or overlapping the bottom and right-hand sides even if they are within the square.

(8) Calculate the average number of corpuscles in a type-C square. Bearing in mind the volume represented by a type-C square, and the dilution factor, calculate the number of red corpuscles per mm³. A typical figure might be 5 million/mm³. How does your figure compare with this?

Assuming that the total volume of blood in the body is 5 dm³, calculate the number of red corpuscles in the entire circulation.

For Consideration
Under what circumstances might you find the red blood cell count to be (1) unusually high, (2) unusually low?

Requirements
Haemocytometer (slide, coverslip, pipette and rubber tubing)
Lancet or sterilized needle
Lens paper
Small beaker

NaCl solution (3 per cent)
Distilled water
Acetic acid
Acetone
Ethanol and cotton wool

Investigation 11.2

Dissection of the mammalian circulatory system

The mammal has a **double circulation** with a completely divided heart. The right side of the heart receives deoxygenated blood from the great veins which it propels to the lungs. The left side receives oxygenated blood from the lungs which is then pumped into the aorta.

The entire system constitutes a dynamic transport device. Both in the structure of its individual units (heart, arteries, capillaries, veins) and in its overall layout, we see a close relation between structure and function.

Procedure

(1) If it has not already been done, open up the thorax as described on p. 58. Remove the thymus gland to expose the heart and pluck away fat from around the great vessels. Push the heart to your left and identify the structures shown in Fig. 11.2A.

Push the heart to your right and identify the point where the **venae cavae** enter the right atrium (Fig. 11.2B). Mentally reconstruct the flow of blood through the heart and great vessels. Carefully trace the right anterior vena cava and innominate artery from the thorax up into the neck, removing all the muscle which now is obscuring these vessels. (Arrows, Fig. 11.3A).

(2) Remove the pectoral muscle and clavicle. This will reveal the origin of the **subclavian** and **internal jugular veins**. In tracing the **innominate artery** forwards notice the origin from it of the **right subclavian**

A   Heart deflected to your left

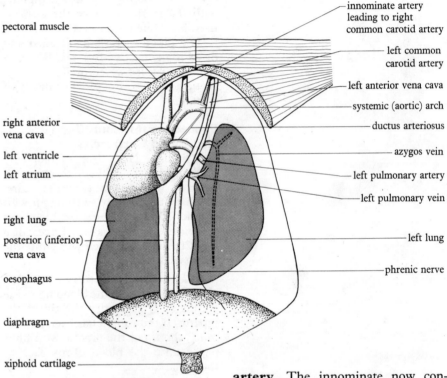

- pectoral muscle
- right anterior vena cava
- left ventricle
- left atrium
- right lung
- posterior (inferior) vena cava
- oesophagus
- diaphragm
- xiphoid cartilage
- innominate artery leading to right common carotid artery
- left common carotid artery
- left anterior vena cava
- systemic (aortic) arch
- ductus arteriosus
- azygos vein
- left pulmonary artery
- left pulmonary vein
- left lung
- phrenic nerve

B   Heart deflected to your right

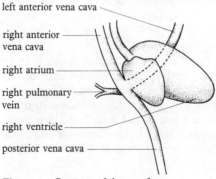

- left anterior vena cava
- right anterior vena cava
- right atrium
- right pulmonary vein
- right ventricle
- posterior vena cava

Fig. 11.2   Contents of thorax of rat.

artery. The innominate now continues forward as the right common carotid artery. Now repeat on the left side of the body. On this side there is no innominate artery, the left subclavian arising from the systemic arch. To see the left subclavian you will probably need to deflect the left anterior vena cava to your right.

Finally trace the arteries and veins forwards to the jaw as shown in Fig. 11.3B.

Identify the structures shown in Fig. 11.3B. What is the function in relation to the circulation of the nerves shown in the inset to Fig. 11.3B?

A   Before removal of pectoral and neck muscles

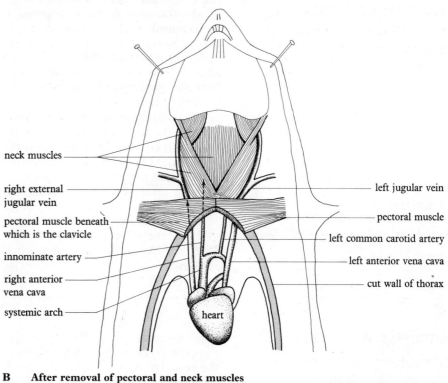

neck muscles

right external
jugular vein

pectoral muscle beneath
which is the clavicle

innominate artery

right anterior
vena cava

systemic arch

left jugular vein

pectoral muscle

left common carotid artery

left anterior vena cava

cut wall of thorax

heart

B   After removal of pectoral and neck muscles

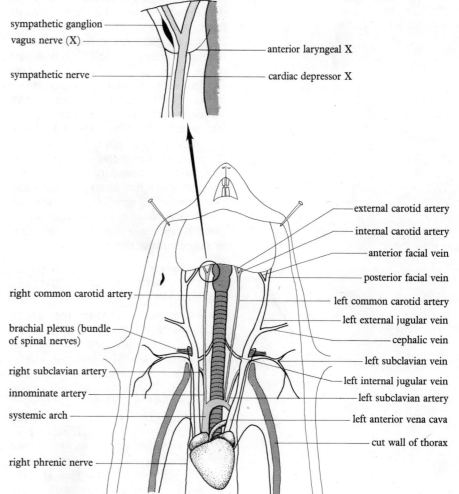

sympathetic ganglion
vagus nerve (X)

sympathetic nerve

anterior laryngeal X

cardiac depressor X

right common carotid artery

brachial plexus (bundle
of spinal nerves)

right subclavian artery

innominate artery

systemic arch

right phrenic nerve

external carotid artery

internal carotid artery

anterior facial vein

posterior facial vein

left common carotid artery

left external jugular vein

cephalic vein

left subclavian vein

left internal jugular vein

left subclavian artery

left anterior vena cava

cut wall of thorax

Fig. 11.3 Dissection of the heart and anterior circulation of the rat.

(3) Follow the **dorsal aorta** through the thorax into the abdominal cavity. In the thorax it gives off a series of **intercostal arteries**. Blood is returned to the heart from the thoracic wall by the **azygos veins**. Trace the azygos veins back to the heart. Also note the **posterior vena cava** running close to the dorsal aorta.

(4) If it has not already been done, open up the abdominal cavity, identify the **coeliac**, **anterior mesenteric** and **posterior mesenteric arteries** and remove the gut (*see* pp. 73–76).

Pluck away the fat obscuring the posterior vena cava and dorsal aorta and investigate the branches of each.

Trace the **iliac** arteries and veins into the hind legs, noting their branches.

Requirements
Dissecting instruments
Hand lens or binocular microscope

Fresh-killed rat

For Consideration
(1) In your dissection you will have noticed that the veins generally have a wider diameter and a darker colour than the arteries. Can you suggest reasons for this difference?

(2) Suggest explanations of the asymmetrical origin of the subclavian arteries on the two sides of the body.

(3) By what series of blood vessels does blood reach (a) the brain, (b) the right forelimb, (c) the liver, (d) the left kidney, and (e) the left hindlimb.

(4) By what series of blood vessels is blood returned to the heart from (a) the sides of the head, (b) the left forelimb, (c) the lungs, (d) the intercostal muscles, and (e) the gut.

Investigation 11.3
Structure and action of the mammalian heart

The mammal has a double circulation, that is, blood flows twice through the heart for every complete circuit of the body. Deoxygenated blood from the tissues enters the right atrium from the great veins (venae cavae). From the right atrium blood flows into the right ventricle whence it is pumped to the lungs via the pulmonary artery. Having been oxygenated in the lungs blood is returned to the heart via the pulmonary veins: after entering the left atrium the blood flows into the left ventricle whence it is pumped to the body via the aorta.

Two main questions arise in connection with the heart: how is the blood propelled from it, and how is the blood kept moving in the right direction. Keep these questions in mind as you examine the heart.

Procedure
For this experiment use the heart of a pig, sheep or ox.

(1) Distinguish between the dorsal and ventral sides of the heart. The ventral side is more rounded (convex) than the dorsal side, and the thick-walled arteries arise from this side (Fig. 11.4). **Note**: right and left **atria**, right and left **ventricles**; pulmonary artery and aorta (systemic arch) arising from right and left ventricles respectively; anterior and posterior venae cavae opening into right atrium; pulmonary veins opening into left atrium; coronary vessels in heart wall (function?).

(2) The following experiment can only be done if the heart and great vessels are still complete.

Insert a tap, or rubber tubing from a tap, into the anterior vena cava. Clamp the posterior vena cava. Run water into the vena cava and note its flow through the heart. From which blood vessel does the water emerge? This is the pulmonary artery. Explain fully.

Now run water into the pulmonary vein and note the vessel from which it emerges. This is the aorta.

(3) Expose the interior of the left ventricle by a longitudinal cut through the ventral wall of the ventricle (*see* right-hand dotted line in Fig. 11.4). At the top of the ventricular cavity to your left: an opening into the aorta guarded by **semilunar (pocket) valves**. To your right: the atrio-ventricular opening guarded by **bicuspid (mitral) valve**; note that the latter has two flaps attached to the ventricular wall by tendinous cords and papillary muscles (function?).

(4) Turn the heart upside down and run water into the ventricle through the slit which you have cut. Notice the impeding action of the bicuspid valve. In life it prevents the backflow of blood from ventricle to atrium.

Now turn the heart the right way up; run water into the cut end of the aorta and note the impeding action of the semilunar valves. What are the functions of the semilunar valves?

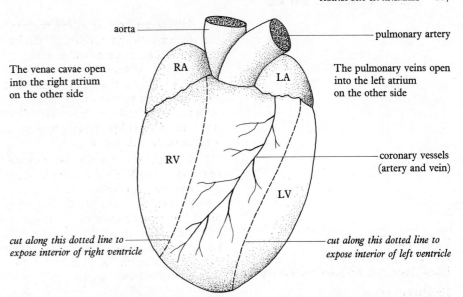

aorta

pulmonary artery

The venae cavae open
into the right atrium
on the other side

The pulmonary veins open
into the left atrium
on the other side

RA

LA

coronary vessels
(artery and vein)

RV

LV

*cut along this dotted line to
expose interior of right ventricle*

*cut along this dotted line to
expose interior of left ventricle*

Fig. 11.4 Ventral view of mammalian heart. This is the more convex side of the heart and shows the attachment of the aorta and pulmonary artery to the ventricles. The veins are on the other side. **RA**, right atrium; **LA**, left atrium; **RV**, right ventricle; **LV**, left ventricle.

(5) Cut open the left atrium and aorta by continuing your ventricular cut upwards. Observe details of the bicuspid valve and the pocket-like semilunar valves. Also notice the opening into the coronary artery from the aorta just above the semilunar valves. This important artery takes oxygenated blood to the wall of the heart.

(6) Expose the interior of the right ventricle by a longitudinal slit through the ventral wall (left hand dotted line in Fig. 11.4). **Note**: the atrio-ventricular opening guarded by **tricuspid valve** (three flaps); **semilunar valves** guarding pulmonary artery. Pour water into the pulmonary artery and note the impeding action of semilunar valves exactly as on the other side of the heart.

(7) Slit open the right atrium and pulmonary artery by continuing your ventricular slit upwards. Observe the valves in detail as you did on the other side of the heart. In addition note the opening of the coronary vein on the left hand side of the atrium (the right as you view it from the ventral side). In the inter-atrial septum you may see a small oval depression, the **fossa ovalis**. This is a relic of the foramen ovale which connects the right and left atria in the embryo. Function of fossa ovalis?

(8) Examine the openings of the pulmonary veins and venae cavae into their respective atria. Are there any valves guarding these openings and what might be their functions?

(9) Take your **pulse rate** by placing your finger over your radial artery at the wrist. How many times is your heart beating per minute? Assuming a constant rate, how many beats will your heart have undergone by the time you are sixty? From these considerations what predictions can you make about the properties of heart (cardiac) muscle?

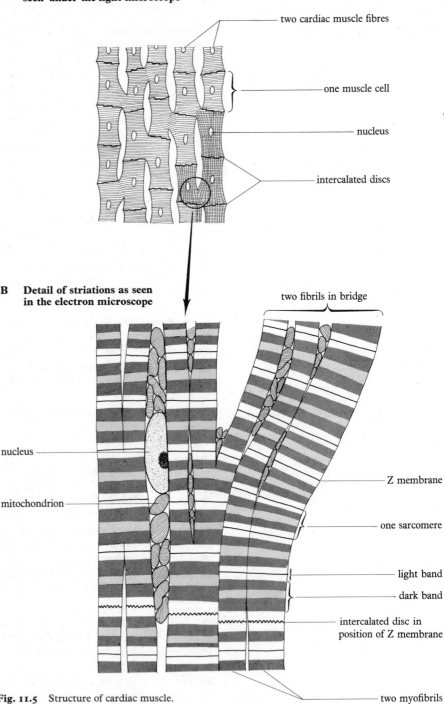

A **Longitudinal section of cardiac muscle as seen under the light microscope**

two cardiac muscle fibres

one muscle cell

nucleus

intercalated discs

B **Detail of striations as seen in the electron microscope**

two fibrils in bridge

nucleus

mitochondrion

Z membrane

one sarcomere

light band

dark band

intercalated disc in position of Z membrane

two myofibrils

Fig. 11.5 Structure of cardiac muscle.

Requirements
Dissecting instruments
Microscope
Tap and rubber tubing
Clamps for sealing of blood vessels

Heart of pig, sheep or ox
Section of cardiac muscle
Electron micrograph of cardial muscle

(10) Examine a section of **cardiac muscle**. Note the network of muscle fibres interconnected by bridges, and the other structures shown in Fig. 11.5A. The muscle fibres consist of chains of muscle cells, each containing a single nucleus. The muscle cells are separated from one another by intercalated discs which the electron microscope shows to correspond in position to the Z membrane (Fig. 11.5B).

For Consideration
(1) Trace the sequence of events that takes place as blood flows through the heart from the venae cavae to the pulmonary artery, and from the pulmonary veins to the aorta.
(2) Can you think of any functional reason why the atrio-ventricular valve has three flaps on the right hand side of the heart but only two flaps on the left?

Investigation 11.4

Structure and properties of blood vessels

Pumped by the muscular action of the heart, the blood is propelled round the body in tubular blood vessels. The latter include, in the order in which blood flows through them, arteries, arterioles, capillaries, venules, veins. Each of these units has certain functions to perform and problems to contend with, and is adapted accordingly. Before beginning this investigation think of the job which arteries and veins have to do and see if you can predict what their structure is likely to be.

Procedure

(1) Using the heart from the previous investigation, cut out a ring from the aorta and another from one of the venae cavae. Suspend each ring from a hook on a stand and attach it to a lever as shown in Fig. 11.6. Load the ring with increasing weights, 10 g at a time. Record the increase in length each time until no further change takes place. Then remove the weights, one at a time, and record the decrease in length each time until no further decrease is given.

How do the artery and vein compare with regard to (a) percentage increase in length on loading; (b) ability to return to their original length on unloading.

How do their elastic properties relate to the stresses and strains which arteries and veins are likely to experience in life?

(2) Examine the webbed foot of an anaesthetized frog under the microscope.

Prepare frog as follows
This is best done by the teacher or technician before the laboratory session. Place the animal in a 3–5 per cent solution of urethane until movement ceases. Then with rubber bands and/or thread attach the animal to a glass plate with the webbed foot spread out as much as possible. Alternatively mount the animal on a piece of cork with a hole in it. Arrange the animal in such a way that the webbed foot is over the hole. Cover the animal with wet cotton wool so as to keep it moist. Pipette frog Ringer's solution (*see* p. 437) onto the foot at frequent intervals to prevent it drying up. Clip to the microscope stage so the webbed foot is visible under low power.

Notice corpuscles flowing through capillaries in single file. Compare the velocity of blood-flow in a capillary with that in an artery. Is there any difference? Explain.

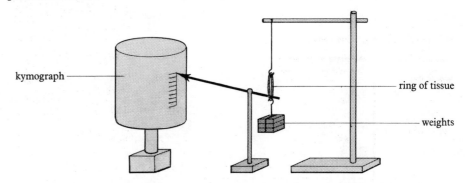

kymograph

ring of tissue

weights

Fig. 11.6 Technique for recording and measuring changes in length of a ring of tissue with increasing load. After adding each weight the new length acquired by the tissue is recorded by manually rotating the kymograph.

A   Complete artery and vein

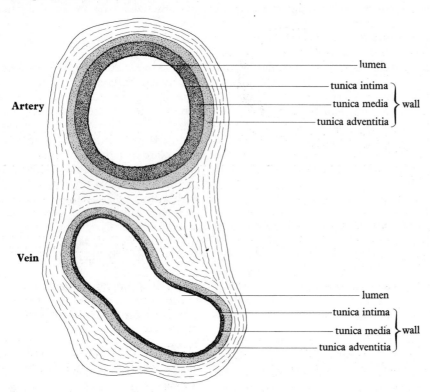

B   Large artery e.g. aorta

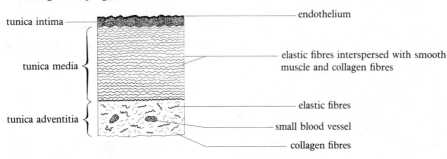

C   Large vein e.g. vena cava

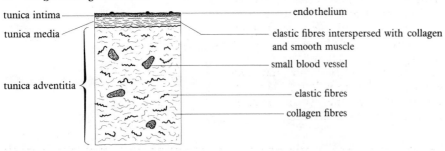

D   Medium sized artery

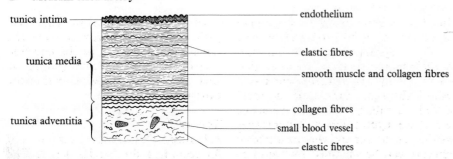

(3) Examine transverse sections of a large artery and vein under the microscope (Fig. 11.7). In each case the wall is made up of three layers:

(a) **Tunica intima**: thin, lined with **pavement** endothelium and composed **of** delicate collagen and elastic fibres **mainly** disposed longitudinally.

(b) **Tunica media**: thicker, consisting of tough circular elastic fibres **and** smooth muscle interspersed with **collagen** fibres. In appropriately **stained** sections the elastic fibres **appear** as wavy lines.
Note: The tunica media of arteries is **much** thicker than that of veins and contains much more smooth muscle.

(c) **Tunica adventitia**: varies in thickness, contains a mixture of irregularly arranged elastic and collagen fibres and blood vessels (vasa vasora).
Note: The tunica adventitia of arteries contains mainly elastic fibres; the tunica adventitia of veins contains mainly collagen fibres.

Compare the artery and vein with regard to the relative amounts of elastic fibres, collagen fibres and smooth muscles in their walls. Do your observations fit in with the results you obtained in the previous experiment? How do living cells in the walls of the artery and vein obtain oxygen? Do your observations agree with Table 11.1?

(4) Examine a transverse section of a smaller artery. How does it differ from the large artery? As arteries get smaller their walls become thinner and contain progressively less elastic tissue relative to smooth muscle. Functional explanation? The smallest arterioles have an endothelium surrounded by a layer of smooth muscle, then a thin layer of connective tissue.

(5) Examine a capillary in a microscopical section of an organ such as lung, kidney or thyroid gland. Notice that the wall consists of a single layer of pavement epithelial cells (Fig. 11.7E). Significance?

Fig. 11.7   Microscopic structure of blood vessels.

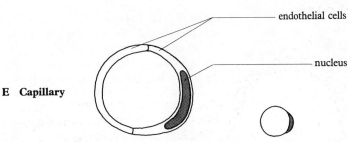

Fig. 11.7 Microscopic structure of blood vessels: E, on the right the capillary as it appears in a typical microscopic section.

Artery	Vein
Smaller overall diameter	Larger overall diameter
Narrower lumen	Wider lumen
Thicker wall with greater overall content of elastic fibres	Thinner wall with smaller overall content of elastic fibres
Tunica intima thicker	Tunica intima thinner
Tunica media thicker and more muscular	Tunica media thinner and less muscular
Tunica adventitia thinner with high elastic fibre content relative to collagen	Tunica adventitia thicker with low elastic fibre content relative to collagen
Vasa vasora situated further away from inner surface of wall	Vasa vasora penetrate closer to inner surface of wall

Table 11.1 Comparison between typical mammalian artery and vein. The vasa vasora are small blood vessels situated in the wall of the artery and vein.

Requirements
Microscope
Kymograph (stationary)
Stand with lever and hook
(*see* Fig. 11.6)
Weights (10 g)

Ring of large artery and vein
TS large artery
TS large vein
TS smaller artery
Section of e.g. lung, kidney or thyroid gland for capillaries

Anaesthetized frog

For Consideration
(1) In Table 11.1 the structure of a typical mammalian artery is compared with that of a vein. On the assumption that your own observations agree with the information given in the table, comment on the functional significance of each of the differences.
(2) Arterioles have an endothelium surrounded by a layer of smooth muscle, then connective tissue. Venules, on the other hand, have an endothelium surrounded by connective tissue, but there is no muscle present. Why the difference?

Investigation 11.5
Action of the frog's heart

In this exercise we will investigate some aspects of how the heart works.

The amphibian heart is basically similar to the mammal's except that there is only one ventricle and blood enters the right auricle via an additional chamber, the sinus venosus. Evidence suggests that the latter serves as the pacemaker and that its intrinsic rhythm can be modified by impulses reaching it from the vagus and sympathetic nerves. The rate at which the heart beats is known as the **cardiac frequency**. In this investigation we shall investigate factors affecting the cardiac frequency.

Technique
A kymograph (*see* Appendix) is used for recording the beating heart. The

heart is attached to a lever which writes on the drum. The drum should revolve at approximately 5 mm/s. If you have a time marker, a time tracing can be made on your drum along with your heart records. If you have not got a time marker, a time scale can be constructed from the time taken for the drum to do a complete revolution. The heart is attached to the lever by a thread with a hook at the end[1].

Preparing the frog

You should be provided with a 'double pithed' frog, i.e. one whose brain and spinal cord have been destroyed. The animal is therefore dead but its heart will remain physiologically active for several hours provided it is kept well doused with Ringer's solution.

Proceed as follows:

(1) Pin the frog, ventral side uppermost, to a piece of cork. Cut away the skin from the pectoral region and identify the xiphisternum (xiphoid cartilage) just behind the bony part of the pectoral girdle (Fig. 11.8A).

(2) Carefully cut round the perimeter of the xiphisternum and remove it. This will create a window in the body wall through which the heart may be seen. *Warning*: there is a large superficial vein just behind the xiphisternum: be careful not to cut it.

(3) Cut away the pericardial membrane surrounding the heart and force the latter out of the window so it is completely exposed. Now insert the hook into the thick muscle at the tip of the ventricle (Fig. 11.8B).

(4) The drum should be as high as possible on its spindle. Attach the thread to the lever by means of a small piece of plasticine. Position the lever on its stand, if necessary counterbalancing it with plasticine, so the heart is fairly well stretched and the lever horizontal. The writing point should give an excursion of about a centimetre every time the heart contracts. The writing point should be directed towards the base of the drum (Fig. 11.8C).

Experiments

NORMAL BEATING OF THE HEART
Record about 12 complete heart cycles at room temperature round the base of the drum. If there is not too much friction between the writing point of the lever and the paper, you should be able to distinguish the contractions of the atria and ventricle (*see* Fig.

11.8C). How do the contractions compare with one another? Explain any differences.

EFFECT OF TEMPERATURE
ON CARDIAC FREQUENCY
Have available Ringer's solution at the following temperatures: ice-cold, 30°C and 40°C. Record about six normal contractions (at room temperature). Then, with the kymograph still running, pipette ice-cold Ringer's solution onto the heart. With a pencil or pen make a mark on the drum corresponding to the moment that the Ringer's is applied. Record the effect until the heart beat returns to normal. Repeat the experiment with Ringer's solution at 30°C and 40°C. Explain your results.

RESPONSE OF THE HEART TO DRUGS
Record about six normal contractions, then pipette a drop $\frac{1}{5,000}$ acetyl choline onto the sinus venosus of the heart. If this produces little or no change try $\frac{1}{1,000}$. When the heart has returned to normal wash it with Ringer's solution and repeat the experiment with $\frac{1}{10,000}$ and $\frac{1}{1,000}$ adrenaline. Does the heart respond differently to acetyl choline and adrenaline? Explain fully.

EFFECT OF HIGH TEMPERATURE ON
VARIOUS REGIONS OF THE HEART
Record normal contractions. Then bring the tip of a blunt rod, heated to 40°C, close to (a) the ventricle, (b) left atrium, (c) right atrium, (d) sinus venosus. In each case record the effect, if any, on the cardiac frequency. Explain your results.

Results

From each set of recordings which you have made, calculate as accurately as possible the cardiac frequency as the number of heart beats per minute.

---

[1] If no kymograph is available the cardiac frequency can be estimated by counting the number of heartbeats per minute. For ease of counting, the heart may be attached to a lever and the beats recorded by making a mark on a sheet of paper every time the lever rises.

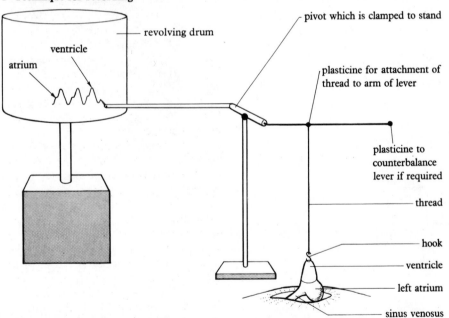

A Exposure of heart

cut skin away along dotted line

xiphisternum: remove to expose heart

B Pushing hook through tip of ventricle

C Technique for recording

revolving drum

ventricle

atrium

pivot which is clamped to stand

plasticine for attachment of thread to arm of lever

plasticine to counterbalance lever if required

thread

hook

ventricle

left atrium

sinus venosus

Fig. 11.8   Recording the heart beat of the frog. In B and C the heart is not to scale.

Requirements
Dissecting instruments
Kymograph set at approximately
  5 mm/s
Time marker (if available)
Heart lever, thread and hook
Recording pen
Plasticine
Blunt rod

Frog Ringer's solution (*see* p. 437)
  at room temperature, ice-cold,
  30°C and 40°C.
Acetyl choline, 1 in 5,000 and 1 in
  1,000
Adrenaline, 1 in 10,000 and 1 in
  1,000

Pithed frog

For Consideration
(1) Did you encounter any particular difficulties, or obtain any unexpected results, in the course of this work? How did you, or would you, cope with such situations?
(2) What light is thrown by the drug experiment on the normal working of the heart? Can you think of a better way of investigating this aspect of the physiology of the heart?
(3) Summarize the results of the experiment in which you investigated the effect of temperature on the cardiac frequency. Relevance to real life?

Questions and Problems

1   In an experiment to investigate factors affecting the carriage of oxygen by blood, the relationship was determined between oxygen tension and the percentage saturation of the blood with oxygen at two different concentrations of carbon dioxide. The results are given below:

Oxygen tension (mm Hg)	Percentage saturation of blood with oxygen	
	20 mm CO_2	80 mm CO_2
0	0	0
10	10	5
20	22	12
30	60	24
40	84	42
60	94	78
80	97	90
100	98	95

Plot the results on graph paper, putting oxygen tension on the horizontal axis and percentage saturation on the vertical axis. State exactly what the graphs show and give an explanation of the results as far as you can.

2   Fig. 11.9 shows the results of an experiment in which the effect of temperature on the oxygen dissociation curve of human blood was investigated. How would you explain these results?

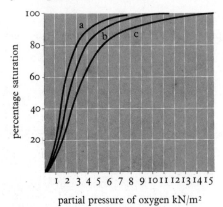

Fig. 11.9 Oxygen dissociation curves for haemoglobin at three different temperatures: **a**, 23°C; **b**, 30°C; **c**, 37°C.

3   Comment on the following:
(a) red blood cells are shaped like biconcave discs and lack a nucleus;
(b) the haemoglobin of the lugworm *Arenicola* has an oxygen dissociation curve well to the left of man's;
(c) the oxygen dissociation curve of the pigeon's haemoglobin is situated well to the right of man's;
(d) breathing coal gas for more than a few seconds is generally fatal;
(e) approximately 95 per cent of the CO_2 released from the tissues in a mammal is returned to the lungs in combination with the red blood cells.

4   The oxygen capacity of an animal's blood is the amount of oxygen carried in the blood when it is saturated. The following table gives the oxygen capacity of the blood of various vertebrates. In all cases the blood contains haemoglobin.

Comment on the figures.

Animal	Oxygen capacity (cm^3 O$_2$/100 cm^3 of blood)
man	20.0
seal	29.3
llama	23.4
crocodile	8.0
frog (*Rana esculenta*)	9.8
carp	12.5
mackerel	15.7
electric eel	19.75
toadfish	6.2

release O$_2$ at higher O$_2$ tension

at high CO$_2$ tension Hb less efficient at taking up O$_2$ — more efficient at releasing

Most invertebrates have blood with low oxygen capacities (generally ranging from 0.1 to 2.5 cm^3 O$_2$/100 cm^3) but there are two notable exceptions: the lugworm *Arenicola* has a capacity of 8.0 and the cuttlefish *Sepia* has a capacity of 7.0 cm^3 O$_2$/100 cm^3. Comment.

CO$_2$ amino groups carbamino compou..

bicarbonate & tissue & oxyhaemoglobin alternate routes

5 Give an illustrated account of the structure of the mammalian heart. Consider what would happen if (a) the right side of the heart were to beat more powerfully than the left side; (b) a blood clot develops inside one of the coronary vessels; (c) the heart is totally denervated.

6 Examine Fig. 11.10 which shows the changes that occur in the aortic, atrial, and ventricular pressures during the human cardiac cycle. What phases in the cycle are represented by A, B and C? Explain what is happening at each of the points D to L in the graphs.

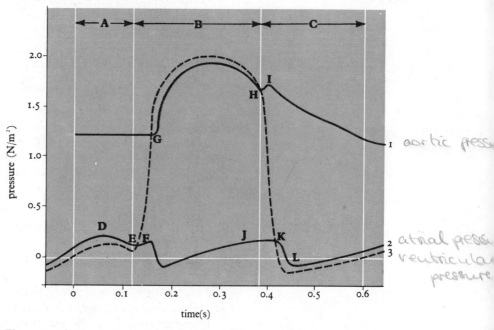

Fig. 11.10 Changes in the aortic, atrial, and ventricular pressures in the course of the human cardiac cycle. Curve 1, aortic pressure; 2, atrial pressure; 3, ventricular pressure.

7 Suggest explanations for each of the following, all of which relate to the human heart:

(a) the beginning of ventricular systole is accompanied by first a slight increase in atrial pressure and then a marked decrease.

(b) the onset of ventricular diastole is accompanied by a sudden slight increase in aortic pressure.

(c) at the commencement of ventricular systole and diastole thud-like sounds can be heard through a stethoscope applied to the chest.

(d) blood leaves the heart at high pressure and with uneven flow but by the time it reaches the capillaries it is at low pressure and flows evenly.

(e) in certain types of disease the ventricles beat at a slower rate than the atria.

8 In the mammalian foetus a hole (the foramen ovale) connects the right and left atria, and an open vessel (the ductus arteriosus) connects the pulmonary and aortic arches. Soon after birth the foramen ovale closes and the ductus arteriosus becomes constricted so that it no longer carries blood.

What is the function of the foramen ovale and ductus arteriosus in the foetus, why are they necessary, and what would be the consequences of either or both of them failing to close at birth?

9 One of the earliest experiments on the initiation of the heart beat was carried out by Stannius on the frog's heart. This is basically similar to the mammalian heart except that there is only one ventricle and the great veins, instead of opening direct into the right atrium, open into a vestibule-like sinus venosus (Fig. 11.11). The heartbeat starts at the sinus venosus and then spreads via the atria to the ventricle. Stannius found that if he tied a ligature round the junction between the sinus venosus and right atrium, the atria and ventricle stopped beating, but the sinus continued to beat at its normal rate. However, after about half-an-hour the atria and ventricle started beating again, but at a slower rate than the sinus. Stannius next tied a second ligature, this time between the atria and ventricle. The result was that the ventricle stopped beating, but the sinus venosus and atria continued to beat at the same rate as before. However, after about an hour the ventricle started beating again extremely slowly, much more slowly than the atria.

Explain Stannius's observations in terms of modern knowledge of how the heart works. How does the cardiac mechanism of the frog compare with that of a mammal?

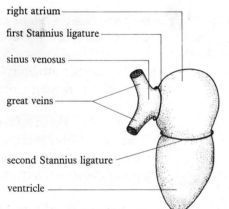

right atrium

first Stannius ligature

sinus venosus

great veins

second Stannius ligature

ventricle

Fig. 11.11 Side view of frog's heart showing the position of the first and second Stannius ligatures.

10 From your knowledge of the structure and functioning of the human heart, lungs and circulatory system, put forward a design for a heart–lung machine that would enable blood to by-pass a person's own heart and lungs, thereby allowing surgical operations to be carried out on these organs. How, precisely, should the machine be connected with the patient?

11 Compare the mechanisms by which blood is returned to the heart from the tissues in a mammal, a fish, and an insect. How does the composition of an insect's blood differ from that of the mammal?

12 Uptake and Transport in Plants

Background Summary

1. Plants require carbon dioxide, water, mineral salts and oxygen. In primitive plants these are transported by diffusion; in higher plants specialized **vascular tissues** are required.

2. In land plants carbon dioxide is taken up through **stomata** which are located mainly on the underside of the **leaves**. The degree of openness or closure of the stomata can be estimated by means of a **porometer** which measures the resistance to airflow of a leaf.

3. Bordered by **guard cells**, the stomata are controlled by an osmotic mechanism which ensures that generally they open by day and close by night.

4. The stomata also permit the escape of water vapour from the plant (**transpiration**). Transpiration is an integral part of the mechanism by which water is taken up into the plant from the soil.

5. In the leaves water evaporates from the surfaces of the spongy mesophyll cells into sub-stomatal air chambers, whence it diffuses through the stomata to the outside.

6. The rate of transpiration can be estimated by measuring the rate of water uptake by a cut leafy shoot, using a **potometer**. The rate of transpiration depends on temperature, relative humidity, air movements, atmospheric pressure, light, and water supply. Plants are adapted in various ways to prevent excessive transpiration.

7. To replace the water transpired from the above-ground parts of a plant water, taken up by the roots, flows through the plant in the **transpiration stream**.

8. The **roots** are adapted for the uptake of water. Movement of water from soil to vascular tissues occurs at least in part by osmosis. The existence of **root pressure** suggests that osmosis may be aided by active transport, possibly in the **endodermis**.

9. The **stem** is well adapted for carrying the transpiration stream from roots to leaves. Water (and mineral salts) are transported in lignified **xylem elements** (**vessels** and **tracheids**) within **vascular bundles** continuous with those in the roots and leaves. Lateral movement is permitted by **bordered pits** which in certain plants contain a **torus**.

10. How water is drawn through a tall stem, and the water columns prevented from breaking or falling back, may be tentatively explained by **adhesion** and **cohesion** forces developed in the anatomical elements involved.

11. Mineral salts are taken up and transported through the same pathway as water. Experimental evidence indicates that ion uptake involves **active transport**.

12. Some plants have special methods of obtaining essential elements, particularly nitrogen—e.g. they may harbour symbiotic **mycorrhiza** or **nitrogen-fixing bacteria** in their roots, or they may adopt the **carnivorous habit**.

13 Organic compounds manufactured in leaves are **translocated** to the rest of the plant in **sieve tubes** in the **phloem** within the vascular bundles. The structure of the sieve elements shows adaptations for this function.

14 The mechanism of translocation is not fully understood but it may take place by means of **mass flow** and/or **active transport**.

Investigation 12.1

Investigation into factors affecting the rate of transpiration

An important factor determining the rate at which water enters a plant is the rate at which it evaporates from the leaves, i.e. is transpired into the surrounding atmosphere. Any external condition which affects the rate of transpiration will be expected to have a corresponding effect on the rate of water uptake.

In this exercise we will use a **potometer** to investigate the effect of various external conditions on the rate at which water enters a leafy shoot. The potometer consists of a length of graduated capillary tubing which can be attached to the cut end of the stalk. The rate at which water traverses the capillary tube is then measured.

Procedure

Two types of potometer are illustrated in Fig. 12.1. In both cases water uptake is measured by timing how long it takes for an air bubble to pass along the graduated capillary tube.

SETTING UP THE POTOMETER
Immerse the potometer in water and make sure it is completely filled. Now put the cut stalk (but *not* the leaves) of your leafy shoot (sycamore is recommended) into the water and cut off the last centimetre of the stalk *under water*. Pause for a moment and then attach the stalk to the potometer as shown in Fig. 12.1. The object is to ensure that the water in the xylem elements of the plant is continuous with the water in the potometer: there must be no air bubble in the system.

Now remove the plant and potometer from the water and mount them in a fixed position. The end of the capillary tube should rest in a beaker of water and any air bubbles in the capillary tube should be expelled by letting in water from the reservoir (type 1) or squeezing the rubber tubing (type 2).

If necessary, smear vaseline on the joints between the stalk and the potometer so as to prevent leakage. If you are using type 2 it may help to wind some copper wire round the rubber tubing in which the stalk is inserted.

USING THE POTOMETER
Perform a trial run with the plant in normal room conditions. Remove the capillary tube from the beaker for a few seconds to allow a bubble of air to enter it. Time how long it takes for the bubble to move along a convenient fixed length of the capillary tube. Then return the bubble to the beginning and repeat the procedure.

In the case of type 1 the air bubble can be returned by carefully letting in water from the reservoir: the same air bubble can then be used again. In type 2 it is necessary to expel the air bubble into the beaker by pressing the rubber tubing: a new one may then be introduced.

Continue making trial runs until your plant settles down to a steady rate of transpiration.

Experiments

Investigate the effect on the rate of transpiration of some or all of the following:
(1) placing the plant in a current of air, created by, e.g. an electric fan;
(2) putting the plant in a humid environment by, e.g. covering the leaves with a polythene bag;
(3) raising the temperature by putting the plant close to a heat source such as a radiator;
(4) blocking the stomata on the lower and/or upper surfaces of the leaves by smearing the leaves with vaseline or silicone grease;
(5) removing some or all of the leaves from the plant.

EXPRESSION OF RESULTS
Express the rate of water uptake in distance moved by the air bubble per unit time (e.g. cm/min).

Type 1 (with reservoir)

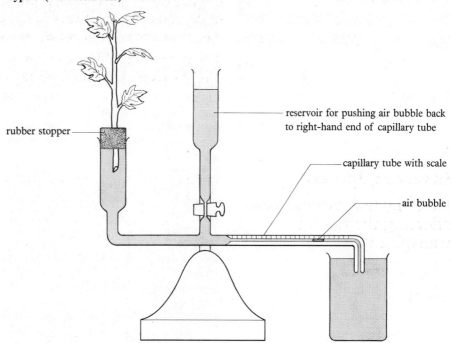

rubber stopper

reservoir for pushing air bubble back to right-hand end of capillary tube

capillary tube with scale

air bubble

Type 2 (without reservoir)

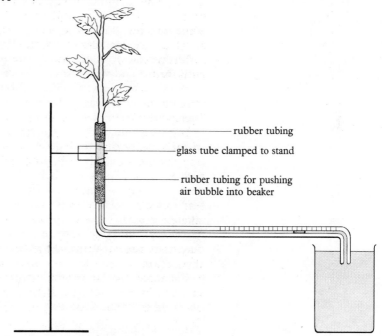

rubber tubing

glass tube clamped to stand

rubber tubing for pushing air bubble into beaker

Fig. 12.1 Two types of potometer.

Requirements
Potometer (*see* Fig. 12.1)
Cutters
Electric fan
Polythene bag
Heat source
Squared paper

Vaseline or silicone grease

Leafy shoot of, e.g. sycamore

If you also know the volume of water that each division of the scale corresponds to, convert your results to volume of water taken up per unit time (e.g. cm³/h).

You can make an estimate of the total leaf area of your plant by removing the leaves and laying them on squared paper. This will enable you to express the water uptake in volume per unit time per leaf area (e.g. $cm^3/h/m^2$).

For Consideration
(1) What conclusions can you draw from your results?
(2) What possible sources of error or ambiguity are there in these experiments and what could you do to overcome them?
(3) Will evaporation alone account for the movement of water through a plant? What other forces might be involved?

Investigation 12.2
Investigation of water loss from the leaves of a plant

The evaporation of water from the leaves of a plant (transpiration) depends on numerous internal and external factors. One of the most important internal factors is the number and distribution of the stomata. In this exercise it will be seen whether there might be a correlation between the rate of transpiration from the leaves of a plant, as determined by weight loss, and the stomatal frequency and distribution.

Cherry laurel is recommended for this experiment. However, the investigation is more meaningful if several different plants are investigated, and the results compared and related to their different ecological situations. For example the situation in a typical **mesophyte** can be compared with a **xerophyte**. A mesophyte is a plant growing in conditions where there is a reasonably plentiful water supply; a xerophyte is a plant growing in conditions where there is a low water supply.

Procedure

(1) Take several leaves with their stalks (petioles), string them together with thread and weigh them[1]. Aim to have a bunch of leaves weighing about 10 g (about four leaves in the case of cherry laurel).

Note the time, then hang your leaves in such a way that they do not touch one another: (a) in still air, (b) in turbulent air from a fan, hair drier or convector heater, and (c) with both leaf surfaces smeared with vaseline or silicone grease.

Reweigh at 15-minute intervals at least four times during the course of the session.

Calculate the percentage decrease in weight at each interval. This can be obtained from the formula:

percentage decrease in weight =

$$\frac{a - b}{a} \times 100$$

where a = initial weight at beginning of experiment
b = new weight

(2) Apply a strip of anhydrous **cobalt chloride** or **cobalt thiocyanate paper** to the upper and lower surface of a leaf on an intact leafy shoot. Hold the paper in place by means of two glass slides held firmly together by elastic bands or paper clips (Fig. 12.2). Anhydrous cobalt paper is blue, but in

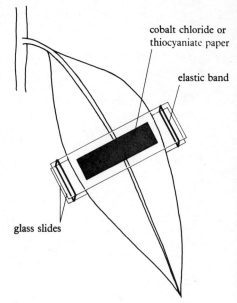

Fig. 12.2 Cobalt chloride or thiocyanate 'sandwich' experiment.

the hydrated state it is pink. Emission of water by the leaf will be shown by the paper turning blue to pink.

Compare the time taken for the colour change to occur in the paper on the upper and lower sides of the leaf. Conclusions?

(3) Estimate the number of stomata in a given area of a leaf by making a polystyrene replica of the leaf surface. Place a small drop of polystyrene cement on the lower surface of the leaf. Spread it out into a thin film with a pin. When dry, peel off the polystyrene film with forceps and place it on a slide, putting on a coverslip to keep it flat. Repeat for upper surface of leaf.

Examine your polystyrene films under low or medium power. Estimate the frequency of stomata per field of view (average of several different areas) for both upper and lower surfaces.

By means of a transparent ruler or calibrated slide estimate the diameter of your field of view and from this the area of the field. (Area of a circle = πr^2, where $\pi = 3.142$ and r is the radius). From this calculate the number of stomata per unit area (e.g. mm^2) on the two sides of the leaf. Do your answers bear any relation to the results of (2) above?

---

[1] If a suitable balance is unavailable, a sensitive microbalance, suitable for weighing single leaves, can be made from a knitting needle, cork and pins. See Nuffield Advanced Biological Science, *Control and Coordination in Organisms*, Penguin 1970.

Requirements

Balance
Electric fan, hair drier, or
  convector heater
Thread
Anhydrous cobalt chloride or
  thiocyanate paper
Glass slides (× 2)
Elastic bands or paper clips
Polystyrene cement
Pin
Forceps
Slide and coverslip

Leafy shoot of cherry laurel, etc.

For Consideration

(1) If you have been able to carry out this investigation on several contrasting plants, do you find a correlation between the number of stomata and rate of transpiration?

(2) What other structural features of leaves besides the number of stomata may influence the rate of transpiration?

(3) In what kind of environments would you expect to find (a) plants with a large number of stomata, and (b) plants with relatively few stomata?

Investigation 12.3
Structure of roots

In addition to providing a firm anchorage, roots absorb water and mineral salts from the soil, and transport these to the stem. The root system of a plant must, therefore, provide an adequate surface area for absorption, and it will also be expected to contain conducting (vascular) tissues. In addition food reserves may be stored in the roots.

In this investigation we shall be concerned only with the primary tissues, namely those that are derived from the dividing cells at the apex of the growing root.

Procedure

EXTERNAL FEATURES
Examine the radicle of a seedling of mustard, cress or pea. The apex of the radicle is smooth, but further back it is covered with **root hairs** (Function?). Further back still it is smooth again. The apex is the yougest part of the root. How do you account for the fact that root hairs are confined to a zone just behind the apex? The radicle develops into the **taproot**, or main root, from which **lateral roots** sprout. Are any lateral roots visible in your seedling?

Mount a radicle in iodine without crushing it and note that each root hair is a single cell.

Examine a maize or wheat seedling. In this case there is no radicle as such, the taproot being replaced by a bunch of **fibrous roots** characteristic of grasses.

INTERNAL STRUCTURE
Cut transverse sections of a young primary root, i.e. one that has not laid down secondary tissues. Buttercup or broad bean are recommended. As the root is flexible it must be held in a firm position while the sections are cut: do this by inserting a short length of the root into a vertical slit made down the centre of a piece of moistened elder pith or carrot tuber (see Fig. 10.4, p. 96). Hold the pith or carrot in one hand and cut your sections rapidly and smoothly with a sharp razor blade or 'cut-throat' razor held in the other hand. Place the sections in a dish of water.

Select one or two thin sections and, transferring them with a fine brush, mount them in a drop of iodine on a slide.

E

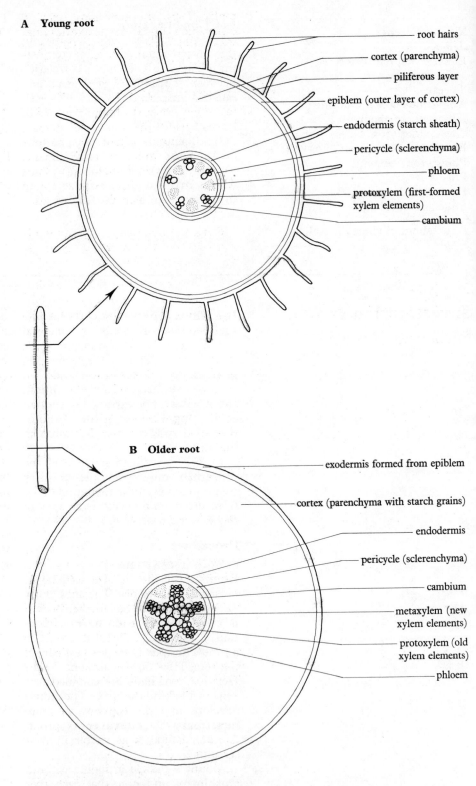

A Young root

root hairs
cortex (parenchyma)
piliferous layer
epiblem (outer layer of cortex)
endodermis (starch sheath)
pericycle (sclerenchyma)
phloem
protoxylem (first-formed xylem elements)
cambium

B Older root

exodermis formed from epiblem
cortex (parenchyma with starch grains)
endodermis
pericycle (sclerenchyma)
cambium
metaxylem (new xylem elements)
protoxylem (old xylem elements)
phloem

C Endodermis cells showing Casparian strip

Casparian strip on radial wall

Fig. 12.3 Transverse sections of dicotyledonous root, based on buttercup *Ranunculus*, **A**, young part of root in the root hair zone just behind the tip; **B**, older part of root further back.

Examine your sections and, using Fig. 12.3 as a guide, identify the following tissues; starting at the outside and working inwards:

Outer cell layer: in the root-hair zone this is the **piliferous layer**; further back in older parts of the root the piliferous layer withers and is replaced by the outermost cortical cells which constitute the **exodermis**. The latter may become corky due to deposition of suberin in the cell walls.

Cortex: extensive area of parenchyma (packing) cells which may contain stored starch grains (stain blue-black with iodine).

Endodermis: single layer of cells marking the inner boundary of the cortex. Note thickened radial walls (Casparian strip) and stored starch grains (the endodermis is also known as the 'starch sheath').

Pericycle: indistinct layer of lignified sclerenchyma immediately inside the endodermis. Function?

Vascular tissues: enclosed within the endodermis and pericycle, water-conducting **xylem tissue** and food-conducting **phloem tissue**. Xylem generally star-shaped, with phloem lying between the 'spokes' of the star. The first elements to form are located in the 'spokes' of the star (protoxylem), xylem elements in the centre of the star develop later (metaxylem). Wedged between protoxylem and phloem are small groups of **cambium cells**. These give rise to secondary vascular tissue in older roots (*see* p. 300).

Examination of the vascular tissues is aided by staining sections in acidified phloroglucinol (eight drops of phloroglucinol plus three drops of concentrated hydrochloric acid) which stains lignin red. Alternatively FABIL may be used (*see* p. 437).

For Consideration
In what ways is the structure of the root adapted to perform its functions?

Requirements
Slides and coverslips
Elder pith and/or carrot tuber
Dish for sections
Fine brush
Section lifter
Safety razor blade or 'cut-throat' razor
Iodine solution
Phloroglucinol
Concentrated hydrochloric acid
FABIL
Seedling of mustard, cress or pea
Seedling of maize or wheat
Fresh roots of, e.g. broad bean or buttercup

Investigation 12.4
Structure of stems

Stems have three main functions: (1) to lift the leaves and flowers into an elevated position, (2) to convey water and mineral salts from the roots to the leaves, and (3) to convey synthesized food materials from the leaves to other parts of the plant. The first function is achieved by various types of strengthening tissue; the second and third functions by conducting (vascular) tissues. In this practical we will first investigate the general organization of a typical dicotyledonous stem and then look at the structure of the vascular tissues in detail.

As with the root we shall be here concerned only with the primary tissues of the stem, namely those formed from the dividing cells at the apex of the growing shoot.

Procedure

EXTERNAL FEATURES
Examine an entire plant so as to appreciate the relationship between the main stem and other structures (Fig. 5.12A). In particular notice that leaves and axillary buds arise at intervals (**nodes**). The region of the stem between two successive nodes is called an **internode**. The stem may be hairy (function?) and/or pigmented. Strip off a piece of the epidermis and examine under the microscope. Anything interesting?

GENERAL INTERNAL ORGANIZATION OF A DICOTYLEDONOUS STEM
For internal structure of the stem, the sunflower, *Helianthus*, is a good plant with which to start. With a sharp scalpel square off the ends of a short length of an internode. The stem and sections must be kept moist at all times. Cut thin transverse sections with a cut-throat razor or *stiff* razor blade (Fig. 12.4A). Hold the piece of stem in one hand and the blade in the other: cut smoothly and rapidly, constantly wetting the blade and surface of the stem with water. As the sections accumulate on the surface of the blade, transfer them to a dish of water with a paintbrush. Select your thinnest sections. Stain for 5–10 minutes in FABIL, then mount in dilute glycerine. Alternatively the sections can be mounted in the stain.

To build up a complete picture of the three-dimensional structure of the stem, and its constituent cells, it is essential to examine longitudinal as well as transverse sections. Longitudinal sections can be cut, stained and mounted in the same way as transverse sections (Fig. 12.4B).

Your investigation of stem structure, by cutting and staining your own sections, can be supplemented by examining prepared slides.

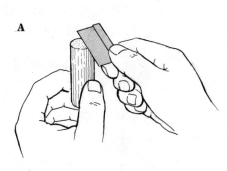

A

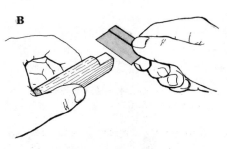

B

Fig. 12.4 Cutting **A** transverse, and **B** longitudinal sections of a plant stem.

A   Stem and associated structures of a generalized plant

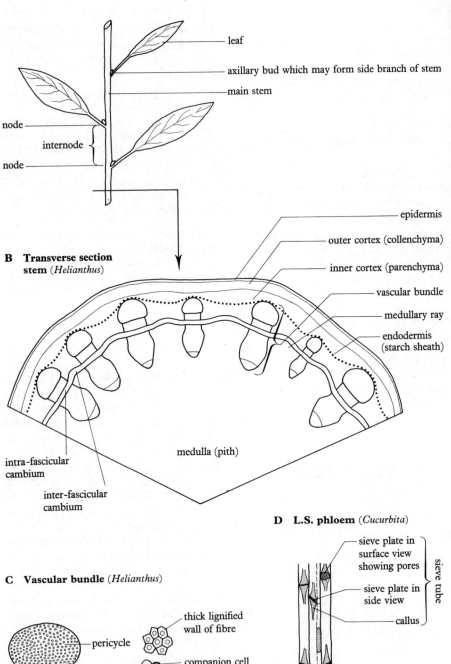

**B   Transverse section
stem** (*Helianthus*)

- leaf
- axillary bud which may form side branch of stem
- main stem
- node
- internode
- node

- epidermis
- outer cortex (collenchyma)
- inner cortex (parenchyma)
- vascular bundle
- medullary ray
- endodermis (starch sheath)

medulla (pith)

intra-fascicular cambium

inter-fascicular cambium

C   Vascular bundle (*Helianthus*)

- pericycle
- phloem
- cambium
- metaxylem (newer xylem elements)
- protoxylem (older xylem elements)

- thick lignified wall of fibre
- companion cell
- sieve tube
- lignified wall of vessel
- thin lignified wall of xylem element

D   L.S. phloem (*Cucurbita*)

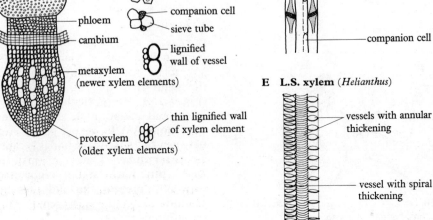

- sieve plate in surface view showing pores
- sieve plate in side view
- callus
- sieve tube
- companion cell

E   L.S. xylem (*Helianthus*)

- vessels with annular thickening
- vessel with spiral thickening

Fig. 12.5   Structure of the stem of a dicotyledon, based mainly on the sunflower *Helianthus*. This particular plant is unusual in having an endo-dermis (starch sheath) in its stem. There is much variation in the detailed arrangement of the peri-cycle fibres and vascular tissues in other dicotyle-dons. Thus the pericycle may be separated from the phloem, it may form a continuous cylinder round the stem, or it may be absent altogether. The vascular tissues, too, may form a complete cylinder instead of dicrete vascular bundles.

Requirements
Slides and coverslips
Dish for sections
Fine brush
Stiff razor blade or cut-throat
  razor

Iodine solution
FABIL
Dilute glycerine (25 per cent,
  aqueous)

Entire plant with leaves and
  axillary buds
Stem of sunflower (*Helianthus*) and
  marrow (*Cucurbita*), fresh
  or in 70 per cent ethanol
Stem of monocotyledon, e.g.
  maize, iris, lily
Macerated stem tissue
TS and LS *Helianthus* and
  Cucurbita
TS monocotyledon

Stem tissue macerated as follows
(Franklin's method):
Cut stem into small pieces about the
size of half a matchstick. Immerse in a
mixture of glacial acetic acid and 20 Vol.
hydrogen peroxide in equal parts.
Maintain at 60°C for 24 h or boil
under reflux for 1 h. Tissues should
then disintegrate with vigorous shaking
or stirring. Wash by decantation,
neutralize with a little ammonium
hydroxide, and store in 70 per cent
alcohol.

Using Fig. 12.5 to help you, identify the following tissues, first in transverse and then in longitudinal section.

Protective and strengthening tissues:
Epidermis: single layer of rectangular cells covering surface of stem.

Collenchyma: about four layers of thick-walled cells immediately beneath epidermis; cells vertically elongated and cellulose walls thickened at corners; constitute outer part of cortex.

Parenchyma: layer of thin-walled packing cells making up the bulk of the cortex, medulla (pith) and medullary rays.

Sclerenchyma: large group of vertically elongated lignified fibres on the immediate outside of each vascular bundle; constitute the pericycle, part of the cortex (N.B. in some stems the pericycle fibres form a continuous ring).

Conducting (transport) tissues— Vascular bundles:
Phloem: vertically elongated sieve tubes, companion cells, and small-celled parenchyma on the immediate inside of the pericycle fibres.

Xylem: mainly vertically elongated xylem elements (vessels) with thick lignified walls: first-formed xylem elements small in cross-section and located in innermost region of xylem (protoxylem); more recently formed elements larger and located further out (metaxylem).

Cambium: several layers of small rectangular cells wedged between xylem and phloem in each vascular bundle; cambium tissue within each vascular bundle (intrafascicular cambium) joined by cambium cells which traverse each medullary ray (interfascicular cambium). Cambium cells divide tangentially to form secondary vascular tissues within the vascular bundles and secondary parenchyma in the medullary rays—*see* p. 300).
N.B. In some stems the xylem and phloem form a continuous ring of vascular tissue.

CONDUCTING CELLS IN DETAIL
Examine xylem vessels of *Helianthus* in longitudinal section under high power. Explore their structure in as much detail as you can. Note: **lignified walls**, **open ends**, **pits**, various forms of **lignified thickening** particularly annular and spiral. (Fig. 12.5E).

Examine sieve tubes in longitudinal sections of either *Helianthus* or, better, *Cucurbita* (marrow). Note: **cellulose walls**, **sieve plate** with **pores**, cytoplasmic slime (**callose**) in the immediate vicinity of sieve plate; **companion cell(s)** with cytoplasm and nucleus (Fig. 12.5D).

In addition to studying cell detail in longitudinal sections, macerated tissue is recommended. Pieces of stem are treated with a chemical agent which causes the cells to separate from one another.

Take a very small quantity of macerated stem tissue and mount in FABIL.

COMPARISON WITH MONOCOTYLEDONOUS STEM
Cut, stain and mount (or examine prepared) transverse sections of the stem of a typical monocotyledon such as maize, iris or lily. The structure of each vascular bundle is basically similar to that of the dicotyledon: **phloem** towards the outside, **xylem** towards the inside. However, there is no cambium. (Consequences?). Notice also that the arrangement of the vascular bundles is different. How would you describe the difference?

For consideration
In what ways is the structure of the stem adapted to perform its function?

Reference
Noel, A. R. A., *Improvements in Botanical Microtechnique. School Science Review*, 1966, **48**, 164, pp. 155–9.

Questions and Problems

1 How would you proceed to measure the transpiration and water absorption of a sunflower plant? These amounts were determined over two-hour periods for plants subject to natural daily variation in the environment and given an adequate supply of water. Comment on the results which are given below and suggest possible explanations.

Process	Grammes per plant and time of day							
	1100—1300—1500—1700—1900—2100—2300—0100—0300							
Transpiration	33	44	52	46	27	16	10	4
Absorption	20	30	41	46	32	22	15	12

2 A potted hydrangea plant was put out for a period of one hour in each of the following conditions in the order given, and the transpiration in each hour measured. The air temperature was 18°C throughout the experiment:

Conditions	Relative humidity (per cent)	Transpiration (g)
(a) Still air, in light shade	70	1.2
(b) Moving air (fan) in light shade	70	1.6
(c) Still air, in bright sunlight	70	3.75
(d) Still air, dark (moist chamber)	100	−0.20

Experimental error: ±0.05 g
Suggest explanations of these results.

(O and C)

3 The rate of transpiration of maize plants was compared with the rate of evaporation of water from a porous-pot atmometer over a 24-hour period with the following results:

Period (h)	Water lost per hour (cm^3)	
	Porous pot	Maize leaves (per m^2)
7–9	3.8	91
9–11	6.6	160
11–13	8.1	218
13–15	9.5	248
15–17	9.4	195
17–19	9.0	179
19–21	6.6	124
21–23	3.8	8
23–1	3.4	18
1–3	1.5	18
3–5	0.7	13
5–7	0.9	23

Explain these results as fully as you can.

(O and C)

4 (a) Explain fully how the properties of water molecules help to explain the passage of water up a tree.

(b) The linear velocity of flow of sap through the xylem of a tree was measured in m/h in the trunk and in one of the small branches at the top of the tree. Measurements were taken at two-hourly intervals during a summer day. The results are shown in Fig. 12.6.

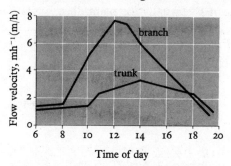

Fig. 12.6 Changes in the linear velocity of flow of sap through the xylem in the trunk and in one of the small branches at the top of a tree.

(i) What difference would you expect in the circumference of the trunk measured at 1400 hours when compared with that measured at 1800 hours? Briefly explain your answer.

(ii) What light do these results shed on the mechanism by which water passes up the tree? (*JMB modified*)

5 Using very sensitive recording equipment, a scientist observed that the diameter of certain very large tree trunks changed from day to day. The graph in Fig. 12.7 shows the results he obtained when he recorded the changes in one of these trunks over a period of about four days. What conclusions would you draw from his results? (*VU modified*)

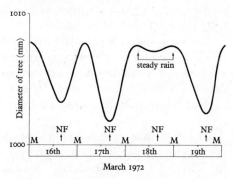

Fig. 12.7 Changes in the diameter of the trunk of a large tree during the course of four days in March 1972. **M**, midnight; **NF**, nightfall.

6 The relationship between oxygen concentration, potassium uptake, and sugar consumption in isolated barley roots was determined. Comment on the results obtained which are given below. (The sugar loss and potassium gain are expressed in arbitrary units.)

	Percentage oxygen in aeration stream					
	0	5	10	15	20	100
Sugar loss	15	20	42	45	45	48
Potassium gain	5	55	70	73	75	70

(*O and C*)

7 A ringing experiment is carried out in which a girdle of bark (corky cells plus phloem tissue) is removed from a woody stem, leaving only the xylem tissues connecting the upper and lower parts of the plant (Fig. 12.8A).
(a) After a time the bark immediately above the ring develops the swollen appearance shown in Fig. 12.8B. Put forward hypotheses to explain this phenomenon. How would you test your hypotheses?
(b) Eventually the plant dies. Explain in detail the reason why death occurs.

A Immediately after cutting the ring

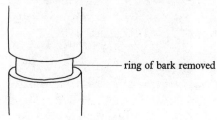

ring of bark removed

B Appearance later

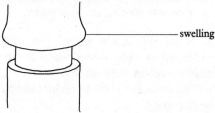

swelling

Fig. 12.8 Ringing experiment on a woody stem.

8 Carbon dioxide whose normal carbon has been replaced by the radioactive isotope ^{14}C is introduced into a leaf half way up a bean plant from which all other leaves have been removed. Estimations of the distribution of the ^{14}C throughout the plant are made at five-minute, two-hour and ten-hour intervals from the start of the experiment.

The experiment is repeated with another similar plant from which a ring of tissue, including the phloem, has been removed just below the leaf. Both plants are exposed to full illumination throughout the experiments. What would you expect the distribution of the ^{14}C to be in each case? Explain your reasoning.

(Use arbitrary units adding up to 10 for the quantities of ^{14}C).

9 Briefly explain each of the following phenomena:
(a) Stomata generally open during the day and close at night.
(b) There are generally fewer stomata on the upper side of a leaf than the lower side.
(c) When transplanting a plant it is advisable to remove some of the leaves.
(d) The stump of a severed tree trunk may exude copious quantities of fluid after cutting.
(e) On warm humid evenings water may be seen to drip from the edges of the leaves of certain plants.

13 The Principles of Homeostasis

Background Summary

1 '*The constancy of the internal environment is the condition for free life.*' This principle was first enunciated by Claude Bernard in 1857. Maintenance of a constant internal environment is **homeostasis**.

2 By 'free life' is meant the ability of an organism to inhabit a wide range of environments.

3 The 'internal environment' is the immediate surroundings of the cells. This consists of **tissue fluid** (also known as **intercellular** or **interstitial fluid**) which bathes the cells.

4 Tissue fluid consists of plasma minus proteins and is formed by ultra-filtration from the capillaries. Excess tissue fluid passes into **lymphatic vessels** where it constitutes **lymph**.

5 The following features of the internal environment must be kept constant:
(a) its chemical constituents, e.g. glucose, ions, etc.
(b) its osmotic pressure
(c) its carbon dioxide content
(d) its temperature.
Nitrogenous waste products and other toxic substances must be eliminated altogether.

6 The principles of homeostasis can be illustrated by the control of blood sugar in the mammal. A rise in the sugar level results in the secretion of **insulin** from the islets of Langerhans in the pancreas and this brings about metabolic disposal of excess blood sugar in the liver. Failure of the pancreas to perform this function results in diabetes melitus.

7 In general homeostatic control processes work as follows: any deviation from the norm (**set point**) sets into motion the appropriate **corrective mechanisms** which restore the norm (**negative feedback**).

8 In abnormal circumstances a deviation from the norm may result in a further deviation (**positive feedback**).

9 One of the most important homeostatic organs in the body is the **liver**. Its homeostatic functions include the regulation of sugar, lipids and amino acids, elimination of haemoglobin from used red blood cells and the production of heat.

10 The structure of the liver, showing an intimate association between the liver cells, blood vessels and bile channels, is admirably adapted to perform its numerous functions.

Investigation 13.1

Homeostasis in action

The efficiency of a homeostatic feedback process depends on the sensitivity of the **detector (receptor)**, the speed with which the **corrective mechanism** is brought into operation, and the time taken for the **effector** to respond to instructions reaching it from the **control centre**. In this practical you will examine some of the properties of homeostatic mechanisms in several quasi-biological situations.

Experiments

THE HUMAN POSITION INDICATOR

Work in pairs. One of you should act as the experimenter, the other as helper. The experimenter makes a chalk mark (which the helper should not be able to see) on the front of a table. The helper should stand on the other side of the table and place his finger on the edge (Fig. 13.1).

The experimenter and helper now form a homeostatic system whose purpose is to keep the helper's finger as close to the chalk mark as possible. The helper should run his finger along the edge of the table to the left or right at constant speed, without stopping it. The experimenter keeps the helper's finger as close to the check mark as possible by saying 'left' or 'right' every time his finger passes the mark. Thus the experimenter serves as the detector and control centre, and the helper as the effector, in a crude homeostatic system.

Investigate the effect on the efficiency of this system of varying two factors:

(a) vary the speed at which the helper moves his finger (i.e. you vary the speed at which the effector fluctuates on either side of the set point).

(b) vary the time lag between seeing the helper's finger pass the chalk mark and saying 'left' or 'right' (i.e. you introduce a variable delay into the control system).

The experimenter should instruct the helper:

(i) as quickly as possible after his finger passes the chalk mark;
(ii) after a delay of 1s;
(iii) after a delay of 2s.

What difference does it make if the experimenter says 'left' or 'right' as soon as the helper's finger passes the mark but the helper introduces a variable delay before responding? What does this tell us about possible sources of delay in a true homeostatic system?

A HUMAN THERMOSTAT

In this experiment, which can be done singly or in pairs, the experimenter himself forms the control mechanism in a homeostatic system whose purpose is to keep the temperature of an aluminium block as constant as possible. The apparatus is shown in Fig. 13.2.

(a) A simple on–off system

Decide on a set point (e.g. 60°C). Switch on the heater until this temperature is reached, then switch it off. When the temperature drops below the set point switch the heater on again, and so on.

Plot the temperature of the block (vertical axis) against time (horizontal axis), at half-minute intervals.

Note the form of your graph. By how much and how quickly did the temperature deviate from the set point? How quickly was it restored? You will probably have found that the temperature reaches the set point quickly and overshoots it, fluctuating widely either side of it. The fluctuations can be dampened by modifying the system in such a way that the power depends on the error.

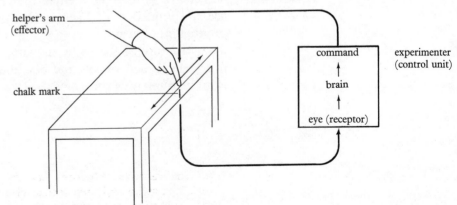

Fig. 13.1 Human position indicator.

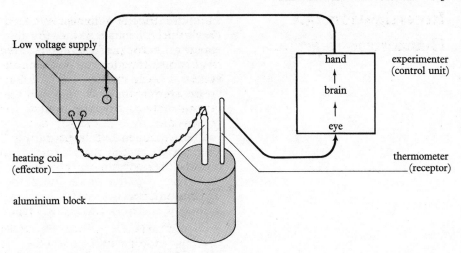

Fig. 13.2 Human thermostat.

(b) An error-dependent system
Starting with the aluminium block cold, over-run the heater slightly by setting the power supply at 16V. When the temperature reaches 58.5°C, reduce the supply to 12V. Then follow the settings shown below:

Temperature of block (°C)	Setting of power supply (V)
Below 58.5	16
at 58.5	12
59	8
59.5	4
60	0

In other words, as the temperature approaches the set point, you reduce the power supply by 8V every time the temperature increases by 1°C. Plot temperature of the block (vertical axis) against time (horizontal axis) at half-minute intervals. Do the temperature fluctuations dampen out now?

Let the block cool to about 50°C, repeat the experiment with a 4V reduction in power per degree rise in temperature as indicated in the settings shown below:

Temperature of block (°C)	Setting of power supply (V)
Below 56.5	16
at 56.5	14
57	12
57.5	10
58	8
58.5	6
59	4
59.5	2
60	0

Plot the results as before. Let the block cool again and then repeat with a 2V reduction in power per degree rise in temperature:

Temperature of block (°C)	Setting of power supply (V)
Below 52.5	16
at 52.5	15
53	14
53.5	13
54, etc.	12

Plot the results. Finally let the block cool to about 40°C and repeat the experiment with a 1V reduction in power per degree:

Temperature of block (°C)	Setting of power supply (V)
Below 45	16
at 45	15
46	14
47	13
48, etc.	12

Plot the results as before. Compare your four graphs, interpret them as fully as you can, and answer these questions:
(1) Why does the temperature tend to overshoot the set point?
(2) How can the overshooting be reduced?
(3) How do the methods used here for reducing fluctuations compare in efficiency in terms of:
(a) their dampening effect on fluctuations, and
(b) the time taken for a deviation from the set point to be corrected?

EFFICIENCY OF THE DETECTOR
The efficiency of a homeostatic mechanism depends in large part on the sensitivity and accuracy of the detector. In experiment (2) the detector consisted of a thermometer in conjunction with the human eye.
Investigate the effect of substituting

the above detector with a less efficient one, namely the human skin. For this experiment a helper is required. The helper forms the detector and control system: he places a finger of one hand on the aluminium block, and uses his other hand to control the output of the power unit.

The experimenter arranges the thermometer so he can see the mercury but the helper cannot. The helper then turns up the power until the temperature reaches 40°C. The experimenter tells him when this temperature is reached. Thereafter, using only his finger as the detector, the helper endeavours to keep the temperature of the block as close to 40°C as possible. The experimenter then records the temperature at half-minute intervals.

Plot temperature against time. How do the fluctuations compare with those obtained in the previous experiments? Conclusions?

LOADED HAND INDICATOR
Suspend a hanger which takes slotted kilogram weights on the end of a cord about 2 m long. Adjust its height above a table or bench so that there is room for you to fit your hand underneath (Fig. 13.3). Tie a loop of string on the hanger so that when your hand is inside the loop and holding a pencil it is attached fairly tightly.

Copy Fig. 13.3 (or some other diagram) onto a piece of paper and fix it to the table immediately beneath the hanger. Put a 1 kg weight on the hanger.

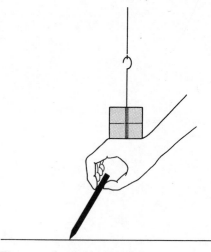

Fig. 13.3 Loaded hand indicator.

With your hand in the loop try with a pencil to follow the line on the paper as quickly and as accurately as possible. Repeat with 2, 3, and 4 kg weights.

Do oscillations occur? What is the effect of the different weights? What is the effect of pressing harder on the paper? Interpret this experiment in terms of homeostasis.

For Consideration

(1) Summarize the potential sources of error and delay in a homeostatic control system.
(2) Think up, design, and carry out other experiments involving the human body which demonstrate the properties and action of a homeostatic control system.

Requirements
Table or island laboratory bench
Chalk
Stop-clock
16V power pack with variable control
Aluminium block with heating coil and thermometer
Slotted weights 1–4 kg
Hanger suspended above table or bench
Sheet of paper

Investigation 13.2

Two organs involved in the regulation of blood sugar: pancreas and liver

Blood sugar is regulated by a homeostatic process involving the pancreas and liver. An increase in blood sugar is detected in the pancreas by the **islets of Langerhans** whose cells secrete the hormone **insulin**. The latter is carried via the bloodstream to the liver whose cells respond by extracting the excess sugar from the bloodstream and breaking it down into carbon dioxide and water (respiration) or building it up into glycogen for storage.

If there is a deficiency of blood sugar insulin ceases to be secreted and the liver cells convert glycogen into glucose and shed it into the bloodstream. The formation of blood sugar from glycogen is claimed to be enhanced by the action of **glucagon**, another hormone secreted by the islets of Langerhans.

The regulation of blood sugar is therefore achieved by a homeostatic mechanism in which the islets of Langerhans serve as the receptor and the liver as the effector. In this control process the effector receives its instructions by hormonal means.

Procedure

MICROSCOPIC STRUCTURE OF
THE PANCREAS
The pancreas is a gland with a dual function:
(1) It secretes digestive enzymes

(which ones?) into the pancreatic duct which conveys them to the duodenum: in this capacity the pancreas is functioning as a ducted **exocrine gland**. The enzymes secreted by the exocrine portion of the gland play no direct part in the homeostatic control of blood sugar.

(2) It secretes the hormones insulin and glucagon into the bloodstream: in this capacity the pancreas is functioning as a ductless **endocrine gland**, or gland of internal secretion.

Knowing these two functions of the pancreas, can you make any predictions about its microscopical structure?

Examine a section of pancreas under low power. The bulk of it is made up of numerous secretory cells, together with blood vessels and branches of the pancreatic duct (Fig. 13.4). If you go over to high power you will see that these secretory cells are arranged in small groups surrounding a narrow duct (Fig. 13.4B). As well as having

prominent nuclei, these cells have a highly granular cytoplasm indicative of their secretory function. They secrete **digestive enzymes** and represent the exocrine portion of the pancreas.

Return to low power and notice that here and there amongst the cells mentioned above are bunches of cells that look different from the rest. These are the **islets of Langerhans**; a typical section might show about six islets. Three features should enable you to pick them out:
(a) They are generally stained differently from the rest of the pancreas; for example in a section stained with haematoxylin and eosin (*see* p. 429) the cytoplasm of the islet cells is a lighter pink than the surrounding enzyme-secreting cells.
(b) The islet cells are arranged differently from the enzyme-secreting cells: in irregular chains rather than small groups (Fig. 13.4C).
(c) The diameter of a typical islet is approximately 200 μm (0.2mm), 6–8 times that of a group of enzyme-secreting cells. The islets of Langerhans secrete the hormones **insulin** and **glucagon** and represent the endocrine portion of the pancreas.

ISLET OF LANGERHANS IN DETAIL
Examine an islet of Langerhans under high power (Fig. 13.4C). Notice irregular chains of **hormone-secreting cells** with lightly-stained cytoplasm and prominent nuclei. Between the chains of secretory cells are **capillaries** lined with pavement endothelium: flattened nuclei of endothelial cells will be seen in places. In suitably stained preparations two different kinds of secretory cell may be detected: β **cells** which secrete insulin and α **cells** which are said to secrete glucagon. The β cells are the more numerous of the two and fill the interior of the islet; the α cells are fewer, larger, more densely stained and occur in the peripheral part of the islet.

In examining an islet of Langerhans notice particularly the intimate association between the hormone-secreting cells and capillaries, a basic requirement of any endocrine organ.

Fig. 13.4 Microscopic structure of the pancreas. The β cells secrete insulin. The α cells secrete glucagon, a hormone whose actions are the opposite of those produced by insulin.

A   General section of pancreas

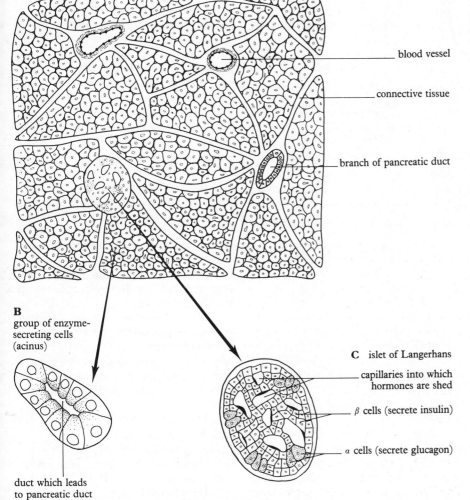

blood vessel

connective tissue

branch of pancreatic duct

B
group of enzyme-secreting cells (acinus)

C   islet of Langerhans

capillaries into which hormones are shed

β cells (secrete insulin)

α cells (secrete glucagon)

duct which leads to pancreatic duct

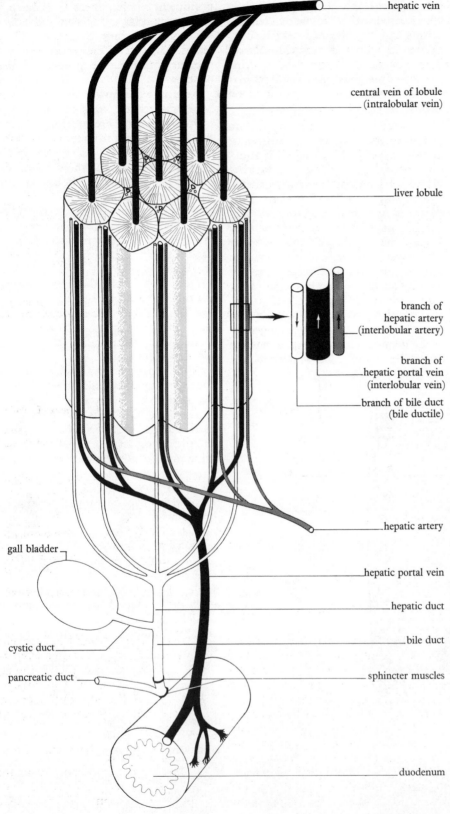

Labels in figure:
- hepatic vein
- central vein of lobule (intralobular vein)
- liver lobule
- branch of hepatic artery (interlobular artery)
- branch of hepatic portal vein (interlobular vein)
- branch of bile duct (bile ductile)
- hepatic artery
- gall bladder
- hepatic portal vein
- hepatic duct
- cystic duct
- bile duct
- pancreatic duct
- sphincter muscles
- duodenum

Fig. 13.5 Diagram to show the relationship between the liver lobules and other structures. The arrows in the inset indicate the direction of flow of materials.

MICROSCOPIC STRUCTURE OF THE LIVER

The liver has numerous functions of which we can single out two principal ones:

(1) It secretes bile into the bile duct which carries it via the gall bladder to the duodenum. Bile plays no direct part in the control of blood sugar, but it is associated with certain other homeostatic functions (which?).

(2) It regulates the amounts of blood sugar, lipids and amino acids by removing them from the bloodstream or adding them to it, according to circumstances.

Knowing these two general functions of the liver, can you make any predictions about its microscopical structure? Before examining it, recall that the liver receives a dual blood supply: oxygenated blood is taken to it via the hepatic artery, blood rich in food substances via the hepatic portal vein from the gut (see p. 76). Blood is drained from the liver via the hepatic vein (Fig. 13.5).

Examine a transverse section of liver under low power (Fig. 13.6). You will see it to consist of numerous closely packed lobules, each roughly 1.0 mm in diameter. A lobule is composed of radiating rows of cells arranged round a central (intralobular) vein, a tributary of the hepatic vein. In pig's liver the lobules are completely encased in connective tissue and can therefore be easily distinguished from each other. In the livers of most other mammals connective tissue is confined to the corners of the lobules.

Now look at the edges of the lobules: here, wedged between one lobule and the next, are various tubes cut in cross-section. In good sections they can be identified by their size and walls when viewed under high power (Fig. 13.6B):
(a) Large lumen, thin endothelial lining: **interlobular vein** (branch of hepatic portal vein).
(b) Much smaller lumen, thicker walls: **interlobular artery** (branch of hepatic artery).
(c) Intermediate in size between (a) and (b), lining of cuboidal or columnar epithelium: **bile ductile** (branch of bile duct).
Blood corpuscles may be seen in (a) and (b).

Now examine the radiating rows of cells inside one of the lobules under high power (Fig. 13.6C). These are the **liver cells**, each row comprising a

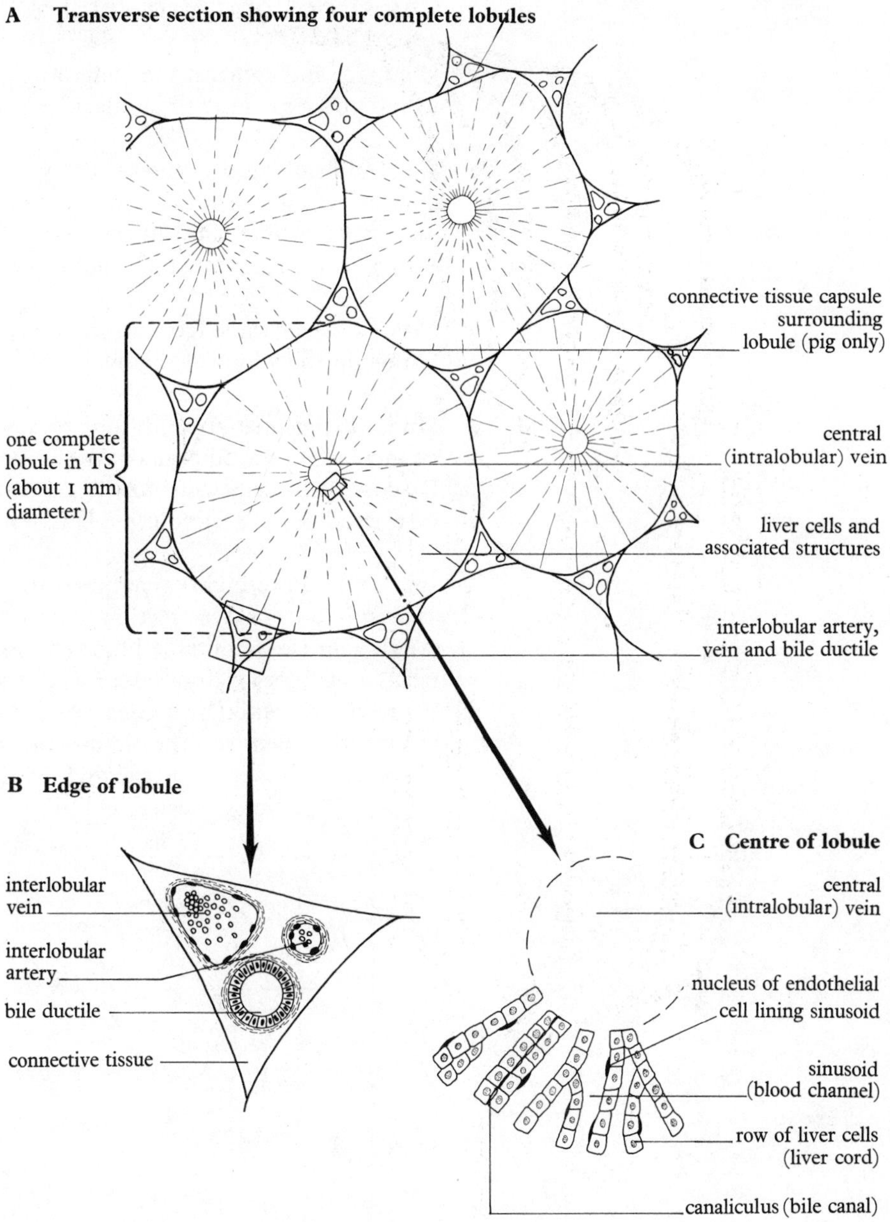

A Transverse section showing four complete lobules

connective tissue capsule
surrounding
lobule (pig only)

one complete
lobule in TS
(about 1 mm
diameter)

central
(intralobular) vein

liver cells and
associated structures

interlobular artery,
vein and bile ductile

B Edge of lobule

interlobular
vein

interlobular
artery

bile ductile

connective tissue

C Centre of lobule

central
(intralobular) vein

nucleus of endothelial
cell lining sinusoid

sinusoid
(blood channel)

row of liver cells
(liver cord)

canaliculus (bile canal)

Fig. 13.6 Microscopic structure of the liver.

liver cord. Notice that the liver cells are of a regular cuboidal shape, have prominent nuclei and granular cytoplasm. Glycogen droplets may be seen in suitably stained sections.

Between the liver cords are clearly visible channels: these are **sinusoids**, small capillaries which connect the interlobular arteries and veins at the edge of the lobule with the intralobular vein in the centre. Blood flows from the former to the latter via the sinusoids and in so doing comes into close association with all the liver cells.

The sinusoids are lined with pavement endothelium whose flattened nuclei may be visible in places. Phagocytic Kupffer cells may be seen.

Also running between the liver cells are much smaller spaces called **canaliculi**. Lacking an endothelial lining, these connect with the interlobular bile ductiles at the edge of the lobule. They are difficult to detect except in very good sections.

For Consideration
(1) How is the microscopic structure of (a) the pancreas and (b) the liver adapted to perform their particular functions?
(2) Homeostatic mechanisms can be analysed in terms of receptors (detectors), control mechanisms and effectors. How do the pancreas and liver fit into such a scheme?

Requirements
Microscope
Section of pancreas (haematoxylin and eosin will do)
If available, section of pancreas stained with, e.g. aldehyde fuchsin to show alpha and beta cells
TS liver (haemalum and Van Gieson's stain recommended)
If available, TS liver stained for glycogen by periodic-acid Schiff technique

Questions and Problems

1 Compare and contrast the composition of blood, tissue fluid, and lymph. How would you explain their similarities and differences?

2 What physiological events would you expect to follow the injection of a small quantity of glucose into the bloodstream of a healthy mammal? What would be the result of injecting glucose into the bloodstream of a pancreatectomized animal, i.e. one from which the pancreas had been removed?

3 Make a list of the effects on the body's homeostatic mechanisms of surgically removing the whole of the liver.

4 Under experimental conditions removal of the liver from a mammal results in an increase in the amounts of amino acids in the blood and the absence of urea. Discuss possible reasons for this and suggest further experiments which might be carried out to test your explanations. (O and C modified)

5 Fig. 13.7 shows highly magnified the relationship between cells and channels as seen in a transverse section of a mammalian liver lobule;
(a) the sinusoid contains blood; what vessels does the blood come from and what vessel does it drain into?
(b) what enters the liver cells from the blood in the sinusoid (arrow 1)?
(c) what is shed into the sinusoid from the liver cells (arrow 2)?
(d) what is shed into the canaliculus from the liver cells (arrow 3)?
(e) what function is performed by the Kupffer cells?

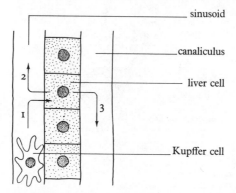

Fig. 13.7 Diagram showing the relationship between cells and channels in a mammalian liver lobule.

6 Discuss the possible circumstances in which:
(a) more blood sugar may be found in the hepatic portal vein than in any other vessel;
(b) more blood sugar may be found in the hepatic vein than in the hepatic portal vein;
(c) the blood sugar level in the general bloodstream may depart dangerously from the normal;
(d) sugar is present in the urine. (O and C modified)

7 How many biological and non-biological situations can you think of that involve control by feedback mechanisms?

8 When a speech is made using a public address system a howl is sometimes heard from the loudspeakers. How would you explain this? Draw an analogy between this and events that sometimes occur in biological situations.

9 Imagine that the mechanism which initiates the human body's cooling processes (sweating, etc.) breaks down completely. Under these circumstances what will be the result of the body temperature being raised by 2°C?

10 What do you understand by an *efficient* homeostatic process? In physiological circumstances homeostatic processes involve the use of a receptor, control centre, and effector. Discuss the extent to which shortcomings in these three components can influence the efficiency of the overall process.

14 Excretion and osmoregulation

Background Summary

1 **Excretion** is the removal from the body of the waste products of metabolism: these are predominantly nitrogenous compounds. **Osmoregulation** is the process by which the osmotic pressure of the blood and tissue fluids is kept constant.

2 In the mammal both functions are performed by the **kidney**, the basic unit of which is the **nephron**. Plasma minus protein passes from the blood vessels of the **glomerulus** into the cavity of Bowman's capsule. Some of the constituents of the glomerular filtrate are then selectively reabsorbed across the walls of the **tubules** before the renal fluid is shed into the ureter, and thence to the bladder.

3 The blood pressure in the glomerulus, and the fine structure of the barrier between the glomerular capillaries and cavity of Bowman's capsule, are consistent with the view that renal fluid is formed by **ultra-filtration**.

4 Analyses of renal fluid at different points along the tubules in the frog indicates that water and glucose are selectively reabsorbed in the proximal convoluted tubule, chloride in the distal tubule. Urea is eliminated altogether and is actively secreted into the proximal tubule from the surrounding blood vessels.

5 Comparable experiments on the mammalian kidney suggest that in mammals chloride, as well as water, glucose and amino acids, are reabsorbed in the proximal tubule. Over two thirds of the water in the glomerular filtrate is reabsorbed in the proximal tubule. Urea is actively secreted into the tubule.

6 Evidence suggests that water is reabsorbed mainly by osmosis, glucose and salts by diffusion and active transport. There is a close association between the epithelium of the tubule and capillaries, and the fine structure of tubule epithelial cells is consistent with its function of active absorption.

7 Further water is reabsorbed from the **collecting ducts** by the high osmotic pressure in the medullary region of the kidney. This high osmotic pressure is achieved by the loop of Henle conserving salts on the principle of a **hair-pin countercurrent multiplier**.

8 Reabsorption of water by the kidney is controlled homeostatically by the osmotic pressure of the blood acting through the intermediacy of **anti-diuretic hormone** (**ADH**) secreted by the posterior lobe of the **pituitary gland**. In conditions of high internal osmotic pressure ADH causes retention of water by the kidney.

9 The body fluids of **marine invertebrates** are isotonic with sea water and such animals have no osmoregulatory devices.

10 **Freshwater and estuarine animals**, including freshwater fishes, counter the osmotic influx of water that inevitably results from their internal osmotic pressure (OPi) being greater than the external osmotic pressure (OPe) by eliminating excess water and actively taking up salts from the external medium.

11 In **marine vertebrates** OPi is less than OPe. Such animals counter the resulting osmotic outflux of water by eliminating excess salts and retaining water.

12 **Migratory fishes** (salmon, eel) present a special problem since in the course of their life cycle they move from one extreme of aquatic environment to the other (fresh water to sea water, and vice versa).

13 **Terrestrial animals** are liable to water loss by evaporation. They have evolved a variety of structural and physiological techniques for preventing excessive water loss.

14 A variety of water-conservation devices, equivalent to those seen in animals, have been developed by plants, particularly **xerophytes** and **halophytes**.

Investigation 14.1

Examination of the mammalian kidney

The kidney is the mammal's principal organ of excretion and osmoregulation. It performs these functions by 'purifying' the blood that flows through it. The blood is first filtered to form **renal fluid**. Useful substances are reabsorbed back from the renal fluid into the bloodstream, while unwanted substances are eliminated as **urine**. The formation and processing of renal fluid are carried out by the **nephrons**, of which each kidney contains about 1.5 million. Each nephron consists of a **Bowman's capsule** containing a bundle of capillaries called the glomerulus, and a long differentiated tubule which leads to a collecting duct. (Bowman's capsule and glomerulus together comprise a **Malpighian corpuscle**).

Blood is filtered across the walls of the glomerular capillaries into the cavity of Bowman's capsule (capsular space). The glomerular filtrate, (renal fluid), consists of plasma minus proteins. As the renal fluid flows along the tubule of the nephron useful substances are selectively reabsorbed back into the bloodstream, whilst harmful substances that escaped filtration are secreted from the blood into the tubule.

Obviously, to achieve this, there must be a close association between the nephron and its blood supply. Bear this in mind as you look at the structure of the kidney.

Coarse Structure of the Kidney

In a dissection of the rat (or other mammal) note that the kidneys are asymmetrically disposed towards the anterior end of the abdominal cavity (Fig. 14.1).

Observe the good blood supply: **renal artery** bringing blood to the kidney from the dorsal aorta runs close to the larger, thinner-walled, **renal vein**, which takes blood from the kidney to the posterior vena cava. Thin thread-like **ureter** leads from the indented inner side of the kidney to the **bladder**.

Remove a kidney from the body (or obtain kidney of pig, sheep or cow from the butcher). Slice through the kidney horizontally (Fig. 14.2A) and notice: light-coloured **cortex** towards

Kidneys and their blood supply in situ (rat)

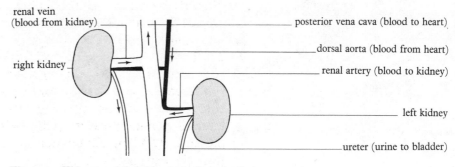

Fig. 14.1 Kidneys and their blood supply *in situ* (rat).

the outside, darker **medulla** towards the inside, much blood, **pyramids** (onto which open the collecting ducts), **pelvis**.

The pelvis is the cavity where the urine collects before flowing down the ureter to the bladder. Fig. 14.2B shows on a distorted scale how the nephrons and collecting ducts are disposed relative to the cortex and medulla.

Microscopic Structure of the Mammalian Kidney

Examine a horizontal or transverse section of mammalian kidney. First hold it up to the light and examine it under a lens and note the demarcation between cortex and medulla.

A Whole kidney showing the different planes in which it may be sectioned

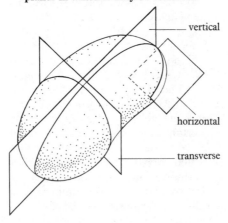

B Horizontal section showing the positions of the nephrons and associated structures

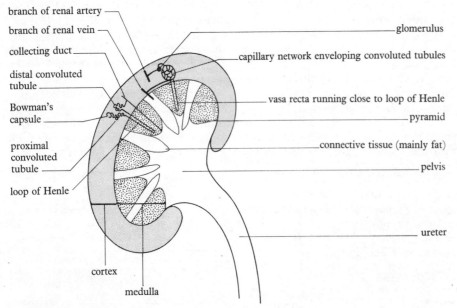

Fig. 14.2 General structure of the mammalian kidney.

CORTEX

If you examine the cortex under low power you will see it contains **Bowman's capsules**, numerous tubules in section, and capillaries (Fig. 14.3A). The capsules contain **glomeruli** and in some cases you may see **afferent** and **efferent arterioles** which take blood to and from them. Notice the gap between the bunch of glomerular capillaries and the outer epithelium of the capsule: this is the capsular space into which blood is ultra-filtered to form renal fluid. The renal fluid then passes along the tubules and so eventually to the collecting ducts.

The tubular structures visible in a section of the cortex are a mixture of proximal and distal convoluted tubules and collecting ducts. The three can be distinguished as follows:

Proximal tubule:
outer diameter about 60 μm,
large lining cells,
cell membranes between adjacent cells not visible,
relatively few nuclei visible in cross section of tubule,
small irregular lumen,
brush border,
three times as long as distal tubule so more plentiful in section.

Distal tubule:
outer diameter 20–50 μm,
smaller lining cells,
cell membranes between adjacent cells not visible,
more nuclei visible in cross section of tubule,
large regular lumen,
no brush border visible,
shorter than proximal tubule so less plentiful in section.

Collecting duct:
outer diameter 25–60 μm,
wall similar to distal convoluted tubule except that cell membranes between adjacent cells are clearly visible.

Pay special attention to the proximal tubule, particularly its **brush border** (function?) and its proximity to capillaries. What blood vessels are the capillaries derived from?

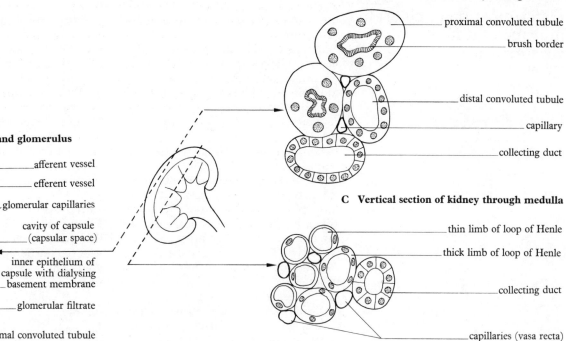

A Bowman's capsule and glomerulus (Malpighian corpuscle)

afferent vessel
efferent vessel
glomerular capillaries
cavity of capsule (capsular space)
inner epithelium of capsule with dialysing basement membrane
glomerular filtrate
proximal convoluted tubule

B Vertical section of kidney through cortex

proximal convoluted tubule
brush border
distal convoluted tubule
capillary
collecting duct

C Vertical section of kidney through medulla

thin limb of loop of Henle
thick limb of loop of Henle
collecting duct
capillaries (vasa recta)

Fig. 14.3 Microscopic structure of the mammalian kidney.

MEDULLA

Now examine the medulla under high power (Fig. 14.3B). This part of the kidney contains the **loops of Henle** which connect the proximal and distal convoluted tubules. Notice the **descending** and **ascending limbs** of the loops of Henle and the long straight capillaries, **vasa recta**, running parallel with them. Why are the tubules and blood vessels long and straight in this part of the kidney?

Descending and ascending limbs and collecting ducts can be distinguished as follows:

Descending limb : (except the lowest third)
outer diameter 14–20 μm,
lining of thin pavement epithelium,
nuclei flattened and bulging into lumen; (known as thin limb).

Ascending limb : (plus lowest third of descending limb)
outer diameter 30–35 μm,
lining of cuboidal epithelium,
cell membranes between adjacent cells not visible,
rounded nuclei,
(known as thick limb).

Collecting duct :
outer diameter 50–200 μm,
lining cells cuboidal to columnar,
cell membranes between adjacent cells clearly visible,
rounded nuclei.

N.B. The wall of the thin limb is confusingly similar to that of the capillaries (vasa recta). However, the capillaries can usually be distinguished by the fact that they contain red blood corpuscles.

To appreciate the vascular nature of the kidney examine a section in which the blood vessels have been injected. In particular notice the **vasa recta** and the intimate association between the proximal convoluted tubule and capillaries.

For Consideration

In what respects is structure of the kidney adapted to perform its functions of excretion and osmoregulation?

Requirements

Microscope
Rat for dissection (demonstration dissection showing kidneys, ureters and blood supply will do)
Fresh kidney from butcher
TS or HS kidney
Section of injected kidney

Investigation 14.2

Determination of chloride in urine in different circumstances

One of the kidney's functions is to regulate the **osmotic pressure** of the blood and tissue fluids. It does this by selectively eliminating varying amounts of water and salts depending on their levels in the body. In this investigation you will estimate the quantity of chloride in the urine in different conditions.

Method

The method involves reacting the chloride present in the urine sample with a known quantity of **silver nitrate**, and then titrating the unprecipitated silver nitrate (i.e. the silver nitrate which has not been used up in the reaction) against **potassium thiocyanate**. From this the amount of chloride in the urine sample can be calculated.

The reactions are as follows:

(1) Chloride in the urine sample reacts with silver nitrate, forming a white precipitate of silver chloride:

$$NaCl + AgNO_3 \rightarrow NaNO_3 + AgCl$$

(2) Potassium thiocyanate from the burette reacts with unprecipitated silver nitrate, forming a white precipitate of silver thiocyanate:

$$KCNS + AgNO_3 \rightarrow KNO_3 + AgCNS$$

(3) The first excess drop of potassium thiocyanate reacts with ferric alum (indicator), forming red ferric thiocyanate:

$$Fe^{3+} \text{ alum} + KCNS \rightarrow$$
$$FeCNS + K \text{ alum}$$

Collecting the Urine Samples

This should be done before the laboratory session. You will need two large test tubes with stoppers.

Urinate about an hour before a meal. With your meal consume plenty of salt but no water. One hour later collect a sample of your urine in one of the tubes: at least 5 cm³ are required. Label this sample 'fed urine'.

Now drink as much water as possible. One hour later collect a further urine sample in the second tube. Label this 'watered urine'.

Add 2–3 drops of toluene to each specimen to prevent bacterial action.

Laboratory Procedure

Measure fed urine (2 cm³) into a porcelain bowl. Add silver nitrate solution (10 cm³), stirring well with a glass rod. Any chloride in the urine produces a white precipitate of silver chloride. Allow the mixture to stand

for five minutes to ensure maximum coagulation of the precipitate.

Add a small spatula-end of ferric alum, the indicator. Titrate any unprecipitated silver with 0·1 M potassium thiocyanate from a burette: run the thiocyanate in carefully, stirring gently all the time until a red colour, permanent for 15 seconds, is produced.

Repeat with another 2 cm³ sample of fed urine. Then do the same for your watered urine. In each case make a note of the volume of thiocyanate required to produce the red colour.

Calculation

From your readings you can calculate the quantity of chloride in fed and watered urine. In each case express your result as mg chloride per 100 cm³ urine.

The calculation is carried out as follows:

Suppose the volume of potassium thiocyanate (KCNS) required to precipitate the unused silver nitrate ($AgNO_3$) is x cm³.

This is equivalent to x cm³ $AgNO_3$ since both are 0.1 M solutions.

Then $10 - x$ cm³ $AgNO_3$ has been precipitated by the chloride in 2.0 cm³ urine.

Now 1,000 cm³ M $AgNO_3$ is equivalent to 35.5 g chloride.

Therefore 1.0 cm³ M $AgNO_3$ is equivalent to $\frac{35.5}{1,000}$ g chloride and 1.0 cm³ 0.1 M $AgNO_3$ is equivalent to $\frac{35.5}{1,000} \times \frac{1}{10}$ g chloride.

So $10 - x$ cm³ 0.1 M $AgNO_3$ is equivalent to $\frac{35.5}{1,000} \times \frac{1}{10} \times (10 - x)$ g chloride.

This is the amount of chloride in 2.0 cm³ urine. Convert to mg/100 cm³.

For Consideration

(1) Compare your results for fed and watered urine and explain the difference between them in terms of the functioning of the kidney.

(2) Suppose that samples of your blood had been taken at the same time as your two urine samples. What would you expect the chloride content of the blood samples to be?

Requirements

Large test tubes (2) with stoppers (for urine samples)
Labels
Burette 25 cm³
Pipette 2 cm³
Pipette 10 cm³
Porcelain bowls (2)
Glass rod

Toluene
50 cm³ silver nitrate (0.1 M in 50 per cent v/v nitric acid)
50 cm³ potassium thiocyanate 0.1 M.
Ferric alum (powdered)

Investigation 14.3

Action of the contractile vacuole of a protozoon

The **contractile vacuole**, found in unicellular protozoons, is a comparatively simple osmoregulatory device. As quickly as it enters by osmosis, water is collected into the vacuole, which swells and, when full, discharges its contents through a temporary pore in the cell membrane.

We will examine the action of the contractile vacuole in the suctorian *Podophrya*. Like *Paramecium* (*see* p. 20), *Podophrya* is a ciliate, but—unlike *Paramecium*—it only possesses cilia in the juvenile (larval) stage. The free-swimming ciliated larva settles down and develops into a non-ciliated, sessile adult. The adult suctorian (Fig. 14.4A), has a typical protozoon 'body' with a single nucleus. It has a non-contractile **stalk** by which it is attached to the substratum, and an array of **tentacles** with sucker-like ends with which it catches prey. If, for example, *Paramecium* touches the tentacles it is held by the suckers and its contents are absorbed (via the tentacles) into the suctorian's body.

external medium should increase the activity of the contractile vacuole and increasing the osmotic pressure of the medium should decrease the activity of the contractile vacuole. In this investigation we will test this prediction.

Procedure

(1) Place two small squares of filter paper on a clean slide as indicated in Fig. 14.4B. Saturate each piece with *Podophrya* culture solution and fill the area between them with the same solution. The purpose of the filter paper is to support the coverslip and provide a reservoir of solution.

(2) With small forceps transfer a short piece of silk thread which has *Podophrya* attached to it into the culture solution between the filter-paper squares. Anchor the silk thread by wrapping it round a short length (about 15 mm) of fine cotton as shown in Fig. 14.4B. Now put on a coverslip so that it rests on the two pieces of filter paper.

A *Podophrya* **in the adult condition**

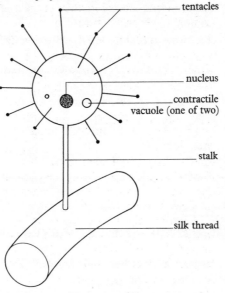

tentacles

nucleus

contractile vacuole (one of two)

stalk

silk thread

B **Method for mounting** *Podophrya*

viewed from above

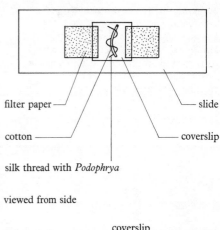

filter paper —

cotton —

silk thread with *Podophrya*

— slide

— coverslip

viewed from side

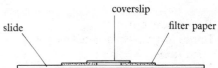

slide

coverslip

filter paper

Fig. 14.4 Investigating osmoregulation in the suctorian *Podophrya*.

Podophrya lends itself well to studies on the contractile vacuole. The two contractile vacuoles, easily identified from their glistening appearance, discharge relatively frequently while the animal, being sessile, will remain steady under the microscope.

If the function of the contractile vacuole is to eliminate excess water from the cell, it stands to reason that decreasing the osmotic pressure of the

Requirements

Filter paper (preferably strips about 15 mm wide)
Fine pipette
Small forceps
Slide and coverslip
Microscope and lamp (*not* substage illumination)
Micrometer eyepiece or transparent ruler
Fine cotton thread

Distilled water
Sucrose solution (0.05 M)

Podophrya on silk thread in culture solution

Prepare *Podophrya* as follows: Tease out short *fine* fibres from a small piece of undyed silk fabric and float on surface of *Paramecium* culture (about 200 cm³) in crystallizing dish. Add culture of *Podophrya*. In 2 or 3 days larvae of *Podophrya* settle on threads and grow rapidly. After a further week fully grown adults should be ready for observation. Maintain culture by feeding with *Paramecium* every 2–3 days.

(3) Examine your slide under medium power of a microscope, arranging the illumination in such a way that there is no chance of the temperature of the *Podophrya* increasing. (N.B. Avoid using a microscope with a substage lamp). Locate *Podophrya* attached to the silk thread, and identify a **contractile vacuole** under high power. Focus on the contractile vacuole and keep watching it for at least five minutes. Can you learn anything about the mechanism of discharge from your observations?

(4) By observing the contractile vacuole over a period of time, estimate its frequency of discharge. While you are doing this keep the filter paper saturated with culture solution.

(5) Now replace the culture solution under the coverslip with distilled water. This can be done by adding distilled water to one of the two pieces of filter paper and withdrawing the solution from the other with dry filter paper. Do this for sufficiently long to ensure that the liquid is completely changed. Now re-estimate the frequency of discharge of the contractile vacuole.

(6) Now replace the distilled water with sucrose solution (0.05 M) by the same technique. Estimate the frequency of discharge as before.

(7) Finally make an approximate estimate of the diameter of the animal's body and of a fully inflated contractile vacuole.

For Consideration

(1) How does the frequency of discharge compare in the culture solution, distilled water, and 0.05 M sucrose solution? Interpret your results as fully as you can.

(2) Assuming that the animal's body and contractile vacuole are spherical, estimate the volume of liquid discharged per unit time in distilled water. How long would it take for a contractile vacuole to discharge a volume of liquid equal to the volume of the animal's body?

Reference
Nuffield Biology Text IV, *Living Things in Action*, Longman Penguin 1966.

Questions and Problems

1 Explain the effect that each of the following will have on the quantity and composition of urine:
(a) drinking a large amount of water;
(b) eating a very salty meal;
(c) a hot dry day;
(d) consuming a large quantity of carbohydrate;
(e) high arterial pressure;
(f) low arterial pressure;
(g) sleep
(h) prolonged muscular exertion;
(i) removal of the pancreas;
(j) destruction of the posterior lobe of the pituitary gland.

2 Comment on this statement: 'Those mammals which can produce the most concentrated urine have the longest loops of Henlé.'

(O and C)

3   In experiments on the mammalian kidney, samples of renal fluid from different regions along the length of the proximal convolution have been withdrawn and analysed. The results, expressed in terms of the renal : plasma ratio for each constituent, are shown graphically in Fig. 14.5. (The renal : plasma ratio is the amount of the particular constituent in the renal fluid divided by the amount of the same constituent in the blood plasma.) The graph also includes a curve for glucose absorption in a phlorizinized kidney, one that has been treated with phlorizin which renders the tubules incapable of absorbing glucose.

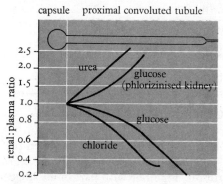

Fig. 14.5 Graphs showing the renal : plasma ratio for various constituents of renal fluid in the nephron of a mammal.

Explain each of the graphs. In what respects do the results of comparable experiments on amphibian kidneys differ from the mammal?

4   The loops of Henle are located in the medulla of the kidney. Blood flowing through this region of the kidney does so in long U-shaped capillaries, the vasa recta, which run close to and parallel with the loops of Henle. Blood flowing through the vasa recta represents only about one per cent of the blood that flows through the kidney. The vasa recta are derived from the efferent arterioles serving the glomeruli.

What can you say about:

(a) the pressure of blood in the vasa recta?
(b) the viscosity of the blood in the vasa recta?
(c) the rate of flow of blood through the vasa recta?
(d) the osmotic pressure of the blood in the vasa recta?
What light is thrown by your answers on the function of the vasa recta?

5   Fig. 14.6 summarizes the results of experiments carried out to determine the composition of the urine and faeces of four different species of mammal. Explain the data as far as you can in physiological and ecological terms.

(SA modified)

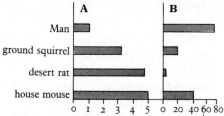

Fig. 14. Summary of data on the composition of the urine and faeces of four different species of mammal. A, urine salt content (arbitrary units); B, water content of faeces (per cent).

6   Sea birds such as the herring gull secrete salts twice as concentrated as in sea water from nasal glands situated on top of their heads, and marine reptiles such as turtles secrete salts from glands close to the eyes. Explain why this should be necessary.

7 The following table shows the composition of plasma, filtrate, and urine in a mammal.

Component	Plasma (g/100 cm^3)	Filtrate (g/100 cm^3)	Urine (g/100 cm^3)
Urea	0.03	0.03	2.00
Uric acid	0.004	0.004	0.05
Glucose	0.10	0.10	0.00
Amino acids	0.05	0.05	0.00
Salts	0.72	0.72	1.50
Proteins	8.00	0.00	0.00

Comment on these data. (*SBT modified*)

8 The diagram in Fig. 14.7 has been reconstructed from electron micrographs of a mammalian kidney tubule. It shows one cell of the proximal convoluted tubule, together with part of an adjacent capillary.

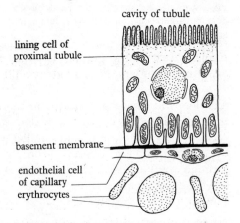

Fig. 14.7 Diagram based on an electron micrograph of the proximal convoluted tubule of a mammal.

Fluid entering the proximal tubule from Bowman's capsule contains water, salts, glucose, amino acids and excretory substances. More than 80 per cent of the water is reabsorbed in the proximal tubule and there is normally no glucose or amino acids, and a greatly reduced sodium content in the fluid which is passed on to the loop of Henle.

Using the information given, together with your own biological knowledge, write short answers on the following:

(a) What do you suggest is the mechanism responsible for the reabsorption of water in the proximal tubule? Explain your answer.

(b) The reabsorption of glucose and amino acids is described as taking place 'against a concentration gradient'. What does this mean?

(c) If the kidney is cooled for a time, glucose and amino acids appear in the urine. How would you explain this?

(d) In what ways is the structure of the tubule cell particularly suited to its function of absorption? (*CL modified*)

9 An experiment was carried out on the Pacific shore crab *Hemigrapsus* to determine the relationship between the osmoconcentration of the blood and that of the surrounding water. Specimens were subjected to different external salinities, and, after being given time to come to equilibrium with the medium, blood samples were taken and their osmoconcentration determined by measuring the depression of the freezing point. The results were as follows:

Osmoconcentration
(as depression of
freezing point °C)

Water	Blood
0.1	0.1
0.15	1.0
0.2	1.1
0.3	1.2
0.4	1.3
0.5	1.35
1.0	1.4
1.5	1.6
2.0	2.0

Plot the results on a graph, putting osmoconcentration of blood on the vertical axis and osmoconcentration of water on the horizontal axis. Interpret the results as fully as you can.

10 Fig. 14.8 shows the relationship between the osmotic pressure of the blood (internal OP = OP_i) and of the external medium (external OP = OP_e) in four different arthropods: the marine spider crab *Maia*, the estuarine shore crab *Carcinus*, the freshwater crayfish *Astacus*, and the larva of the mosquito *Aedes detritus* which inhabits salt marsh pools.
(a) Explain what each curve shows.
(b) Relate the information which the curves show to the habitat of each animal.
(c) Discuss the anatomical and physiological basis of the curves.

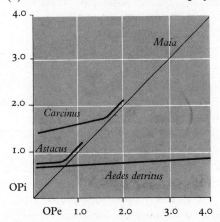

Fig. 14.8 Relationship between the osmotic pressure of the blood (OPi) and the external medium (OPe) for four different arthropods. The osmotic pressure is expressed as depression of the freezing point in °C.

11 To what extent are the methods employed by terrestrial animals and plants to prevent excessive water loss comparable with those seen in plants?

12 (a) Why do miners eat salt tablets or drink salted beer?
(b) Sea water contains three per cent salt, the salt content of the blood is about one per cent, and, as a maximum, the kidneys can excrete only a two per cent salt solution. Give a brief explanation for the serious consequences that occur after drinking a large quantity of sea water.

13 You are given the following information concerning the kidney of three different animals, together with their respective environments, the nature of their nitrogenous waste, and the type and quantity of urine produced. Study the information and then describe how the structure and product of the kidney is adapted to the osmotic environment in which the animal lives:

Animal	Environment	Kidney		Nitrogenous waste	Urine relative to blood
		Glomerulus	Tubule		
Trout	freshwater	large	long	ammonia	hypotonic, copious
Herring	sea	nil	short	trimethylamine oxide	isotonic, scanty
Lizard	land	small	very long	uric acid	hypertonic and scanty

How would you characterize the structure and product of the human kidney on the above chart?   (*AEB modified*)

15 Temperature Regulation

Background Summary

1 Organisms cannot withstand fluctuations in body temperature beyond that which is compatible with the functioning of their enzymes.

2 On the basis of their ability to regulate their body temperature animals are classified into **endothermic** and **ectothermic**.

3 Heat is lost or gained by **radiation**, **evaporation**, **conduction** and **convection**. The problem in temperature regulation is to overcome, control, or make use of these physical processes.

4 Endothermic animals (e.g. mammals) have various structural and physiological ways of coping with excessive cold and heat. Physiological responses are controlled by a **thermo-regulatory centre** in the **hypothalamus** of the brain which responds to changes in the temperature of the blood.

5 In endotherms physical (non-metabolic) mechanisms maintain a constant body temperature when the environmental temperature ranges between a high and low **critical temperature**. This is the body's **efficiency range**. Above the high critical temperature, and below the low critical temperature, the metabolic rate rises.

6 It has been found that the low critical temperature is significantly lower for arctic than for tropical animals. Man's low critical temperature is about 27°C, that of the arctic fox −40°C.

7 Many animals regulate their body temperature by **behavioural means**. In ectothermic animals this is the only method. **Migration** and **hibernation** may be regarded as ways of avoiding unfavourable environmental temperatures.

8 A few animals and many plants are able to tolerate wide temperature fluctuations. To an extent plants are cooled by **transpiration**, in some cases markedly so.

Investigation 15.1

Microscopic structure of skin

Though important in **temperature regulation**, the skin performs many other functions as well. For example, it contains numerous sensory devices and is therefore important in **reception of stimuli**, informing the body of environmental changes at the surface; its toughness affords the body **physical protection**; the fact that hairs, nails and the keratinized outer layer of the epidermis contain protein means that the skin contributes towards **nitrogenous excretion**; and its impermeability makes it an effective water-proofing layer, thus giving it a passive rôle in **osmoregulation**. The fact that man can go swimming in water of any salinity without untoward osmotic effects bears witness to this last function. The skin is therefore important in protection against, and adjustment to, changing external conditions.

Procedure

Examine a vertical section of hairy skin (for example, human scalp). First distinguish between the superficial epidermis, the deeper, more extensive dermis, and (strictly not part of the skin) the hypodermis (subcutaneous tissue) (Fig. 15.1). Now examine each in turn.

EPIDERMIS

The epidermis is composed of stratified epithelium (*see* p. 15). Cells formed by proliferation of the basal **Malpighian layer** get pushed outwards, flattening as they do so (**stratum spinosum**, so-called because under high power the cells sometimes have a 'prickly' appearance); eventually they become granulated and die (**stratum granulosum**), then clear (**stratum lucidum**) and finally the cells become converted into scales of keratin (**stratum corneum**) which flake off (**stratum disjunctum**). The strata granulosa and lucida represent stages in the keratinization of the cells. The stratum corneum is impermeable, thus conferring on the epidermis its protective properties.

In your section of skin, the nuclei of the Malpighian layer and stratum spinosum should be clear, but the cell membranes will probably be difficult to make out. The stratum granulosum should be apparent, but the stratum lucidum, unless it stands out as a clear bright line, may be difficult to see. Cells of the Malpighian layer may be pigmented, containing **melanin** towards their outer surface. Function?

DERMIS

First notice that the surface of the dermis is folded to form a series of ridges, the **dermal papillae**. Receptors sensitive to touch are located in the dermal papillae, but these will only be seen with special staining techniques.

The rest of the dermis is made up of loose connective tissue, mainly collagen fibres, but some elastic fibres too. **Note**: **capillaries** (red blood corpuscles may be seen in them), **sweat glands** and their **ducts**, **hairs** in **hair follicles** with **erector pili muscles** and **sebaceous glands**. Functions? What sort of glands are the sweat and sebaceous glands?

Few sections will be as complete as the one drawn in Fig. 15.1. Seldom will a hair follicle be cut throughout its full length: more often it will be cut tangentially as shown on the right of the drawing. Similarly sweat ducts seldom appear complete in a section. The organization of the dermis must therefore be reconstructed by searching the entire section for clues and, if necessary, examining several different sections.

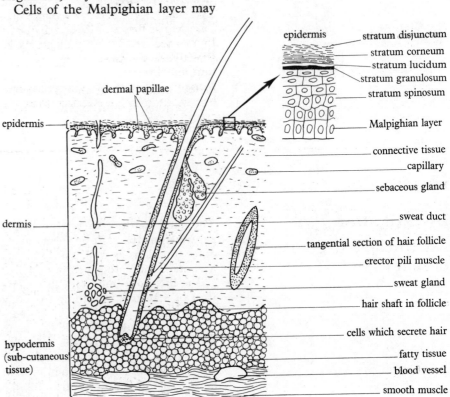

Fig. 15.1 Microscopic structure of human skin (scalp).

HYPODERMIS (SUB-CUTANEOUS TISSUE)

Hypodermis means 'beneath the dermis'; it is not strictly part of the skin. It consists of a layer of adipose tissue (**subcutaneous fat**) of variable thickness, beneath which is a layer of smooth muscle and blood vessels.

The adipose tissue as well as providing an insulating layer against heat loss, allows the skin to move freely on underlying structures.

RECEPTORS IN THE SKIN

Here is a simplified summary of the distribution and functions of skin receptors:

In epidermis:

Free nerve endings derived from branched nerve fibre—sensitive to pain.

In dermal papillae:

Meissner's corpuscles:
Branched nerve endings embedded in connective tissue—sensitive to touch.

In rest of dermis:

Krause's end bulbs:
Bundle of branched nerve endings enclosed in connective tissue capsule—thought to be sensitive to cold.

Ruffini's endings:
Tree-like system of nerve endings terminating as flattened discs, supported by connective tissue—thought to be sensitive to warmth;

Free nerve endings wrapped round base of hair follicle—sensitive to movements of hair;

Pacinian corpuscle:
Unbranched nerve ending encapsulated in thick multi-layered covering of connective tissue—sensitive to pressure. Only the Pacinian corpuscles can be seen properly without special staining methods. The others are visible in methylene blue or silver preparations.

COMPARISON WITH THE SKIN OF AN ECTOTHERM

Observe, and feel, the skin of a live frog. What structures would you predict to be present (and not present) in the skin?

Now examine a vertical section of the skin of the frog. Are your predictions correct? In what respects is the frog skin (a) similar to, and (b) different from mammalian skin? Explain the differences.

For Consideration

(1) How much of the surface of human skin can be removed without feeling pain and without bleeding?

(2) What areas of the skin would you expect to: (a) lack hair follicles, (b) have a particularly thick layer of subcutaneous fat, (c) have little or no subcutaneous fat, (d) have a particularly thick stratum corneum, (e) have numerous Meissner's corpuscles, (f) have a particularly large number of sweat glands?

(3) In sunbathing, what causes the skin to develop: (a) a pink colour; (b) a brown tan; (c) blisters?

(4) Why is it a good thing for sunlight (in moderation) to fall on the skin? (Refer to Table 5.3, p. 42 if you do not know the answer).

Requirements

VS hairy skin (e.g. scalp)
VS skin stained to show receptors (if available)
Common frog
VS frog skin

Investigation 15.2

Effect of temperature on heartbeat of *Daphnia*

Daphnia, the 'water flea', is a small freshwater crustacean which lacks physiological methods of maintaining a constant body temperature. This means that if the environmental temperature changes, its body temperature follows suit. This being so, its metabolic rate will be expected to rise or fall accordingly. In this investigation we shall test the hypothesis that as the environmental temperature rises, the metabolic rate rises too. We shall use the rate at which the heart beats (**cardiac frequency**) as a measure of the metabolic rate. Fortunately *Daphnia* is comparatively transparent, which enables the workings of its internal organs, including the heart, to be seen without dissection.

Procedure

SETTING UP THE EXPERIMENT

(1) Select a large specimen and transfer it by means of a pipette to the centre of a small petri dish. Remove excess water from around the specimen, so it is temporarily stranded. Now smear a little silicone grease onto the floor of the petri dish and, using a mounted needle, push the posterior end of the animal into the grease so it is firmly anchored in position. Now fill the petri dish with water.

A Side aspect of Daphnia as seen under low power

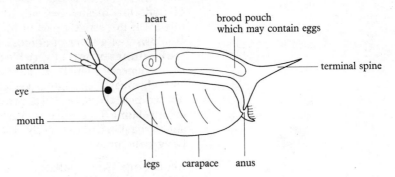

B View of set-up from front

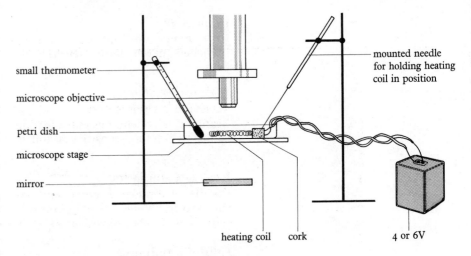

C Petri dish and heating coil viewed from above

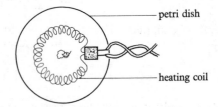

Fig. 15.2 Experiment on the effect of temperature on the heartbeat of *Daphnia*.

(2) Place the petri dish on the stage of a microscope and observe the animal under low power. The beating **heart** is located on the dorsal side just above the **gut** and in front of the **brood pouch** (Fig. 15.2A). Do not confuse the beating of the heart with the flapping of the **legs** which lie within the carapace on the ventral side.

(3) Surround the animal with a circular heating coil and clamp it in position. Also clamp a thermometer in position so that its bulb is in the water (Fig. 15.2B).

ESTIMATING THE CARDIAC FREQUENCY

A convenient way of doing this is to time, by means of a stopwatch, how long it takes for the heart to beat 50 times. The heart may be beating sufficiently slowly for every pulsation to be counted. If, however, it is beating too rapidly, count by making a mark on a sheet of paper at every tenth beat. Do several practice runs to get used to the technique.

When you feel ready to proceed with the experiment, replace the water in the petri dish with ice-cold water. Estimate the cardiac frequency and note the temperature. Now connect

Requirements

Small petri dish
Mounted needle
Heating coil plus stand and clamp
Small thermometer plus stand and
  clamp
4V or 6V battery
Stopwatch
Pipette (10 or 25 cm³)
Rubber pipette
Iced water
Silicone grease

Daphnia

the heating coil with a 4 or 6 volt battery so as to gradually heat up the water in the dish. Estimate the cardiac frequency at 5°C intervals, noting the temperature each time.

If the temperature of the water rises too quickly, disconnect the heating coil from the battery and, if necessary, add a few ice chippings.

Present your results in a table, recording the cardiac frequency at each temperature.

Plot your results on a graph, temperature along the horizontal axis and cardiac frequency along the vertical axis.

For Consideration

(1) What conclusions do you draw from your results?
(2) Would you conclude from your results that *Daphnia* has no means of controlling its body temperature?
(3) Did you reach the upper lethal temperature in your experiment? If not, what would you expect it to be? Explain fully.
(4) Would you expect to get the same sort of results with any 'cold-blooded' animal?
(5) How would you carry out the same investigation on a human subject? In what respects would you expect the results to differ from those obtained with *Daphnia*?

Questions and Problems

1   Comment on the graph shown in Fig. 15.3 and discuss reasons for the difference in the three curves.

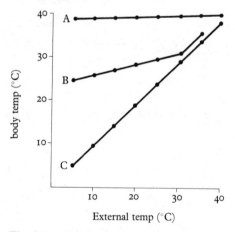

Fig. 15.3 Relationship between the external temperature and body temperature for three different animals. **A**, cat; **B**, spiny anteater *Echidna*; **C**, lizard.

2   Why is it that: (a) the skin is red in hot weather, white in cold weather, and blue in *very* cold weather; (b) in prolonged exposure to severe cold the living cells at the tips of the fingers may die ('frostbite'); (c) for bodily warmth a 'string' vest is particularly effective; (d) during a fever a body temperature of less than about 40°C is treated by covering the patient with extra blankets, but if the temperature exceeds about 40°C ice packs are placed in contact with the patient?

3   It has been observed that in many animals the veins that bring blood back from exposed extremities (such as the hands and feet in man) run close to, and parallel with, the artery that takes blood to these structures. In some cases, such as the flippers of whales, the artery may be completely surrounded by the veins. What do you think is the significance of this arrangement?

4 Graph A in Fig. 15.4 shows how the rate of heat production by a naked human body varies with the temperature of the surrounding air. Graph B shows how the rate of heat loss from a naked human body varies with the temperature of the surrounding air. Explain the form of each graph. (*JMB*)

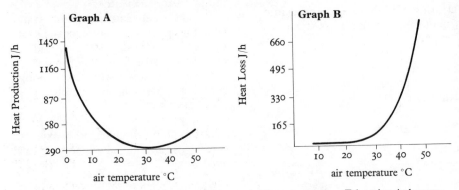

Fig. 15.4 Relationship between air temperature and **A** heat production, **B** heat loss in human.

5 Fig. 15.5 shows the results of an experiment carried out by Pugh and Edholm on two human volunteers, A and B. A is fat (i.e. has much subcutaneous adipose tissue), whereas B is thin. Both subjects had their body temperatures recorded at intervals while immersed in water at 16°C. This was done first with the subjects lying still, and then while the subjects were swimming. Interpret the results as fully as you can.

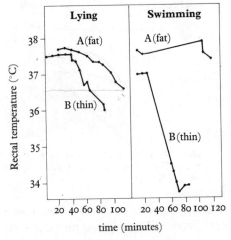

Fig. 15.5 Results of Pugh and Edholms' experiment on the body temperature of two human subjects, one thin, the other fat, immersed in water at 16°C.

6 Within a given species of animal, the average size of individuals tends to be smaller in warmer climates and larger in colder climates (Bergmann's principle), whereas the extremities, such as ears and tail, tend to be longer in warmer climates and shorter in colder climates (Allen's principle). Can you explain these generalizations? Would it be justified to call these two generalizations laws?

7 On a still sunny day it often happens that the tar on roads melts in the sun, and car roofs become too hot to touch. Comment on the fact that in these circumstances the leaves of plants are usually within 12°C of the air temperature, even in full sunlight. (*CCJE*)

8 In man there is little evidence for acclimatization to adverse temperatures. Man has mastered the climate 'by the creation of a *milieu intermediaire* rather than by physiological adjustment'. (*HE Lewis*) Discuss. (*CCJE*)

9 The graph in Fig. 15.6 shows the effect of environmental temperature on the metabolic rate of (a) a monkey and (b) a polar bear cub. Comment on these results.

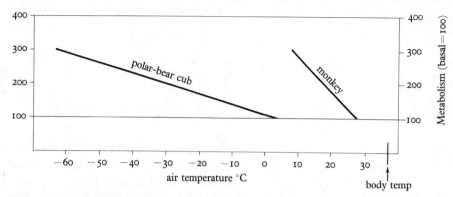

Fig. 15.6 Effect of environmental temperature on metabolic rate of monkey and polar-bear cub. Metabolism is expressed as oxygen consumption (cm^3 O_2/h).

10 The graph in Fig. 15.7 shows the oxygen consumption of a humming bird and a shrew over a period of 24 hours. Explain the differences between the two curves.

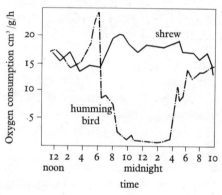

Fig. 15.7 Oxygen consumption of a humming bird and a shrew over a 24-hour period. (*After* Pearson, 1957).

11 Two groups of fish of the same species were kept for some time at different temperatures. One group was kept at 25°C, and the other at 35°C. At the end of the period of time batches of fish from each group were exposed to different temperature conditions and the number of survivors counted. The results of this experiment are shown in Fig. 15.8. Discuss them as fully as possible.

(*SA modified*)

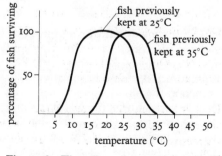

Fig. 15.8 The effect of temperature on the survival rate of two groups of fish, one previously kept at 25°C, the other at 35°C.

16 Control of Respiratory Gases

Background Summary

1   The importance of maintaining a constant level of oxygen and carbon dioxide in the blood is demonstrated by the fact that an alteration in the level of these gases results in an appropriate change of ventilation rate and cardiac frequency.

2   It has been found that carbon dioxide is the most important stimulus initiating circulatory and respiratory changes.

3   These changes are co-ordinated by **cardio-vascular** and **respiratory centres** in the **medulla** of the brain. These are informed of the level of the carbon dioxide by **chemoreceptors** located in the **carotid bodies** and elsewhere.

4   The cardio-vascular centre also responds to changes in blood pressure which are monitored by stretch receptors in the walls of the **carotid sinuses**.

5   The higher centres of the brain, a variety of reflexes, and the hormone adrenaline are also involved in the initiation of respiratory and cardio-vascular responses.

6   The effects of a slowly diminishing oxygen supply, such as occurs in ascending a mountain, are offset by **acclimatization**, a series of long-range responses to low oxygen tension.

7   A more rapid response to changing conditions is seen during and after a bout of heavy muscular exercise. The respiratory, circulatory, nervous and endocrine systems all co-operate to bring about appropriate adjustments.

8   When facing total oxygen deprivation many animals, particularly diving mammals and birds, undergo **bradycardia**: the cardiac frequency falls and blood is redistributed to the vital organs.

Investigation 16.1

Effect of lack of oxygen and accumulation of carbon dioxide on the ventilation rate in man

It is well known that during muscular exertion the rates of respiration and heart beat increase. Is this caused by a temporary decrease in the amount of oxygen in the blood, an increase in the level of carbon dioxide, or some other change resulting from muscular activity? The purpose of this investigation is to test whether a shortage of oxygen or an increase in the carbon dioxide tension is the most effective stimulus initiating appropriate adjustments. The rate of respiration is expressed in terms of the **ventilation rate**, which is the total volume of air inspired per minute.

Method and Procedure
A **spirometer** and **kymograph** are required. For a description of the apparatus and procedure for carrying out experiments *see* p. 112. The kymograph should be set at a speed of approximately 15 mm/min.

Work in pairs, one student acting as subject, the other as experimenter. In this investigation the subject should sit as relaxed as possible at all times

EFFECT OF LACK OF OXYGEN

Fill the spirometer with oxygen. The carbon dioxide absorber (canister of soda lime) should be in position, so the subject will inspire pure oxygen from the spirometer chamber. Connect the subject, who should be sitting comfortably, to the spirometer. Record his respirations on the kymograph drum for as long as possible. As the oxygen supply gets used up, changes will occur in the subject's respiratory pattern, and he may begin to feel faint. *Do not go too far!* The experimenter should watch the subject carefully and as soon as his lips or cheeks begin to go blue (cyanosed) he should disconnect the subject.

EFFECT OF REBREATHING EXPIRED AIR

Allow the subject to recover from the previous experiment. To investigate the effect of increasing the carbon dioxide level, remove the carbon dioxide absorber from the spirometer, so the subject will rebreathe his own expired air. This will become increasingly rich in carbon dioxide as the experiment progresses.

Repeat the procedure which you followed for the first experiment, but without the carbon dioxide absorber. Record the subject's respirations for as long as possible. It is important that you should take the same precautions as before: *disconnect the subject as soon as he shows the slightest signs of cyanosis.*

Requirements

Recording spirometer with pen
Kymograph set at speed of about
15 mm/min
Eosin (or non-clogging ink)
Kymograph paper calibrated for
  volume
Oxygen cylinder

(For further details *see* p. 52).

Results

Calculate the ventilation rate (*VR*) during each minute from the beginning to the end of each set of recordings. The *VR* is the total volume of air inspired per minute. It can be worked out from the fact that:

Ventilation rate = frequency × depth

where frequency is the total number of inspirations carried out during the one minute period, and depth is the average volume of air inspired at each breath during the same one minute period. Express the *VR* in dm^3/min.

Now plot your results graphically, ventilation rate for *both* experiments on one sheet of graph paper, so the curves can be readily compared. The time, in minutes, should be on the horizontal axis.

How do the results of the two experiments compare with one another? What conclusions do you draw as to the most effective stimulus initiating changes in the ventilation rate?

For Consideration

(1) For a respiratory gas such as carbon dioxide to influence the ventilation rate there must be receptors sensitive to the amount of gas in the body. Where are these receptors?

(2) You have probably found that carbon dioxide provides a very effective stimulus bringing about changes in the ventilation rate. Does this rule out the possibility that in natural circumstances, such as muscular activity, other factors besides carbon dioxide might help to bring about such adaptive changes? What might these other factors be?

Investigation 16.2

Effect of various factors on human blood pressure and pulse rate

Blood pressure depends on (a) the volume of blood expelled by the heart per unit time, i.e. the **cardiac output**, and (b) the resistance offered by the capillaries and arterioles, i.e. the **peripheral resistance**. Cardiac output and peripheral resistance themselves depend on various factors including posture and exercise.

The cardiac output is determined by the **stroke volume** (volume of blood pumped into the aorta per minute) and the **cardiac frequency** (number of heart beats per minute). The latter can be determined by measuring the pulse rate.

The speed at which blood reaches the tissues is determined by the pressure. Blood in the arteries is at high pressure, which is maintained by the pumping action of the heart. However, the arterial pressure is by no means steady, but fluctuates according to the phases of the heart beat (**cardiac cycle**). Pressure is highest when the heart contracts (**systole**), and lowest when it relaxes (**diastole**). In this investigation we shall be concerned with the maximum (systolic) pressure.

Method

Work in pairs, one of you acting as subject, the other doing the experiments. While not taking exercise the subject should be as relaxed as possible: he should sit comfortably in a chair with his arm resting on a table at about the level of the heart.

The **pulse rate** is measured by placing a finger immediately over the radial artery on the median side of the wrist (Fig. 16.1).

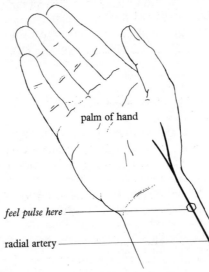

palm of hand

feel pulse here ———

radial artery ———

Fig. 16.1 How to take a subject's pulse rate.

The blood pressure can be measured in the brachial artery by means of a **sphygmomanometer**, an inflatable rubber armlet enclosed within an inextensible cloth covering. The armlet is connected to a pump and mercury manometer. The pump is fitted with a release valve to allow slow deflation of the armlet.

The experimenter should wrap the armlet snugly (but not tightly) round the subject's upper arm well above the elbow: begin with the broad end and tuck the narrow end into the turns already made. The outlet valve should be closed to allow inflation: the milled head should be turned clockwise. Feel the subject's radial pulse with your finger. Inflate the rubber armlet to a pressure of about 180 mm Hg (239×10^2 N/m^2) and notice that the pulse disappears. (Why?). Now open the release valve slowly and gradually decrease the pressure until the pulse *just* reappears. This is the **systolic pressure**.

With the subject relaxed, take several readings of the pulse pressure until you have got used to the apparatus. Also practice taking the subject's pulse rate

by counting the number of throbs in a one minute period.

EFFECT OF POSTURE ON ARTERIAL PRESSURE AND PULSE RATE

The subject should lie down quietly for five minutes after which his pulse rate and arterial pressure should be determined. The subject should then stand up for a further five minutes after which his pulse rate and arterial pressure should again be taken. Explain the difference (if any) between the two readings.

EFFECT OF BREATHING CARBON DIOXIDE

With his nose clipped, the subject should breathe in and out of an inflatable bag fitted with a mouthpiece (Fig. 16.2). The experimenter should take readings of his pulse rate and arterial pressure at one minute intervals. The subject should disconnect himself from the bag when he feels he can continue no longer, and the experimenter should go on recording his pulse rate and arterial pressure until they return to their normal values.

Plot a graph of the subject's pulse rate and arterial pressure (vertical axis) against time (horizontal axis). Explain your results.

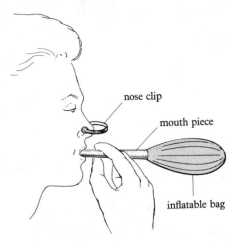

nose clip

mouth piece

inflatable bag

Fig. 16.2 For Investigation 16.2 the subject breathes in and out of an inflatable bag fitted with a mouthpiece.

EFFECT OF HYPERVENTILATION ON PULSE RATE

In this experiment it is necessary to obtain continuous readings of the subject's pulse over a period of seven minutes. This is best done by counting the number of pulses every 15 seconds and converting these to pulses per minute afterwards.

Requirements
Sphygmomanometer
Inflatable bag fitted with mouth-piece (a polythene bag connected by rubber tubing to a spirometer mouthpiece will do).

When the subject is relaxed and breathing normally, the partner should take his pulse rate continuously for two minutes. The subject should then engage in **forced breathing** for 30 seconds: breathe as deeply and frequently as possible, but with the minimum of muscular effort to the body as a whole. The subject now breathes normally for a further five minutes during which the partner records his pulse rate every 15 seconds. Does the pulse rate fall below normal at any point? Explain.

EFFECT OF EXERCISE
The subject should take severe exercise for five minutes. His partner should then determine his pulse rate and arterial pressure at one minute intervals until they return to normal. Plot the results, pulse rate and arterial pressure on the vertical axis, time on the horizontal axis. Compare with other members of your class.

For Consideration
To what extent do your results correspond to those obtained in experiment 16.1?

Questions and Problems

1 Explain as precisely as you can why:
  (a) an excess of carbon dioxide if inspired is dangerous;
  (b) the rate of an athlete's heart beat (cardiac frequency) increases just before he starts a race;
  (c) violent exercise is dangerous for elderly people;
  (d) carbon monoxide is poisonous.

2 It is commonly stated that the main stimulus responsible for the faster rate of breathing that occurs during muscular exercise is the increased carbon dioxide tension in the blood. However, it has been found that the ventilation rate achieved by a human subject during severe exercise is considerably greater than the maximum response shown by the same subject breathing carbon dioxide at rest. Explain.

3 The graph in Fig. 16.3 shows the pulse rate of an athlete during three-and-a-half months training on a bicycle joulometer (bicycle ergometer). The work load was kept constant throughout the training period at 14 715 watts. Comment on these measurements.

   How would you expect his cardiac output (volume of blood expelled from the heart per minute), respiratory frequency (inspirations per minute) and ventilation rate (volume of air inspired per minute) to change during this period? Explain fully.

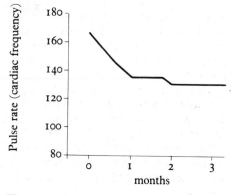

Fig. 16.3 Graph showing change in pulse rate of an athlete during three and a half months regular training on a bicycle joulometer.

4 Outline the physiological changes that take place in the human body in:
(a) performing a 100-metre sprint;
(b) a long-distance race;
(c) ascending to a height of 9,000 metres in an aeroplane without oxygen apparatus;
(d) climbing a mountain 9,000 metres high without oxygen apparatus.

5 The Olympic Games in 1968 were held in Mexico City and competing national teams gathered there for a much longer period in advance of the events than in previous Games. Discuss the necessity for this long period.

(O and C modified)

6 The following table shows the blood flow in cm^3/min to various regions of the human body while it is at rest, and also during different states of physical activity:

Region	At rest	Light exercise	Fairly strenuous exercise	Maximum exertion
Heart muscles	250	350	750	1000
Skeletal muscles	1200	4500	12 500	22 000
Kidneys	1100	900	600	250
Gut	1400	1100	600	300
Skin	500	1500	1900	600
Brain	750	750	750	750
All other regions	600	400	400	100
TOTAL	5800	9500	17 500	25 000

Explain these figures as far as you are able.

(JMB)

7 What enables certain species of whales to remain submerged under water for as much as an hour?

8 The term hypoxia usually refers to a condition in which the availability or utilization of oxygen is depressed. The data listed below illustrate four different types of hypoxia compared with the state of a 'normal' person breathing fresh room air. (Assume that the weight, sex, and age of all subjects are the same.)

Subject		Haemoglobin (g Hb per 100 cm^3 blood)	O_2 Content of Arterial Blood (cm^3 O_2 per 100 cm^3 blood)	O_2 Content of Venous Blood (cm^3 O_2 per 100 cm^3 blood)	Cardiac Output (litres per minute)
A	Normal	15	19	15	5.0
B	Hypoxia	15	15	12	6.6
C	Hypoxia	8	9.5	6.5	7.0
D	Hypoxia	16	20	13	3.0
E	Hypoxia	15	19	18	no information

(handwritten note: increased rate of breathing)

Suggest explanations of the cause of the hypoxia in subjects B to E. Subject B has an increased rate of breathing. Briefly describe the physiological mechanism that is responsible.

(SA modified)

9 When a diver returns to the surface after being submerged at a depth of about 25 metres he may suffer from decompression sickness, commonly known as 'the bends'. Bubbles of nitrogen form in his blood, restricting his circulation and causing partial paralysis. Great pain in the middle of his body causes him to bend over—hence the name given to this condition.

(a) Explain why the nitrogen bubbles are formed in his blood.

(b) What factors would be expected to increase the severity of the condition?

(c) What precautions might be taken to prevent it happening?

(d) Why is it that fat people are generally more susceptible to decompression sickness than thin people?

(e) Decompression sickness also occurs when a person ascends rapidly in an aeroplane, but the height through which he can rise before suffering from 'the bends' is very much greater than for a diver (approximately 6,000 metres as against 25 metres). Why the difference?

17 Defence Against Disease

Background Summary

1   The destruction of pathogenic micro-organisms, and/or the neutralization of toxic substances produced by them, is an important aspect of homeostasis.

2   The body of an organism is defended from pathogenic micro-organisms by preventing their entry and/or destroying them after they have entered. In both cases man has augmented the body's natural defence mechanisms with artificial ones.

3   **Preventing entry** is achieved by means of barriers, e.g. skin and clotting of blood; rejection, e.g. coughing; and destruction, e.g. lysozyme in external secretions. Artificial methods include public health measures, antisepsis, and asepsis.

4   Destruction of microbes, once they are in the bloodstream, is carried out by **phagocytosis** and the **immune response**, both functions of the white blood cells (leucocytes). The immune response involves the production of specific **antibodies** in response to microbe-borne **antigens**. The antibodies include antitoxins, agglutinins, precipitins, and lysins.

5   **Immunity** may be conferred on an animal by active or passive means depending on whether the animal is stimulated to produce its own antibodies or receives ready-made antibodies from an external source. Active artificial immunity may be conferred by injecting antigens (**vaccine**) into the body.

6   There are two hypotheses accounting for the production of antibodies: the **instructive hypothesis** and **clonal selection hypothesis**. Evidence favours the latter, which has the merit of explaining why antibodies are not formed in response to an animal's own antigens.

7   The immune reaction can be seen when blood belonging to two incompatible groups is mixed, as might occur in a blood transfusion. With respect to the ABO group system, a **universal donor** (group O) can give blood to a recipient of any group without agglutination; a **universal recipient** (group AB) can receive blood from a donor of any group without agglutination.

8   The immune reaction is also seen when Rhesus positive and Rhesus negative bloods come into contact. In certain situations this occurs during pregnancy, resulting in haemolytic disease of the newborn (erythroblastosis foetalis).

9   An unfortunate aspect of the immune response is that foreign tissue introduced into a recipient in surgical transplantation is usually rejected. Compatibility is possible only between genetically identical individuals, otherwise the normal immune response must be prevented.

10 Man's other endeavours to combat disease include the use of **chemotherapeutic agents**, e.g. penicillin and sulphonamide drugs. **Interferon**, a cell protein that prevents multiplication of viruses, may offer possibilities in the future.

Investigation 17.1

The different types of white corpuscles in human blood

Blood is not just a fluid. It is a complex circulating tissue consisting of **red** and **white corpuscles** and **platelets** suspended in a fluid medium, **plasma**. No staining is required to see the red corpuscles (*see* p. 101), but staining is necessary to distinguish the platelets and five types of white cells.

Procedure

We will stain a blood smear with **Leishman's stain** or **Wright's stain** which consist of a mixture of eosin and methylene blue. Proceed as follows:

(1) Clean two slides thoroughly with acid ethanol and dry them. Using a sterilized lancet or needle, draw some of your own blood by pricking a finger or thumb. Sterilize your skin first with ethanol.

(2) Quickly place a drop of blood at the very end of one of the slides. Invert that slide, apply it at an angle of about 45° to the surface of the other slide about a centimeter from the end, and draw it towards the other end so as to make an even smear.

(3) Dry the smeared slide by waving it in the air. (Do not heat it). Now pipette onto the smear eight drops of stain and leave for about 45 seconds. Then add to the stain eight drops of distilled water and mix by rocking the slide. Leave for six to ten minutes.

(4) Remove excess stain in a stream of distilled water for five seconds. Blot gently with clean, dust-free filter paper. When dry, the stained smear can be viewed without a coverslip, but if you wish to keep your preparation it is advisable to put on Canada balsam and a coverslip so as to protect the smear.

IDENTIFICATION OF WHITE CORPUSCLES

White blood corpuscles can be immediately distinguished from red ones by the fact that the former have a nucleus. This will have stained purple or blue. Platelets will appear as small groups of purple dots.

There are five types of white corpuscles, which can be identified by reference to Fig. 17.1. Distinguish between the **granulocyte (polymorph)** with its granular cytoplasm and lobed nucleus (which looks like a string of sausages), and the **agranulocyte** with its non-granular cytoplasm and large spherical or bean-shaped nucleus.

As can be seen from Fig. 17.1, the most common type of granulocyte is the **neutrophil** which is manufactured, along with red blood corpuscles, in the bone marrow. The most common type of agranulocyte is the **lymphocyte** which is manufactured in the nodes of the lymphatic system.

Fig. 17.1 Summary of the different types of white blood cell found in human blood. The relative proportions are given as percentages of the total number of white blood corpuscles.

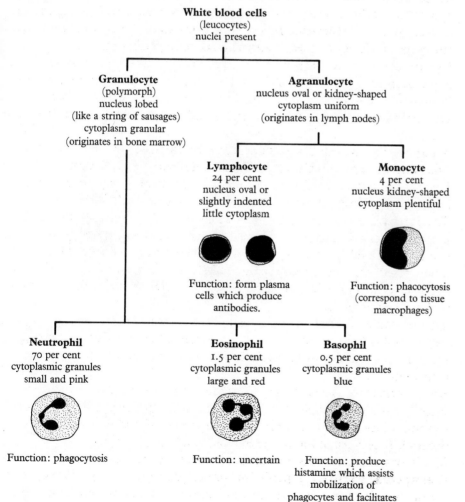

White blood cells
(leucocytes)
nuclei present

Granulocyte
(polymorph)
nucleus lobed
(like a string of sausages)
cytoplasm granular
(originates in bone marrow)

Agranulocyte
nucleus oval or kidney-shaped
cytoplasm uniform
(originates in lymph nodes)

Lymphocyte
24 per cent
nucleus oval or
slightly indented
little cytoplasm

Function: form plasma
cells which produce
antibodies.

Monocyte
4 per cent
nucleus kidney-shaped
cytoplasm plentiful

Function: phacocytosis
(correspond to tissue
macrophages)

Neutrophil
70 per cent
cytoplasmic granules
small and pink

Function: phagocytosis

Eosinophil
1.5 per cent
cytoplasmic granules
large and red

Function: uncertain

Basophil
0.5 per cent
cytoplasmic granules
blue

Function: produce
histamine which assists
mobilization of
phagocytes and facilitates
tissue repair and healing

EXAMINATION OF LYMPH NODE

Numerous **lymphocytes** and their precursors can be seen in a section of a **lymph node**. A lymph node is a swelling in the course of a lymphatic vessel. Its principal functions are to produce lymphocytes, and 'filter' the lymph, removing toxic substances from it as it flows through the node. A diagram of a generalized lymph node is shown in Fig. 17.2.

Careful examination of a stained section of a lymph node shows it to consist of an extensive network of fibres to which are attached numerous **reticulo-endothelial cells**. The latter give rise to lymphocytes by mitosis. The reticulo-endothelial cells may be distinguished from the lymphocytes by the fact that they are larger, more elongated and are capable of phagocytosis. The filtering function of the lymph node is due to the ability of the reticulo-endothelial cells to phago-cytose small particles as they flow through it.

For Consideration

(1) Which are the most numerous of the five types of white cell in your own smear? Is any type absent altogether as far as you can see? What might be the explanation of its absence?

(2) Leishman's stain and Wright's stain consist of a mixture of acidic eosin (pink) and alkaline methylene blue (blue). Can you explain why the three types of granulocyte stain differently with these stains, and why they are given their respective names?

(3) Where in the human body, apart from the bloodstream, would you expect to find white blood corpuscles or their derivatives? How do they get there?

(4) There are in the body certain lymph nodes which receive lymph from the lungs. It has been observed that the cytoplasm of the reticulo-endothelial cells in these particular nodes contain numerous carbon particles in people who live in, or near, large cities. Explain.

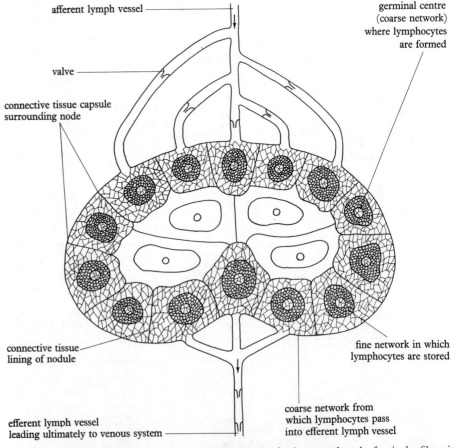

Fig. 17.2 Diagram of a lymph gland. The gland is made of a dense meshwork of reticular fibres in which lymphocytes are formed. It also filters the lymph flowing through it, removing pathogenic micro-organisms and foreign particles.

Requirements

Slides (× 2)
Large coverslip
Sterilized lancet or needle
Clean dust-free filter paper

Acid ethanol
Ethanol and cotton wool
Leishman's or Wright's stain
Distilled water
Canada balsam

Section of lymph node

Investigation 17.2

The number of white corpuscles in human blood

As with red blood corpuscles, white corpuscles can be counted by employing a sampling technique using a **haemocytometer**. Such estimations are important clinically, for an abnormally low or high white cell count may be characteristic of certain diseases.

APPARATUS

For details of the haemocytometer *see* p. 25. Examine the haemocytometer slide under low power and identify the type-**B** squares. These are the squares you should use for your white cell count. Each type-**B** square has an area of $\frac{1}{25}$ mm^2, and the volume represented by a type-**B** square is 0.004 mm^3.

The pipette for white blood corpuscles is the one with the mark 11 above the bulb. Make sure the slide and pipette are clean.

Procedure for Counting White Blood Corpuscles

(1) Obtain a large drop of blood by pricking the end of your thumb with a sterilized lancet or needle. Sterilize your skin first with ethanol. Do not squeeze out the blood too vigorously, for this will force out a lot of tissue fluid which will dilute the corpuscles.

(2) With the rubber tubing attached to the pipette, suck the blood up to the '1' mark. Quickly dry the tip of the pipette. If the '1' mark has been passed, touch the tip of the pipette with filter paper until the blood drops back to the '1' mark.

(3) Now suck up **Turk's solution** until the contents of the pipette reach the '11' mark. If the '11' mark is passed, the pipette must be emptied and the procedure started again. (Turk's solution renders the red blood corpuscles invisible (how?) and stains the white corpuscles).

(4) Remove the rubber tubing from the pipette. Close the two ends of the pipette with thumb and finger and rock the pipette for at least a minute. When the blood and Turk's solution are thoroughly mixed, blow out six drops so as to expel excess Turk's solution. Your blood sample is now diluted 1 in 10.

(5) Place the coverslip in the centre of the slide. Put one drop of diluted blood from the pipette onto the slide alongside the coverslip in the area between the two deep grooves. The blood should be drawn under the coverslip by capillary action. Wait five minutes to allow the corpuscles to settle. If the blood flows into the grooves, clean the slide and put on another drop.

(6) Place the slide under the microscope and adjust the illumination so the grid and corpuscles can be clearly seen. If the corpuscles are unevenly distributed, clean the slide and start again.

(7) Count the white blood corpuscles in at least ten type-**B** squares. Record your results by ruling out the appropriate number of squares on a piece of paper and writing the number of corpuscles in each square.

Note: In each square count all the corpuscles which lie entirely within it, plus those that are touching or overlapping the top and left hand sides. Do not include those touching or overlapping the bottom and right hand sides even if they are within the square.

(8) Calculate the average number of corpuscles in a type-**B** square. Bearing in mind the volume represented by a type-**B** square, and the dilution factor, calculate the number of white corpuscles per mm^3. A typical figure might be 6,000/mm^3. How does your result compare with this?

Assuming that the total volume of blood in the body is 5 dm^3, calculate the number of white corpuscles in the entire circulation.

For Consideration

(1) In what circumstances would you expect there to be, temporarily or permanently, an abnormally large number of white blood corpuscles in the bloodstream?

(2) In what circumstances would you expect there to be an abnormally low white cell count?

Requirements

Haemocytometer (slide, coverslip, pipette, and rubber tubing)
Sterilized lancet or needle
Lens paper

Ethanol and cotton wool
Turk's solution
Distilled water
Acetic acid
Acetone

Turk's solution made up as follows :
Distilled water (100 cm^3)
Glacial acetic acid (1 cm^3)
1 per cent aqueous solution of gentian violet (1 cm^3)

Investigation 17.3

Determination of human blood groups

The human population can be divided into four groups on the basis of the reaction between the blood of different individuals when mixed together. These groups are called **A**, **B**, **AB**, and **O**. The capital letters denote types of **antigens** present in the person's red blood cells. Corresponding **antibodies** in the plasma are designated **a**, **b**, **ab** and **o**. If a person has a particular antigen (say **A**) in his red cells, he cannot have the corresponding antibody (in this case **a**) in his plasma, otherwise agglutination will occur.

So the blood groups of different individuals can be summarized as follows:

Blood group	Type of antigens in red blood corpuscles	Type of antibodies in plasma
A	A	b
B	B	a
AB	A and B	nil
O	nil	a and b

When the bloods of different individuals are mixed, as in a transfusion, no reaction takes place provided that the recipient's blood does not contain antibodies corresponding to the donor's antigens. If it does, agglutination occurs. (Normally it does not matter if the donor's antibodies are incompatible with the recipient's antigens. Why?).

In addition to the ABO system, the red blood cells of most people contain an antigen called the **Rhesus factor** (**Rh** factor). Such people are **Rh positive**. People lacking this factor are **Rhesus negative**. The plasma of Rh negative people does not contain Rhesus antibodies but it may be induced to develop them in certain circumstances. (What circumstances?).

It is obviously of medical importance that a person's blood group should be known, so in the event of his or her requiring a transfusion blood of a compatible group can be used.

Method

An individual's blood group can be determined by mixing a sample of his blood with a series of sera each containing a reagent corresponding to a specific type of antibody. The reagents contained in the sera are **anti-A** (corresponding to type **a** antibodies), **anti-B** (corresponding to type **b** antibodies), and **anti-D** (corresponding to **Rhesus** antibodies, also called **anti-Rho**).

The reagents are contained in test panels on an **Eldoncard** (Fig. 17.3A). Each panel consists of a cellulose strip on which the specific reagents have been deposited and dried.

Procedure

(1) Using the standard pipette supplied, place a full drop of tap water on the reagent in the **anti-A** test panel. Dissolve the reagent in the water by mixing with the flat end of the plastic stick provided. Confine the mixing to a small area in the centre of the panel.

(2) Clean the stick thoroughly with cotton wool and repeat the above procedure for the **anti-B**, **anti-D** and **control** panels. Clean the stick thoroughly with fresh cotton wool between each.

A Eldoncard

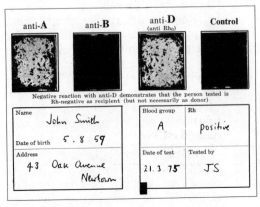

B The drop of blood on the flat end of the plastic stick should be just the right size —a hemisphere as shown below:

Fig. 17.3 The use of an eldoncard for determining blood group. (Eldoncard reproduced by permission of Nordisk Insulinlaboratorium, Gentofte, Denmark and Phillip Harris Biological Limited).

(3) Obtain a drop of blood by pricking your thumb or finger with a sterilized lancet or needle. Squeeze the blood onto the flat end of the plastic stick. The blood should form a hemisphere as shown in Fig. 17.3B.

(4) Mix the blood with the dissolved reagent in the **anti-A** panel, and then spread it over the whole panel.

(5) Clean the stick and repeat steps (3) and (4) with the **anti-B, anti-D** and **control** panels. Clean the stick thoroughly between each.

(6) Tilt the card backwards and forwards and from side to side, holding it vertically for about 10 seconds in each direction. This helps to mix the test solutions, but do not let any of them run over the boundaries of the panels.

Results

If agglutination occurs the blood sample in the test panel will form red streaks and blotches. The results can be interpreted as follows:

Agglutination in the **anti-A** panel means you belong to group **A**.
Agglutination in the **anti-B** panel means you belong to group **B**.
Agglutination in both **anti-A** and **anti-B** panels means you belong to group **AB**.
Agglutination in neither **anti-A** nor **anti-B** panels means you belong to group **O**.
Agglutination in the **anti-D** panel means you are **Rhesus positive**.
No agglutination in the **anti-D** panel means you are **Rhesus negative**.

Work out the percentage of students belonging to each blood group in your class and compare with the national frequencies given below:

Requirements
Fresh Eldoncard (sealed) (includes dropping pipette and plastic stick)
Beaker for tap water
Cotton wool for cleaning plastic stick
Sterilized lancet or needle
Ethanol and cotton wool

Blood group	England	Scotland
	% population	
O	42	52
A	46	35
B	9	10
AB	3	3
Rh-positive	84	83
Rh-negative	16	17

For Consideration
(1) Can you account for the fact that blood groups **O** and **A** are so much more common than **B** and **AB**?
(2) Can you explain why **Rh-positive** individuals are so much more common than **Rh-negative** individuals?
(3) Why the differences between England and Scotland?
(4) What is the relevance of blood groups to the mechanisms by which the body defends itself against disease?

Investigation 17.4

Effect of antibiotics on bacteria

Antibiotics are substances, produced by certain living organisms, which kill, or prevent the reproduction of, microorganisms belonging to other species, including bacteria. Such substances, which include the well-known antibiotic, penicillin, are of great medical importance. This investigation is designed to demonstrate the antibacterial action of certain antibiotics.

Procedure

(1) Obtain a stoppered bottle or test tube containing 15 cm^3 of sterile nutrient **agar**. Melt the agar by placing the bottle in a water bath at 100°C. Most agars melt at about 95°C.

(2) Remove the bottle and cool to about 45°C. The agar should remain in liquid form, but be cool enough to hold against your cheek. Most agars solidify at about 42°C.

(3) Obtain a test tube containing a culture of non-pathogenic bacteria on an agar slope. Suitable bacteria include *Escherichia coli* and *Bacillus cereus*. The tube should be kept plugged with sterile cotton wool.

(4) The object of the next step is to transfer a sample of bacteria from the agar slope in the test tube to the liquid agar in the bottle. To avoid contamination the whole operation should be carried out as quickly as possible under sterile conditions. Proceed as follows (Fig. 17.4):

Quickly sterilize an inoculation loop by heating it in a flame, then cooling it in sterile water. Remove the cotton wool bung from the test tube containing the bacterial culture, sterilize the mouth of the tube by passing it through a flame several times. Now remove the cap from the bottle of liquid agar, flame the mouth of the bottle and with the sterilized loop transfer one loopful of bacterial culture from the agar slope to the liquid agar in the bottle. Dip the loop right into the bottle so as to thoroughly mix the bacteria with the agar.

(5) Without delay pour the liquid agar containing the bacteria into a warm, sterilized petri dish. In doing this, avoid contamination by raising the lid of the petri dish only just enough to pour in the agar. (Warming the petri dish prevents water condensing on the lid as you pour in the agar.) After replacing the lid, ensure even distribution of the agar by gently moving the petri dish from side to side on a flat surface.

(6) After the agar has hardened, for each antibiotic you wish to test place on the surface of the agar a small disc of filter paper, which has been soaked in the **antibiotic** then allowed to dry. The disc should be transferred with sterile forceps. Suitable antibiotics for investigation include **penicillin** and **aureomycin**. A control disc, i.e. one that has been soaked in water then allowed to dry, should also be placed on the agar.

A convenient alternative to the above procedure is to use a '**multodisk**'. This consists of a central con-

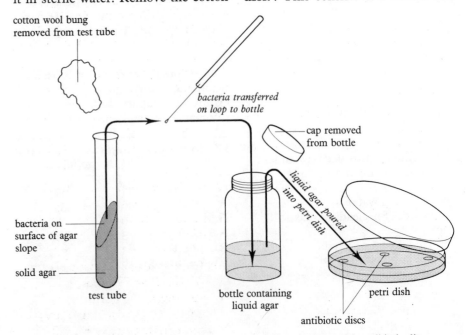

cotton wool bung
removed from test tube

*bacteria transferred
on loop to bottle*

cap removed
from bottle

*liquid agar poured
into petri dish*

bacteria on
surface of agar
slope

solid agar

test tube

bottle containing
liquid agar

petri dish

antibiotic discs

Fig. 17.4 Preparation of bacterial culture in petri dish for treatment with antibiotic discs.

trol disc with side-arms leading to small discs each impregnated with a different antibiotic.

(7) Turn the petri dish upside down to avoid water of condensation accumulating on the surface of the agar. Incubate at 37°C. Examine at intervals over several days for bacterial growth.

Results
Draw a diagram of the agar surface showing the areas where bacterial growth has, or has not, occurred. How do these areas relate to the positions of the antibiotic and control discs?

Further Experiments
VIEWING OF BACTERIA
Should you wish to examine your bacteria under the microscope proceed as follows:

Make a thin smear of the bacterial culture on a microscope slide. Onto the smear pipette a few drops of polychrome methylene blue or some other suitable bacterial stain. Leave for the appropriate time (about one minute in the case of methylene blue). Wash the slide in a gentle stream of water. Dry with blotting paper. Examine under high power, preferably with oil immersion.

THE SOURCE OF PENICILLIN
Penicillin is obtained from the fungus *Penicillium*. With a sterile needle mix spores of *Penicillium notatum* with a drop of water on a slide.

Obtain a sterile petri dish of solid agar. Raise the lid and with a sterile needle make a smear of your spore suspension across the agar to one side of the dish (Fig. 17.5). Incubate for 2–3 days to allow the fungus to establish itself.

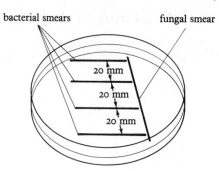

Fig. 17.5 Plating of fungal and bacterial smears.

Two-to-three days later inoculate the agar with several different types of bacteria, e.g. *Bacillus cereus*, *Escherichia coli*, *Serratia marcenscens*, and *Sarcina lutea*. Using a sterile inoculation loop, smear the bacteria onto the agar at right angles to the fungal smear as shown in Fig. 17.5. Do not inoculate more than four types of bacteria, leaving at least 20 mm between each.

Incubate at 37°C for several days and examine at intervals. Interpret your results.

For Consideration
(1) Why was it necessary to ensure sterile conditions in this investigation?
(2) When, by whom, and in what circumstances, was penicillin discovered?
(3) Suggest hypotheses to explain how antibiotics (in general) might prevent the development of bacteria.
(4) What other agents, besides those tested in this investigation, might be effective against bacteria? How would you test their efficiency?
(5) Antibiotics have been extracted from various species of fungus and soil micro-organisms. Do you think they have any use to the organisms which produce them?

Requirements
Microscope
Water bath at 100°C
Incubator at 37°C
Bunsen burner or spirit lamp
Petri dish of diameter 88 mm (× 2)
Inoculation loop
Forceps

Sterile nutrient agar (15 cm³) in stoppered bottle
Sterile water
Methylene blue and/or other bacterial stains
Antibiotic discs (e.g. penicillin, aureomycin) or 'multodisk'

Pure culture of non-pathogenic bacteria, e.g. *Escherichia coli* or *Bacillus cereus* on agar slope in test tube
Spores of *Penicillium*

To make inoculation loop
Insert the end of a 3 cm length of platinum or 'nichrome' wire into the end of a glass rod softened over a bunsen flame. The loop itself should have a diameter of approximately 2 mm.

To prepare antibiotic discs
Cut out a disc of filter paper and soak in a solution of the antibiotic. Dry and store in a sealed bottle.

Questions and Problems

1 Write short explanatory notes on lysozyme, rhesus factor, interferon, and penicillin.

2 Explain the difference between:
(a) antisepsis and asepsis;
(b) fibrinogen and fibrin;
(c) active and passive immunity;
(d) homografts and heterografts.

3   In 1952 two patients, whom we can call Mr A and Mr B, were in the same ward in University College Hospital, London. Both appeared to suffer from haemophilia, a condition in which the blood takes an abnormally long time to clot. At this time haemophilia was known to be caused by the absence of a chemical factor, anti-haemophilic globulin, necessary for the conversion of disintegrated platelets into thromboplastin.

In treating Mr A it was found that the normal clotting time could be restored by injections of anti-haemophilic globulin; however, Mr B did not respond to this treatment. It was further found that transferring plasma from Mr B into Mr A's bloodstream restored Mr A's clotting time to normal, and transferring plasma from Mr A to Mr B restored Mr B's clotting time to normal.

What conclusions could be drawn from these observations?

In transferring plasma from one patient to the other the doctors did not need to ensure that both shared the same blood group. Why was this unnecessary?

4   Give a brief account of how antibodies are thought to be produced in the mammalian body. Bearing in mind that antibodies are too small to be seen under a microscope, how could you determine the relative amounts of a specific antibody in an individual at different times?

Fig. 17.6 shows the relative amounts of a specific antibody formed in the bloodstream after (a) a first injection of an antigen, and (b) a later injection of the same antigen. Discuss the difference between the two curves.

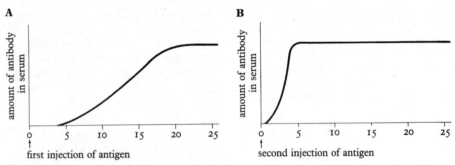

Fig. 17.6   Graphs showing the amounts of an antibody formed in the bloodstream after (A) first injection of an antigen, and (B) after a second injection of the same antigen. The time scale on the horizontal axis is in days.

5   Why is transplantation surgery beset with so many difficulties?

6   Here are the results of three experiments on skin transplantation in mice:
Experiment 1: A piece of skin from mouse A is transplanted to mouse B. After 11 days the transplant (graft) is rejected (sloughed off).

Experiment 2: A second piece of skin from mouse A is then transplanted to mouse B. This second graft is rejected after six days.

Experiment 3: A piece of skin from mouse C is transplanted to mouse D, the latter having received injections of cellular material from mouse C before birth. The graft is not rejected.
Answer the following questions as concisely and clearly as you can:
(a) Explain the results of experiments 1 and 2.
(b) Put forward one or more hypotheses to explain the result of experiment 3.
(c) By what means, apart from the procedure outlined in experiment 3, might rejection of grafts be prevented?
(d) What would you conclude about mice A and B if no rejection had occurred?

(O and C modified)

7   Do not be put off by the length of this question; it is not as bad as it looks.

In mammals, including man, the thymus gland is prominent at the time of birth, but gradually diminishes after birth. Removal of the thymus (thymectomy) from a four-week-old mouse produces no harmful consequences, but removal of the thymus at birth results in a wasting disease: illness with a much retarded growth rate sets in after about five weeks and death occurs by the eighth week; the lymph nodes are severely reduced in size and there is an abnormally small number of lymphocytes in the blood; the immune response does not develop, plasma cells and antibodies failing to be produced. Before reading further, suggest hypotheses to explain the results of this experiment.

In an attempt to discover the role of the thymus gland the following experiment was carried out by R. H. Levey:

Special plastic capsules were prepared. Each capsule was filled with thymus tissue obtained from a newborn mouse. The walls of the capsules contained pores too small to let cells through, but large enough to let through all chemical products of the cells.

Five groups of mice were treated as follows:

Group A: thymectomized at birth, capsules filled with thymus tissue implanted under skin.
Group B: thymectomized at birth, no capsules implanted.
Group C: thymectomized at birth, empty capsules implanted under skin.
Group D: thymectomized at birth, uncapsulated thymus tissue implanted under skin.
Group E: unthymectomized (normal mice).

The results, in summarized form, were as follows:

Group A: normal development: no wasting disease, normal lymph nodes, and lymphocyte count, normal immune response develops with full production of plasma cells and antibodies.
Group B: wasting disease.
Group C: wasting disease.
Group D: exactly like group A.
Group E: exactly like group A.

Now answer the following questions:
(a) Suggest a full explanation for these results.
(b) What is the point of setting up groups B to E?
(c) How do you think the experimenter tested the immune responses of the mice?
(d) It was known before these experiments were performed that at birth lymphocytes migrate in large numbers from the thymus gland to the lymph nodes. Is this observation compatible with the results of the experiments described above?

18 Nervous and Hormonal Communication

Background Summary

1 The **nervous system** provides the fastest means of communication within the body. In most animals the nervous system consists of a **central nervous system** (**CNS**) and **peripheral nerves**.

2 **Nerve cells** (**neurones**) vary in their structure but in general they possess a cell body (centron) from which arise a variable number of dendrons and dendrites, together with one or more axons. The axon generally has a **myelin sheath**.

3 Nerve cells are broadly classified into **sensory** (**afferent**), **intermediate** (also known as **connector** and **internuncial**), and **motor** (**efferent**).

4 By recording the potential difference between the inside and outside of giant axons of the squid, it has been shown that at rest the inside is negative with respect to the outside (**resting potential**), but during passage of a nerve impulse this situation is momentarily reversed (**action potential**).

5 The resting and action potentials can be explained by the distribution of ions on the two sides of the nerve membrane. At rest Na^+ ions are actively extruded (**sodium pump mechanism**), but during passage of the impulse the membrane becomes depolarized and Na^+ ions enter the axon.

6 The size of an action potential is independent of the strength of stimulation (**all-or-nothing law**).

7 For a brief period immediately after it has transmitted an impulse, the axon is totally inexcitable (**absolute refractory period**). This is followed by a slightly longer period during which it is partially excitable (**relative refractory period**).

8 Transmission speeds vary from 0.5 m/s to over 100 m/s. High speeds of transmission are achieved by having a **myelin** (**medullary**) **sheath** with **nodes of Ranvier** (vertebrates), or **giant axons** (certain invertebrates).

9 Contiguous nerve cells are connected by **synapses**. Typically, a nerve cell is covered with hundreds of **synaptic knobs**. Transmission at the synapse is achieved by a chemical transmitter substance, **noradrenaline** or **acetylcholine**, being liberated into the synaptic cleft. In sufficient quantity, this depolarizes the membrane of the post-synaptic cell. The nerve-muscle (neuromuscular) junction operates in the same way.

10 Important properties of synapses include **spatial** and **temporal summation** (**facilitation**), **inhibition** and **fatigue** (**accommodation**). Synapses account for the actions of many drugs and poisons and they play a major part in control processes within the CNS.

11 Nerve cells are frequently organized into **reflex arcs**. A typical reflex arc consists of receptors, sensory (afferent), intermediate, and motor (efferent) neurones, and effectors. Reflex arcs provide the anatomical basis of **reflex action**. Successive reflex arcs are interlinked by longitudinal neurones.

12 The vertebrate CNS consists of **brain** and **spinal cord**. The brain is subdivided into **forebrain**, **midbrain** and **hindbrain**, the forebrain being further divided into **endbrain** and **'tweenbrain**. These fundamental divisions are readily seen in primitive vertebrate-like fishes where they are associated with comparatively rigid localization of function. Such distinctions are less well seen in the mammalian brain, particularly on the functional level.

13 In primitive vertebrates integrative functions, including motor co-ordination, are carried out by the hindbrain, the forebrain and midbrain being mainly for sensory relay. In mammals integrative functions are performed by the forebrain.

14 The **peripheral nerves** are divided into two types: **spinal** and **cranial**. The former follow a segmental pattern, the latter are less regular due to the development of specialized head structures. The basic arrangement of vertebrate cranial nerves is best seen in the dogfish.

15 Involuntary activities are controlled by the **autonomic nervous system**. This is subdivided into the **sympathetic** and **parasympathetic** systems whose actions are for the most part antagonistic.

16 Certain primitive invertebrates, notably coelenterates, have a **nerve net** and lack any trace of a CNS. Such primitive nervous systems show **interneural** and **neuromuscular facilitation**. Transmission speeds are slow.

17 The first glimpses of **through-conduction** appear in sea-anemones, but this phenomenon is more fully developed in higher invertebrates such as the earthworm where it is associated with the development of **giant axons**.

18 The brain is an integral part of the **head**. In the evolution of the animal kingdom the head becomes progressively more elaborate (**cephalization**). Starting as little more than a sensory relay centre, the brain gradually takes over the functions of integration and motor co-ordination.

19 **Hormones**, chemical messengers secreted into the bloodstream by **endocrine organs** (**ductless glands**), can be compared with nerve impulses. They provide an additional means of communication within the body.

20 The basic requirement of any endocrine organ, namely a close association between the secretory cells and bloodstream, is admirably illustrated by the thyroid gland. This secretes thyroxine which controls the basal metabolic rate.

21 There is a close structural and physiological connection between the endocrine and nervous systems. This is well demonstrated by the adrenal medulla and pituitary gland. The latter contains **neurosecretory cells** which, in addition to transmitting nerve impulses, produce hormones which flow along their axons.

Investigation 18.1

Microscopic structure of nervous tissue

The principal function of nerve cells is to transmit messages around the body. In order to transmit the correct messages each nerve cell must be connected with adjacent nerve cells. It must also have the means to transmit messages quickly over long distances where appropriate. Keep these functions in mind as you examine nervous tissue under the microscope.

Procedure

THE CELL BODY OF A MOTOR NEURONE

Examine a transverse section of mammalian spinal cord. Locate the cell body of a motor neurone in the ventrolateral region of the grey matter (see Fig. 18.5) and examine under high power. **Note**: **cell body (centron)** with nucleus, Nissl's granules in cytoplasm, **dendrons** possibly showing dendritic branches, beginning of long **axon** (Fig. 18.1). Silver preparations may show the swollen ends (terminal buttons) of dendrites of neighbouring nerve cells in contact with the membrane of the cell body: these are synapses.

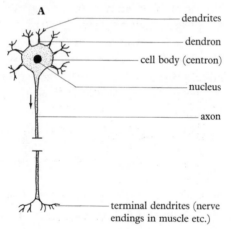

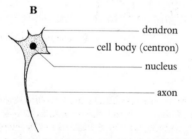

Fig. 18.1 Motor neurone in grey matter of spinal cord: **A**, diagram; **B**, as the neurone appears in a typical section.

MYELINATED AXONS

Dissect out part of the sciatic nerve from the leg of a frog (see p. 209). With needles tease out one end of the nerve on a slide, add a drop of salt solution (0.75 per cent) and put on a coverslip. Note nerve fibres (**axons**) which have a double contour on account of their **myelin (medullary) sheath**. Irrigate with a drop of osmic acid; after 15 minutes observe **myelin sheath**, **nodes of Ranvier**, nuclei of **Schwann cells**.

Examine a prepared longitudinal and transverse section of a whole myelinated nerve, e.g. sciatic. Use Fig. 18.2 to help you interpret its structure.

The axons of such nerves are responsible for innervating effectors like skeletal muscle. For example, what particular muscles are innervated by axons of the sciatic nerve? Examine a prepared longitudinal section of skeletal muscle showing **nerve endings**. (Fig. 18.3). What do you know of the ultra-structure of such nerve endings?

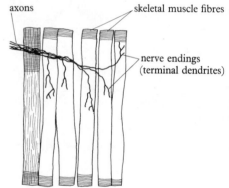

Fig. 18.3 Longitudinal section of skeletal muscle showing nerve endings.

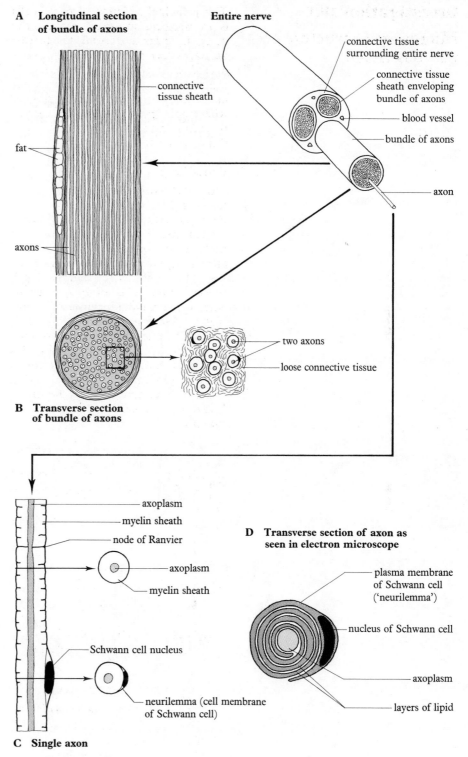

A   Longitudinal section of bundle of axons

Entire nerve

connective tissue surrounding entire nerve

connective tissue sheath enveloping bundle of axons

blood vessel

bundle of axons

connective tissue sheath

fat

axons

axon

two axons

loose connective tissue

B   Transverse section of bundle of axons

axoplasm

myelin sheath

node of Ranvier

axoplasm

myelin sheath

D   Transverse section of axon as seen in electron microscope

plasma membrane of Schwann cell ('neurilemma')

nucleus of Schwann cell

Schwann cell nucleus

axoplasm

layers of lipid

neurilemma (cell membrane of Schwann cell)

C   Single axon

Fig. 18.2   Microscopical structure of myelinated nerve. **A–C** are based on light microscope preparations, **D** on an electronmicrograph. In **D** note that the meylin sheath is multi-layered, the layers of lipid being formed by the Schwann cell wrapping itself round the axon as shown. In the electron microscope the axoplasm is seen to contain numerous longitudinally orientated microtubules, known as neurotubules, which are thought to assist transport of materials from the nerve cell body to the far end of the axon.

PYRAMIDAL CELLS IN CEREBRAL CORTEX

Examine a vertical section of the cerebral cortex and notice the layer of **pyramidal cells**, so called because of their characteristic pyramid shape (Fig. 18.4A). Observe a pyramidal cell under high power. Numerous **dendrites** towards the surface connect with other nerve cells. Its long **axon** extends via the lower parts of the brain to the spinal cord where it joins the descending motor tracts taking impulses to effectors.

A   Pyramidal cell in cerebral cortex

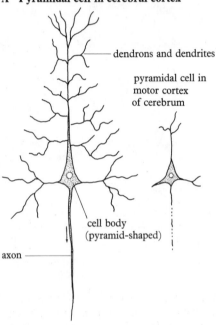

dendrons and dendrites

pyramidal cell in motor cortex of cerebrum

cell body (pyramid-shaped)

axon

B   Purkinje cell in cerebellum

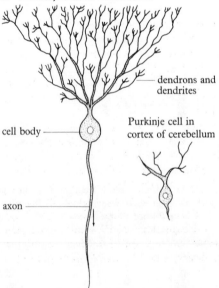

dendrons and dendrites

cell body

Purkinje cell in cortex of cerebellum

axon

Fig. 18.4 Two types of nerve cell from the brain. To the right of each diagram is the cell as it actually appears in a typical microscopic preparation.

PURKINJE CELLS IN CEREBELLAR CORTEX

In a vertical section of the cerebellum, observe **Purkinje cells** (Fig. 18.4B). Note numerous **dendrites** ramifying towards surface, and long **axons** extending downwards. The cerebellum controls fine movements, impulses in the Purkinje cells initiating or inhibiting motor activities. They receive, via their dendrites, impulses from other parts of the CNS, including the cerebral cortex.

SPINAL CORD

Return to the transverse section of the spinal cord and examine it under low or **medium power**. Identify as many of the structures shown in Fig. 18.5 as you can. Nerve cell bodies are confined to the central **grey matter**; longitudinal axons of ascending and descending tracts to the peripheral **white matter**. The small **central canal** is a reminder of the fact that the CNS of all chordates is hollow. This contrasts with the solid nerve cord of invertebrates such as the earthworm (*see* p. 228).

Note the **meninges** surrounding and protecting the spinal cord: thick **dura mater** and thin **pia mater**, separated by the vascular **arachnoid**.

NERVOUS AND HORMONAL COMMUNICATION · 177

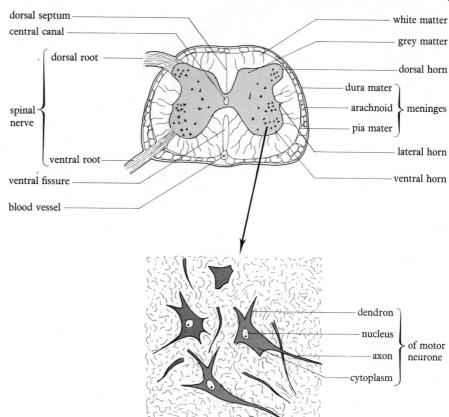

Fig. 18.5 Microscopical structure of spinal cord (based on cat). Above, transverse section of spinal cord; below, detail of grey matter (ventral horn).

Requirements

Slide and coverslip

Salt solution (NaCl, 0.75 per cent)
Osmic acid (1 per cent)

TS spinal cord
LS myelinated nerve
TS myelinated nerve
LS skeletal muscle with nerve ending
VS cerebral cortex
VS cerebellar cortex
Frog's leg in Ringer's solution

Investigation 18.2

The nerve impulse and reflex action in the earthworm

Nerve impulses can be recorded, and the properties of transmission investigated, using a **cathode ray oscilloscope**. In the present experiment you will record impulses from **giant axons** in the ventral nerve cord of the earthworm. These impulses are responsible for eliciting the earthworm's escape response, a rapid contraction of the longitudinal musculature which, in normal life, draws the worm into its burrow.

The earthworm commends itself for this purpose for two reasons. First, impulses recorded from the giant axons are unusually large and relatively little amplification is required to make them show up distinctly on the oscilloscope screen. Secondly, the giant axons are intermediate neurones in a relatively simple reflex arc, whose properties can be easily investigated.

Apparatus

OSCILLOSCOPE
The cathode ray tube contains a fila-

ment which, when heated, emits a stream of electrons. These are focused to form a narrow beam travelling towards the fluorescent screen. At the point where the beam strikes the screen a spot of light is formed. This can be moved from side to side by applying voltages to plates on either side of the beam: the 'X' deflection plates. The beam can be moved vertically up and down by plates above and below it: the 'Y' deflection plates.

To observe the small and rapid potential changes that occur during transmission of a nerve impulse, the spot is swept rapidly at constant speed from left to right forming the time base, and the action potentials, greatly amplified, are applied to the 'Y' plates. Every time this happens the spot is deflected vertically. If the spot is traversing the screen at a sufficiently high frequency, each action potential appears as a momentary stationary wave on a horizontal line of light across the screen.

The length of the wave depends on

the frequency of the time base; its height depends on the amount by which the action potential is amplified before it is applied to the 'Y' plates.

The controls which you may need to alter are as follows:

Brilliance: controls the brightness of the spot on the screen.

Focus: controls the sharpness of the spot.

Y shift: moves the time base up and down; set it so the trace is set in the middle of the screen.

X shift: moves the time base from side to side; set it so the leading edge is at the extreme left of the screen.

X gain: controls the length of the time base; set it so the time base just fits onto the screen.

Variable time/cm: controls the speed and frequency of the time base. You will want to alter this in the course of your experiments: start by setting it so that separate spots are just not discernible, the time base appearing as a continuous line across the screen.

Volts/cm: controls the amplification (Y gain); set it so the action potentials are approximately 2 cm high.

RECORDING ELECTRODES
A pair of platinum electrodes, their ends bent upwards to form hooks, should be clamped to a stand and connected by coaxial cable to the input of the oscilloscope.

Making the Preparation
Lightly anaesthetize an earthworm with, e.g. MS 222 solution. Leave it in the solution for long enough to stop it responding violently to manipulation. Proceed as follows:

(1) With the worm on a piece of cork, its ventral surface towards you, hold the worm firmly about 1 cm behind the clitellum ('saddle').

(2) With á pair of small scissors, make a *mid-ventral* slit through the body wall about 1.5 cm long. The front end of your slit should be about 1 cm behind the back of the clitellum. Be careful not to insert your scissors too far as the nerve cord is immediately beneath the body wall.

(3) Pin the worm, ventral surface uppermost, to the cork as shown in Fig. 18.6A.

(4) Separate the body wall from the underlying septa (the connective tissue partitions between adjacent segments). To do this, grasp the body wall with forceps and cut the septa with small scissors or a sharp scalpel.

(5) Pin back the body wall on either side, so the nerve cord can be seen lying on the ventral side of the intestine (Fig. 18.6B).

A Opening up the worm

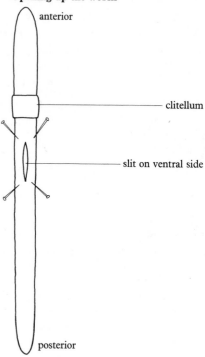

B Deflecting the body wall

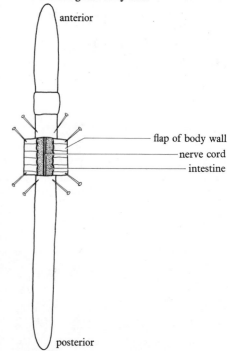

Fig. 18.6 Preparing an earthworm for recording nervous impulses from the giant axons in the ventral nerve cord.

C The final set-up

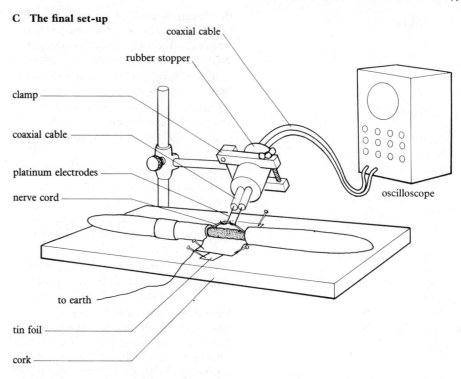

(6) Free the nerve cord from the underlying intestine by lifting it gently with a seeker and cutting through any connective tissue beneath it.

(7) Gently lift the nerve cord onto the hooked recording electrodes (Fig. 18.6C). The cord must not be stretched too much, but it should be lifted clear of the gut. There must be no film of water between the nerve cord and gut. If there is, blow it away.

Keep your preparation, particularly the nerve cord, moistened with Ringer's solution.

(8) Finally, earth the preparation by sliding a piece of metal foil under the operated region of the worm and connecting this to, e.g. a gas tap.

(9) Test your preparation by touching the anterior end of the worm with the point of a *wooden*-handled mounted needle. Set the gain of the oscilloscope (volts/cm) so the action potentials recorded from the giant axons are about 2 cm high. Set the time base frequency (time/cm) so the width of each impulse is approximately 1 mm.

Experiments

It is best to work in pairs, one student stimulating the worm, while the other examines the action potentials on the oscilloscope screen.

PROPERTIES OF THE ACTION POTENTIALS

Increase the time base frequency so as to examine the wave form of the action potentials recorded by touching (a) the anterior, and (b) the posterior end of the worm. With a pair of dividers, determine the exact height and width of the recorded action potentials. From the known amplification (gain) and time base frequency, calculate the magnitude in millivolts, and duration in milliseconds, of the action potentials.

RESPONSES TO TACTILE STIMULI OF DIFFERENT INTENSITIES

Re-set the time base frequency so the width of the action potentials is about 1 mm. With a needle touch the skin (a) very lightly, and (b) more sharply. How many action potentials are recorded in each case? Is there a correlation between the intensity of tactile stimulation and the number of impulses generated? Explain fully.

RESPONSES TO REPETITIVE TACTILE STIMULATION

Touch one end of the worm repeatedly at half-second intervals and observe the response to each stimulus: one of you should do the stimulating and the partner should watch the screen, counting and recording the number of action potentials elicited by each stimulus. Explain your results.

Requirements

Cathode ray oscilloscope
Mounted platinum electrodes
  connected to coaxial cable
Stand and clamp for above
  (*see* Fig. 18.6C)
Small dissecting instruments
Needle with *wooden* handle
Camel-hair brush
Pins
Sheet of cork (approx.
  10 cm × 18 cm)
Piece of metal foil (approx.
  1 cm × 3 cm)
Source of bright light
Pair of dividers

Earthworm Ringer's solution
  (*see* p. 437)
Anaesthetic (MS 222, 1 g in
  400 cm³ Ringer's solution
  recommended)
Range of solutions of different pH

Live earthworms in dish of
  Ringer's solution

Clearly a 'block' develops on the afferent side of the reflex. What experiments could you perform to locate the precise point in the reflex where this block develops? What do you think might cause the development of such a block?

Allow the worm to recover and repeat the experiment noticing, this time, the muscular responses which are given to each volley of impulses. Explain your observations.

DIFFERENCES BETWEEN ANTERIOR AND POSTERIOR STIMULATION

Are the action potentials recorded by stimulating the anterior end of the worm identical with those recorded by stimulating the posterior end? If not, how exactly do they differ from each other? Can both types of action potential be recorded together if the anterior and posterior ends of the worm are stimulated simultaneously? What sort of action potentials are recorded if the middle region of the worm (e.g. just behind the clitellum) is stimulated? What conclusions about the structural organization of the nervous system can you draw from your observations? How might you confirm them?

RESPONSES TO DIFFERENT TYPES OF STIMULATION

Perform experiments to find out if impulses can be elicited in the giant axons by other kinds of stimuli, e.g. gentle stroking with a camel hair brush, intense light, vibration, air currents, change in pH, chemicals. Explain your results.

For Consideration

What light is thrown by this investigation on (1) the nature of the nerve impulse, (2) the functioning of receptors, (3) the anatomy of the earthworm's nervous system, and (4) the behaviour of earthworms in their natural environment?

Investigation 18.3

Spinal reflexes in frog and man

Many of our activities depend on **reflexes**. Most of these involve the brain, but some of them use only the spinal cord and will be given even when the brain is completely inactivated. If an animal's brain is destroyed whilst the spinal cord is left intact, its behaviour and reflexes can be examined in order to determine how much of the animal's activity is independent of the brain and controlled only by the spinal cord. Any reflexes observed are called **spinal reflexes**.

Procedure

You should be given a pithed frog, i.e. one whose brain has been destroyed. First allow it to recover from spinal shock which always follows sudden destruction of the brain. Then suspend the frog by the lower jaw by pinning it to a cork held in a clamp on a retort stand. Note the position of the limbs: they take up a definite *posture*, indicating that some tone is present in the flexor muscles. **N.B.** You may

find that for some observations it is best to have the frog lying ventral side downwards on a flat surface. The reflexes *fatigue* very rapidly, so do not stimulate too frequently.

SPINAL REFLEXES OF FROG

(1) **Flexor reflex**: Gently pinch one of the webbed feet with a pair of forceps and note what happens. The extent of the response depends on the intensity of stimulation: start by pinching very gently and then repeat with a firmer pinch.

Reconstruct the sequence of events which takes place in this reflex starting with the **receptor** and finishing with the **effector**.

(2) **Scratch reflex**: Place on one flank a small piece of filter paper (about 2 mm²) soaked in acetic acid (5 per cent). The foot of the same side will probably be raised to rub off the irritant. What happens if that foot is held down? Explain your observations.

(3) **Crossed-extensor reflex**: This is a difficult reflex to induce, but you may be lucky. If one leg is flexed slightly the other may extend. This reflex fatigues very rapidly and is, therefore, difficult to demonstrate. However, when it does work it shows that crossed-extension does not require an intact brain.

Crossed-extension is the basis of locomotion by alternate flexion and extension of first one leg and then the other. This can be demonstrated by gently pulling your spinal frog along the top of the bench by its arms so that its legs drag along the bench. Unless the preparation is badly fatigued the hind legs will start 'walking'. Although a spinal frog will show this activity it is better demonstrated by a toad. Why?

SPINAL REFLEXES OF MAN
In man there are a number of reflexes which only involve the spinal cord. Two of these can be demonstrated quite easily.

(1) **Knee jerk**: One of you should sit with the right thigh crossed loosely over the left knee in such a way as to slightly stretch the extensor muscle of the leg. If your partner now taps the right-knee tendon (just below the knee cap), a sharp extension of the leg results.

This is one of the simplest reflexes known in the human. It involves the lumbar region of the spinal cord. Reconstruct the reflex arc involved and trace the sequence of events which takes place in the course of the reflex.

(2) **Ankle jerk**: Kneel on a chair and let one foot dangle loosely. If your partner now taps the tendon at the back of the foot there should be a sudden extension of the foot. This reflex involves the sacral part of the spinal cord. Reconstruct the reflex arc involved.

Repeat stimulating at approximately two taps per second. Does the response decline or disappear on repetition? Explain.

For Consideration
(1) From your observations on the behaviour of a spinal frog, what part does the brain play in locomotion?
(2) What experiments could you do to demonstrate the role of different regions of the brain in reflex action and locomotion in a primitive vertebrate like the frog?
(3) How does the part played by the CNS in locomotion of the frog compare with an earthworm, insect, and mammal?
(4) What other reflexes (spinal or otherwise) are shown by the human? Speculate on the part(s) of the CNS involved in each case.

Requirements
Stand and clamp
Cork
Pins
Forceps
Filter paper

Acetic acid (5 per cent)

Pithed frog (only the brain should be destroyed)

Investigation 18.4
Dissection of the cranial nerves and brain of the dogfish

The dogfish is a good animal in which to dissect the brain and cranial nerves. Being a comparatively primitive vertebrate, the brain is readily divisible into its four fundamental regions (endbrain, 'tweenbrain, midbrain, and hindbrain), and the pattern of cranial nerves is relatively simple. Moreover, since the skeleton is entirely cartilaginous, the brain and nerves are easy to get at. The arrangement of the cranial nerves is fundamentally similar to the mammal, though of course there are many detailed differences.

The dogfish has ten cranial nerves, numbered I to X, some of which have numerous branches.

Procedure

REMOVAL OF THE EYE AND EXAMINATION OF THE ORBIT: IDENTIFICATION OF NERVES II–VII

(1) Skin the top and left side of the head from the snout to the most posterior gill slit.

(2) Cut away the conjunctiva and loose connective tissue surrounding the **eye** in the orbit (eye socket), taking care not to cut any nerves, particularly on the ventral side.

A   Removal of eye

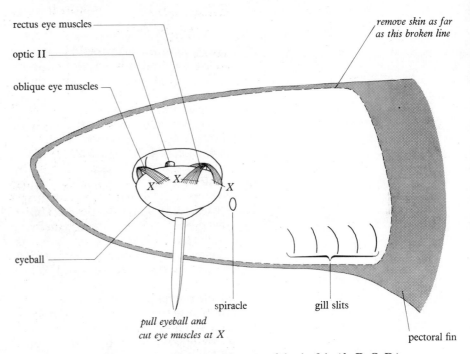

rectus eye muscles

optic II

oblique eye muscles

*remove skin as far
as this broken line*

eyeball

X   X   X

spiracle

gill slits

pectoral fin

*pull eyeball and
cut eye muscles at X*

Fig. 18.7   Steps in the dissection of the cranial nerves of the dogfish. (**A, B, C, D.**)

(3) Grip the front of the eye with large forceps and pull it towards you. Peer down behind the eyeball and identify the **eye muscles** and **optic nerve (II)** (Fig. 18.7A). Cut all six eye muscles, and the optic nerve, as close to their attachment to the eye as possible. *Be careful not to cut any other structures.* Lift the eye from the orbit.

(4) In removing the eye, you will have cut into the orbital blood sinus (part of the anterior cardinal vein) which lies immediately behind it. There will, therefore, be blood in the orbit. Wash this away and identify all the structures shown in Fig. 18.7B. The nerves emerge from holes (foramina) in the cartilage at the back of the orbit and then run across the back, sides, or floor of the orbit to their destination. To see **pathetic (trochlea) IV** it is necessary to look behind the superior oblique eye muscle; and to see **hyomandibular VII** you must look behind the posterior (external) rectus eye muscle. **Ophthalmic V** and **VII** may be a little difficult to see: they run in grooves in the cartilage at the back of the orbit. **Oculomotor III** will be easy to see, but you will probably see little or nothing of **abducens VI** since it usually enters the posterior rectus muscle before emerging from the back of the orbit.

DISTRIBUTION OF NERVES V AND VII

(5) You have identified cranial nerves II–VII. It is now necessary to trace nerves V and VII (trigeminal and facial) to their destinations.

Start with **ophthalmic V** and **VII**. By slicing horizontally, remove the cartilage lying above these two nerves. In so doing you will expose the brain. Anteriorly the ophthalmic nerves come very close to the surface, so be careful not to cut them. They lie immediately above the olfactory organ and serve the snout (function?). Cut away the cartilage covering the olfactory organ and olfactory lobe of the brain. The **olfactory nerve (I)** connects the olfactory lobe with the olfactory organ.

(6) Next follow out **maxillary** and **mandibular V** and **buccal VII**. What structures do they serve and what are their functions? **Palatine VII** need not be traced out any further. It serves the palate.

(7) Now explore the distribution of **hyomandibular VII**. To do this examine the side of your fish just behind the spiracle and notice two superficial nerves very near the surface (Fig. 18.7B). These are **external mandibular** and **hyoidean VII**, both branches of hyomandibular VII. Note that dorsally these two nerves unite just before plunging into the muscle behind the spiracle. Because of its

Fig. 18.7 B Identification and dissection of nerves in the orbit

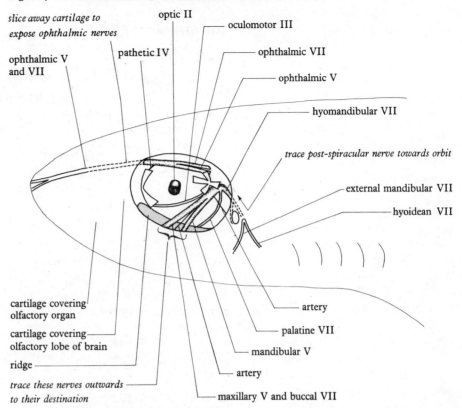

slice away cartilage to expose ophthalmic nerves

optic II

oculomotor III

ophthalmic VII

ophthalmic V

pathetic IV

ophthalmic V and VII

hyomandibular VII

trace post-spiracular nerve towards orbit

external mandibular VII

hyoidean VII

artery

palatine VII

mandibular V

artery

maxillary V and buccal VII

cartilage covering olfactory organ

cartilage covering olfactory lobe of brain

ridge

trace these nerves outwards to their destination

position behind the spiracle this common nerve is called the **post-spiracular**. Trace it back towards the orbit, cutting upwards towards you with a sharp scalpel, *not* downwards towards the nerve. En route, it gives off a **pre-spiracular** branch to the spiracular muscle (function?). The origin of the pre-spiracular is variable.

EXPOSURE OF NERVES VIII, IX AND X

(8) The branches of cranial nerves IX and X (glossopharyngeal and vagus) run across the floor of the anterior cardinal sinus which is located on the median side of the gill pouches. To get at nerves IX and X you must, therefore, open up the sinus.

First locate the **post-orbital groove** in the cartilage immediately behind the orbit. This groove carries the narrow vein which connects the orbital sinus with the anterior cardinal sinus. Its position relative to the spiracle and other landmarks is indicated in Fig. 18.7C.

Run a blunt probe along the groove and gently insert it into the posterior cardinal sinus. Now cut vertically through the roof of the sinus so as to open up the sinus as shown in Fig. 18.7C. Stretch out the floor of the sinus and note the nerves. Five nerves crossing the floor of the sinus obliquely should be clearly visible.

Fig. 18.7 C Cutting open anterior cardinal sinus

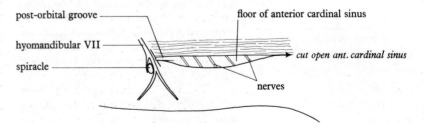

post-orbital groove

floor of anterior cardinal sinus

hyomandibular VII

cut open ant. cardinal sinus

spiracle

nerves

Fig. 18.7   D   Dissection of cranial nerves IX and X

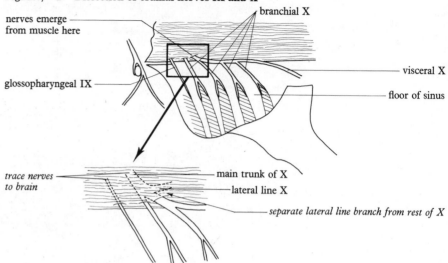

(9) The five nerves serve the gill pouches. The first is the **IXth nerve**, the others are all **branchial branches** of **X, the vagus**.

The main part of the vagus nerve, from which the branchial branches arise, is located on the median wall of the sinus. It is in fact compounded of several nerves which you should endeavour to separate from one another (Fig. 18.7D). As the vagus nerve emerges from the muscle at the anterior end of the sinus it gives off dorsally a **lateral line branch**, and ventrally the five **branchial branches**. Between the lateral line and branchial branches is the **visceral branch** which enters the abdominal cavity and supplies the gut, etc. Between lateral line X and visceral X spinal nerves will be seen.

(10) Trace the Xth and IXth nerves back to the brain. This necessitates the removal of all the muscle and cartilage which at present lie above these nerves. At the same time slice away the cartilage of the auditory capsule so as to expose **auditory VIII**. This is situated deep down, is stumpy and has three short branches, one serving each of three semicircular canals.

Cut away the cartilage covering the posterior regions of the brain.

(11) Clean up your dissection, removing all superfluous connective tissue, muscle and cartilage, so as to display the whole of the brain and the ten cranial nerves to maximum advantage. The origin and distribution of each nerve should now be clear. Can you identify all the structures shown in Fig. 18.8?

For Consideration

(1) What are the functions of impulses transmitted in each of the cranial nerves?

(2) What functions are performed by the olfactory lobes, optic lobes, and cerebellum?

(3) What important component of the 'tweenbrain is not fully visible in your dissection? What is its function?

(4) Explain in detail how the six eye muscles move the eye in the orbit.

(5) In the course of your dissection you have cut into the anterior cardinal sinus which is part of the venous system. Why should the veins of the dogfish take the form of large sac-like sinuses?

(6) What part of the nervous system does the vagus (Xth) cranial nerve belong to?

(7) To what extent do the cranial nerves appear to show a segmental arrangement like the spinal nerves? What in your view appears to have obscured the regular segmental pattern typical of spinal nerves?

Requirements
Dissecting board
Dissecting instruments
Dissecting awls

Dogfish for dissection

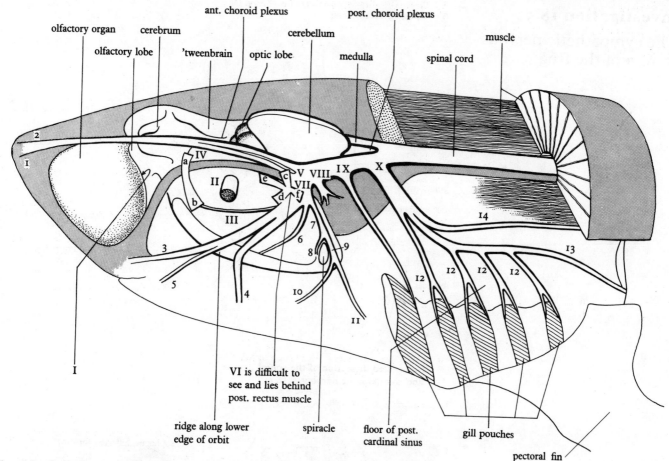

Fig. 18.8 Completed dissection of the brain and cranial nerves of the dogfish. The main cranial nerves are numbered and named as follows: I, olfactory; II optic; III, oculomotor; IV pathetic (trochlea); V, trigeminal; VI, abducens; VII, facial; VIII, auditory; IX, glossopharyngeal; X, vagus.

The branches are as follows: 1 & 2, ophthalmic V & VII; 3, maxillary V; 4, mandibular V; 5, buccal VII; 6, palatine VII; 7, hyomandibular VII; 8, prespiracular VII; 9, post-spiracular VII; 10, external mandibular VII; 11, hyoidean VII; 12, branchial X; 13, visceral X; 14, lateral line X. The eye muscles are as follows: a, superior oblique; b, inferior oblique; c, superior rectus; d, inferior rectus; e, anterior rectus; f, posterior rectus.

Investigation 18.5

The sympathetic nervous system of the frog

A   Opening up the frog

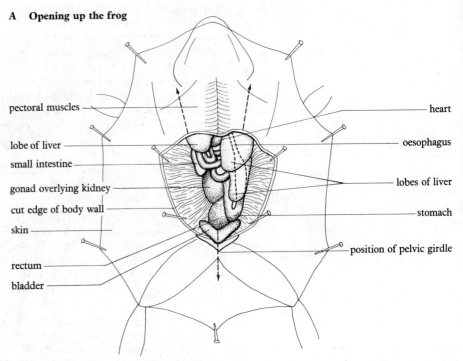

pectoral muscles

heart

lobe of liver

oesophagus

small intestine

lobes of liver

gonad overlying kidney

stomach

cut edge of body wall

skin

position of pelvic girdle

rectum

bladder

B   Completed dissection of the spinal nerves and sympathetic nervous system

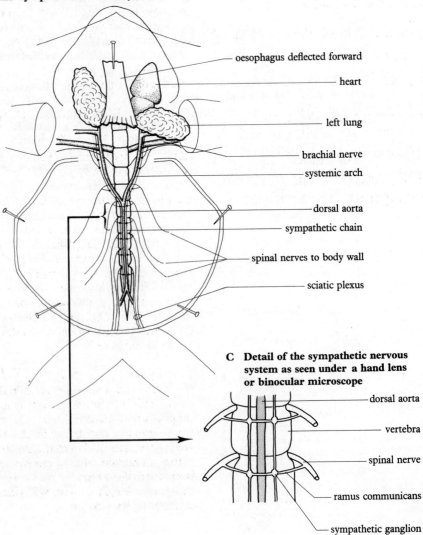

oesophagus deflected forward

heart

left lung

brachial nerve

systemic arch

dorsal aorta

sympathetic chain

spinal nerves to body wall

sciatic plexus

C   Detail of the sympathetic nervous system as seen under a hand lens or binocular microscope

dorsal aorta

vertebra

spinal nerve

ramus communicans

sympathetic ganglion

Fig. 18.9   Dissection of the frog to show the spinal nerves and sympathetic nervous system.

A good animal in which to observe the sympathetic nervous system is the frog. The foreshortening of the body reduces the amount of dissection, the sympathetic chain follows the course of the dorsal aorta and systemic arch, which helps identification, and the relationship of the sympathetic ganglia to the spinal nerves is clear.

Procedure

(1) Open up the frog on the ventral side and pin back the skin and body wall as shown in Fig. 18.9A.

(2) Remove the central portion (coracoid) of the pectoral girdle by cutting on either side of the mid-line as shown by the two forward-pointing arrows in Fig. 18.9A. In lifting out the coracoid be careful not to cut the systemic arch beneath.

(3) Cut through the pelvic girdle as shown by the backward-pointing arrow in Fig. 18.9A, and pull the legs apart.

(4) Cut through the oesophagus just in front of the stomach, and the rectum just in front of the anus. Remove the gut. Also remove the liver, kidneys, gonads and genital ducts, fat bodies, and veins (including the large posterior vena cava between the two kidneys). In removing the kidneys notice the elongated light-coloured adrenal glands on their ventral surfaces. What is the relevance of the adrenal glands in a dissection of the sympathetic nervous system? Draw the oesophagus, heart and lungs forward so as to see the **brachial plexus** (nerves serving the arms). Do not remove the systemic arches and dorsal aorta, for the sympathetic chain follows the course of these arteries.

(5) Identify the structures shown in Fig. 18.9 B and C. The ten **spinal nerves** are easy to identify, but the sympathetic nervous system is a little more difficult. A hand lens or binocular microscope will help. Observe the **sympathetic chain** with **ganglia**, and the **rami communicantes** linking each ganglion with its adjacent **spinal nerve**. Can you see any sympathetic nerves extending out from the ganglia?

For Consideration

(1) What are the general functions of the sympathetic nervous system?

(2) By what means are responses mediated by the sympathetic nervous system (a) enhanced, and (b) inhibited?

Requirements

Dissecting dish
Dissecting instruments
Pins
Binocular microscope or hand lens

Preserved or fresh-killed frog for dissection

Investigation 18.6

Dissection of the nervous system of the earthworm

The general pattern of the earthworm's nervous system is typical of bilaterally symmetrical invertebrates. A pair of **cerebral ganglia** (brain) on the dorsal side of the pharynx is connected by circumpharyngeal connectives to a sub-pharyngeal ganglion on the ventral side of the pharynx. From this a ventral ganglionated **nerve cord** extends the length of the body. From the cerebral, sub-pharyngeal and segmental ganglia **peripheral nerves** extend out to receptors and effectors in the body wall.

Procedure

(1) Make an incision on the mid-dorsal side of the worm about 4 cm behind the clitellum ('saddle'). With small scissors cut through the body wall along the mid-dorsal line to the extreme anterior end. Be careful, particularly at the anterior end, not to dig down too deeply or you will damage underlying structures.

(2) Cut through the septa between adjacent segments; deflect the body wall and pin it out on either side right up to the extreme anterior end.

(3) Identify the **cerebral ganglia** (**brain**) as a pair of small white bodies lying above the buccal sac. Behind the clitellum deflect the intestine to one side to reveal the **ventral nerve cord** beneath it.

(4) Now cut through the intestine posteriorly; grasp hold of it with forceps and fillet it away from underlying structures, starting at the posterior end and working forward. Be careful not to damage the ventral nerve cord beneath the gut. Continue filleting away the gut until you get to the cerebral ganglia. At this point pull the gut back and transect it cleanly immediately behind the cerebral ganglia. Be careful not to damage the circumpharyngeal connectives which run from the cerebral ganglia above the gut to the ventral nerve cord which lies beneath it.

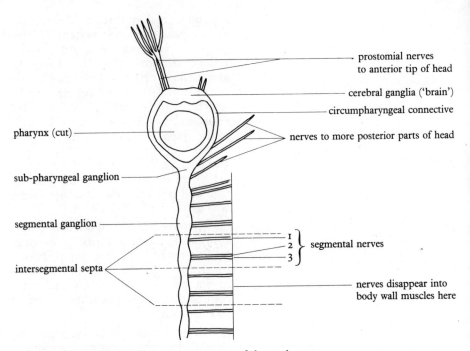

prostomial nerves
to anterior tip of head

cerebral ganglia ('brain')

circumpharyngeal connective

pharynx (cut)

nerves to more posterior parts of head

sub-pharyngeal ganglion

segmental ganglion

1
2
3 } segmental nerves

intersegmental septa

nerves disappear into
body wall muscles here

Fig. 18.10 Anterior end of the nervous system of the earthworm.

Requirements
Dissecting dish
Dissecting instruments
Pins
Binocular microscope or hand lens

Freshly killed earthworm for
dissection

(5) Identify the following structures
using a lens or binocular microscope as
necessary (Fig. 18.10): **cerebral ganglia** with nerves; circum-pharyngeal connectives; sub-pharyngeal ganglion with nerves; **ventral nerve cord**, swollen in each segment to form a segmental ganglion from which three pairs of **segmental nerves** pass outwards to the muscular body wall. The second and third segmental nerves are very close to one another.

For Consideration
How does the general plan of the earthworm's nervous system compare with that of the dogfish?

Investigation 18.7

Microscopic structure of two endocrine organs: the adrenal medulla and thyroid gland

Communication between different parts of the body is carried out by **hormones** as well as nerve impulses. The hormone is manufactured, and sometimes stored, in an **endocrine organ** whence it is secreted into the bloodstream. An endocrine organ would, therefore, be expected to show an intimate and extensive association between secretory cells and blood vessels. To see if this is so, examine sections of the thyroid gland and adrenal medulla.

Procedure

ADRENAL MEDULLA
Examine the central part of a median section of the adrenal (suprarenal) gland (Fig. 18.11A). This is the **adrenal medulla** which secretes the hormone **adrenaline**.

First notice numerous **capillaries**. These have a thin lining of pavement endothelium (note occasional flattened nuclei). Red blood cells may be seen in them.

The blood vessels are surrounded by columnar cells elongated at right angles to the walls of the blood vessels. They are called **chromaffin cells** because the granules in their cytoplasm stain brown with chrome salts. A brown colour is also given if adrenaline is mixed with chrome salts or chromic acid in a test tube. This, and other evidence, suggests that these cells secrete the hormone.

Observe chromaffin cells under high power. Note their prominent nuclei, cytoplasmic granules (distribution within the cell?) and proximity to blood vessels.

A Adrenal medulla

B Thyroid gland

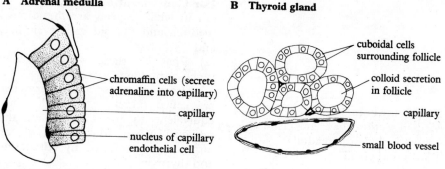

chromaffin cells (secrete adrenaline into capillary)

capillary

nucleus of capillary endothelial cell

cuboidal cells surrounding follicle

colloid secretion in follicle

capillary

small blood vessel

Fig. 18.11 Detail of two representative endocrine organs to show the relationship between the secretory cells and the blood vascular system.

What other types of cell can you see in the adrenal medulla? Account for any observations you make.

THYROID GLAND

The thyroid differs from the adrenal medulla in that the thyroid hormone (**thyroxine**) is stored in the gland before being shed into the bloodstream.

Examine a section of thyroid gland under medium power (Fig. 18.11B). Notice numerous follicles, each surrounded by a single layer of cuboidal cells. The cavity (lumen) of each follicle contains a colloid secretion which is usually visible in sections.

The follicle cells secrete thyroxine, but precisely how they do so is uncertain. One hypothesis, supported by various lines of evidence, is shown in Fig. 18.12. According to this scheme the functions of the follicle cells would be to absorb iodide, sugars, tyrosine and other amino acids from the bloodstream; to shed thyroglobulin (synthesized from these raw materials) into the lumen of the follicle for temporary storage; to absorb thyroglobulin from the lumen and convert it into thyroxine, and secrete the thyroxine into the bloodstream. In addition the cells respond to thyrotrophic hormone (secreted by the anterior lobe of the pituitary gland) in the bloodstream by increasing the rate of absorption of iodide and production of thyroglobulin and thyroxine.

It follows that the follicle cells must have an intimate association with the bloodstream on the one hand and the lumina of the follicles on the other. Is this the case in your section? Capillaries can be recognized by their pavement endothelium with flattened nuclei, and the fact that they contain red blood corpuscles.

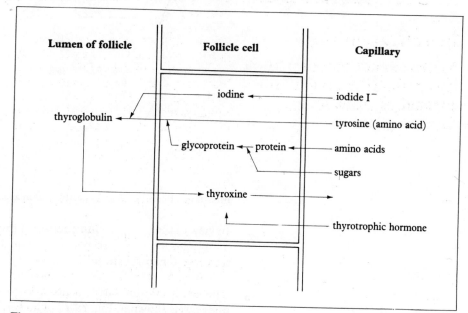

Lumen of follicle   Follicle cell   Capillary

iodine ← iodide I⁻

thyroglobulin ← tyrosine (amino acid)

glycoprotein ← protein ← amino acids

sugars

thyroxine

thyrotrophic hormone

Fig. 18.12 Proposed sequence of events leading to the secretion of the hormone thyroxine into the bloodstream from a follicle cell of the thyroid gland.

For Consideration

(1) In what respects are the adrenal medulla and thyroid glands similar, and dissimilar?

(2) What tells these two glands how much hormone to secrete?

(3) Organs that respond to hormones are called target organs. They are equivalent to the effectors of the nervous system. Make a list of the target organs affected by adrenaline and thyroxine.

Requirements

Median section of adrenal gland
  (must include the medulla)
TS thyroid gland

Questions and Problems

1   Fig. 18.13 shows an oscillograph record obtained when one end of an axon is stimulated with a single shock, and electrical events are recorded from the other end. The recording electrodes, connected to an oscilloscope, are arranged as shown in the diagram: one is inserted into the axon, the other is located outside the membrane enveloping the axon.

Explain as precisely as possible phases A to E in the oscillograph record.

What would you expect the oscillograph record to look like if both recording electrodes were placed side by side on the outside of the nerve membrane as shown at the foot of the diagram? Explain fully.

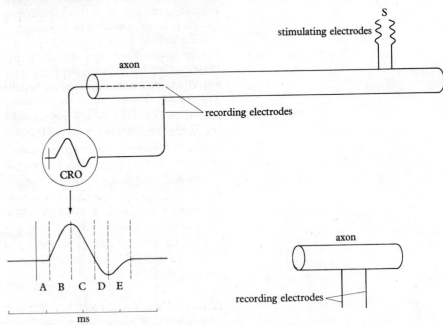

Fig. 18.13   Recording electrical activity in an axon.

2   In view of the all-or-nothing nature of the nerve impulse, how would you explain the graded muscular responses of which our bodies are clearly capable when reacting to natural stimuli?

3   The effect of stimulating a nerve with repetitive stimuli at two different frequencies is shown in Fig. 18.14. The vertical line preceding each action potential marks the moment of stimulation. Comment fully on these results.

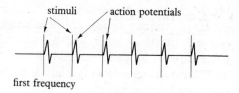

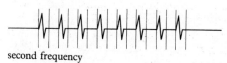

first frequency

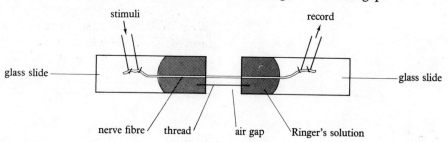

second frequency

Fig. 18.14 Effect of stimulating a nerve with repetitive stimuli at two different frequencies. The time scale is the same for both sets of recordings.

4 In an experiment to investigate the function of the myelin sheath, a single myelinated nerve fibre was placed on two moist glass slides separated by a short air gap. The nerve fibre was arranged so that the gap was bridged by an inter-nodal stretch, that is a length of the nerve fibre in between two consecutive nodes of Ranvier. Each slide was covered with a film of Ringer's solution, and the gap was bridged by a moist thread as shown in Fig. 18.15. After being set up, the sheath of the nerve fibre soon dried in the region of the air gap.

Fig. 18.15 Experiment on the transmission properties of an excised myelinated nerve fibre.

A pair of fine platinum electrodes was now placed in contact with each end of the nerve fibre. Through one of the two pairs of electrodes stimuli were delivered to the nerve fibre. Through the other pair of electrodes action potentials were recorded by means of an oscilloscope.

It was found that every time a stimulus was applied to one end of the nerve fibre, an action potential was recorded from the other end of the nerve fibre provided that the air gap was bridged by the moist thread. If the thread was removed, or allowed to dry, transmission ceased.

Explain these observations as fully as you can. What would you predict would be the result if the portion of the nerve fibre spanning the air gap included a node of Ranvier?

5 Acetylcholine, the chemical transmitter at the terminals of cholinergic nerves, is a derivative of acetyl coenzyme A. Explain why lack of vitamin B_1 in the diet results in paralysis. (If necessary refer to Table 5.2, p. 45.)

6 Comment on each of the following statements:
   (a) Synapses prevent impulses going in the wrong direction.
   (b) It is possible to suppress a reflex such as withdrawing one's hand from a hot object.
   (c) Curare, the poison used by certain natives on arrow tips, produces paralysis.
   (d) Certain drugs have hallucinogenic effects in which sensations are abnormal and distorted.

7 Fig. 18.16 shows the electrical activities in a motor neurone in response to applying three separate volleys of stimuli to an afferent nerve fibre. The first volley is small, the third large, and the second intermediate between the other two. Electrical responses are recorded from within the cell body of the motor neurone (intracellular electrodes, R_1) and from the surface of its axon (R_2). Explain the results as fully as you can.

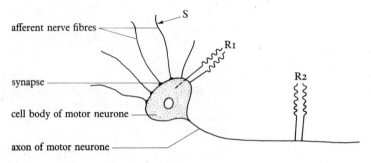

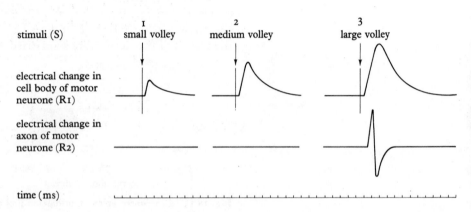

Fig. 18.16 Recording electrical events in a motor neurone.

8 'It is questionable if the human brain will ever succeed in understanding itself.' Discuss.

9 What do you understand by the term 'head'? Discuss in functional terms the sequence of evolutionary steps which you believe may have led to the development of a head of the kind seen in the vertebrates.

10 Compare concisely the nervous and endocrine system as means of communication within the body of an animal.

11 When frightened, the human body responds as follows: the face goes white, the pupils dilate, the heart beats faster, blood pressure increases, the mouth goes dry, and—in extreme cases—urination and defecation may occur. Explain these responses.

12 The responses of sea anemones can be investigated by the technique shown in Fig. 18.17. A series of electrical stimuli is delivered to the base of the animal and the animal's muscular contractions are recorded by means of a thread attached to a spring lever which writes on a revolving drum (kymograph). A representative recording is shown below the diagram. Explain the results in terms of the working of the nervous system.

What other primitive features does the sea anemone's nervous system possess besides the one demonstrated by this experiment?

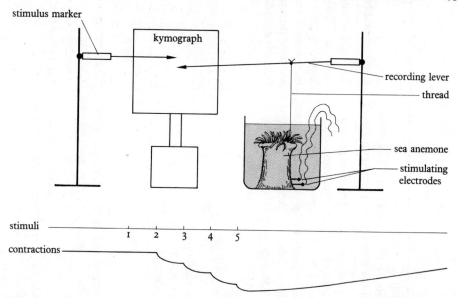

stimuli

1 2 3 4 5

contractions

Fig. 18.17 Experimental set-up for recording the muscular responses of a sea anemone to electrical stimulation.

13 Endocrine glands do the secreting in the endocrine system; neurones do the secreting in the nervous system. Discuss this statement.

14 Using the information in Table 18.1 and/or any other information available to you, answer the following questions, all of which relate to the mammalian endocrine system.

(a) Which hormones have the effect of raising the metabolic rate?

(b) Give a brief account of the hormones which play a part in lactation (milk secretion).

(c) Give one example of a situation where opposing effects are produced by different concentrations of one and the same hormone. Briefly discuss the survival value, if any, of your example.

(d) Which hormones control the levels of Ca^{2+} and Na^+ in the body?

(e) Which hormone appears to oppose the action of melanophore-stimulating hormone (MSH)?

(f) Which hormones affect the retention, or otherwise, of water in the body? Briefly explain how they achieve their effects.

(g) Which hormone or hormones bring about a rise in blood pressure? Briefly explain how they achieve their effects.

(h) Very briefly explain the role of hormones in the body's defences against disease.

(i) Is there anything in the table to suggest how in general hormones may exert their action within the cells of their target organs?

(j) Which hormones play a part in the chemical (enzymatic) digestion of food?

(k) Name an endocrine organ which is an exocrine as well as an endocrine gland.

(l) A trophic hormone is one which stimulates another endocrine gland to secrete. Make a list of such hormones found in the mammalian body and in each case state its target organ.

(m) Explain the term 'negative feedback' as used in Table 18.1.

(n) The pituitary is often described as the 'master gland' of the endocrine system. Briefly justify this title.

(o) Construct a flow diagram to illustrate the different compounds into which sugar can be converted in the human body, and show the role of the various hormones in these conversions.

Table 18.1   Hormones secreted by endocrine glands (mammals unless otherwise stated)

Gland	Secretion (hormone)	Functions	Result of deficiency (hyposecretion)	Result of excess (hypersecretion)	Controlled by
Thyroid (on either side of trachea just below larynx)	Thyroxine (iodine-containing ring compound attached to an amino acid)	Controls basal metabolic rate probably facilitating action of respiratory enzymes. Promotes action of pituitary growth hormone, diuresis, breakdown of protein, milk-production and metamorphosis in amphibians. Enhances action of adrenalin and sympathetic nervous system	Hypothyroidism: simple goitre (swollen neck) decreased metabolic rate, physical and mental lethargy—cretinism in children, myxoedema in adults. (N.B. Endemic goitre is due to lack of iodine in diet)	Hyperthyroidism: exopthalmic goitre (swollen neck, protruding eyeballs) increased metabolic rate, excitability:	Thyroid-stimulating hormone (TSH) (thyrotrophic hormone) from anterior pituitary (negative feedback)
	Calcitonin (polypeptide)	Opposes actions of parathormone (see below)			
Parathyroids (on either side of thyroid gland)	Parathormone (polypeptide with about 80 amino acids)	Controls concentration of Ca^{2+} in blood by: 1 regulating exchange of Ca^{2+} and PO_4^- between blood and bone 2 regulating excretion of Ca^{2+} and PO_4^-	Ca^{2+} concentration in plasma falls and PO_4^- rises. Increased excitability of nerve and muscle; muscle undergoes tetany	Ca^{2+} concentration in plasma rises and PO_4^- falls. Weakness, nausea, softening of bones	Concentration of Ca^{2+} in bloodstream (negative feedback)
Thymus (ventral side of heart)	Hormone not yet characterized	Renders lymphocytes (many of which are found in the thymus itself and then migrate to the lymph nodes) capable of developing into antibody-producing plasma cells. This happens immediately after birth, after which the thymus gradually regresses	Removal of thymus from mouse at birth destroys immune response. If thymus is removed after weaning age no ill effects result		
Pancreas (islets of Langerhans)	Insulin (protein) secreted by β cells	Suppresses blood sugar by causing respiratory breakdown of sugar, conversion to glycogen or fat in liver, and inhibiting gluconeogenesis (formation of glucose from protein). These effects probably achieved by insulin decreasing permeability of cell membranes to glucose	Blood sugar rises (hyperglycaemia) resulting in diabetes melitus. Sugar appears in urine (glycosuria)	Blood sugar falls (hypoglycaemia) resulting in nervous convulsions and failure of muscles to contract properly	Blood-sugar level (negative feedback)

Gland / source	Function (normal action)	Deficiency	Excess	Controlled by / stimulus
noradrenaline (ring compounds with short side chain) secreted by adrenal medulla	...broken down to lactic acid, cardiac frequency increases, blood vessels serving vital organs dilate, blood vessels serving non-vital organs constrict, spleen (blood reservoir) contracts, smooth muscle in walls of bronchioles relax so breathing is facilitated, pupils of eye dilate, sweating occurs, erector-pili muscles contract so hair stands up vertically, fatigued skeletal muscle contracts more vigorously; gut muscles relax but sphincters contract (except when adrenaline level is abnormally high in which case sphincters relax)			system. (Adrenal medulla is in effect a modified sympathetic ganglion)
Adrenal cortical hormones (steroids) secreted by adrenal cortex	*Mineralocorticoids* control ionic balance, specifically relative concentration of Na^+ and K^+	Kidney fails to reabsorb Na^+ and therefore water, resulting in reduction in volume of plasma and tissue fluid, fall in blood pressure and retention of K^+ in blood	Increased reabsorption of Na^+ and therefore water by kidney resulting in increased volume of plasma and tissue fluid, rise in blood pressure and loss of K^+ via kidney	Adreno-cortico trophic hormone (ACTH) from anterior pituitary (negative feedback)
	Glucocorticoids control carbohydrate metabolism by inhibiting cell respiration and encouraging gluconeogenesis, thereby raising blood sugar level. Latter promotes adaptation to stress. (N.B. Cortisone causes rise in blood pressure, shrinkage of thymus and lymph nodes, and reduction in number of white blood cells. This prevents body's anti-infection reaction from being too pronounced and widespread, i.e. it is an anti-stress hormone)	Fall in blood sugar increases susceptibility to stress	Increase in blood sugar increases resistance to stress	ACTH
	Sexcorticoids control development of gonads and secondary sexual characteristics	Weakness, circulatory and renal failure accompanied by pigmentation of skin: Addison's disease	Sexual precocity in young male. Sexual atrophy in adult male and female. Masculinization (virilism) in male	ACTH
Pineal body (roof of 'tween-brain)	**Melatonin** (hydroxy-indol compound) Causes concentration of melanin in pigment cells of frog. Function unknown in man but may promote sexual development in male	Removal causes spreading of melanin in frog pigment cells		
Wall of stomach	**Gastrin** (polypeptide) Causes secretion of gastric juice by gastric glands in stomach wall			Presence of food in stomach

Gland	Secretion (hormone)	Functions	Result of deficiency (hyposecretion)	Result of excess (hypersecretion)	Controlled by
Wall of duodenum	Cholecystokinin (long-chain polypeptide)	Causes bile to flow from gall bladder to duodenum by causing contraction of gall bladder and relaxation of sphincter muscle where bile duct joins duodenum			Presence of food in duodenum
	Secretin (long-chain polypeptide)	Causes secretion of pancreatic juice by pancreas			Presence of food in duodenum
Testes (interstitial cells = Leydig cells)	Androgens (androsterone and testosterone) (steroids)	Promote growth and activity of male accessory reproductive organs and secondary sexual characteristics by activating protein synthesis through intermediacy of RNA. Also stimulate spermatogenesis	Before puberty: failure of accessory reproductive organs and secondary sexual characteristics to develop (resulting in eunuch). After puberty: inhibition of spermatogenesis; also atrophy of accessory glands and failure of them to secrete	Before puberty: premature development of accessory reproductive organs and secondary sexual characteristics. Also increase in size of bones and skeletal muscles	Gonadotrophic hormones (FSH and ICSH) from anterior pituitary (feedback)
Ovaries During pregnancy placenta takes over production of hormones	Oestrogens (Oestradiol, oestrone, oestriole, hydroxyoestrone) (steroids)	Promote development of female accessory reproductive organs and secondary sexual characteristics, repair of uterus following menstruation, development of ducts (but not secretory cells) of mammary glands. Normally inhibit lactation, and make uterus more sensitive to oxytocin		Loss of appetite and feeling of nausea	FSH from anterior pituitary
	Progesterone (steroid) secreted by corpus luteum	Promotes proliferation of uterine mucosa and thickening of muscle, inhibits ovulation, promotes development of secretory cells of mammary glands but thought to inhibit lactation, make uterus less sensitive to oxytocin			LH from anterior pituitary
Pituitary (floor of 'tweenbrain) Pars dist...	Thyrotrophic hormone (thyroid-stimulating hormone, TSH)	Causes thyroid gland to secrete thyroxine	Less thyroxine secreted	More thyroxine secreted	Thyroxine in blood (negative feedback)

	Hormone	Function	Effect of hyposecretion	Effect of hypersecretion	Secretion controlled by
	(ACTH) (protein)				
	Growth hormone (protein)	Stimulates growth by promoting protein synthesis; also increases level of blood sugar by inhibiting action of insulin and stimulating action of glucagon	In young: dwarfism	In young: gigantism In adult: enlargement of extremities (acromegaly)	
	Prolactin (protein)	Stimulates mammary glands to secrete milk			Oestrogens and progesterone (inhibition)
	Gonadotrophic hormones: follicle-stimulating hormone (FSH) (glycoprotein)	Causes spermatogenesis in male. Causes development of Graafian follicle and secretion of oestrogens in female	Damage to seminiferous tubules		Oestrogens and progesterone (inhibition)
	Luteinizing hormone (LH) or **interstitial cell-stimulating hormone** (ICSH)	Causes secretion of androgens in male. Causes ovulation and development of corpus luteum in female			Oestrogens (stimulation) and progesterone (inhibition)
Pars intermedia (part of posterior lobe)	**Melanophore-stimulating hormone** (MSH) (polpeptide with 18 amino acids)	Expansion of melanin pigment in chromatophores (melanophores) in skin, particularly effective in, e.g. amphibians			
Pars nervosa (part of posterior lobe) Secretions produced by neuro-secretory cells in hypothalamus whence they flow down axons into pars nervosa	**Anti-diuretic hormone** (ADH) (**vasopressin**) (polypeptide with 8 amino acids)	Causes reabsorption of water in kidney with concomitant rise in blood pressure	Urine copious and dilute (water diuresis), tissue fluids concentrated, low blood pressure. Permanent hyposecretion results in diabetes insipidus	Urine concentrated, tissue fluids diluted, high blood pressure	Osmotic pressure of blood
	Oxytocin (**pitocin**) (polypeptide with 8 amino acids)	Causes contraction of uterus at parturition (birth). Also causes expulsion of milk from mammary glands by making smooth muscle surrounding secretory cells contract	Parturition delayed	Premature parturition	Oestrogen and progesterone

19 Reception of Stimuli

Background Summary

1 Receptors may consist of isolated **sensory cells**, or the cells may be compacted together to form a **sense organ**. They can be classified according to the type of stimuli they respond to.

2 Individual sensory cells are classified into **primary** and **secondary receptors** according to whether the sensitive device is the terminal of an afferent neurone or a specially adapted epithelial cell.

3 In general sensory cells, when stimulated, develop a local **generator potential** which, if it builds up sufficiently, elicits **action potentials** in an afferent neurone. The frequency of discharge depends on the size of the generator potential.

4 If a stimulus is maintained, the generator potential usually declines and the action potentials decrease in frequency until they cease altogether (**adaptation**).

5 For repetitive stimuli to be detected separately the generator potential produced by each stimulus must fall below the firing threshold before the next stimulus is delivered. If the frequency is too high for this to happen, **stimulus fusion** occurs.

6 Evidence suggests that the link between the stimulus and the development of a generator potential is a chemical process. This is known to be the case in photo-receptors and may apply to other receptor cells as well.

7 Receptor cells are often subject to inhibitory influences either from neighbouring sensory cells, as in the case of **mutual inhibition** in the compound eye, or through efferent nerves, as in the case of the lateral line receptors of fishes.

8 Individual receptor cells are often extremely sensitive, even to the slightest stimulus. The effective sensitivity of groups of sensory cells, e.g. the rods in the eye, may be increased by **convergence** and **summation**.

9 Sensory precision, e.g. of the cones in the eye, results from the receptor units being closely packed and by having a one-to-one relationship between the sensory cells and afferent neurones.

10 The mammalian eye and ear can be taken to illustrate the structure and functioning of two representative sense organs. In both cases the individual sensory cells, structurally and functionally integrated, are enclosed within a complex ancillary apparatus which protects them and ensures that they receive and respond to the appropriate stimuli.

11 In the case of the eye the receptor cells, **rods** and **cones**, are located in the **retina** on which light rays are brought to a focus by an adjustable **lens**. The cones, which are particularly concentrated in the foveal region of the retina, are responsible for high acuity colour vision in conditions of good illumination (i.e. **daylight vision**), the rods for black-and-white vision at low levels of illumination (i.e. **night vision**). These respective functions, imposed on the rods and cones by their structural and physiological properties, can be related to a wide range of everyday experiences.

12 The **compound eye** of arthropods contrasts sharply with the vertebrate eye and illustrates how in the course of evolution the same physiological problem has been solved in two contrastingly different ways.

13 The mammalian ear performs two functions: hearing and balance. Hearing is dealt with by sensory cells of the **organ of Corti** in the **cochlea**, to which sound waves are transmitted via a series of membranes, ossicles, and fluid-filled canals. The properties of the cochlea permit the ear to discriminate between sounds of different intensity and pitch.

14 **Balance** is dealt with by the **ampulla organs**, **utricle**, and **saccule** of the **vestibular apparatus**. The ampulla organs, associated with the **semicircular canals**, are sensitive to movements of the head, whilst the utricle and saccule contain **otolith organs** sensitive to the position of the head relative to the force of gravity.

Investigation 19.1

Analysis of human skin as a receptor

Requirements
For each pair of students:
Ruler
Ball-point pen with fine point
Flexible bristle mounted in wooden holder
Wooden-handled mounted needle whose sharp end has been sawn off about 0.5 cm from the tip; the cut surface must be flat and smooth
Mounted needle with sharp point
Pair of dividers
Bunsen burner
Beaker
Ice-cold water and/or acetone

The efficiency of a receptor in monitoring changes at the surface of the body depends on (a) the sensitivity of the individual sensory cells, (b) the variety of stimuli to which the receptor responds, (c) its ability to distinguish between, i.e. resolve, two stimuli applied simultaneously, and (d) the rapidity with which the sensory cells adapt when stimulated continuously.

Investigating (a) requires an elaborate set-up (*see* p. 177), but (b), (c), and (d) can be investigated with a minimum of apparatus. In this case your own skin will be examined as a receptor. You will need to work in pairs.

Experiments

THE DIFFERENT STIMULI TO WHICH THE SKIN RESPONDS
With a fine ball-point pen, rule a grid of not less than 25 squares on the back of your partner's hand. The sides of the squares should be 2 mm long so the area of each one will be 4 mm^2.

Explore each square in turn for its sensitivity to **touch, heat, cold,** and **pain.** For touch use a flexible bristle mounted in a wooden holder. For heat and cold use a flat metal surface approximately 1 mm in diameter (a large pin with the end sawn off will do): to investigate heat reception warm the pin in hot water; to investigate cold reception cool it in acetone or ice-cold water. For pain use a sharp needle.

Draw your grid on a large scale on a sheet of paper and indicate within each square which, if any, of the four modalities listed above it is sensitive to. In this way you can map the distribution of receptors sensitive to the different kinds of stimulation in the skin.

The experiment can be repeated in different parts of the body so the distribution of receptors may be compared.

THE RESOLVING POWER OF THE TACTILE RECEPTORS
A receptor's resolving power is the minimum distance required between two simultaneously applied stimuli so that they can be detected as two separate stimuli rather than a single stimulus. One usually thinks of resolving power in connection with the eye, but it also applies to the sense of touch.

Use a pair of dividers to apply two simultaneous tactile stimuli to your partner's arm. Vary the distance between the two points of the dividers and determine the minimum distance by which the two stimuli must be separated for the subject to feel both of them.

Repeat the experiment on other parts of the body including the finger tips and thigh. Record and explain your results.

ADAPTATION
With a needle wiggle one of the hairs on your partner's hand until the subject ceases to feel it. In this way estimate how long it takes for the receptors at the base of the hair to **adapt** to continual stimulation. Is the time the same for all hairs investigated? Are there regional differences over the body surface?

Interpret the results of these experiments in terms of the structure, neural connections, and physiology of receptor cells.

Investigation 19.2
Structure of the mammalian eye

The eye is an elaborate sense organ consisting of a layer of light-sensitive cells, the retina, in front of which is a transparent cornea and adjustable lens for focussing the light.

Procedure

DISSECTION OF THE EYE

Use the eye of, e.g. sheep or ox. The posterior part of the eye will probably have fat clinging to it. Note the transparent **cornea** continuous with the tough, white **sclera (sclerotic layer)**. Also notice the **conjunctiva** covering the surface of the cornea; it is continuous with the mucous membrane of the eyelids. The cornea is devoid of blood vessels: what are the consequences of this?

Proceed as follows:

(1) Remove the fat from the eyball so as to expose the **optic nerve** and extrinsic **eye muscles** (function?).

(2) Make an *almost* complete circular cut round the edge of the cornea, in front of where it joins the sclera (dotted lines 1, Fig. 19.1A). The watery fluid which emerges is **aqueous humour**. Deflect the cornea forwards like a lid. Note the **iris** (brown), **pupil**, front of **lens**. With a blunt needle push the lens backwards and forwards and from side to side. What restricts its movement?

(3) Make a circular cut all round the eyeball so as to cut the eye into anterior and posterior halves (dotted line 2, Fig. 19.1A). Remove the gelatinous **vitreous humour** with large forceps.

Now examine the interior of the eye (Fig. 19.1B). In the posterior half observe the **retina** (grey), **choroid** (black) and thick **sclera** (white). Notice blood vessels radiating from the point where the optic nerve is attached to the retina (**blind spot**). To one side of the blind spot the **fovea centralis** (yellow spot) should be visible.

(4) Now look at the front half of the eyeball. The retina is continuous with the **ciliary body** which forms a circular black band round the lens. With a blunt needle gently move the lens and observe that it is attached to the ciliary body by a delicate **suspensory ligament**.

(5) Make two radial cuts in the wall of the anterior half of the eye and remove the portion of the wall in between (dotted line 3, Fig. 19.1A). Leave the lens attached to the rest of the wall. This should enable you to see more clearly the relationship between the lens, ciliary body and iris. Use a hand lens or binocular microscope if necessary. Compare with the diagram in Fig. 19.2A.

(6) How would you describe the appearance of the lens and how does it feel to the touch? Explain.

A Mammalian eye: exterior

anterior half: exterior

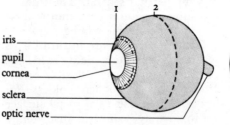

B. Interior view

anterior half · posterior half

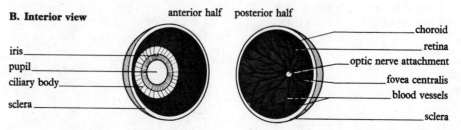

Fig. 19.1 Dissection of mammalian eye.

A Median vertical section of mammalian eye

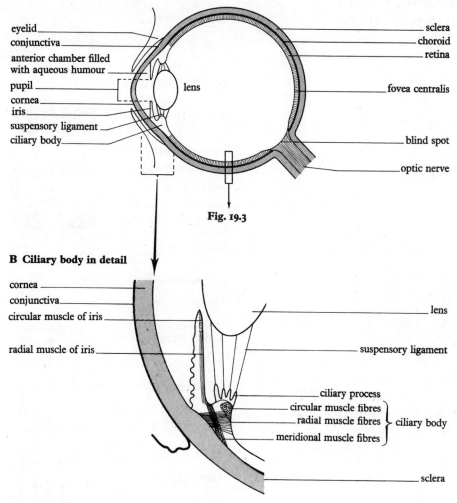

Fig. 19.3

B Ciliary body in detail

Fig. 19.2 General microscopic structure of mammalian eye.

MICROSCOPIC STRUCTURE OF THE CILIARY BODY AND IRIS

Examine the ciliary body and iris in a median section of the eye (Fig. 19.2B). The **ciliary body** contains circular, radial, and meridional muscle fibres (the **ciliary muscle**) responsible for changing the shape of the lens during **accommodation** to near or far objects. When the circular and radial muscles contract the tension on the lens is relaxed: the lens assumes a more rounded shape and its front bulges forwards. How does the disposition of these muscles enable them to achieve this effect?

When the circular and radial muscles relax the reverse happens and the lens is pulled out into a more flattened shape. This is further aided by contraction of the meridional muscles. What triggers these various muscles to contract and relax?

Now turn your attention to the **iris**. This contains circular and radial muscle fibres which constrict and dilate the **pupil** respectively. Under what circumstances do these muscles contract and relax?

A Vertical section of retina as seen in a typical microscopic preparation

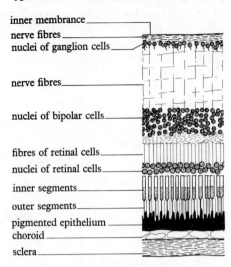

inner membrane
nerve fibres
nuclei of ganglion cells

nerve fibres

nuclei of bipolar cells

fibres of retinal cells
nuclei of retinal cells
inner segments
outer segments
pigmented epithelium
choroid
sclera

B Diagram of retina in detail

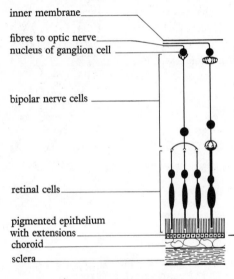

inner membrane

fibres to optic nerve
nucleus of ganglion cell

bipolar nerve cells

retinal cells

pigmented epithelium with extensions
choroid
sclera

MICROSCOPIC STRUCTURE OF THE RETINA

(1) Examine a vertical section of the **retina** (not the foveal region) under high power. Observe the various layers shown in Fig. 19.3A, using Fig. 19.3B to help you interpret them. Note the **choroid** and **sclera** (**sclerotic layer**) outside the retina. The choroid is vascular its outer membrane being pigmented. The sclera is composed of connective tissue: mainly collagen fibres with some elastic fibres.

(2) Examine a vertical section of the **fovea**. How does it differ from the non-foveal part of the retina?

(3) Can you tell the difference between the two types of retinal cells: the **rods** and **cones**? This may be difficult because they are densely packed and in certain parts of the retina they look alike. In general cones are fatter than rods (Fig. 19.3D). The centre of the fovea (**fovea centralis**) contains only cones, densely packed and unusually slender. Moving outwards from the fovea centralis, the cones become fatter, and rods are present in ever increasing proportions. The extreme periphery of the retina contains only rods. What is the functional significance of these facts?

(4) Examine an individual **retinal cell** (rod or cone) in as much detail as you can using Fig. 19.3D to help you. The dark part of the inner segment contains numerous densely packed mitochondria.

(5) Compare the retina of a light adapted and dark adapted eye. How do they differ with regard to the distribution of pigment in the pigmented epithelium at the base of the retina? Explain the reason for the difference.

For Consideration
In what ways is the structure of the eye related to its functions?

C Detail of pigmented epithelium

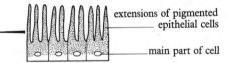

extensions of pigmented epithelial cells

main part of cell

D Rod and cone compared

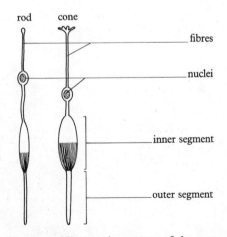

rod   cone

fibres

nuclei

inner segment

outer segment

Fig. 19.3 Microscopic structure of the mammalian retina and associated structures.

Requirements
Binocular microscope or hand lens
Dissecting instruments
Eye of, e.g. sheep or ox
MS eye (for ciliary body and iris)
VS non-foveal part of retina, light and dark adapted
VS foveal part of retina

Investigation 19.3
Structure of the mammalian ear

The ear is a complex sense organ responsible for hearing and balance. In this investigation we will confine our attention to the ear's auditory function.

Procedure

(1) First study Fig. 19.4A or a model of the ear, or a vertical section of the whole ear. The ear is sub-divided into an air-filled **outer** and **middle ear**, and a fluid-filled **inner ear**. The middle ear contains the three **ear ossicles** (malleus, incus, and stapes) which are held in position by delicate muscles. The inner ear contains the **membranous labyrinth** which consists of the **cochlear duct** (median canal) and the **vestibular apparatus** (semicircular canals, ampullae, utricle, and saccule). The whole of the membranous labyrinth is filled with fluid (**endolymph**) and surrounded by fluid (**perilymph**). The perilymph is enclosed within the **bony labyrinth** which includes the vestibular and tympanic canals of the cochlea (*see* below).

(2) Examine a vertical section of the **cochlea** under low power. Observe the coiled cochlea tube in section. Refer to Fig. 19.4C to see how your section relates to the ear as a whole.

Notice that the cochlear tube is subdivided by membranes into three fluid-filled canals:

(a) **Vestibular canal** (scala vestibuli)

(b) **Middle canal** (scala media, cochlear duct)

(c) **Tympanic canal** (scala tympani).

A Vertical section of mammalian ear: diagrammatic

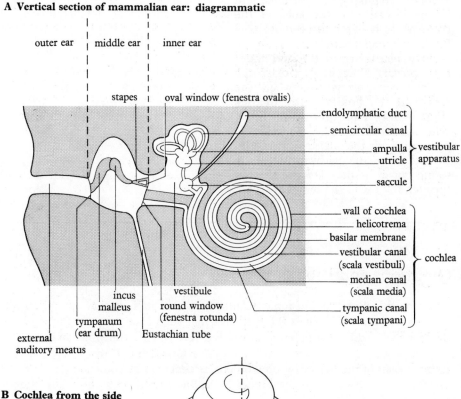

B Cochlea from the side

Fig. 19.4 Structure of mammalian ear.

C Vertical section of cochlea

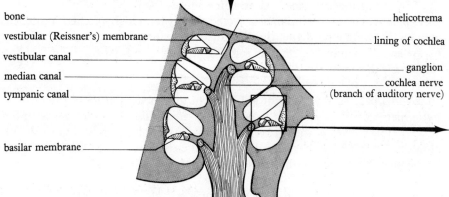

D Organ of Corti in detail

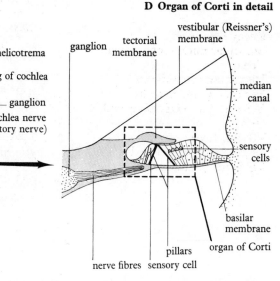

What parts of the inner ear does each of these canals connect with? In addition to the canals observe the other structures shown in Fig. 19.4. What do you make of the **helicotrema**, the small orifice linking the vestibular and tympanic canals at the apex of the cochlea? Is it likely to be more of a hindrance than a help?

Examine the **organ of Corti** under high power. Note in particular the sensory cells wedged between the **tectorial** and **basilar membranes** (Fig. 19.4D). How are these sensory cells stimulated? What function is performed by the **pillars**? Use Fig. 19.4 to trace the pathway taken by sound waves as they are transmitted through the ear to the organ of Corti.

For Consideration
In what ways is the structure of the ear related to its auditory function?

Requirements
Diagram, model, or VS whole ear
VS cochlea

Questions and Problems

1 Describe the electrical and ionic events that occur in a sensory cell when the latter is stimulated. What is the evidence that such events occur, and how do they compare with what happens in a nerve fibre when it transmits an impulse?

2 Explain each of the following phenomena as concisely as you can:
(a) Wearing a coarse shirt produces a tickling sensation at first but after a period of time this sensation ceases.
(b) Treating sensory cells with an ATP inhibitor results in the receptor failing to respond to stimulation.
(c) When reading a book the words you are looking at directly are sharply defined whereas surrounding words are blurred.
(d) It is impossible to see a very faint star by looking directly at it, but it is possible to do so if you look slightly to one side of it.
(e) The flicker on a cinema screen can be detected if one looks at the screen out of the corner of one's eye, but not if one looks directly at it.
(f) If you go into a dimly lit room from bright daylight, it is at first difficult to see anything, but gradually objects become visible.
(g) In very dim light it is impossible to distinguish between different colours, everything appearing as black, white, or various shades of grey.
(h) When pipetted into the eye, atropine causes the pupil to dilate.
(i) A person who is continually subjected to very loud high-pitched sounds may eventually become permanently deaf to such sounds.
(j) After a few minutes on the rotor in a fairground a person suffers from temporary dizziness.

3 In myopia parallel rays of light are brought to a focus in front of the retina; in hypermetropia the rays are focused behind the retina. What are the symptoms and possible causes of these two conditions? Astigmatism is another refractive abnormality in which there is a difference of curvature in the various meridians of the eye. What will be the consequences of this condition?

4 How do you think the brain causes a specific muscle, for example the biceps, to exert differences in power according to its load? (*AEB*)

5 Look at Fig. 19.5 from a distance of at least 30 cm. Close your left eye and focus on A with your right eye. Now slowly move the book towards you. What happens to B as you do this? Explain fully.

Repeat the process, but this time focus on B and see what happens to A. What conclusions can be drawn?

Fig. 19.5

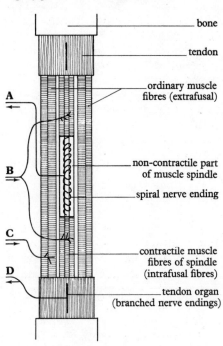

Fig 19.6   Schematic diagram of muscle spindle and tendon organ.

6 Muscle spindles and tendon organs are associated with skeletal muscles. Fig. 19.6 is a highly schematic diagram of these two types of receptor and their nerve supply. The following are the results of experiments carried out to investigate their physiology:

(a) The whole muscle is stretched. Result: a train of nervous impulses is discharged in nerve fibre **A**, the frequency of impulses being proportional to the stretching force. Only very slight stretching is required to elicit this response.

(b) If **B** is stimulated the same result is produced as in (a).

(c) **C** is stimulated while impulses are being discharged in **A**. Result: the frequency of impulses in **A** decreases.

(d) The tendon is stretched. Result: impulses are discharged in **D**, but this only occurs if the stretching is fairly severe.

(e) If the nervous system is intact, stretching the tendon may result in a decrease in the frequency of impulses in **C**.

(f) With prolonged stretching of the muscle and tendon, discharge of impulses in **A** and **D** continues for a long time without any diminution in frequency.

What conclusions would you draw from these experiments as to the role and mode of functioning of the muscle spindle and tendon organ in the living animal?

7 What do receptors have in common?

8 Outline how you would assess the sensory capacity of a newly discovered animal?

(CCJE)

9 Give a brief description of the following sensory functions in man, relating each to the structure of the organ involved:
   (a) detection of angular acceleration of the head;
   (b) detection of sound waves by the inner ear;
   (c) reception of visual images of different light intensity. (*JMB*)

10 It is claimed that bees can distinguish between yellow and blue. How would you test whether or not this is true? Describe experiments you would carry out, paying attention to any special considerations you would need to take into account in designing them.

20 Effectors

Background Summary

1 Effectors are structures which respond directly or indirectly to stimuli. The body's principal effectors are **muscles** and **glands**.

2 Vertebrate muscles are classified into **skeletal** (voluntary, striated), **visceral** (involuntary, non-striated, 'smooth'), and **cardiac** (heart) muscle. Each possesses certain characteristic properties.

3 Generally muscles contract when impulses reach them through the nervous system but sometimes, as in the case of cardiac muscle, contractions are **myogenic**, i.e. they arise within the muscle tissue itself.

4 When a skeletal muscle is excited, either directly with a single electrical stimulus or by a single impulse through the nerve that innervates it, it responds—after a brief latent period—by giving a **twitch**.

5 On repetition at sufficiently high frequencies muscle twitches can **summate** to produce a **tetanus**, this being the basis of normal graded responses in the body. In contrast to visceral and cardiac muscle, skeletal muscle contracts rapidly but fatigues comparatively quickly.

6 Contraction of a muscle fibre is initiated by an electrical impulse (**action potential**) which has the same ionic basis as the nerve impulse.

7 The muscle action potential obeys the **all-or-nothing law** and is followed by an absolute and relative **refractory period**.

8 A skeletal muscle is subdivided into **fibres**, and the fibres into **myofibrils**. Each myofibril consists of alternating sets of thick myosin and thin actin **filaments** whose arrangement gives the muscle its striated appearance.

9 Evidence strongly supports the theory that when skeletal muscle contracts the thick and thin filaments slide between one another, possibly propelled by cross bridges acting as ratchets. Energy is supplied by ATP from mitochondria situated between adjacent myofibrils.

10 Little is known about how a muscle action potential is linked with the contraction process but there is evidence that the **sarcoplasmic reticulum** is involved in initiating the splitting of ATP.

11 Other effectors, besides muscle, include **chromatophores** (pigment cells), **electric organs**, **light-producing organs**, and **nematoblasts** (stinging cells). The last is an example of an **independent effector**.

Investigation 20.1

Action of skeletal muscle: experiments on a nerve-muscle preparation

The physiological properties of skeletal muscle can be investigated by applying electrical stimuli either to the muscle direct or to the nerve that supplies it. In the latter case the muscle and its nerve are removed from a leg to give a nerve–muscle preparation. A convenient preparation is the **sciatic nerve** and **gastrocnemius (calf) muscle** of the frog. The nerve is placed in contact with a pair of electrodes through which electrical shocks can be delivered; the muscle is attached to a lever which records its contractions on a revolving drum (kymograph).

Apparatus

This consists of (1) the stimulating equipment, (2) the muscle bath, and (3) the kymograph.

For details of the **kymograph** see p. 434. As with other experiments involving the kymograph it is necessary to know the speed at which the drum rotates. If this is not already known, a time tracing should be made with either a tuning fork or an electrical time marker.

Stimulating equipment comes in various forms. One commonly used arrangement is described on p. 435. It consists of two connected circuits for delivering brief induction shocks. The primary circuit incorporates a 2-volt battery, operator key, and the primary coil of an inductorium; the secondary circuit consists of the secondary coil of the inductorium, a short-circuit key, and the electrodes. Single or repetitive shocks can be delivered.

Muscle baths vary in their specifications. Generally the stimulating electrodes are located at one end; at the other end there is a hook, connected with the recording lever, to which the muscle should be attached. In the floor of the muscle bath between the electrodes and the hook is a hole or piece of cork into which a pin can be stuck for firmly anchoring the nerve–muscle preparation.

Making the Nerve–Muscle Preparation

You will be provided with the hind leg of a frog. To keep the tissues alive keep the leg well moistened with Ringer's solution. Remember this is no ordinary dissection: the tissue has got to function properly after you have finished butchering it.

Proceed as follows:

(1) Skin the leg. This can be done by grasping the skin at the top of the leg with a pair of forceps and stripping off the skin like pulling off a stocking.

(2) Place the leg, posterior surface uppermost, on a piece of cork, and identify the structures shown in Fig. 20.1A. (The posterior surface can be easily recognized by the glistening white tendon just above the knee).

(3) With your thumbs firmly pull apart the two muscles in the thigh. This will expose the ribbon-like **sciatic nerve** which lies between them. Pin the muscles down on either side. (Fig. 20.1B.)

(4) Follow the nerve up as far as you can, i.e. to the cut end of the leg. Grasp the cut end of the nerve with small forceps and carefully separate the nerve from surrounding tissue down to the knee. Be careful not to nick the nerve.

(5) With a scalpel blade free the **Achilles tendon** from the foot and cut it *below* the sesamoid bone (Fig. 20.1B).

(6) With ligature thread tie a double knot round the Achilles tendon just *above* the sesamoid bone; then make a small loop by tying a second double knot approximately 2 mm from the first as shown in Fig. 20.1B. Cut one of the two loose ends of the thread but leave the other about 4 cm long.

(7) Take hold of the thread, and gently free the **gastrocnemius muscle** from the rest of the leg all the way up to the knee.

(8) Lay the nerve and the muscle to one side of the leg as shown in Fig. 20.1C. Now cut through the leg immediately above and below the knee, making sure you do not damage either the gastrocnemius muscle or the sciatic nerve. You have now completed making your nerve–muscle preparation which should look like Fig. 20.1D.

A Posterior surface of leg after removing skin

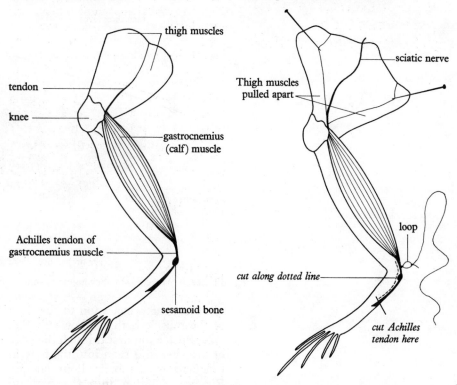

thigh muscles

tendon

knee

gastrocnemius (calf) muscle

Achilles tendon of gastrocnemius muscle

sesamoid bone

B Exposure of sciatic nerve and tying of loop round Achilles tendon

sciatic nerve

Thigh muscles pulled apart

loop

cut along dotted line

cut Achilles tendon here

C Detachment of nerve–muscle preparation from rest of leg

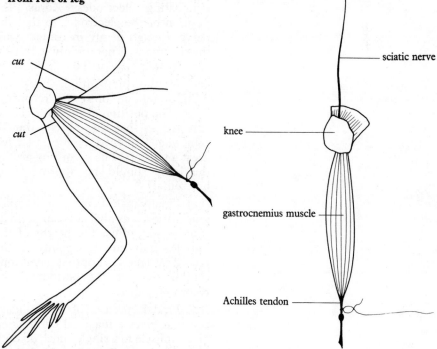

cut

cut

D Completed nerve–muscle preparation

sciatic nerve

knee

gastrocnemius muscle

Achilles tendon

Fig. 20.1 The making and setting up of a nerve-muscle preparation using the sciatic nerve and gastrocnemius muscle of frog.

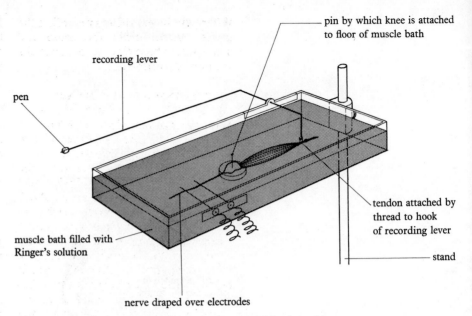

pen

recording lever

pin by which knee is attached
to floor of muscle bath

tendon attached by
thread to hook
of recording lever

stand

muscle bath filled with
Ringer's solution

nerve draped over electrodes

Fig. 20.2   Muscle bath showing nerve–muscle preparation in position.

(9) Fix your preparation to the floor of the muscle bath as shown in Fig. 20.2. Push a pin first through the edge of the knee and then into the hole in the floor of the bath. Pour enough Ringer's solution into the bath to cover the preparation.

(10) Drape the sciatic nerve over the stimulating electrodes. Adjust the position of the electrodes so that the nerve, though firmly in contact with the electrodes, is not unduly stretched.

(11) By means of the loop at the other end of the preparation, hook the Achilles tendon onto the recording lever. The position of the lever can be adjusted so that the muscle, though somewhat stretched, is not under too great a tension. The recording arm of the lever should be approximately horizontal.

(12) Finally make sure that the kymograph drum is as high as it can be on its spindle and adjust the height of the muscle bath so the point of the recording lever is approximately one centimetre from the bottom of the recording paper.

(13) Find the position of the secondary coil that gives a maximal contraction of the muscle to a single shock on both make and break. Procedure for delivering a single shock: with the short-circuit key open, close the key in the primary circuit (make shock), and then release it (break shock). You are now ready to proceed with the experiments.

Experiments

(1) RESPONSE TO A SINGLE STIMULUS: THE SIMPLE TWITCH

Set the kymograph at about 400 mm/s. Bring the tip of the recording lever into contact with the paper about one centimetre from the base of the drum. Record a base line by allowing the drum to make one complete revolution.

Connect the primary key to terminal **S** at the end of the primary coil of the inductorium. Now record the response of the muscle to a single make or break shock. If possible record the moment of stimulation.

This can be done in one of two ways:

(a) By using a stimulus marker. This should be incorporated into the primary circuit between the primary key and the inductorium.

(b) By using the contact key at the base of the kymograph as the key in the primary circuit. This key is operated by an arm that projects from the spindle of the kymograph so the circuit is made or broken (or both) every time the drum revolves. Ensure that when the key is operated the muscle only responds to make or break, not both. (There are several ways of achieving this; what are they?). When you have recorded the response of the muscle, mark the moment of stimulation by slowly revolving the drum by hand until the shock is just sent in to the nerve and a single twitch is recorded on a stationary drum.

When you have finished recording the single twitch, lower the drum and make a new base line in readiness for the next experiment.

(2) RESPONSE TO TWO STIMULI IN SUCCESSION: SUMMATION

With the kymograph set at about 100 mm/s, record the response to two shocks in quick succession (make followed by break). Start with the interval between the two shocks such that the response to the second shock starts well after the muscle has begun to relax following the first contraction. Then repeat the procedure, gradually decreasing the interval between the two shocks until there is no distinction between the two responses which become smoothly summated to give a single contraction larger than the twitch obtained in the first experiment.

(3) EFFECT OF HIGH-FREQUENCY STIMULI: TETANUS

Disconnect the primary key from its connection to terminal **S** at the end of the primary coil of the inductorium and reconnect it to terminal **R**. Set the interrupter at the end of the primary coil so that it vibrates on make and ceases to do so on break of the primary circuit.

Set the kymograph at about 20 mm/s. Make a base line round the drum. Now record the response of the muscle to approximately three seconds of repetitive stimulation. Then let the muscle relax until the point of the lever returns to the base line. (There is no need to record for the full duration of the muscle relaxation but note how long it takes to return to the base line).

(4) EFFECT OF REPETITIVE STIMULI FOR A LONG PERIOD: FATIGUE

Slow the drum to about 0.5 mm/s, connect the primary key to the **S** terminal again, and make a new base line. Now stimulate the preparation repeatedly at two stimuli per second until no recordable response is given by the muscle. When the preparation is thus fatigued shift the electrodes so that they are in contact with the muscle, one on each side of it. Now stimulate again at two stimuli per second. Record the responses.

The paper can now be removed from the kymograph drum.

Presentation of Results

Cut out your recordings and stick them in your laboratory notebook. In the case of the second experiment (summation of contractions) display the recordings in sequence *one above the other* showing the way the second contraction fuses with the first as the interval between the two stimuli is gradually decreased.

Beneath each recording, or set of recordings, give a time scale. For experiments 1–3 draw a horizontal line corresponding to 0.5 seconds; for experiment 4 (fatigue) a line corresponding to five seconds would be more appropriate.

For a single twitch (experiment 1) measure the following:
(a) Latent period (time delay between the moment of stimulation and the onset of the muscular response) in milliseconds.
(b) Contraction time in seconds.
(c) Relaxation time in seconds.
(d) Duration of twitch, i.e. (a) + (b).
(e) Contraction height in mm.
Measure the height of the largest completely summated response obtained in experiment 2 and of the tetanus obtained in experiment 3.

Finally measure the time taken for no recordable contractions to be given in experiment 4.

For Consideration

(1) What is the latent period due to?
(2) What light is shed by the results of experiments 2 and 3 on the normal functioning of nerves and muscles in the intact animal?
(3) Do the results of experiment 4 enable you to say whereabouts in a nerve–muscle preparation fatigue occurs in the course of repetitive stimulation?
(4) What may fatigue be due to?

Requirements

Kymograph
Muscle bath with recording lever and electrodes
Recording pen
Stimulator or stimulating circuit
Stand for muscle bath
Pipettes (2)
Pins
Thread
Dissecting instruments
Piece of cork approximately 12 × 12 cm
Eosin
Frog Ringer's solution (500 cm³)

Frog's leg, cut from the body as high up as possible, and kept in Ringer's solution

Investigation 20.2
Structure of skeletal muscle

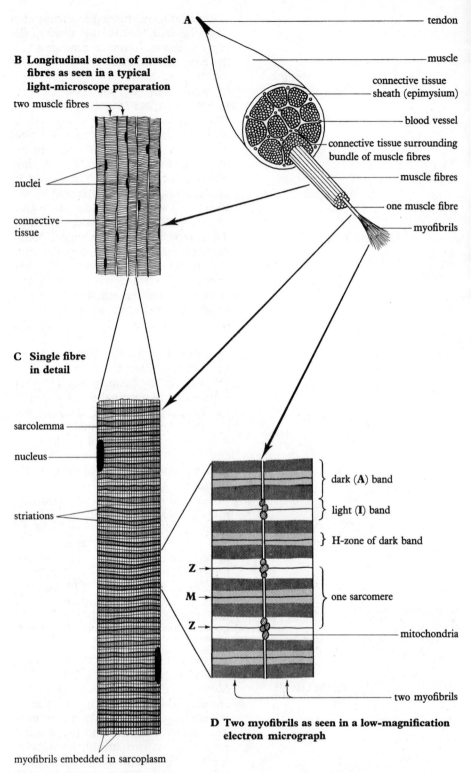

B Longitudinal section of muscle fibres as seen in a typical light-microscope preparation

two muscle fibres

nuclei

connective tissue

A — tendon

— muscle

connective tissue sheath (epimysium)

— blood vessel

connective tissue surrounding bundle of muscle fibres

— muscle fibres

— one muscle fibre

— myofibrils

C Single fibre in detail

sarcolemma

nucleus

striations

myofibrils embedded in sarcoplasm

dark (**A**) band

light (**I**) band

} H-zone of dark band

Z →
M →
Z →

one sarcomere

— mitochondria

— two myofibrils

D Two myofibrils as seen in a low-magnification electron micrograph

Fig. 20.3 Structure of skeletal muscle. The detail of the striations shown in **B** is visible in good light-microscope preparations.

Vertebrate muscle is classified into **visceral** (non-striated, smooth), **skeletal**, and **cardiac muscle**. In each case the cells are adapted for contraction. The close relationship between structure and function is particularly apparent in the case of skeletal muscle.

Procedure
(1) Remove one of muscles from the leg of a frog by freeing it from neighbouring muscles and cutting the tendons at its origin and insertion. *See* p. 208 for instructions on the removal of the gastrocnemius (calf) muscle.

The muscle fibres are surrounded by a sheath of connective tissue of the **epimysium**. Refer to Fig. 20.3A to see how the muscle fibres are organized within the sheath. With needles break through the sheath and tease out a few fibres on a glass slide. Crush some of them with a bristle, add a drop of sodium chloride (0.75 per cent) and put on a coverslip. Observe **striations**, **sarcolemma** (in crushed region). Irrigate with acetic acid to show scattered oval **nuclei**.

(2) Examine a longitudinal section of skeletal muscle under high power. In a good section, appropriately stained and illuminated, you should be able to see all the structures shown in Fig. 20.3B and C. What is the explanation of the striations in each fibre? Notice that the nuclei which occur along the edge of each fibre are not separated by intervening cell membranes. A striated muscle fibre is, therefore, a multinucleate structure (**syncytium**).

Examine a low magnification electron micrograph of skeletal muscle fibrils in longitudinal section and note the structures shown in Fig. 20.3C. Herein lies the explanation of the striations seen in the light microscope. (3) Permanent preparations of skeletal muscle fibres can be made by staining them with haematoxylin and eosin (*see* p. 437).

For Consideration
How does the microscopic structure of skeletal muscle reflect the way it works?

Requirements
Frog for dissection (can be shared between several students)
Dissecting instruments and bristle
Microscope
Slides and coverslips

NaCl (0.75 per cent)
Acetic acid

LS skeletal muscle
Electron micrograph of skeletal muscle fibrils (LS)
(Materials for permanent preparations using haematoxylin and eosin).

Questions and Problems

1 Comment on Fig. 20.4 which shows (a) the action potential and (b) the tension recorded from an isolated muscle fibre of the frog at 20°C.

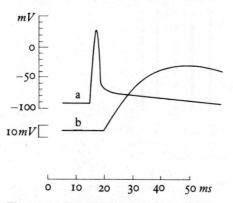

Fig. 20.4 Simultaneous recording of action potential and tension in an isolated muscle fibre of frog. (*After* Hodgkin and Horowicz, 1957).

2 In suitable conditions isolated segments of mammalian intestine will contract spontaneously and rhythmically even after all their nerve plexuses have been inactivated by drugs such as nicotine or cocaine. How would you account for this?

It has been found that spontaneous movements of the mammalian intestine are enhanced (i.e. increased in frequency and amplitude) by acetylcholine, but suppressed by adrenaline. Explain.

3 Draw up a table in which skeletal, visceral and cardiac muscle are compared from both a structural and functional point of view.

4 What is the all-or-nothing law? The nerve of a nerve–muscle preparation (*see* p. 209) was stimulated with a series of single shocks of gradually increasing intensity. The responses of the muscle, recorded on a slowly revolving drum, are shown in Fig. 20.5. How would you reconcile these results with the all-or-nothing law? What would you expect the results to be if (a) the whole muscle and (b) a single muscle fibre were stimulated direct?

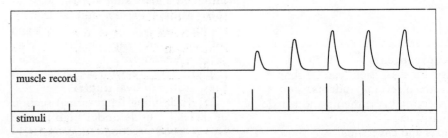

muscle record

stimuli

Fig. 20.5 Kymograph recording of responses of the muscle of a nerve–muscle preparation to stimulation of the nerve with a series of shocks of gradually increasing intensity. The relative intensity of the stimuli is indicated by the heights of the vertical lines in the lower record.

5 Why is it that in some longitudinal sections of skeletal muscle, viewed in the electron microscope, the filaments are arranged as in Fig. 20.6A whereas in other sections taken from the same muscle they are arranged as in Fig. 20.6B.

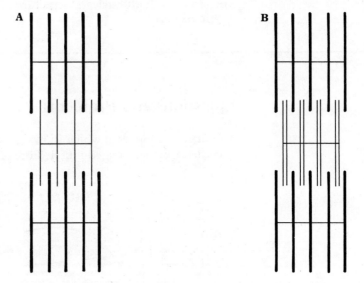

Fig. 20.6 The thick and thin filaments of a skeletal muscle fibril as they appear in the electron microscope.

6 What would you expect Fig. 20.6A to look like if the muscle was fully contracted? Explain in as much detail as you can how the change from the relaxed to the contracted state is brought about.

7 Comment briefly on each of the following statements:
(a) The flatfish *Paralichthys albiguttus* can change the colour and pattern of its markings to suit its background.
(b) In the flight muscles of the bee mitochondria are particularly numerous and their cristae are very close together.
(c) If a tentacle of the sea anemone *Anemonia* is stroked repeatedly with a fine glass rod the withdrawal response spreads to neighbouring tentacles, but the discharge of nematoblasts is confined to the tentacle which is stimulated.
(d) Bats and swamp-dwelling fishes both have to find their way about in the dark.

8 What factors influence the time a muscle takes to contract fully to a single impulse received from its nerve? The following table shows the contraction times of various muscles in the cat:

Muscle	Contraction time
Internal rectus	7.5 ms
Inferior oblique	18.7 ms
Gastrocnemius	39 ms
Diaphragm	480 ms
Stomach	2.2 s

Discuss these figures.

9 Give an account of the appearance of skeletal muscle as seen in longitudinal section under a good optical microscope. Explain its histological appearance in terms of its fine structure as deduced from electron micrographs.

21 Locomotion

Background Summary

1 Locomotion is generally brought about by a **musculo-skeletal system**. The skeleton may be an **endoskeleton**, **exoskeleton**, or **hydrostatic skeleton**. The skeleton is operated by sets of **antagonistic muscles**.

2 In considering the locomotion of any animal three things must be taken into account: **propulsion**, **support**, and **stability**. These can be conveniently dealt with in relation to the medium in or on which the organism moves: **water** (e.g. fishes), **land** (e.g. tetrapods), or **air** (e.g. birds).

3 In fishes propulsion is generally achieved by side-to-side movement of the tail (or sometimes the whole body), achieved by differential contraction of the **myotomes**. The mechanical principles involved in this kind of propulsion also apply to spermatozoa, many aquatic invertebrates, and flagella.

4 In cartilaginous fishes (elasmobranchs) support is achieved by the large **pectoral fins** and the **heterocercal tail fin**. In bony fishes (teleosts) support is achieved by the buoying effect of the **swim bladder**, the tail fin being **homocercal**.

5 Stability in fishes is achieved by the various fins collectively counteracting **yawing**, **pitching**, and **rolling**.

6 Some aquatic animals propel themselves by a mechanism akin to that of the breaststroke of a swimmer. This also applies to the action of cilia.

7 Although terrestrial locomotion can be achieved by undulatory movements, it generally involves the use of **limbs** acting as **levers**. For effective action the limbs must be operated by appropriate muscles acting across well lubricated **joints**. The various muscles are conveniently classified according to the effects they produce.

8 In tetrapods support is achieved by the limbs acting as **struts** and the **vertebral column** as the span of a **cantilever bridge**.

9 Stability in a moving tetrapod is maintained by its **diagonal locomotory pattern** aided by **vestibular reflexes** and reflexes arising from various **proprioceptors**, notably muscle spindles.

10 Locomotion in air depends on the development of **flight**. The same aerodynamic principles apply to all flying animals but the structural and physiological basis of flight varies from one group to another. This can be appreciated by comparing the flight mechanisms of birds and insects.

11 At the cellular level movement is poorly understood. In the case of **amoeboid movement** it involves changes in the physical properties of the cytoplasm, though the exact mechanism is disputed.

Investigation 21.1

Types of musculo-skeletal system

Skeletons may be conveniently classified, on the basis of their topographical relationship to the muscles that operate them, into **endoskeletons** where the skeleton is internal to the muscles, and **exoskeletons** where the skeleton is external to the muscles. The endoskeleton is typical of vertebrates, the exoskeleton of arthropods. To these two types of skeleton may be added a third, the **hydrostatic skeleton**, where the muscles surround a fluid-filled cavity. This is found in various soft-bodied invertebrates.

Procedure

ENDOSKELETON

As an example of an endoskeleton, examine the tail of a dogfish. Cut a hand-section a short distance behind the pelvic fins and remove the skin from one side.

In the end-on view notice that the **vertebral column** is completely enveloped by blocks of muscle, the **myotomes** (Fig. 21.1A). The individual muscle fibres have been cut in cross section so it is apparent that they are orientated longitudinally.

In side-view notice the connective tissue sheets, the **myocommata**, between successive myotomes. Tease out some of the **muscle fibres** and confirm that they are orientated longitudinally.

The fibres of each myotome run longitudinally from one vertebra and myocomma to the next. The vertebrae, myotomes and myocommata are all serially repeated along the body, i.e. they are **metamerically segmented**. What is the significance of this? Why are the myotomes ⋛-shaped when viewed from the side?

The effectiveness of the myotomes in generating propulsive movements can be realized by watching fish in an aquarium tank.

Examine the **vertebral column** in side-view, noting its segmented pattern and the structures shown in Fig. 21.1B. (The right-hand diagram shows a single vertebra in end-on view). Why are the neural spines so small? (Compare with mammal, p. 221).

A Transverse section of tail of dogfish

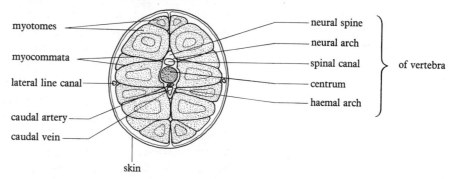

B Vertebral column of dogfish in side view

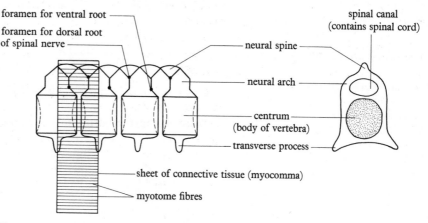

Fig. 21.1 The locomotory apparatus of the dogfish. The haemal arch shown in **A** is formed by the union of the left and right transverse processes. It envelopes and protects the caudal artery and vein.

H

Examine a transverse section from an embryo dogfish (caudal region) to see details of the vertebra, myotomes and myocommata. Notice that the muscle fibres are orientated longitudinally.

EXOSKELETON

Remove one of the largest legs from a crab (the shore crab *Carcinus* is recommended on account of its large size). With a scalpel blade cut a rectangular window in the side of the largest segment of the limb (i.e. the fourth from the end), taking care not to damage the internal structures.

When you have cut your window you will see the white **flexor** and **extensor** muscles inside the limb. Note that the muscles are enclosed *within* the hard cuticle (exoskeleton). Remove the muscles (which are soft), leaving the tendon-like **apodemes** to which they are attached. With small forceps grasp each of the two main apodemes in turn and pull gently. Note that one of the apodemes flexes the leg, the other extends it.

Investigate the joint, as fully as you can, using a hand lens or binocular microscope. How do you think it works? The crab's leg may seem a clumsy affair but it is by no means ineffective as a cursory examination of a living specimen will show. For a more efficient example of an exoskeleton in action observe a cockroach running across the floor of a cage.

HYDROSTATIC SKELETON

Observe an earthworm crawling forwards on damp filter paper in a dish. Observe successive 'peristaltic' bulges which travel from the head backwards. These are accompanied by protraction of the **chaetae** (bristles) so the combined effect results in an effective propulsion.

The hydrostatic skeleton consists of the fluid-filled **body cavity** surrounded by a muscular **body wall**. The fluid is kept under pressure by contraction of the surrounding muscles.

Make a short longitudinal slit in the dorsal body wall just in front of the clitellum. Observe the slit as the worm crawls. When is the slit opened and closed in relation to the passage of a 'peristaltic' bulge? Repeat these observations on a transverse slit. What conclusions can you draw regarding the muscles in the body wall?

With small scissors make a mid-dorsal longitudinal incision starting just behind the clitellum and extending posteriorly for approximately 30 segments. The incision must cut right into the body cavity. Notice that fluid issues from the cut. What effect does this operation have on locomotion? How does the appearance of the worm in the region of the slit differ from its appearance elsewhere? Open the slit with forceps and notice intersegmental **septa** occurring between successive segments. What does all this tell us about the muscles and skeleton?

Examine a transverse and longitudinal section of the earthworm under low power. Notice the large body cavity and thick muscular body wall. The body-wall muscles are in two layers: outer layer of **circular muscles** and an inner layer of **longitudinal muscles** whose fibres are neatly arranged in bundles. Which is the thickest layer, and why? Will these muscles account for your earlier observations on the live worm?

The effectiveness of the hydrostatic skeleton can be seen by placing a worm on the surface of damp soil in a jar. If one side of the jar is shielded from light it will probably burrow into the soil close to the glass. Watch carefully.

Compare with what happens if the worm is placed on a sheet of wet glass. Conclusions?

For Consideration

Compare the relative merits and demerits of the endo-, exo-, and hydrostatic skeletons from a structural and functional point of view.

Requirements
Microscope
Hand lens or binocular microscope
Dissecting instruments
Jar of damp soil
Sheet of glass

Dogfish tail (several students can share)
Vertebral column of dogfish
Fish (any kind that uses tail propulsion) in tank
Largest leg of crab (e.g. *Carcinus*)
Live crabs in dish
Cockroaches in cage
Earthworm on damp filter paper in dish
TS and LS earthworm (preferably intestinal region)
TS dogfish embryo (tail region)

Investigation 21.2

Functional analysis of the muscles in the hind leg of the frog

The purpose of this exercise is to investigate the functions performed by the muscles in the hind limb of a vertebrate. At one time such an analysis would have been strictly anatomical, each muscle being carefully dissected away from the skeleton and named according to its position. We shall adopt a more functional approach, the muscles being described in terms of their functions.

Method

A satisfactory way of analysing the leg muscles functionally is to skin one of the hind legs of a double-pithed frog and stimulate individual muscles electrically.

A **stimulator** capable of delivering high-frequency shocks of different intensities is required. The inductorium described on p. 435 is satisfactory: the interruptor at the end of the primary coil should be used with terminals **C** and **R**. Connect the secondary coil to a pair of stimulating electrodes: these should be wooden-handled mounted needles (Fig. 21.2).

Stimulate the muscles with one electrode; the other may be placed in contact with any part of the animal, or with the moist surface on which the animal is lying.

If a stimulator is not available, single pulses can be achieved by connecting the electrodes directly to a 4-volt battery.

For part of the investigation you will want to have the preparation in a horizontal position, dorsal or ventral side uppermost depending on which muscles you are investigating: pin or hold it to the floor of a clean dissecting dish.

On occasions you will need to have it in a vertical position: pin it through the jaws to a cork held in a clamp on a stand.

Procedure

(1) You should be provided with a double-pithed frog, that is, one whose brain and spinal cord have been destroyed. (Why is it necessary for the spinal cord as well as the brain to be destroyed?)

(2) Skin both legs and the lower part of the trunk. Keep the muscles well moistened with Ringer's solution.

(3) With the preparation horizontal or vertical as appropriate, stimulate each individual muscle with *weak* repetitive shocks. The shocks must be

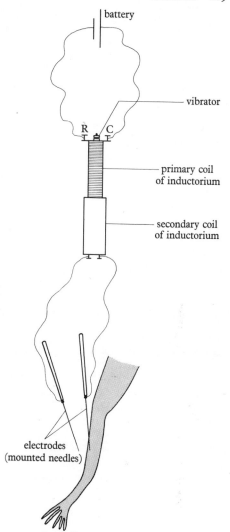

Fig. 21.2 Apparatus for stimulating the muscles in the hind leg of a frog.

as weak as possible otherwise stimulation will spread to adjacent muscles producing confusing results. It is not easy to see the boundary between neighbouring muscles, but the functional distinction between them can be readily appreciated by moving the stimulating electrode here and there and observing the different responses. Sometimes a very slight shift in the position of the electrode evokes a completely different movement of the leg, indicating that a different muscle is being brought into action.

(4) First systematically stimulate all the **superficial muscles** in the leg. Cover the dorsal, ventral, inner, and outer surfaces, and include the muscles at the extreme base of the leg in the pelvic region as well as those at the distal end of the foot. Build up a diagram as you go along, labelling the muscles according to what they do (*see* classification overleaf).

(5) When you have completed your analysis of the superficial muscles, dissect them away, taking care not to damage the muscles beneath. Then stimulate the **deeper muscles** and determine their actions in the same way as you did for the superficial muscles.

DISSECTION OF PRESERVED FROG

It is much easier to see the individual muscles in a preserved frog than in a fresh preparation. It is, therefore, helpful to have a preserved frog alongside you and to systematically dissect its muscles as you proceed with your analysis. This will help you to be certain that the diagrams you build up are anatomically, as well as functionally, correct.

NAMING THE MUSCLES

In your diagrams label the muscles according to the following classification, adding in each case any necessary qualifications:

Protractor: draws femur forwards

Retractor: draws femur backwards

Abductor: draws femur outwards

Adductor: draws femur inwards

Knee flexor: draws femur and tibia-fibula towards each other, i.e. closes knee joint

Knee extensor: draws femur and tibia-fibula away from each other, i.e. opens knee joint

Ankle flexor: draws foot towards tibia-fibula, i.e. closes ankle joint

Ankle extensor: draws foot away from tibia-fibula, i.e. opens ankle joint

Inward thigh rotator: rotates femur inwards

Outward thigh rotator: rotates femur outwards.

Also make clear which muscles are **intrinsic** and which **extrinsic**, and which ones are **antagonistic**.

For Consideration

(1) How many muscles have you accounted for in the hind leg of the frog?

(2) How many of the muscles belong to each of the groups listed above?

(3) Which of the muscles do you consider to be the most important in normal locomotion of the frog?

(4) Although numerous muscles are responsible for operating the hind leg, there is only one nerve (the sciatic nerve) in the leg. How is it that this one nerve can co-ordinate the actions of all the muscles?

Requirements

Stimulating equipment
Wooden-handled mounted
   needles (2)
Clean dissecting dish
Large cork
Stand and clamp
Flex

Frog Ringer's solution
Double-pithed frog
Preserved frog

Investigation 21.3

The mammalian skeleton

For convenience the skeleton is broken-down as follows:

(1) **Axial**
  (a) Skull
  (b) Vertebral column (backbone)
  (c) Ribs
  (d) Sternum (breastbone)

(2) **Appendicular**
  (a) Limbs
    (i) Forelimb (arm)
    (ii) Hindlimb (leg)
  (b) Girdles
    (i) Pectoral (shoulder) girdle
    (ii) Pelvic (hip) girdle

Examine a mounted rabbit skeleton and identify the various parts (Fig. 21.3). It is composed almost entirely of bone, with cartilage only at the articular surfaces. (The microscopic structure of bone and cartilage are dealt with on p. 18). What functions are performed by the skeleton as a whole?

Examine individual bones and relate them to the mounted skeleton. Pay particular attention to:

 (i) **articulating surfaces**: try fitting bones together, estimating how much freedom of movement they have.

 (ii) **projections**: these are usually for the attachment of tendons and ligaments for which they are often adapted by being flattened.

When examining the vertebrae start with a **lumbar vertebra** and then compare it with the others.

In studying the skeleton see if you can answer the following questions:

(1) Why do the lumbar vertebrae have particularly large centra?

(2) What function might be performed by the prominent lateral processes which project from the sides of the lumbar vertebrae, but are not found elsewhere in the vertebral column?

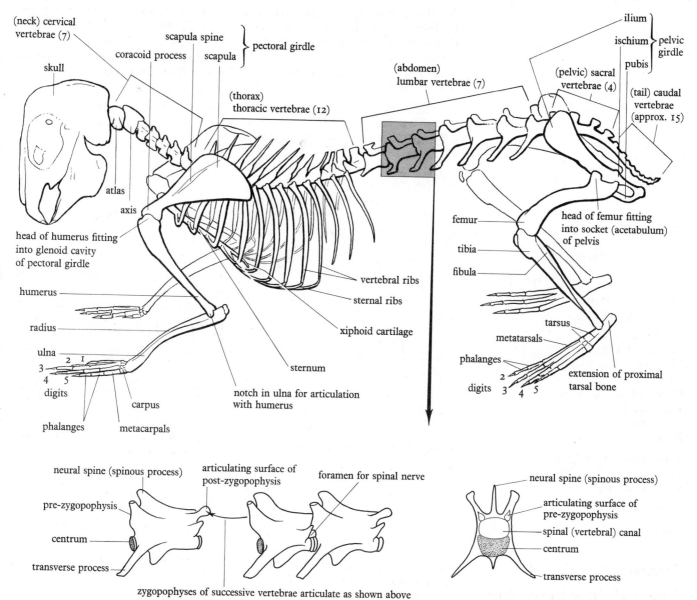

(neck) cervical vertebrae (7)

skull

coracoid process

scapula spine

scapula

pectoral girdle

ilium

ischium

pubis

pelvic girdle

(abdomen) lumbar vertebrae (7)

(thorax) thoracic vertebrae (12)

(pelvic) sacral vertebrae (4)

(tail) caudal vertebrae (approx. 15)

atlas

axis

head of humerus fitting into glenoid cavity of pectoral girdle

humerus

radius

ulna

3
4 5
2 1

digits

phalanges

metacarpals

carpus

vertebral ribs

sternal ribs

xiphoid cartilage

sternum

notch in ulna for articulation with humerus

head of femur fitting into socket (acetabulum) of pelvis

femur

tibia

fibula

tarsus

metatarsals

phalanges

digits

2
3 4 5

extension of proximal tarsal bone

neural spine (spinous process)

articulating surface of post-zygopophysis

foramen for spinal nerve

pre-zygopophysis

centrum

transverse process

zygopophyses of successive vertebrae articulate as shown above

neural spine (spinous process)

articulating surface of pre-zygopophysis

spinal (vertebral) canal

centrum

transverse process

Fig. 21.3 Diagram of the skeleton of rabbit. Below the complete skeleton are four lumbar vertebrae, in side view (left) and anterior view (right).

(3) Why are the neural spines of the lumbar and thoracic vertebrae relatively long compared with most of the cervical vertebrae?

(4) Why are the transverse processes of the atlas vertebra, and the neural spine of the axis vertebra, relatively large and flat?

(5) You will notice that the ribs are double-headed. With what specific parts of the thoracic vertebrae do the heads articulate? What is the point of this double articulation?

(6) Why are the four sacral vertebrae fused together?

(7) In life the ilium is firmly, but not immoveably, attached by ligaments to the first sacral vertebra. Functional significance?

(8) On the other hand the scapula is more loosely attached to the vertebral column. Why?

(9) What function(s) would you ascribe to the pubis and why are the two halves of the pelvic girdle fused together?

(10) Why are the ilium and scapula flattened?

(11) What is the scapula spine for?

(12) Why is the coracoid greatly reduced compared with more primitive tetrapods?

(13) What functions would you ascribe to the clavicles and sternum?

(14) Some of the limb bones possess bumps, crests and grooves. What are these for?

(15) How does the articulation of the forelimb with its girdle compare with that of the hindlimb?

(16) How does the hip joint compare with the other joints of the hindlimb? Functional significance? What about the forelimb?

(17) What explanation would you suggest for the lower part of each limb consisting of *two* parallel bones?

(18) What function do you think is performed by the backward extension of the tarsus in the hindlimb? Is there a functional equivalent in the fore-limb? If so, where?

(19) Any suggestion as to why the hindlimb has only four digits?

(20) Compare the relative sizes of the different components of the forelimb and hindlimb. What significance would you attach to the differences?

For Consideration
To what extent does the rabbit skeleton conform to D'Arcy Thompson's canti-lever analogy?

Requirements
Mounted skeleton of rabbit
Individual bones of rabbit

Investigation 21.4

Formation and structure of a limb bone

Bone is formed as a result of the activities of embryonic **mesenchyme cells**. Mesenchyme cells may develop directly into membrane (dermal) bones (e.g. scapula, skull roof, etc.), or into cartilage which is then turned into bone (ossified). Limb bones are always pre-formed in cartilage.

Procedure
Examine a longitudinal section of a limb bone of a mammalian embryo. The main body of the limb bone is called the **shaft**, the ends are known as the **epiphyses**. To understand its microscopic structure it is necessary to know how it develops. The sequence of events can be summarized as follows (Fig. 21.4):

(1) Mesenchyme cells arrange themselves into three layers (Fig. 21.4A). The outer layer gives rise to **fibroblasts** which form connective tissue (collagen), the middle layer gives rise to **osteoblasts** which are destined to form bone, and the inner core gives rise to small **chondroblasts** which form small-celled cartilage.

(2) An area of **large-celled (hypertrophic) cartilage** appears in the centre of the shaft (Fig. 21.4B). The chondroblasts are arranged in columns which lie parallel to the long axis of the shaft.

(3) The cartilage in the centre of the shaft is invaded by **blood vessels** and **osteoblasts** which migrate inwards from the periphery and work their way towards the two ends of the shaft (Fig. 21.4C), laying down bone as they do so.

(4) A cavity develops in the centre of the shaft, the **marrow cavity**. This contains soft fatty tissue and numerous blood sinuses where red blood cor-puscles and certain types of white blood corpuscles (which types?) are formed. This is the **bone marrow** (Fig. 21.4D).

Examine your section under low power and identify the areas shown in Fig. 21.4D. Now concentrate on the region where cartilage is being replaced by bone (Fig. 21.4D$_1$). The innermost hypertrophic cartilage cells secrete an enzyme which causes the cartilage to calcify. This isolates the cells from their source of nourishment, so they die and disrupt. As this happens they are invaded by osteoblasts which pro-ceed to secrete bone matrix. The bone formed at this stage is permeated by numerous cavities containing blood vessels, and the osteocytes (as the bone cells are now called) are somewhat irregularly arranged. This type of bone is called **spongy** or **cancellous bone**.

Find a cartilage cell which is being invaded by osteoblasts and examine it in as much detail as you can (Fig. 21.4D$_2$).

Examine spongy bone under high power (Fig. 21.4D$_3$). Notice the numerous cavities and the stellate form of the osteocytes.

At birth the cartilage in the epiphyses becomes ossified. However, the bands of hypertrophic cartilage persist, thus allowing the bone to continue growing in length. Increase in girth of the shaft takes place by the addition of new bone by osteoblasts just beneath the periosteum.

As new bone is laid down, old bone is resorbed at the centre. This is achieved by **osteoclasts**, large multi-nucleate cells formed by the fusion of certain osteoblasts. Normally resorp-tion of old bone keeps pace with the formation of new bone, and the marrow cavity expands accordingly.

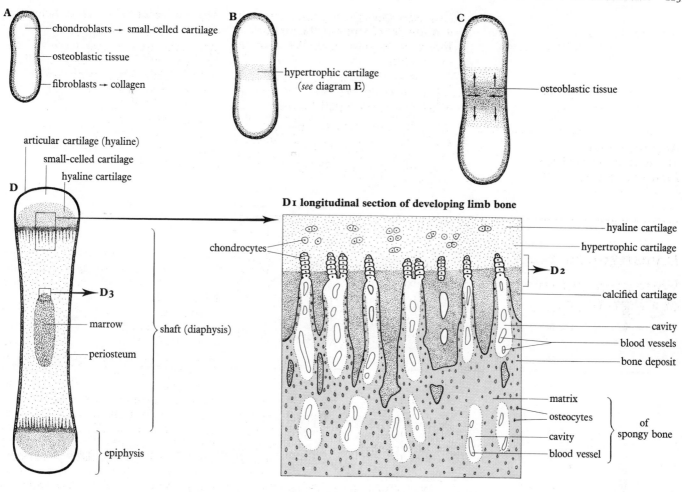

A
chondroblasts → small-celled cartilage
osteoblastic tissue
fibroblasts → collagen

B
hypertrophic cartilage (*see* diagram **E**)

C
osteoblastic tissue

D
articular cartilage (hyaline)
small-celled cartilage
hyaline cartilage
marrow
periosteum
shaft (diaphysis)
epiphysis

D1 longitudinal section of developing limb bone
chondrocytes
hyaline cartilage
hypertrophic cartilage
D2
calcified cartilage
cavity
blood vessels
bone deposit
matrix
osteocytes
cavity
blood vessel
of spongy bone

D2 Hypertrophic cartilage in detail

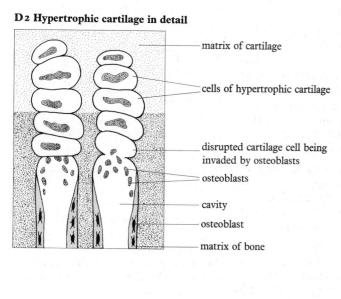

matrix of cartilage
cells of hypertrophic cartilage
disrupted cartilage cell being invaded by osteoblasts
osteoblasts
cavity
osteoblast
matrix of bone

D3 Spongy bone and osteoclasts in detail
osteocyte
cavity containing blood vessels
matrix of bone
osteoclast
bone marrow

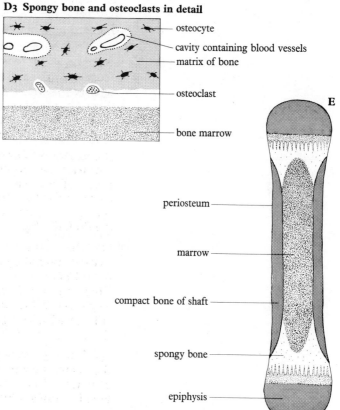

E
periosteum
marrow
compact bone of shaft
spongy bone
epiphysis

Fig. 21.4 Development of mammalian limb bone.

Examine spongy bone in the immediate vicinity of the marrow cavity. Can you see any osteoclasts? (Fig. 21.4D₃).

Osteoclasts are important throughout the time the bone is developing. Their eroding activities enable the bone to be continually changed and re-modelled to meet the stresses and strains to which the limb is subjected. They are also responsible for cutting channels into new bone thereby enabling the invasion of blood vessels. These channels become the **Haversian canals** of mature **compact bone.** The sides of the shaft of a fully formed limb bone are composed of compact bone (Fig. 21.4E). The epiphyses consist of a superficial layer of compact bone with spongy bone beneath.

Examine compact bone in transverse and longitudinal section (*see* p. 18). Notice the Haversian pattern.

Requirements
LS developing limb bone
TS and LS compact bone

In what respects is this organization suited to the carrying of heavy loads?

For Consideration
How is the formation and structure of a limb bone related to the functions that it has to perform?

Questions and Problems

1 'An animal propels itself forward by pushing backwards against its surroundings.' Discuss.

A science-fiction film depicted an organism from another planet as moving by means of a rotating propeller. Do you think this is feasible?

2 Devise a method for measuring the force exerted by an earthworm moving forward on a horizontal surface? How is its movement achieved?

3 An eel swims forward by throwing its body into waves of undulation which pass from head to tail. What mechanical forces are set up during this process and what is their musculo-skeletal basis?

The marine ragworm *Nereis* swims forward by the same method as the eel, but the waves, instead of passing from head to tail, travel from tail to head. The only external structural difference between an eel and *Nereis* is that the latter has paddles projecting from the sides of the body (Fig. 21.5). Why should the possession of these paddles make it necessary for the propulsive wave to travel from tail to head?

Fig. 21.5 The ragworm *Nereis* swims by waves of undulation which pass forward along the body from tail to head. (*From* Gray, J., *How Animals Move*, Cambridge University Press).

4 In what respects do the fins of a bony fish such as a cod differ from those of a shark? How would you explain the differences?

5 To what extent do the stabilizing mechanisms of fishes compare with those of aircraft?

In his investigations into the stabilizing mechanisms of the dogfish, J. E. Harris built an accurate model of this fish with detachable fins. He suspended the model in a wind tunnel and measured the effect of air currents on the model when turned in various directions with or without certain fins (Fig. 21.6). In one series of experiments he swung the model through a fixed angle about the centre of gravity, either to the left or right, and he measured the resulting forces exerted against the fish. His results are summarized graphically in Fig. 21.6C. Explain them, and draw such conclusions as you can.

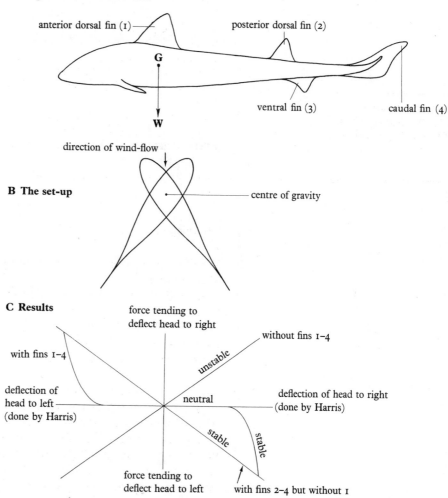

A Dogfish model in side-view

anterior dorsal fin (1)

posterior dorsal fin (2)

G

ventral fin (3)

caudal fin (4)

W

direction of wind-flow

B The set-up

centre of gravity

C Results

force tending to deflect head to right

without fins 1–4

with fins 1–4

unstable

deflection of head to left (done by Harris)

neutral

deflection of head to right (done by Harris)

stable

stable

force tending to deflect head to left

with fins 2–4 but without 1

Fig. 21.6 Harris' experiments on stability of a model dogfish in a wind tunnel. **A** shows the model dogfish in side view with the positions of the various detachable fins; the weight of the fish **W** acts through the centre of gravity **G** about which the model is rotated in the wind tunnel as shown in **B**. **C** shows the results in summarized graphical form.

Do you consider that experiments on models in wind tunnels can form a basis for drawing valid conclusions as to what occurs in the real animal?

6 Give an example, with brief explanatory notes, of each of the following:
(a) an animal which has a hydrostatic skeleton as well as an exoskeleton;
(b) an animal which swims by a 'jet-propulsion' mechanism;
(c) an animal which moves sideways;
(d) a fish that walks;
(e) an animal which uses its hind legs for jumping and swimming;
(f) a mammal that flies;
(g) a fish that glides.

7 How is the structure of bone, articular cartilage, a tendon and a ligament suited to the tasks which these structures have to perform?

8 Fig. 21.7 shows three different types of lever. Give examples, with diagrams, of specific situations in the human musculo-skeletal system where each type of lever occurs.

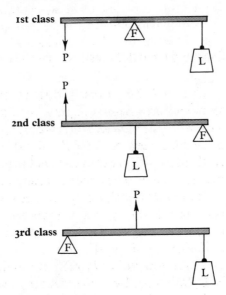

Fig. 21.7 Three classes of lever. **P**, power; **F**, fulcrum; **L**, load.

9 'A cow has four legs, the milking stool has only three.' Discuss.   (*OCJE*)

10 Discuss in physiological terms the problem of learning to ride a bicycle.

11 Compare the flight mechanism of birds and insects.

22 Behaviour

Background Summary

1 Techniques for investigating animal behaviour include kymographs, slow-motion cinematography, time-lapse and multiple-flash photography, various types of activity recorder, and direct observation.

2 Behaviour can be broadly divided into **species-characteristic** and **individual-characteristic** behaviour. Both involve certain fundamental processes, e.g. reflex action, orientation and learning.

3 **Reflex action** includes the escape responses of many invertebrates, which, because of their comparative simplicity, lend themselves readily to experimental analysis. Such is the case with the giant-axon reflexes of the earthworm and squid.

4 **Orientation behaviour**, commonly divided into **kinesis** and **taxis**, is well illustrated by the responses of small organisms such as flatworms to food and light.

5 Given that the animal is appropriately motivated, species-characteristic behaviour is initiated by **releasing stimuli** and ended by **terminating stimuli**. This is seen in numerous cases of territorial behaviour, threat and appeasement displays. It is apparent that in many instances a **stimulus-filtering mechanism** must be involved.

6 In many cases of species-specific behaviour, particularly those involving sexual activity, synchronization between the participating individuals is essential. This is illustrated by the stickleback in whose **chain behaviour** the importance of reciprocal **sign stimuli** has been demonstrated. In other animals external chemical stimuli (**pheromones**), and **hormones**, may influence the animal's motivational state. The latter are closely associated with the hypothalamus.

7 Under certain circumstances species-characteristic behaviour may entail aberrant behaviour, as is illustrated by **displacement** and **vacuum activity**.

8 **Learning** can be conveniently, if somewhat artificially, divided into **habituation, associative learning, imprinting, exploratory learning** and **insight learning**. Associative learning can be distinguished into the **conditioned reflex** and **trial and error learning**.

9 At the physiological level learning can be interpreted on a purely neural basis, or (more controversially) in biochemical terms as a change in the nucleic acid content of the brain cells. The two theories, which are by no means incompatible, are both supported by experimental evidence.

10 Insight learning can provide a clue as to the meaning of the term **intelligence**, a difficult concept to define and even more difficult to analyse at the physiological level.

Investigation 22.1

Analysis of the behaviour of the earthworm

Because higher animals are so complex and difficult to work with, lower animals, like the earthworm, lend themselves more readily to analysis of behaviour. Here comparatively simple acts of behaviour can be analysed in terms of the workings of receptors, nerves and effectors. Moreover, operations can be readily carried out on the worm for it has no pain receptors and regenerates quickly afterwards. (Investigations 18.2, 18.6, and 21.1 should also be consulted.)

Experiments

(1) Observe a worm crawling on damp absorptive paper. Note that forward locomotion is achieved by **peristaltic swellings** passing from the anterior to posterior end.

(2) Allow the worm to crawl forward. Stimulate the head by stroking it gently with a camel-hair brush. What happens to the peristaltic swellings? Stimulate more strongly until **anti-peristalsis** occurs. What happens if and when a peristaltic wave meets an anti-peristaltic wave? What sort of nervous system might be involved in these responses?

(3) Apply a sharp stimulus to the head and note that this results in rapid shortening followed by anti-peristalsis. This constitutes the animal's **escape reaction** and is particularly important when the worm comes out on to the surface of the soil on warm, wet nights. The effectiveness of this reaction is increased by the **chaetae**. Notice what happens to the chaetae when (a) the head, and (b) the posterior end are stimulated.

Rapid shortening of the whole body is evoked by a sharp stimulus applied to either end. This escape response is mediated by impulses which are quickly transmitted along the length of the body by **giant axons** in the ventral nerve cord.

(4) Grasp the extreme posterior end of a worm with a pair of large forceps. Grip the body firmly but do not crush it. You may observe that the animal breaks in two close to where it is being grasped. This response is known as **autotomy**. What is its significance in the animal's life? Where precisely does the body break in relation to the point where it is being grasped? How do you think this response is brought about?

(5) Allow a worm to burrow in a jar of damp soil. After some searching it will probably do this close to the glass, especially if this is shielded from the light. Follow the process closely. Burrowing may take place in one of two ways:

(a) by the inserted head end swelling and forcing open a passage; the swollen head and anterior chaetae then act as an anchor whilst the rest of the body is drawn in;

(b) by the soil being eaten by the worm, passed through the gut and cast out (as 'worm castings') at the rear.

(6) What sort of structures are necessary for all this to take place? Examine transverse and longitudinal sections of the earthworm and notice especially the **longitudinal** and **circular muscles**, **chaetae** in their muscular sacs, **ventral nerve cord** (and possibly **segmental nerves**). The three **giant axons** (one median and two laterals) will be seen towards the dorsal side of the nerve cord.

(7) The ventral nerve cord is visible through the mid-ventral surface of the worm. Cut the cord about half-way down the body with a pair of fine scissors and make the animal crawl (a) on damp absorptive paper, i.e. on a rough surface, and (b) on wet glass, i.e. on a smooth surface. Note what happens.

(8) Now cut the worm in half and sew the two halves together with thread so they are about 10 mm apart. Repeat your observations of locomotion on rough and smooth surfaces. Can you draw any conclusions from this experiment? Do the results tell us anything about the mechanism by which the peristaltic wave is transmitted along the body?

(9) Do the impulses responsible for the propagation of the peristaltic waves travel the full length of the nerve cord or do they peter out as they are transmitted along the cord? Your findings in experiments (7) and (8) suggest what the answer *may* be, but you can only prove it by performing another experiment. Devise a suitable experiment to answer this question, and then carry it out.

(10) Remove the first four segments of a worm, thereby depriving it of its brain. Determine the effect of this on (a) normal peristaltic locomotion and (b) the escape response. What does this tell us about the role of the brain?

Requirements
Microscope
Large sheet of absorptive paper
Glass plate
Jar of damp soil
Dissecting dish or piece of cork
Dissecting instruments
Camel-hair brush
Mounted needle
Needle and thread
Slide and coverslip

Live earthworms (5) in Ringer's
 solution
Preserved earthworm

TS and LS earthworm

For Consideration
The behaviour of the earthworm can broadly be divided into (a) normal peristaltic locomotion, and (b) the rapid escape response.

(1) Summarize the part played by the muscles and nervous system in bringing about normal peristaltic locomotion.

(2) Reconstruct the reflex arc involved in the escape response including the receptors and effectors.

(3) To what extent do the results of your experiments relate to the behaviour of earthworms in their natural habitat?

Investigation 22.2

Orientation and feeding behaviour of flatworms

Although reflex action plays an important part in the behaviour of lower animals, more elaborate behaviour patterns are also shown. Comparatively simple behaviour patterns are seen in the way free-living flatworms orientate themselves in relation to environmental stimuli, notably light and food.

Experiments
(1) Watch the activities of up to ten flatworms in a dish in semi-darkness. Then switch on a lamp placed 60 cm above the dish. Observe their behaviour carefully and note any changes. Now bring the lamp closer to the dish by 15 cm and repeat your observations. Continue to do this until the lamp is only 15 cm from the dish. Now switch the lamp off and observe what happens. Interpret your observations. Could you make this experiment quantitative? Do so if you can.

(2) Place your lamp 20 cm above the dish and cover half the dish with a piece of opaque cardboard. Record what happens and interpret your results. Can you find possible **light receptors** by looking at a specimen under a hand lens or binocular microscope?

(3) Gently push the flatworms to one end of the dish and place a small piece of fresh liver at the other end. Observe what happens and make notes on the behaviour of the animals. Does the type of behaviour change as the animals get closer to the food? Describe fully and explain. What sort of receptors are involved and where are they?

(4) Induce a specimen to feed on a piece of fresh liver. Record its behaviour and observe the actions of the pharynx.

(5) Remove the **pharynx** from one of your larger light-coloured specimens. This can be done by placing the specimen on a slide and pressing down on it with a coverslip. The pressure usually causes the eversible pharynx to pop out and come adrift. Place the isolated pharynx in a watchglass of water and put some food in contact with it. Watch under the lowest power of the microscope. Explain what happens in terms of muscle action and nervous control.

For Consideration
In this investigation you have been observing the behaviour of a simple animal. You should now try to put forward a physiological explanation of your observations. Your explanation will be incomplete if it does not include reference to receptors (eyes, chemoreceptors, etc.), nervous coordination (brain, peripheral nerves, etc.) and locomotory equipment (muscles, cilia, etc.).

Requirements
Binocular microscope or hand lens
Microscope
Slide and coverslip
Dish (20 × 25 cm approx.)
Lamp
Opaque cardboard

Flatworms (up to 10, starved)
Fresh liver

Investigation 22.3

Orientation responses of woodlice to humidity

Assemble a humidity **choice-chamber**. A simple way to do this is shown in Fig. 22.1: Place two square or oblong dishes side by side, touching. Put water in one, and anhydrous calcium chloride in the other. On top of these two dishes place a plastic dish whose bottom has been replaced by a 1 mm mesh net held on with a rubber band.

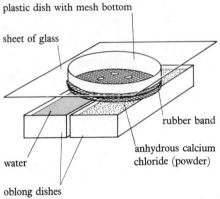

plastic dish with mesh bottom

sheet of glass

rubber band

anhydrous calcium chloride (powder)

water

oblong dishes

Fig. 22.1 A simple choice-chamber suitable for woodlice.

Put up to 20 woodlice on the net. Cover the plastic dish with a lid or sheet of glass. After allowing the woodlice to settle down, record how many are above the water dish and how many above the calcium chloride dish.

Now cover the whole assemblage with a cardboard box perforated by an observation slit. There should be only just enough light for you to see the woodlice.

Examine at intervals and record the percentage of woodlice over the water. Plot the percentage against time.

Requirements

Choice-chamber (*see* above)
Cardboard box with observation slit

Anhydrous calcium chloride

Woodlice (up to 20)

Further Experiments

(1) Devise an apparatus in which woodlice are presented with a humidity or temperature gradient. By means of such an apparatus investigate the range of humidity or temperature for which they show a preference.

(2) In their natural environment woodlice may not be confronted with a simple 'choice' of the kind that you have presented them with in the above experiment. It is possible that they may in reality have to choose between, say, a region with a favourable temperature but unfavourable humidity and a region with a favourable humidity but unfavourable temperature.

Devise, and if possible carry out, experiments to investigate the preferential behaviour of woodlice in relation to various combinations of the following conditions: humidity, temperature, illumination and contact with solid objects.

For Consideration

In their natural environment woodlice tend to congregate under logs, stones, etc. What particular feature of this situation do you think they find agreeable, and how is their behaviour adapted to enable them to seek out, find and finally settle in such regions?

Investigation 22.4

Orientation responses of blowfly larvae to light

Obtain a large sheet of non-white paper and draw on it a circle of diameter 24 cm. Divide this circle into 30° sectors.

With as little surrounding light as possible, position two lamps at approximately 90° to each other on each side of the circle (Fig. 22.2).

Place up to 10 blowfly larvae in the centre of the circle and switch one of the lights on. Record the number of blowflies in the sector of the circle leading directly to the light.

When all the larvae are moving in a definite direction relative to the light, switch that lamp off and the other one on. Record what happens.

Try different variations on this experiment. In all cases record the responses of the larvae, if possible quantitatively, in relation to light.

Place a few blowfly larvae on the surface of, e.g. damp sawdust or bran, in a beaker. What relevance does their observed behaviour have to the above experiments?

For Consideration

Relate these experiments to blowfly larvae in their natural habitat.

A Plan

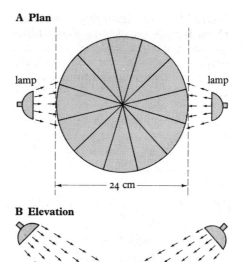

B Elevation

Requirements
Sheet of non-white paper at least
25 cm square (*see above*)
Lamps (2)
Beaker of damp sawdust or bran

Blowfly larvae (up to 10)

Fig. 22.2   Arrangement for Investigation 22.4 on the responses of blowfly larvae to light.

Investigation 22.5
Habituation in fanworms

Fanworms are polychaetes, a group of marine annelids. At the anterior end is an array of delicate **tentacles**, called the **branchial crown** (Fig. 22.3). This is used for gaseous exchange and also for filter feeding, the tentacles possessing numerous cilia whose beating carries food particles towards the mouth.

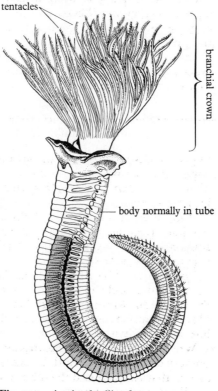

Fig. 22.3   A tube-dwelling fanworm.

In normal circumstances the branchial crown projects from the opening of a protective **tube**. The tentacles are very sensitive, responding to touch and in some cases to shadows falling on them.

Procedure
(1) Very gently brush the branchial crown with a mounted needle until the worm withdraws slowly. What sort of response is this, and through what kind of nerves is it mediated?

(2) Stimulate the branchial crown more strongly until the worm jerks back rapidly into its tube.

This rapid response is mediated through a **giant-axon system**. If available, examine a transverse section of a fanworm and note the **giant axon(s)** in the ventral nerve cord. Also notice the extensive **longitudinal musculature** in the body wall.

(3) Stimulate the branchial crown repeatedly at a regular frequency. Your stimuli should be sufficiently strong to elicit the rapid response. What happens to the rapid responses? Are slow responses still given after the rapid responses have ceased?

(4) Repeat (3) by stimulating a single tentacle with a needle in exactly the same place each time. After the responses have declined shift the point of stimulation to either another part of the tentacle or a different tentacle.

Record what happens. Conclusions?

(5) If stimulation of a single tentacle fails to evoke a rapid response, what happens if you stimulate two, three, or more tentacles simultaneously? Conclusions?

For Consideration

(1) List the possible sites in the giant-axon reflex pathway where transmission failure might be occurring in the decline of the rapid response.

(2) What further experiments might be done to locate the site exactly?

(3) What might be the adaptive value of the rapid response?

(4) What information do these experiments provide on the structural organization of the nervous system of this animal?

Requirements
Microscope
Apparatus for delivering single electrical stimuli (*see* p. 435)
Needle

Fanworm (e.g. *Sabella, Myxicola, Branchiomma, Eudistylia*). The worm should be in its tube in *cold* sea water.
TS fanworm

Investigation 22.6
Effect of thwarting on the behaviour of rats

It is often observed that if an animal is in a state of conflict, or in some way frustrated or thwarted, it may perform acts of behaviour which are unusual or inappropriate.

In this investigation the occurrence of such activities are examined in the rat.

Procedure
You should be provided with a laboratory rat in a cage. The rat has been starved for 48 hours. It should be provided with water but not food. It is advisable to work in pairs, one student observing the behaviour of the rat, the other making a record of the observations.

First spend at least 10 minutes observing your rat. This will allow the animal to get used to you, and will give you an opportunity to watch its various activities. You will probably find that it performs the following types of behaviour: **drinking**, **scratching**, **grooming**, **exploring**, and **inactivity** (**rest**). Learn to recognize each type of behaviour and to estimate the time spent on each.

When you have done this, proceed as follows:

NORMAL BEHAVIOUR
Record as accurately as possible the time (in seconds) spent on each of the above types of behaviour during the course of 20 minutes. Record the times in such a way as to make it clear in which five-minute period, in the course of the 20 minutes, each type of behaviour occurs.

FRUSTRATION BEHAVIOUR
Now present the rat with food in such a way that it can see and smell the food but cannot get at it.

Observe the animal's behaviour carefully. Record the time spent on the five different types of behaviour during the next 20 minutes. As before, make it clear in which five-minute period, in the course of the 20 minutes, each type of behaviour occurs.

FEEDING BEHAVIOUR
When the second 20 minutes are up, give the food to the rat and observe its behaviour. (There is no need to record the time spent in different activities.)

Results
Calculate the total time spent on each type of behaviour (a) without visible food, and (b) with visible food. Any significant differences? How do your results compare with those obtained by the rest of your class?

Next, calculate the time spent on each type of behaviour within each five-minute period within the 20 minutes. Do this for both situations (i.e. without visible food, and with visible food). Plot graphs showing the time spent on each type of behaviour (vertical axis) against the five-minute periods (horizontal axis).

Requirements

A plastic cage (30 × 45 cm is
  recommended)
Food pellets in glass dish
Laboratory rat in cage

The cage should have a wire top with
sunken compartments for water bottle
and food pellets.

The rat should be supplied with
water but deprived of food for 48
hours before the laboratory session.

The glass dish containing the food
pellets should be small enough to fit
into the food compartment of the
cage.

For Consideration

(1) Which particular type(s) of behaviour appear to increase in the
thwarting situation? How would you
explain the increase?

(2) Is the occurrence of the above
behaviour uniform during the thwarting situation, or does it show a
tendency to fall? Explain.

(3) In general an animal, when
thwarted, exhibits either displacement activity or redirected activity.
The former is irrelevant behaviour
which is completely out of context;
the latter is relevant behaviour but it
is directed towards the wrong object.

Of the types of behaviour which
show an increase in the thwarting
situation in your rat experiment, which
do you consider to be displacement
activity, and which is redirected
activity?

(4) Can you think of possible examples
of displacement and redirected activities in man?

(5) What functions, if any, do you
think are fulfilled by these kinds of
behaviour patterns?

Investigation 22.7

Learning in lower animals

It is claimed by many observers that
lower animals such as flatworms and
earthworms show associative learning.
In this type of behaviour the animal
learns to associate a specific stimulus
with a particular situation. In this
investigation you are invited to test
this claim for yourself.

Method

It is suggested that you should try to
teach the animal to associate a particular situation with an unpleasant
stimulus, which in this case is a weak
electric shock. For this you require an
apparatus capable of delivering repetitive stimuli at a high frequency. The
inductorium described on p. 435 is
satisfactory: the interruptor at the
end of the primary coil should be used
with terminals **C** and **R**. Connect the
secondary coil to a pair of stimulating
electrodes as shown in Fig. 21.2, p. 219.
For the stimulating electrodes use two
blunt seekers bound together with insulating tape. When stimulating the
animal, switch on the stimulator and
then touch the animal briefly with
the electrodes. The intensity of the
stimuli should be just sufficient to
evoke an observable response. Before you start training the animal you
must determine the correct intensity
of stimulation. To do this, start by
stimulating with weak shocks, then
gradually increase the intensity of the
stimuli until a clear response is given.
Don't over-stimulate: the object of
this experiment is to train the animal,
not to electrocute it.

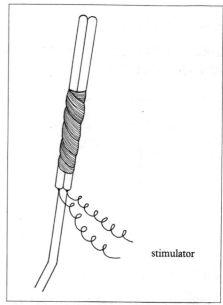

Fig. 22.4 The stimulating electrodes for Investigation 22.7. Two wooden-handled seekers
(probes) are bound together with tape and connected to the stimulator as shown here.

Procedure

(1) Try training a flatworm (e.g.
Planaria sp.) to associate a light stimulus with an electric shock. Place the
flatworm in a dish of water in semi-
darkness. When it has settled down,
shine a bright light on it and then,
about one second later, give it a brief
electric shock. The animal should give
no visible response to the light, but it
should respond to the electric shock.
Repeat at regular intervals to see if
eventually the animal responds as soon
as the light is shone on it, i.e. *before*
the shock is administered.

Record your results and if possible present them in the form of a graph.

(2) Try training an earthworm to turn left in a simple T-shaped maze. Start by placing the worm in the stem of the 'T' (Fig. 22.5). Encourage it to move forwards by gently stroking the posterior end. When it comes to the junction the worm should turn left or right. If it turns left and reaches the end of the cross-piece of the 'T', return it to the beginning and start again. If it turns right, stimulate it with a brief electric shock when it reaches the end, then return it to the beginning and start again.

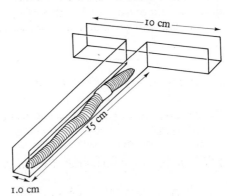

Fig. 22.5 A simple T-maze for training experiments with the earthworm. The maze can be conveniently constructed out of perspex. The passages should be sufficiently wide for the animal to fit comfortably and snugly into them, and the walls should be sufficiently high to prevent the worm climbing out.

Repeat the procedure to see if eventually the animal learns always to turn left so as to avoid the electric shock. The animal should be kept moist at all times.

Record your results and, if possible, present them in the form of a graph.

For Consideration
(1) What difficulties have you encountered in this investigation? How might they be overcome?
(2) Assuming you have been successful, the flatworm and the earthworm both show associative learning in these two experiments. However, the *type* of associative learning differs in the two cases. How do they differ?
(3) Can you think of any situations in their natural habitats where associative learning might be useful to these two animals?
(4) What further experiments might be carried out on learning in these animals?

Requirements
Stimulating equipment (*see* Fig. 21.2, p. 435)
Electrodes (*see* Fig. 22.4)
Dish
Lamp capable of producing bright light
'T' maze (*see* Fig. 22.5)

Flatworm (e.g. *Planaria* sp.)
Earthworm

Investigation 22.8
Learning in rats and gerbils

Many animals can learn to find their way through mazes of varying degrees of complexity, particularly if encouraged by a reward. This experiment involves training rats and/or gerbils to master a comparatively simple **maze**.

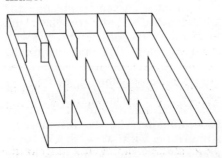

Fig. 22.6 Plan of maze suitable for use with rats and gerbils. The maze should be of approximately 50 cm side and at least 10 cm high.

Procedure
Use a maze of the type shown in Fig. 22.6. This can be constructed from a cardboard or wooden box fitted with hardboard or polystyrene partitions. The whole maze should have a transparent cover.

Put non-odorous food in the food box at the end of the maze and place an uninitiated gerbil or rat at the starting point. The animals should be deprived of food for up to 48 hours before the start of the experiment.

Now run a trial. This consists of letting the animal find its way to the food box. Time how long it takes and score the number of **errors** it makes in doing so. You must decide what constitutes an error: for practical purposes the animal may be considered to have made an error if its head passes a correct opening.

When the animal reaches the food box reward it with food and give it three-quarters of a minute to eat it. Then transfer it to the starting point and run another trial, recording the time taken to complete the maze, and the number of errors, as before.

If time permits continue the experiment until the animal completes the maze without making any errors.

Requirements
Maze of the type shown in Fig. 22.6
Processed rat or gerbil food (available from supplier)
Rat or gerbil, starved for 48 h before experiment

Results
Plot (a) the time taken to complete the maze, and (b) the number of errors (vertical axis), against the trials. Can any conclusions be drawn regarding the ability of rats or gerbils to learn?

For Consideration
(1) What bearing do your results have on the behaviour of these animals in their natural habitats?
(2) Can you suggest any factors that might speed up the process by which these animals learn to master the maze?
(3) What experiments might be carried out to compare the maze-learning abilities of small mammals with, e.g. cockroaches and toads?

Questions and Problems

1 To what extent can an animal's behaviour be explained by the capabilities of its receptors?

2 How far can the behaviour of animals be explained in terms of reflex action?

3 In an experiment on woodlouse behaviour a perspex-walled chamber 40 centimetres long with a constantly maintained humidity gradient from one end to the other was prepared. The chamber was kept under conditions of constant temperature and illumination for 48 hours. One hundred adult woodlice were placed in the centre of the chamber at 10 00 hours (Time 0). Observations were made at intervals and the distribution of woodlice was found to be as shown in the Table.

Relative humidity %	No. of woodlice			
	3 h	14 h	24 h	48 h
0–10	0	2	0	8
10–20	0	15	0	12
20–30	0	16	0	9
30–40	2	6	0	11
40–50	0	20	1	7
50–60	3	17	5	14
60–70	10	8	9	8
70–80	25	6	35	10
80–90	35	4	30	11
90–100	25	6	20	10

Comment on these results. Suggest an hypothesis to explain them. What experiments would you carry out to attempt to confirm your hypothesis?

(OCJE)

4 Discuss the meaning and usage of the terms 'instinct' and 'intelligence' in the context of animal behaviour.

5 Do plants behave?

6 The graphs in Fig. 22.7 show habituation to touch as seen in the tube worm *Branchiomma*. In each the responses of a group containing 17–21 worms were tested by brushing the protruding branchial crown. Two separate experiments were carried out. In the first experiment the worms were stimulated so gently that they only responded by withdrawing slowly into their tubes. In the second experiment the stimuli were sufficiently strong to evoke a rapid response from the worms. In Fig. 22.7 Graph A shows the results of the first experiment (slow withdrawals), and Graph B shows the results of the second experiment (rapid responses). Comment fully on these results.

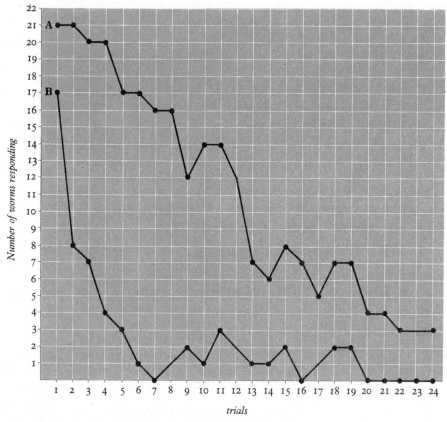

Fig. 22.7 Habituation to touch by the tube worm *Branchiomma*. **A**, slow withdrawals; **B**, rapid responses. (After Krasme).

7 Biologists are not agreed as to the precise meaning of the word 'intelligence', but it is generally held to represent the degree to which an animal can respond adaptively to new environmental situations. Bearing this in mind, outline the tests which you would make to investigate the intelligence of an animal.

8 In the behaviour of the British newt *Triturus* there is an elaborate and sustained encounter (lasting about 20 minutes) involving courtship display in which the female stimulates the male by prodding him. He responds by depositing a spermatophore which she then picks up with her cloaca. In the newt *Euproctus* which lives in fast-flowing streams in the Pyrenees, the male waits in hiding and pounces on the female as she passes. He then wraps his tail round her and transfers sperm to her.

Comment on these behaviour patterns. Discuss how each is adapted to its respective environment. Which do you think is the more primitive?

9 Devise an experiment involving associative learning which could show whether (a) a monkey was colour blind, or (b) a dog could differentiate between two high-pitched sounds.

23 Cell Division

Background Summary

1 Cell division is responsible for **reproduction** and **growth**. In both cases the chromosomes must be correctly distributed between the daughter cells.

2 Two main types of cell division are recognized: **mitosis** and **meiosis**. The latter is involved principally in the formation of **gametes**, or—in certain plants—**spores**.

3 Mitosis can be observed, and filmed, in plant endosperm tissue. Meiosis may be seen in developing spores, pollen grains, embryosacs, ovaries or testes.

4 A cell normally contains two of each type of chromosome, the **diploid state**. Mitosis preserves this condition. Meiosis, however, results in the daughter cells containing only one of each type of chromosome, the **haploid state**.

5 Both types of cell division involve the orderly movement of chromosomes on a **spindle apparatus**. The process can be conveniently, if artificially, divided into five stages.

6 In **interphase** the cell prepares for division, the genetic material replicating; in **prophase** the chromosomes make their appearance as distinct bodies; in **metaphase** they arrange themselves on the equator of the spindle; in **anaphase** they separate and move towards opposite poles of the spindle; and in **telophase** the cell divides into two.

7 In mitosis homologous chromosomes do not associate with one another. At metaphase they arrange themselves independently on the spindle, and at anaphase the sister chromatids of each chromosome part company independently of its homologue. Thus is the diploid state preserved.

8 In meiosis there are two successive divisions. In the first, homologous chromosomes come together (associate) and subsequently segregate into the daughter cells thus reducing the number of chromosomes from the diploid to the haploid state. In the second division the sister chromatids of each chromosome segregate into the daughter cells.

9 During prophase of the first meiotic division the chromatids of homologous chromosomes usually become attached at certain points called **chiasmata** where breakage and rejoining may occur. This permits **crossing over** to take place, a process with important genetic and evolutionary consequences.

Investigation 23.1

Observation of stages of mitosis in root tip

If the chromosomes are selectively stained with a dye such as **acetic orcein**, stages in mitosis can be observed in tissues where cell division occurs. An example of such a tissue is the **meristematic tissue** located in the apical meristem, or zone of cell division, in the tip of a growing root.

Procedure
(1) Cut off the apical 5 mm from the tip of a growing lateral root of, e.g. broad bean, or sunflower.

(2) Place the root tip in a watch glass containing acetic orcein stain and M HCl in the approximate proportions of ten parts of stain to one part of acid.

(3) Warm, but *do not boil*, for five minutes on a hotplate. The acid helps to macerate the tissue (why is that desirable?)

(4) Place the stained root tip on a clean microscope slide. Cut it in half transversely and discard the half furthest from the apex.

(5) Add two or three drops of acetic orcein to the root tip on the slide.

(6) Without interfering too much with the arrangement of the cells, break the root tip up with a needle so as to spread it out as thinly as possible.

(7) Put on a coverslip, cover it with filter paper and squash gently.

(8) Warm the slide on a hotplate for about ten seconds to intensify the staining. (The slide should be very warm, but not too hot to touch).

(9) Examine for stages in mitosis.

Requirements
Microscope
Slide and coverslip
Hotplate
Filter paper
Mounted needle
Razor blade
Watch glass

Acetic orcein
M HCl

Lateral root of, e.g. broad bean
LS root tip of, e.g. *Allium*
Slides of *Ascaris* for mitosis

Broad bean seeds should be germinated in damp peat 10 days before the laboratory session. When the radicle is approximately 12 mm long, cut off the tip to stimulate growth of lateral roots.

Recording your Observations
Divide a page of your laboratory notebook into eight boxes, labelled as shown below:

Mitosis in root tip of _____	
(1) Interphase	(2) Early prophase
(3) Late prophase	(4) Metaphase
(5) Early anaphase	(6) Late anaphase
(7) Early telophase	(8) Late telophase

Using Fig. 23.1 to help you, identify as many stages from mitosis as you can. Make annotated sketches in the appropriate boxes, showing the arrangement of the chromosomes. What can you say about the chromosome number of the plant from which the root was obtained?

Supplement the information obtained from your own slide by observing mitotic figures in a prepared longitudinal section of the root tip of, e.g. onion (*Allium* sp.). Compare with mitosis in an animal, e.g. *Ascaris*.

For Consideration
In what situations apart from those studied here would you expect to find mitosis taking place in animals and plants?

A) Interphase

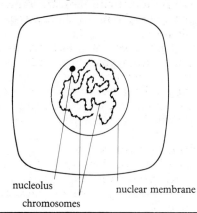

nucleolus   nuclear membrane

chromosomes

B) Early prophase

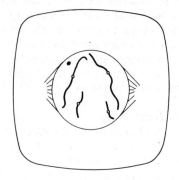

C) Late prophase

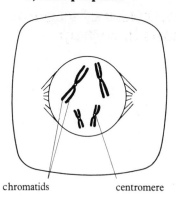

chromatids   centromere

D) Early metaphase

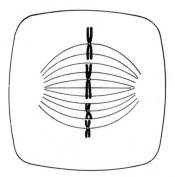

E) Late metaphase

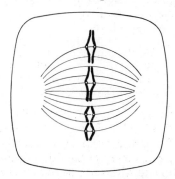

F) Early anaphase

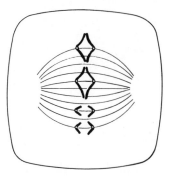

G) Late anaphase

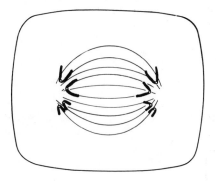

H) Early telophase

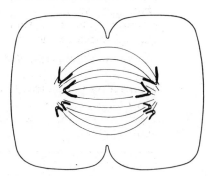

I) Late telophase

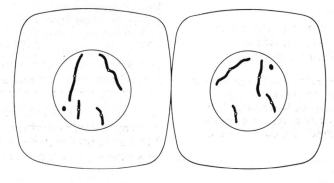

Fig. 23.1 Diagram illustrating the behaviour of chromosomes during mitosis. For simplicity only two pairs of chromosomes are shown, one short pair and one long pair.

Investigation 23.2

Observation of stages of meiosis in anther

Meiosis can be observed in immature anthers that are still enclosed inside the flower bud. Within such anthers diploid **pollen mother cells** may be found dividing meiotically to form haploid **pollen grains**. Anthers displaying a pale creamy colour generally have meiotic stages in them.

Hyacinth is recommended. A hyacinth bulb contains a dormant inflorescence of flower buds. Each flower bud contains six stamens. The technique involves squashing an anther and staining it with acetic orcein.

Procedure

(1) Take a hyacinth bulb and remove the enveloping leaves so as to expose the inflorescence. The flower buds at the base of the inflorescence are the most advanced, those at the apex are the youngest. Make slides of both, and of intermediate buds between the two extremes.

(2) With a needle and forceps open up a bud on a white tile. Using a hand lens or binocular microscope, identify the anthers which, in the more advanced flowers from the base of the inflorescence, will probably be distinctly yellow.

(3) Remove an anther and place it on a clean microscope slide. Add two drops of acidified acetic orcein (ten parts stain to one part M HCl).

(4) Squash with a glass rod and leave for one minute to allow the stain to penetrate the tissue.

(5) Put on a coverslip, cover it with filter paper and squash gently.

(6) Warm the slide on a hotplate for about ten seconds to intensify the staining. (The slide should be very warm, but not too hot to touch).

(7) Examine for stages in meiosis.

Recording your Observations

Using the same technique as for Investigation 23.1, and using Fig. 23.2 to help you, identify as many stages of meiosis as you can.

For simplicity the diagrams in Fig. 23.2 do not show chiasmata. Examples of these are shown separately in Fig. 23.3. Can you see chiasmata in your preparation, and what effect do they have on the appearance of the chromosomes during prophase and metaphase of meiosis I? Can you estimate the chromosome number of the plant?

Supplement the information gained from your own preparation by observing stages of meiosis in prepared sections of the anthers of, e.g. lily. Compare with prepared sections of mammalian testis.

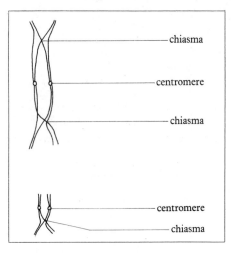

Fig. 23.3 Diagrams illustrating the appearance of bivalents towards the end of prophase I of meiosis. At this stage the two centromeres begin to repel each other so the two constituent chromosomes of the bivalent become clearly visible. Each chromosome may be seen to consist of a pair of chromatids which are in contact at certain points along their length. These points are the chiasmata.

For Consideration

In what situations apart from the ones studied here would you expect to find meiosis taking place?

Requirements

Microscope
Hand lens or binocular microscope
Slides and coverslips
White tile
Hotplate
Filter paper
Mounted needle
Forceps
Glass rod

Acetic orcein
M HCl

Hyacinth bulb
TS or LS anthers of, e.g. lily
TS mammalian testis

A) Interphase

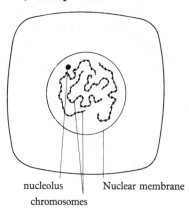

nucleolus

chromosomes

Nuclear membrane

B) Early prophase I

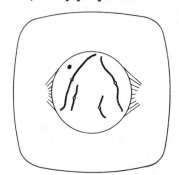

C) Mid prophase I

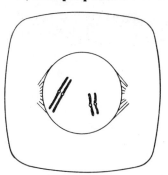

D) Late prophase I

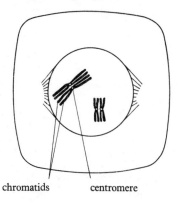

chromatids   centromere

E) Metaphase I

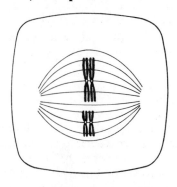

F) Anaphase I

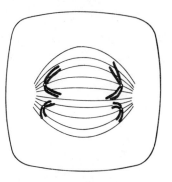

G) Telophase I

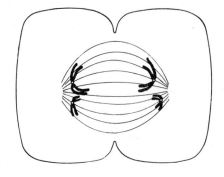

H) Prophase II

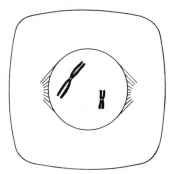

I) Metaphase II

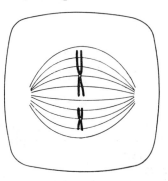

Fig. 23.2 Diagrams illustrating the behaviour of chromosomes during meiosis. As in Fig. 23.1, only two pairs of chromosomes are shown, one short pair and one long pair. Normally chiasmata would be visible at stages D and E but for simplicity these are omitted from the diagrams. Formation of chiasmata results in the chromosomes (bivalents) assuming the kind of configurations shown in Fig. 23.3.

J) Anaphase II

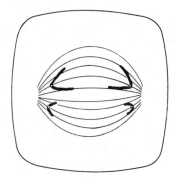

K) Telophase II

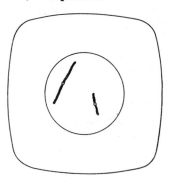

Investigation 23.3

Observation of stages of meiosis in testis of locust

In animals meiosis always takes place in the gonads. A convenient place where meiotic figures can be seen is the testis of the locust. It is necessary to use adult males that are within 7–14 days of their last moult, or fifth instar males that are almost ready to moult into adults.

You can use either the desert locust (*Schistocerca gregaria*), or the migratory locust (*Locusta migratoria*). In the former the testes are easier to dissect out, due to the presence of less fat.

Procedure

(1) Decapitate a male locust and pin it out, dorsal side uppermost, under water in a dish. Pin out the wings on each side.

(2) Open up the abdomen by a mid-dorsal longitudinal cut. Pin back the body wall.

(3) Using a hand lens or binocular microscope, identify the testes. Together with fat, these make up an oval body lying above the gut in abdominal segments 5 and 6 (Fig. 23.4). Transfer the testes to a microscope slide.

(4) Remove as much of the fat (yellow) as you can, leaving only the white tubules of the testes. Two or three tubules are sufficient.

(5) Gently squash the tubules with a glass rod so as to spread out the tissue.

Remove excess water from the slide with filter paper.

(6) Add several drops of acetic orcein stain and put on a coverslip. Cover with filter paper and squash gently, tapping the coverslip with a blunt instrument. This helps to spread the chromosomes.

(7) Warm the slide on a hotplate for about 10 seconds to intensify the staining. (The slide should be very warm, but not too hot to touch).

(8) Examine for stages in meiosis and compare with the results of Investigation 23.2.

For Consideration

Where and when would you expect to find mitosis and meiosis taking place in a female locust?

Requirements

Microscope
Hand lens or binocular microscope
Slide and coverslip
Small dissecting dish
Dissecting instruments
Glass rod
Filter paper

Acetic orcein

Male locust (either young adult or fifth instar nymph that is about to moult).

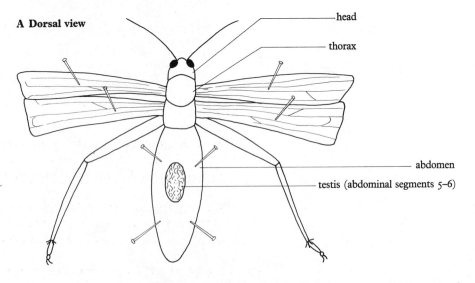

A Dorsal view

head
thorax
abdomen
testis (abdominal segments 5–6)

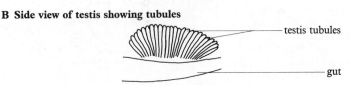

B Side view of testis showing tubules

testis tubules
gut

Fig. 23.4 Diagrams showing the testis in the abdominal cavity of the locust.

Questions and Problems

1   Make clear the difference between (a) chromosome and chromatid; (b) homologous chromosomes and sister chromatids; (c) diploid and haploid; (d) centrosome and centromere.

2   What functions are performed by mitosis? Briefly summarize the *principles* behind each of the stages in mitosis.

3   Speculate on the following:
    (a) What factors might trigger the onset of mitosis?
    (b) What might cause the nuclear membrane to disappear when it does?
    (c) What might cause simultaneous movement of the chromatids during anaphase?
    (d) What initiates the formation of a new cell membrane at the end of mitosis?

4   The graph in Fig. 23.5 illustrates the movement of chromosomes within a cell during mitosis. Curve A shows the changes in the distance between the centromeres of the chromosomes and the poles of the spindle. Curve B shows changes in the distance between centromeres of sister chromatids. On the time scale, zero (o) marks the beginning of the time when the chromosomes line up on the equator.

    Describe what is happening to the chromosomes 15 to 20 minutes after time o. Why does curve A fall slightly during the first 5 minutes?

    (NSW modified)

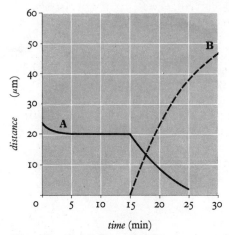

Fig. 23.5   Graph illustrating the movement of chromosomes in a cell during mitosis.

5   Sketch the arrangement of the chromosomes of a plant cell with a diploid number of 6 (a) at early anaphase of mitosis, and (b) at early anaphase of a first meiotic division. What is the significance of the differences between them?

6   Without going into the mechanisms of either in detail, outline the essential differences between mitosis and meiosis. Summarize the importance of these two types of cell division in the lives of organisms.

7   To what extent do you consider (a) mitosis to be a homeostatic process, and (b) meiosis to be random?

8 Fig. 23.6 A to D shows reproductions of photomicrographs of four dividing cells. Interpret each picture as fully as possible.

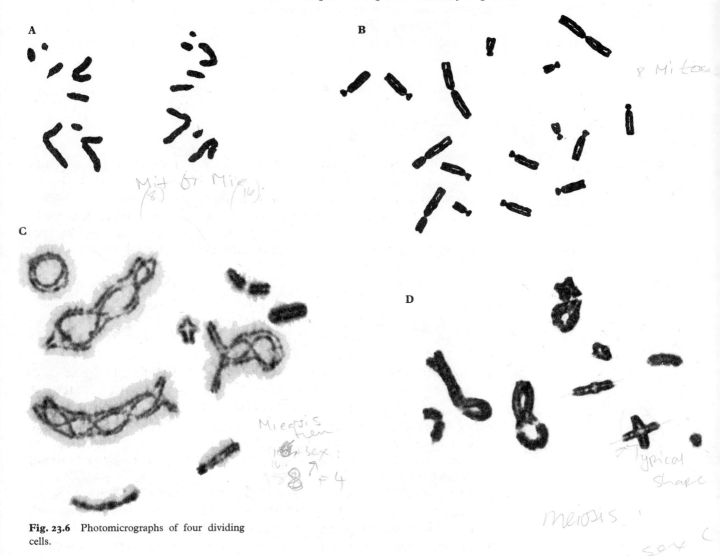

Fig. 23.6 Photomicrographs of four dividing cells.

24 Reproduction

Background Summary

1 Reproduction can be **asexual** or **sexual**. Asexual reproduction generally achieves rapid proliferation of the species. Sexual reproduction, though often less prolific, confers genetic variation on the species.

2 Reproduction is often associated with survival over unfavourable periods.

3 The genetic significance of sexual reproduction is demonstrated by bacterial conugation in which transfer of genetic material occurs without the use of gametes.

4 Most organisms employ **gametes**. These may be **isogametes**, **anisogametes** or **heterogametes**. Organisms with isogametes are generally distinguishable into **plus** and **minus strains**, foreshadowing the development of eggs and sperm in higher forms. Gametes are haploid, the diploid condition being restored when they fuse.

5 The structure of **eggs** and **spermatozoa** is clearly related to their functions, the spermatozoon as a motile vehicle for the male's genetic material, the egg for receiving genetic material from the sperm.

6 In the process of **fertilization** the spermatozoon, aided by its **acrosome reaction**, penetrates the egg membrane, its haploid set of chromosomes (**paternal chromosomes**) uniting with those of the egg (**maternal chromosomes**).

7 Many aquatic animals show **external fertilization**. Most terrestrial animals have **internal fertilization** which in mammals is combined with **placentation**, **viviparity** and extensive **care of the young**.

8 In man sexual reproduction starts with **gametogenesis**: **spermatogenesis** in the testes of the male, **oögenesis** in the ovaries of the female. In both cases the sequence of events, involving meiosis, is essentially the same.

9 The microscopic structure of the **testis** and **ovary** is directly related to their functions of producing large numbers of sperm and eggs respectively.

10 The anatomy of the male and female **reproductive systems** is geared towards bringing sperm into contact with an egg and subsequently protecting and nourishing the developing embryo into which the zygote develops.

11 In man, the female **sexual cycle** follows a monthly pattern, **ovulation** alternating with **menstruation**. The sequence of events is controlled by complex interactions between **gonadotrophic hormones** from the pituitary gland and **ovarian hormones** from the gonads themselves.

12 In the event of fertilization and successful implantation, the hormonal situation is altered in such a way that menstruation is temporarily suspended and the woman becomes **pregnant**. The **placenta** takes over the function of secreting the ovarian hormones which, together with other hormones produced by the pituitary, ensure that **parturition** and **lactation** occur at the appropriate time.

13 Some mammals, like the human, can reproduce at any time of the year; others have specific **breeding seasons** the timing of which is controlled by a combination of environmental and hormonal factors.

14 In the human male there is no sexual cycle as such but sexual activities are controlled by hormones comparable to those found in the female.

15 It is instructive to compare mammalian reproduction with that of flowering plants. In the latter the reproductive apparatus is embodied in the **flowers**. From a functional standpoint the male gamete nuclei in the **pollen grains** produced in the anthers are equivalent to sperm, and the **egg cell** contained in the ovule is equivalent to the ovum. Both are haploid.

16 **Pollination**, the process by which pollen is transferred from the male to the female parts of flowers, is equivalent to copulation and is achieved in a variety of ways. Fertilization is achieved by the formation and growth of a **pollen tube** which conveys the male nuclei to those of the female.

17 In flowering plants a **double fertilization** takes place, one male nucleus fusing with the egg cell to form a zygote, the other with the polar nuclei to form the primary endosperm nucleus.

18 After fertilization the zygote develops into the **embryo**, the primary endosperm nucleus into the **endosperm tissue**, the ovule into the **seed**, the integuments into the **seed coat**, and the ovary (carpel) into the **fruit**.

19 Sometimes an organism develops from an unfertilized egg, a phenomenon called **parthenogenesis.** Two types of parthenogenesis are recognized, diploid and haploid, and—if occurring naturally, as they do in certain animals—they can be used to achieve rapid proliferation without the participation of males.

20 Most animals have separate sexes (**dioecious**), but some animals and the majority of plants are hermaphroditic (**monoecious**). On account of its genetic disadvantages, most monoecious organisms have mechanisms for reducing the chances of self-fertilization, though exceptions occur.

21 Many organisms, particularly plants and lower animals, can reproduce **asexually**. In most cases asexual reproduction occurs in addition to (but, in a very few cases, instead of) sexual reproduction.

22 Asexual methods of reproduction include binary and multiple **fission**, **spore-formation**, **budding**, **fragmentation** and **vegetative propagation**.

23 An essential part of the reproductive processes of many organisms, particularly slow-moving or sessile ones, are mechanisms for **dispersal**. An organism may be dispersed in the form of **spores**, **seeds**, **fruits** or **larvae**, according to the species in question. Agents of dispersal include animals, wind, and water.

Investigation 24.1
Observation of fertilization

It is not easy to study fertilization, but by using suitable material in the right conditions the process can sometimes be observed. For obvious reasons it is best to use organisms which have external fertilization. Such organisms include sea urchins, marine worms, and brown seaweeds.

Sea Urchin
Male and female urchins produce sperms and eggs in March or April. In some species spawning can be induced by injecting them with 0.5 M KCl solution (1 cm^3) close to the mouth. Eggs and sperm shed through pores by the mouth should be collected in separate petri dishes. Eggs form an orange fluid, sperms a white fluid.

Transfer some eggs to a drop of sea water on a clean *cavity* slide. Gently lower a coverslip onto the drop of fluid.

Locate eggs under low power. With a fine pipette transfer a drop of the milky sperm suspension to the edge of the coverslip. Look down the microscope for sperm to swim into the field of view. Observe as much as possible of fertilization. Look particularly for the lifting of the fertilization membrane.

Marine Worm
The marine worm *Pomatoceros* produces eggs and sperm at any time of the year. The animal lives in a **calcareous tube** attached to rocks and stones. It is commonly found on rocky shores between the tide marks.

With a scalpel chip away the narrow tail end of the tube, and do the same with the broader head end. Then push the worm out of the tube from the rear with a blunt needle. Place the worm in a watchglass of sea water.

Repeat for several worms, placing males in one watchglass and females in another. The sexes can be distinguished by the fact that the males are yellow or orange and females purple.

The worms release their eggs and sperms as soon as they are removed from their tubes. With a pipette transfer a drop of sea water from the watchglass containing female worms to a cavity slide. Observe eggs, if present, under low power.

Add a drop of sea water from the watchglass containing males, quickly put on a coverslip and observe under the microscope. Alternatively, put on the coverslip *before* adding the sperms

and then introduce the sperms at the edge of the coverslip as in the sea urchin technique described above. Observe fertilization.

Seaweed
Separate a mass of male and female plants of the dioecious seaweed, *Fucus serratus* (toothed wrack). This is a brown alga which lives between the tide marks. Egg-producing **oögonia** and sperm-producing **antheridia** are formed in flask-like **conceptacles** which open onto the surface of the fronds.

Mature fronds exude mucilage from their conceptacles. The mucilage is coloured orange in the case of the male conceptacles, due to the presence of numerous orange-coloured sperms, and green in the case of the female conceptacles due to the presence of dark-green ova.

Take some of the mucilage from one plant and place it in a drop of sea water on a slide. Take a drop of mucilage from a plant of the other sex and place it in another drop of sea water on the same slide. The two drops should be close but not touching. Place the slide under low power and with a clean needle connect the two drops. Watch the process of fertilization under the microscope.

For Consideration
(1) What events, too small to see down your microscope, take place during the process of fertilization?
(2) What important functions are fulfilled by fertilization?

Requirements
Microscope
Slides and coverslips
Cavity slides
Mounted needle
Pipette
Petri dishes
Watchglasses
Syringe

0.5 M KCl

Sea urchin, e.g. *Echinus* or
  Arbacia
Tube worm, *Pomatoceros triqueter*
Toothed wrack, *Fucus serratus*

All the above specimens are available from Marine Biological Laboratories.

Investigation 24.2

Dissection of the reproductive system of the rat

Procedure

(1) Before commencing the dissection identify the urinary and genital openings on the ventral side (Fig. 9.1, p. 73). Note that in the male there is just one **urino-genital opening** at the end of the penis, whereas in the female there are separate **urinary** and **genital openings**, the former being located just anterior to the latter. In the male, notice also the **scrotal sacs** which contain the **testes**, and in the female identify the **nipples**.

(2) Open up the abdominal cavity as instructed on p. 73, cutting *round* the urinary and genital openings (dotted line in Fig. 9.1). Pin back the skin and cut back the body wall in the usual way. In the female notice the **mammary glands** adhering to the inside of the skin. Now ligature the hepatic portal vein and remove the gut as instructed on p. 75.

You are now in a position to proceed with your dissection of the reproductive apparatus.

First notice that in both sexes the tubes leading to the urino-genital openings are covered by a layer of muscle beneath which lies the pubis, part of the pelvic girdle (Fig. 24.1A for male; 24.2A for female).

MALE

(3) Cut open one of the **scrotal sacs** (dotted line on left-hand side of Fig. 24.1B). With forceps grasp hold of the **testis**, or the fat attached to it, and draw it forwards. Note that this pulls up the bottom of the scrotal sac as a result of the **cauda epididymis** being attached to it by the **gubernaculum**, a strand of muscle whose contraction was responsible for the descent of the testis during development. Identify the structures seen in Fig. 24.1C, if necessary using a hand lens or binocular microscope to help you.

(4) Remove the muscle overlying the pubis of the pelvic girdle. With large scissors cut through the pelvic girdle on either side of the mid-line and remove the pubis (dotted lines in Fig. 24.1A). This will expose the **urethra** which leads into the penis.

(5) Identify the **epididymis** and **vas deferens** leading from the testis. Follow the vas deferens to the mid-line and ascertain that it, along with its fellow from the other side, opens into the anterior end of the urethra. Also opening into the urethra at this point is the duct of the **bladder**, which receives the **ureters** from the **kidneys**.

Now identify the glands associated with the reproductive tract: a gland associated with the inner end of each vas deferens; a pair of large **seminal vesicles** and **coagulating glands**; **prostate glands** lying ventral to the urethra just behind the bladder; a pair of **Cowper's glands** at the base of the penis, and a pair of **preputial glands** towards the tip of the penis. What are the functions of these glands?

Note that in the male the urethra serves as a common **urino-genital duct**.

Finally, investigate the blood vessels serving the reproductive organs, particularly the **spermatic artery** and **vein**, derived (usually) from the dorsal aorta and posterior vena cava respectively. Together the spermatic artery and vein constitute the **spermatic cord**.

(6) Cut open the scrotal sac of a fresh-killed male rat and expose the testis. Cut into the testis and release some of the milky fluid onto a slide. Add a drop of 0.9 per cent salt solution and put on a coverslip. Observe sperm under low and high power.

Requirements
Hand lens or binocular microscope
Dissecting instruments
Thread
Cotton wool
Slide
Coverslip
NaCl solution (0.9 per cent)

Male and female rats for dissection

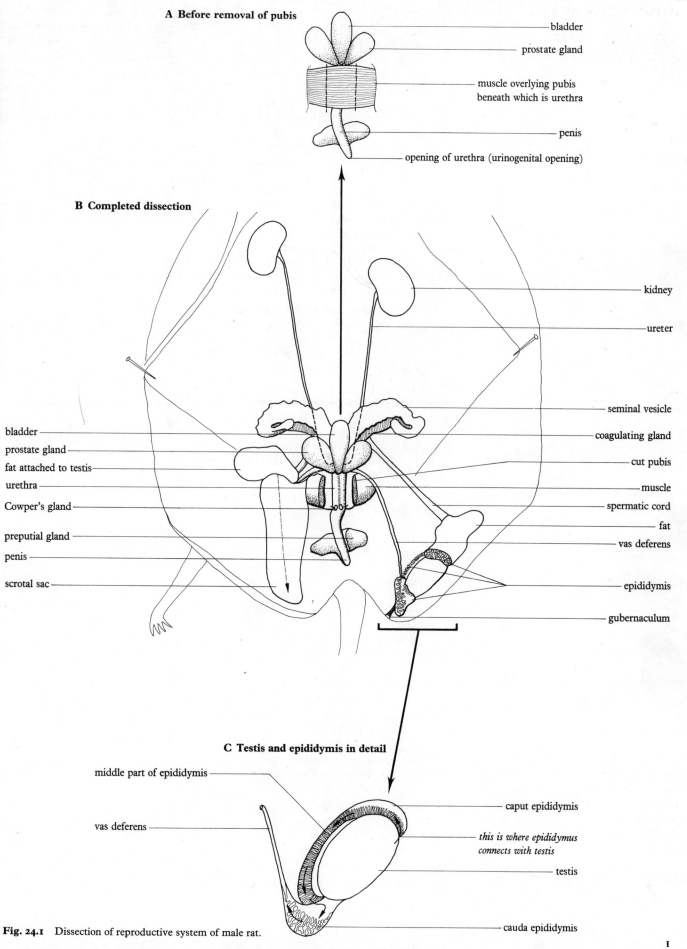

A Before removal of pubis

— bladder

— prostate gland

— muscle overlying pubis beneath which is urethra

— penis

— opening of urethra (urinogenital opening)

B Completed dissection

— kidney

— ureter

— seminal vesicle

— coagulating gland

— cut pubis

— muscle

— spermatic cord

— fat

— vas deferens

— epididymis

— gubernaculum

bladder —

prostate gland —

fat attached to testis —

urethra —

Cowper's gland —

preputial gland —

penis —

scrotal sac —

C Testis and epididymis in detail

middle part of epididymis —

vas deferens —

— caput epididymis

— *this is where epididymus connects with testis*

— testis

— cauda epididymis

Fig. 24.1 Dissection of reproductive system of male rat.

FEMALE

(6) Remove the pubis as in the male (dotted lines in Fig. 24.2A). This will expose the narrow **urethra** beneath which is the more bulbous **vagina**. Identify the small **ovaries** and—with the aid of a lens or binocular microscope—the short, coiled **oviduct** (Fig. 24.2C). The oviducts on each side lead to a long V-shaped **uterus**, the two horns of which unite in the midline to form the **vagina** (Fig. 24.2B). As in the male the bladder, receiving the ureters from the kidneys, opens into the anterior end of the urethra. A pair of **preputial glands** will be seen at the end of the urethra. Note that in the female the urethra (urinary duct) and vagina (genital duct) are separate.

(7) Identify the **ovarian** and **uterine arteries** and **veins**. With which major blood vessels do these connect? The uterine blood vessels are especially large and prominent during pregnancy.

(8) If the rat is pregnant the uterus will be seen to contain a variable number of **embryos** which can be removed and examined. In removing an embryo notice the **umbilical cord** and **placenta**.

For Consideration

(1) What are the principal similarities and differences between the male and female reproductive systems?

(2) You will have observed that in the male the urethra serves as a common urinogenital duct conveying both urine and sperm to the exterior, whereas in the female the urinary and genital systems have separate openings to the exterior. How would you explain this difference between the two sexes and what are its physiological consequences?

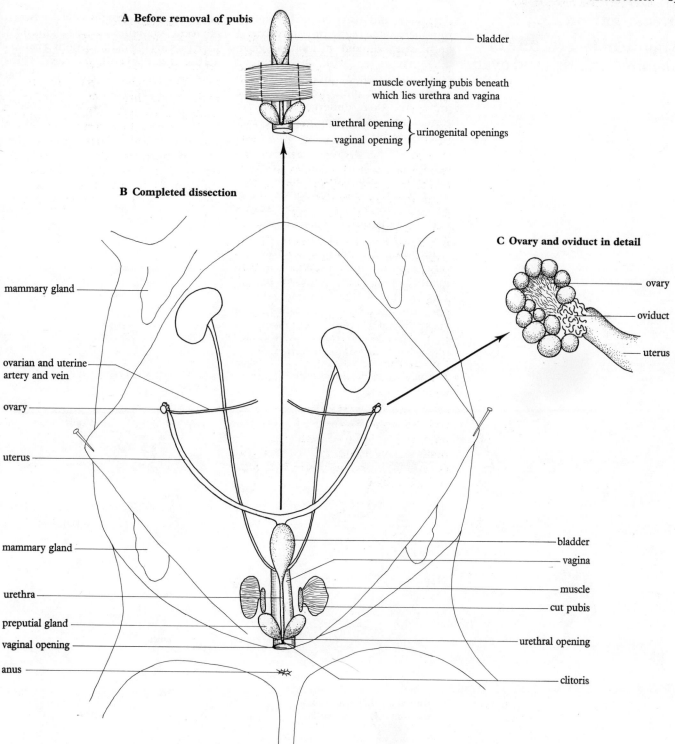

A Before removal of pubis

bladder

muscle overlying pubis beneath which lies urethra and vagina

urethral opening ⎱ urinogenital openings
vaginal opening ⎰

B Completed dissection

C Ovary and oviduct in detail

ovary

oviduct

uterus

mammary gland

ovarian and uterine artery and vein

ovary

uterus

mammary gland

urethra

preputial gland

vaginal opening

anus

bladder

vagina

muscle

cut pubis

urethral opening

clitoris

Fig. 24.2   Dissection of reproductive system of female rat.

Investigation 24.3

Microscopic structure of mammalian testis and ovary

The testis and ovary contain developing sperm and eggs respectively. In both cases diploid primordial germ cells, associated with the germinal epithelium, divide mitotically to form spermatogonia and oögonia which grow and then undergo meiosis to form, ultimately, spermatozoa and ova respectively.

Testis

Examine a prepared section of mammalian testis under low power (Fig. 24.3A). Observe numerous **seminiferous tubules** cut in various planes.

Examine the wall of a seminiferous tubule under high power. Can you identify the structures shown in Fig. 24.3B? Trace the development of **spermatozoa** from **spermatogonia** via **spermatocytes** and **spermatids**. Can you see any stages in meiosis? Can you detect **Sertoli cells**?

Now examine the tissue between adjacent seminiferous tubules under high power (Fig. 24.3C), noting **interstitial (Leydig) cells** and **capillaries** embedded in connective tissue. What is the function of the Leydig cells and why are they located close to capillaries?

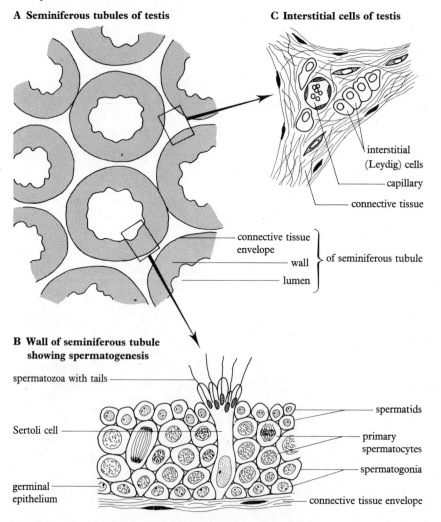

A Seminiferous tubules of testis

C Interstitial cells of testis

interstitial (Leydig) cells

capillary

connective tissue

connective tissue envelope

wall

lumen

⎱ of seminiferous tubule

B Wall of seminiferous tubule showing spermatogenesis

spermatozoa with tails

Sertoli cell

germinal epithelium

spermatids

primary spermatocytes

spermatogonia

connective tissue envelope

Fig. 24.3 Microscopic structure of mammalian testis. Secondary spermatocytes are short-lived and do not normally appear in mictoscopic preparations.

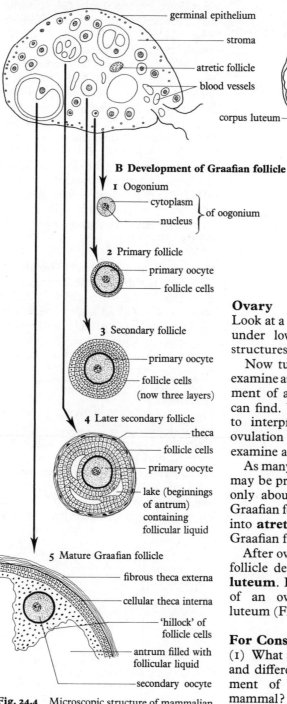

A Section of ovary showing follicles

germinal epithelium

stroma

atretic follicle

blood vessels

C Section of ovary with corpus luteum

corpus luteum

B Development of Graafian follicle

1 Oogonium

cytoplasm ⎫
⎬ of oogonium
nucleus ⎭

2 Primary follicle

primary oocyte

follicle cells

3 Secondary follicle

primary oocyte

follicle cells
(now three layers)

4 Later secondary follicle

theca

follicle cells

primary oocyte

lake (beginnings
of antrum)
containing
follicular liquid

5 Mature Graafian follicle

fibrous theca externa

cellular theca interna

'hillock' of
follicle cells

antrum filled with
follicular liquid

secondary oocyte

Fig. 24.4 Microscopic structure of mammalian ovary. In **B**, the first meiotic division occurs just before ovulation, the second meiotic division occurs just after fertilization.

Requirements
Microscope

Section of mammalian testis
Section of mammalian ovary
showing developing Graafian
follicles
Section of mammalian ovary
showing corpus luteum

Ovary
Look at a section of mammalian ovary under low power and identify the structures shown in Fig. 24.4A.

Now turn over to high power and examine as many stages in the development of a **Graafian follicle** as you can find. Use Fig. 24.4B to help you to interpret the stages. How does ovulation take place? If available, examine a slide which shows this.

As many as 400,000 primary follicles may be present at birth, but normally only about 400 develop into mature Graafian follicles. The rest degenerate into **atretic follicles**. How often are Graafian follicles formed?

After ovulation the hollow Graafian follicle develops into a solid **corpus luteum**. Examine a prepared section of an ovary containing a corpus luteum (Fig. 24.4C).

For Consideration
(1) What are the essential similarities and differences between the development of sperm and eggs in the mammal?
(2) How are the developmental events, which you have observed to occur in the testis and ovary, controlled?

Investigation 24.4
Structure of flowers

First examine a plant such as bluebell or willowherb and notice the relationship between the flowers and the stem (Fig. 24.5). The flowers together constitute the **inflorescence**. Note that the **flower stalk (pedicel)** is borne in the angle between a small leaf (**bract**) and the main stalk of the inflorescence (**peduncle**).

If the plant is at a sufficiently advanced stage of development, notice the developmental transition from apex to base of the inflorescence. Towards apex: unopened **flower buds**; further back: open **flowers** with stamens and carpels in various states of maturity; further back still: **fruits**.

Structure of a Typical Flower
The flower is the reproductive part of the plant. It consists of a series of **whorls** of modified leaves which collectively produce, protect, and ensure the union of, the gametes. The whorls of structures are attached to a **receptacle**, the expanded end of the flower stalk. The whorls are, from the outside inwards:

Calyx (**sepals**): usually small, green and leaf-like.

Corolla (**petals**): often coloured, scented and with nectaries (sac containing sugary solution) towards base for attracting insects.

Androecium (**stamens**): each stamen consists of an anther and fila-ment. Each anther contains four pollen sacs in which the pollen grains develop.

Gynoecium (**carpels**): each carpel consists of stigma, style and ovary. The ovary contains one or more ovules in each of which is an egg cell.

Examine a flower of, e.g. buttercup, and note the four whorls of structures listed above. Remove a representative component of each whorl and examine its parts under a hand lens or binocular microscope (Fig. 24.6).

Variations in Floral Structure
Most flowers possess the structures described above. Flowers vary mainly in the numbers, arrangement, and degree of fusion between the component parts. These are some of the main variations that may be observed:

Sepals: number variable; show varying degrees of fusion; may be green and leaf-like, or coloured and petal-like (petaloid).

Petals: number variable; like sepals, show varying degrees of fusion, in extreme cases forming tube (corolla tube); nectaries may be absent; colour and scent varies widely; sepals and petals may be fused or indistinguishable from one another, in which case they form, together, the **perianth**.

Stamens: number may be large and variable or smaller and fixed; anthers vary in size; filaments may vary in length and in their mode of attach-

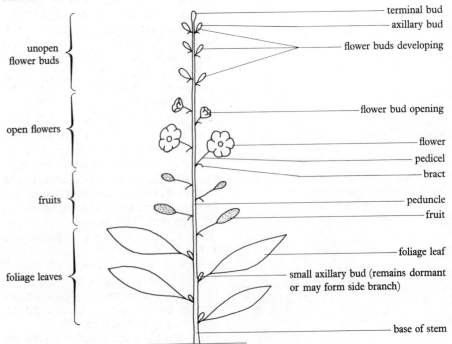

unopen flower buds

open flowers

fruits

foliage leaves

terminal bud
axillary bud
flower buds developing

flower bud opening

flower
pedicel
bract

peduncle
fruit

foliage leaf
small axillary bud (remains dormant or may form side branch)

base of stem

Fig. 24.5 Diagram of a generalized plant to show the arrangement of the flowers and related structures.

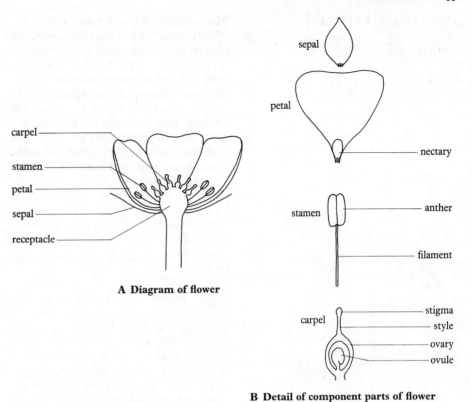

A Diagram of flower

B Detail of component parts of flower

Fig. 24.6 Structure of a generalized flower. Flowers of the buttercup family (Rannuculaceae) conform closely to the organization illustrated here.

ment to the anther; base of filaments may be attached to the petals or peri-anth rather than the receptacle; show varying degrees of fusion, in extreme cases forming a tube round the carpels.

Carpels: number may be large and variable or smaller and fixed; styles vary in length; may be separate or joined; if joined there is much variation in the method of fusion; the ovary may stand proud upon the receptacle (**superior ovary**) or be sunk down into the receptacle (**inferior ovary**).

In addition to the above variations, the flower may be radially symmetrical (**actinomorphic**) or bilaterally symmetrical (**zygomorphic**). A zygomorphic flower can be cut in only one plane to give two equal and opposite halves; an actinomorphic flower, however, may be cut in more than one plane to give two equal and opposite halves.

Examples of actinomorphic flowers include buttercup, lily, tulip and bindweed (*Convolvulus*). Zygomorphic flowers include sweet pea, deadnettle, snapdragon and orchids.

As already mentioned, flowers are usually **hermaphrodite**, but sometimes they are **unisexual**. In the latter either the stamens or carpels are absent. Male and female flowers may be found on one and the same plant, and then the plant is described as **monoecious**; or male and female flowers may be on different individuals, in which case the plant is **dioecious**. Examples of monoecious plants include hazel, sycamore, oak, beech and maize. Dioecious plants include poplar and willow.

Recording Floral Structure

There are three recognized ways of recording the structure of a flower:

DRAWING THE HALF-FLOWER

This provides an elevation of the flower. Cut the flower along the median plane, i.e. the plane in line with the main stem (Fig. 24.7). In zygomorphic flowers cutting along this plane will give two equal and opposite halves.

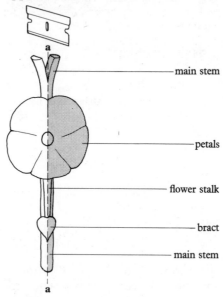

Fig. 24.7 To prepare a half-flower cut the floral structures along the median plane **a–a** as shown in this diagram.

Draw a half-flower (Fig. 24.8A) and/or construct a simple diagram of the cut surface (Fig. 24.8B).

FLORAL DIAGRAM

This provides a plan of the flower as viewed from above, and should look like a diagrammatic transverse section (Fig. 24.8C).

In constructing a floral diagram hold the flower so the bract faces you and the peduncle is furthest away.

Observe the following conventions: if the petals, sepals or stamens are joined, link them with simple brackets; if the stamens arise from the petals link them with radial lines.

FLORAL FORMULA

This is the simplest and quickest way of representing the structure of a flower. Written in coded form, it enables a person to work out the structure of the flower.

The floral formula is given by writing capital letters corresponding to the whorls as follows: **K** for **calyx**; **C** for **corolla**; **A** for **androecium** and **G** for **gynoecium**. Each letter is followed by a figure denoting the number of units in the whorl. If the number is large and variable it is expressed as infinity (∞). A floral formula is given in Fig. 24.8D.

A Half-flower

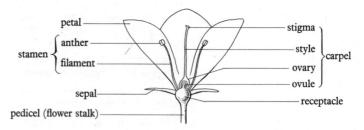

B Vertical section

C Transverse section (floral diagram)

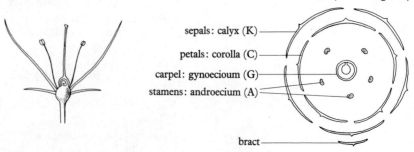

sepals: calyx (K)

petals: corolla (C)

carpel: gynoecioum (G)

stamens: androecium (A)

bract

D Floral formula $\bigoplus$ K5 C5 A5 G$\underline{1}$

Fig. 24.8 Representation of floral structure.

Write the floral formulae of the flowers which you have been studying. Observe the following conventions:

Actinomorphy is designated by writing $\oplus$ at the beginning of the formula;

Zygomorphy is designated by writing $\uparrow$ at the beginning of the formula;

A **superior ovary** is designated by putting a line below the gynoecium number, e.g. $G\underline{5}$;

An **inferior ovary** is designated by putting a line above the gynoecium number, e.g. $G\overline{5}$;

If units are joined, put their number in brackets, e.g. $K(5)$;

If some units are joined but others free the former only are put in brackets, e.g. $A(4) + 2$;

If one whorl is united with another, their symbols should be tied, e.g. $\widehat{C5}\,A5$;

If the sepals and petals are replaced by a **perianth**, the symbol used is **P**.

Examine the flowers of different plants. In each case sketch the half-flower and floral diagram and give the floral formula. Start with a simple actinomorphic flower which lacks any special complications. Then examine other flowers, both actinomorphic and zygomorphic, displaying the variations summarized above. In each case consider the possible functional significance of the variations, particularly those that might promote cross-pollination by wind or insects.

Arrangement of Flowers

Flowers may be solitary or massed together into an inflorescence. What are the possible advantages to the plant of having an inflorescence?

The many different types of inflorescence found in nature can readily be seen by examining different plants. How do the variations arise? They do so mainly by the peduncle branching in various ways, and by the pedicles varying in length (they may indeed be absent altogether, the flowers arising directly from the peduncle).

Examine a range of plants and sketch the arrangement of the flowers in each case. What can you say about the possible functional significance of each pattern?

One particularly specialized inflorescence is found in the Compositae, the family which includes the dandelion. Here the apex of the peduncle is expanded and flattened and bears numerous small flowers. This type of inflorescence is called a **capitulum**.

For Consideration

(1) From your observations what basic features do virtually all flowers have in common?
(2) Which flowers of the ones that you yourself have examined conform most closely to the basic pattern, and which ones deviate most widely from it?
(3) How would you explain the main differences between the reproductive apparatus of flowering plants and mammals?

Requirements

Binocular microscope or hand lens
Needles
Small forceps
Razor blade
White tile

Plants displaying a range of actinomorphic and zygomorphic flowers, some solitary and others forming different kinds of inflorescence

---

Investigation 24.5

Pollination, pollen germination and microscopic structure of stamen and carpel

The flower is part of the **sporophyte** and produces **spores** (*see* p. 273). The spores are of two kinds. Male spores, called **microspores** on account of their small size, are formed by meiosis inside the anthers: the microspores are the **pollen grains** each of which produces two **male nuclei**. Female spores, called **megaspores** because of their comparatively large size, are formed by meiosis inside the ovaries: the megaspore is the **embryosac** and it contains, amongst other inclusions, the **egg cell** or **ovum**.

Structure of Pollen Grains

Collect pollen grains from the anthers of the plants provided. Mount them in water and examine under high power. Can you deduce from the structure of the pollen grains whether the plant is insect or wind-pollinated? Are your deductions supported by the structure of the flower from which the pollen grains were obtained?

Clear some pollen grains in a drop of chloral hydrate or phenol and mount in iodine or methyl green in acetic acid. Examine under high power noting two-layered **wall** and two **nuclei**. What functions are fulfilled by the nuclei?

Germination of Pollen

When a flower is pollinated, pollen grains, usually from another plant, adhere to the stigma. The latter produces a sugary secretion, varying in concentration from about 2 to 45 per cent, which holds the pollen grains and facilitates their germination.

With a paint-brush transfer a few pollen grains of, e.g. nasturtium, deadnettle, chickweed or shepherd's purse, into a drop of sucrose solution (0.4 M) in the central depression of a cavity slide. Put on a coverslip.

Place your slide in a dark place at 20°–30°C and examine at intervals for one-to-two hours. Observe **pollen tubes**.

Stain by irrigating your slide with a drop of acetocarmine or neutral red. Look for **tube nucleus** and two **male nuclei**.

How could this experiment be extended to investigate the effect of various factors, e.g. temperature, on pollen germination?

Examine prepared longitudinal sections of carpels which have been pollinated. Make sketches to illustrate the growth of the pollen tube into the stigma and down the style to the ovary.

Flower Bud

Examine a transverse section of a flower bud of lily and identify the **stamens** and **carpels** (Fig. 24.9A). If the section has been cut at the right level, the part of the stamens visible in your section should be the anthers, and the part of the carpels visible should be the ovaries.

Anther

Examine an anther under high power (Fig. 24.8B). Note the four **pollen sacs** and examine their contents. The contents depend on the state of maturity of the anther:

(1) If immature, the pollen sacs will be full of closely packed **pollen mother cells**.

(2) If more mature they will contain pairs or tetrads of cells (**pollen tetrads**) resulting from meiotic division of the pollen mother cells. Chromosomes may be observed, telling you at what stage of meiosis the cells have been fixed.

(3) If completely mature they will contain fully formed **pollen grains**. The pollen grain has a single haploid nucleus to start with, but this soon divides mitotically into a generative and tube nucleus (Fig. 24.9C). The former divides again to form the two male nuclei.

Observe the wall of the pollen sac, noting the inner **tapetal layer** which nourishes the developing pollen grains, and the **middle** and **fibrous layers**. Drying out of the cells of the fibrous layer creates tension which causes the anther to split open at the **stomium**, a line of weakness running longitudinally along each side of the anther.

Ovary

Examine an ovary under high power and observe the structures shown in Fig. 24.9D. particularly the **ovule**. Focus onto the contents of the **embryosac** in the centre of the ovule. How many **nuclei** can you see in the embryosac? The answer will depend on its state of maturity and the level of the section.

The embryosac starts by having a single haploid nucleus which is formed by meiosis from an **embryosac mother cell**. The haploid nucleus of the embryosac then undergoes three successive mitotic divisions to give a total of eight nuclei, two of which (the polar nuclei) will fuse to form a central **fusion nucleus**. Each of the seven resulting nuclei becomes surrounded by membranes to give the cells shown in Fig. 24.9E. The **antipodal cells** at the chalazal end of the embryosac are thought to provide nourishment for the embryosac, the **synergids** at the micropyle end are nonfunctional eggs and degenerate. In the act of fertilization the **egg cell** (**ovum**) fuses with one male nucleus to give the zygote, and the diploid fusion nucleus fuses with the other male nucleus to form the triploid **primary endosperm nucleus**. The latter gives rise to the endosperm tissue which envelops and nourishes the embryo.

Examine the rest of the ovary noting the structures shown in Fig. 24.9D.

For Consideration

Compare the gametes, and the way they develop, in the flowering plant and mammal.

Requirements

Microscope
Cavity slide and coverslip
Paintbrush
Needle

Sucrose (0.4 M)
Chloral hydrate
Phenol
Iodine solution
Methyl green in acetic acid
Acetocarmine
Neutral red

Flowers of a variety of insect- and wind-pollinated plants.
Flowers of nasturtium, deadnettle, chickweed or shepherd's purse. (in both of the above the stamens must be ripe)
TS flower bud of lily
LS pollinated carpel

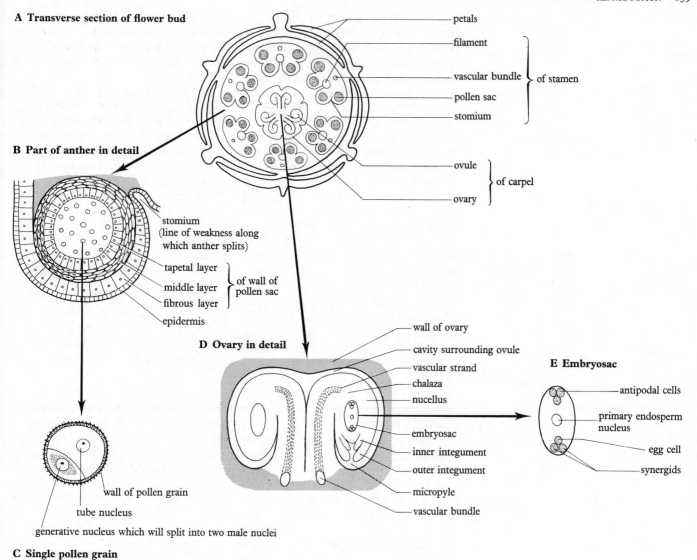

A Transverse section of flower bud

petals
filament
vascular bundle — of stamen
pollen sac
stomium

ovule — of carpel
ovary

B Part of anther in detail

stomium (line of weakness along which anther splits)
tapetal layer
middle layer — of wall of pollen sac
fibrous layer
epidermis

D Ovary in detail

wall of ovary
cavity surrounding ovule
vascular strand
chalaza
nucellus

embryosac
inner integument
outer integument
micropyle
vascular bundle

E Embryosac

antipodal cells
primary endosperm nucleus
egg cell
synergids

wall of pollen grain
tube nucleus
generative nucleus which will split into two male nuclei

C Single pollen grain

Fig. 24.9 Microscopic structure of the flower of the lily.

Investigation 24.6
Pollination mechanisms

In the course of evolution various structural and physiological mechanisms have arisen which tend, with varying degrees of efficiency, to promote **cross-pollination** and prevent (or at least reduce the chances of) **self-pollination**.

Procedure
Examine a wide range of flowers out-of-doors and in the laboratory and find out as much as you can about their pollination mechanisms. In doing so try to answer the following questions:
(1) Is the flower pollinated by wind or animals?
(2) If by animals, is it pollinated by flies, bees, butterflies, birds, elephants, or what?

(3) In what ways is the flower adapted to promote cross-pollination?
In answering this last question the following general summary may help you:

FEATURES FAVOURING
CROSS-POLLINATION
(1) The anthers may be situated below the level of the stigma(s) so there is no possibility of pollen falling onto the stigma of the same flower, e.g. yellow toadflax and many other plants.

(2) The stamens, whole flower or inflorescence may hang downwards so as to ensure wide scattering of pollen grains away from the parent plant, e.g. hazel catkins.

(3) The anthers may be borne on long flexible filaments favouring wide scattering of pollen grains, e.g. grasses.

(4) **Brightly coloured sepals, petals, perianth or bracts** may attract insects. Flowers are often grouped together into a **dense inflorescence** to make a bright splash of colour. Experiments on insect vision have shown that blue and yellow are the most effective colours for attracting insects.

(5) Production of **scent** to attract insects. Some insects have a sensitive sense of smell. Scent production is particularly marked in flowers pollinated at night by moths, e.g. honeysuckle, night-scented stock, butterfly orchid.

(6) **Nectar as a bait**. This is a sugary solution secreted by nectaries, sac-like glands at the bases of the petals. (N.B. Some bee-pollinated flowers have only pollen as bait).

(7) The nectaries are often located deep down, often at the base of a corolla tube, so the insect has to probe right into the flower.

(8) The petals may be so arranged as to make it easy for the insect to alight on the flower and insert its proboscis into it. For example, one of the petals may be adapted as a **landing platform**, e.g. orchids, deadnettle, snapdragon.

(9) **Nectar-guides**, spots or markings on one or more of the petals, may direct the insect to the nectaries, e.g. orchids, yellow toadflax, wild pansy.

(10) When an insect alights on the flower and inserts its proboscis the stamens are jerked in such a way that pollen is deposited on the insect's body, e.g. sage.

(11) A variety of mechanisms, some involving flap-like **valves**, ensure that the stigma is exposed when the insect enters, but is covered when it withdraws, e.g. violet, iris, orchids.

(12) The pollen may be placed on the insect's body (e.g. proboscis, head, etc.) in such a position as to ensure that it is deposited on the stigma when the insect visits another flower, e.g. orchids, deadnettle, yellow toadflax, sage.

(13) A specific application of (11) is seen in **heterostyly**, the condition in which the styles of different flowers may differ in length, e.g. primrose, purple loosestrife.

(14) The pollen grains may be sticky or sculptured, thus enabling them to cling to the body of the insect, or light and dry, enabling them to be carried by wind.

(15) The stamens and carpels within a given flower may mature at different times (**dichogamy**). Stamens ripen before carpels in, e.g. ivy, geranium, and sage (**protandrous condition**). Carpels ripen before stamens in, e.g. plantain, horse chestnut, arum lily (**protogynous condition**).

(16) The stamens and carpels may be located in different flowers, i.e. the **flowers are unisexual**. This comparatively unusual situation is found mainly amongst wind-pollinated plants.

Monoecious condition: male and female flowers on same plant, e.g. hazel, beech, oak, sycamore.

Dioecious condition: the male and female flowers on separate plants, e.g. holly, willow, poplar.

(17) **Self-sterility**. In some plants pollen will not develop on, and indeed may poison, a stigma of the same plant, e.g. certain orchids.

(18) A variety of highly specialized mechanisms which vary according to the plant in question—e.g. arum lily. Other examples may be found in textbooks of botany.

For Consideration

(1) Consider each of the flowers that you have examined and try to decide how efficient it is at securing cross-pollination and preventing self-pollination.

(2) Although cross-pollination is genetically desirable, it is not necessary that it should happen invariably. For this reason the various mechanisms listed above are by no means foolproof. Indeed, some plants have mechanisms which specifically encourage self-pollination. Can you suggest what these mechanisms might be?

(3) Do any animals show structural features or mechanisms comparable to those that promote cross-pollination in plants?

Requirements
Binocular microscope or hand lens
Needles
Small forceps
Razor blade
White tile

Flowering plants in season

Investigation 24.7
Structure and dispersal of fruits and seeds

A **fruit** in the strict sense of the word is formed from the **ovary**. Generally after fertilization the ovary expands, enclosing and protecting the seed(s) which are formed from the ovules. The wall of the fruit, known as the **pericarp**, is derived from the wall of the ovary. As well as protecting the seeds, the fruit is adapted in various ways to promote **dispersal**.

In practice a number of other floral structures besides the ovary may contribute to the formation and dispersal of the fruits. These include the style, receptacle, sepals, and bracts (Fig. 24.10).

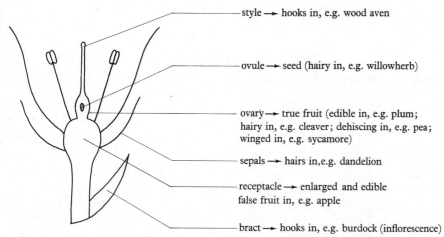

style → hooks in, e.g. wood aven

ovule → seed (hairy in, e.g. willowherb)

ovary → true fruit (edible in, e.g. plum; hairy in, e.g. cleaver; dehiscing in, e.g. pea; winged in, e.g. sycamore)

sepals → hairs in, e.g. dandelion

receptacle → enlarged and edible false fruit in, e.g. apple

bract → hooks in, e.g. burdock (inflorescence)

Fig. 24.10 Schematic diagram summarizing the contribution made to fruit formation and dispersal by floral structures other than the ovary. In some cases the ovary is situated above the receptable as shown here; in other cases the ovary is sunk down into the receptacle. There may, of course, be more than one carpel present, and in some cases the whole inflorescence may enter into the formation of the fruit.

Fruits can be classified on a structural basis, or according to their methods of dispersal. In the following survey the latter system is adopted. However, there are no hard and fast distinctions between the different methods of dispersal. A great many fruits and seeds are dispersed accidentally, for example on the hooves of animals and the wheels of vehicles.

Fruits Dispersed by Clinging to Animals
Such fruits usually have hooked projections enabling them to cling to the fur of animals. The hooks are formed from various parts of the floral apparatus.

HOOKED STYLE
This is common in single-seeded fruits where the fruit is formed without much modification from the carpel. Such a fruit is called an **achene** (Fig. 24.11A). Sometimes, after fertiliza-

tion, the style becomes woody and its tip hooked.
Example: wood avens (*Geum urbanum*).

Examine an achene with a hooked style. Test its ability to cling to your clothes.

HAIRY PERICARP
Here the pericarp of the fruit bears hooked hairs for attachment to animal fur.
Example: cleavers (*Galium aparine*).

Examine a fruit with a hairy pericarp and test its ability to cling to your clothes.

HOOKED BRACTS
Imagine a capitulum (*see* p. 257) in which the compact inflorescence is surrounded by a ring of overlapping bracts. After fertilization, the bracts become woody and each develops a hooked tip. The result is a **bur** which is very efficient at clinging to the fur of animals.
Example: burdock.

Examine a bur of burdock and test its ability to cling to your clothes. Cut it vertically and observe the single-seeded fruits each of which is crowned with a tuft of hairs developed from the sepals.

Fruits Dispersed by being Carried by Animals
Most **nuts** are dispersed this way. A nut is generally a single-seeded fruit with a woody pericarp. The fruit sits in a cup-shaped **cupula** formed from bracts. Dispersal largely depends on the inadvertent dropping of the nuts by animals which collect them, e.g. squirrels.
Examples: acorn, hazelnut, chestnut, beechnut.

Examine a representative nut, take it to pieces and note the features mentioned above.

Fruits Dispersed by being Eaten by Animals
These fruits, often fleshy, juicy and brightly coloured, are eaten by birds and other animals, which digest the fruit, but leave the seeds unharmed. The latter pass out with the faeces.

DRUPE
Usually fleshy, succulent fruits developed from single carpel containing a single seed (Fig. 24.11B).

Pericarp three-layered: outer **epicarp** ('skin') is protective (against

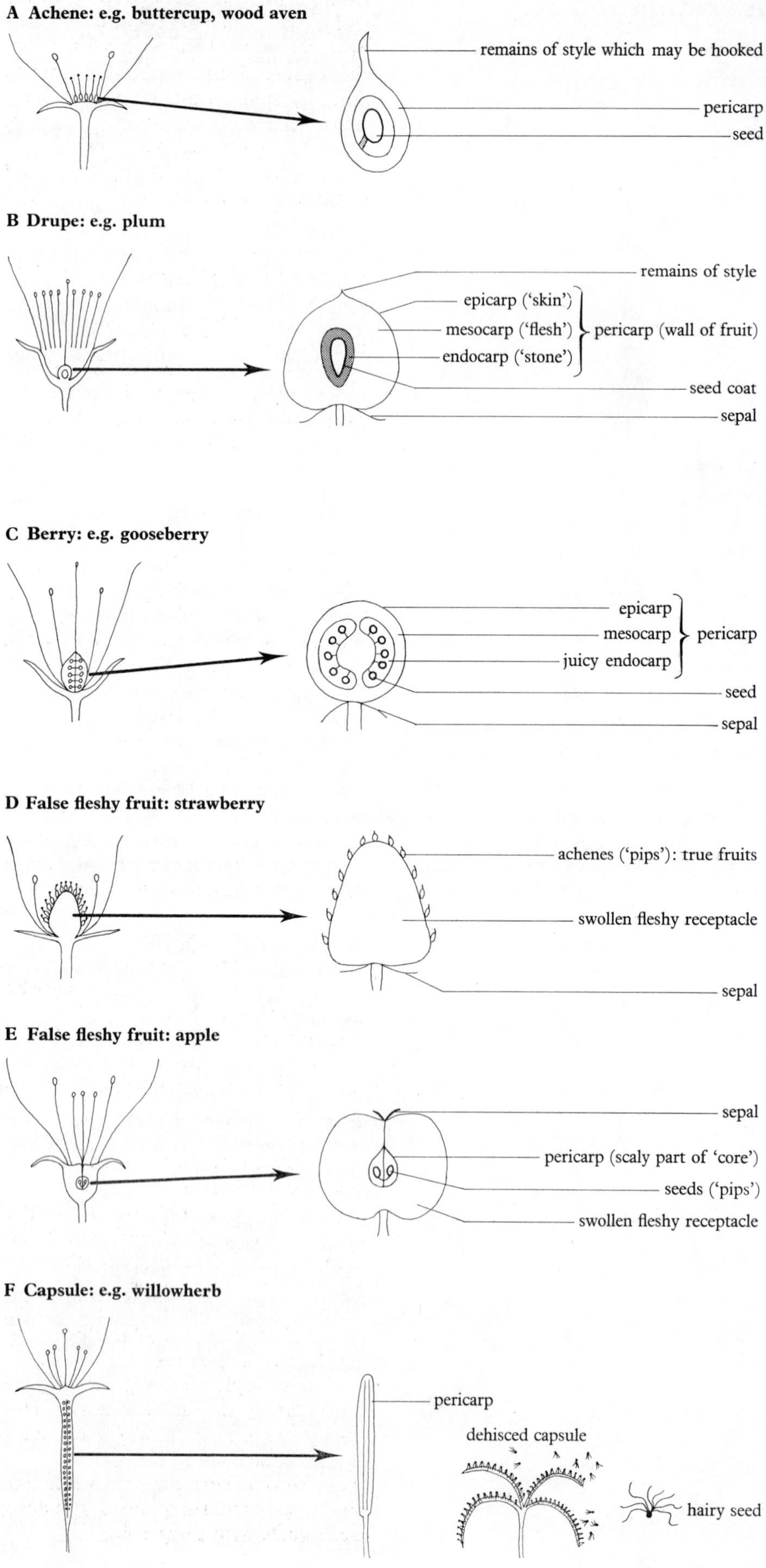

A Achene: e.g. buttercup, wood aven

- remains of style which may be hooked
- pericarp
- seed

B Drupe: e.g. plum

- remains of style
- epicarp ('skin')
- mesocarp ('flesh') } pericarp (wall of fruit)
- endocarp ('stone')
- seed coat
- sepal

C Berry: e.g. gooseberry

- epicarp
- mesocarp } pericarp
- juicy endocarp
- seed
- sepal

D False fleshy fruit: strawberry

- achenes ('pips'): true fruits
- swollen fleshy receptacle
- sepal

E False fleshy fruit: apple

- sepal
- pericarp (scaly part of 'core')
- seeds ('pips')
- swollen fleshy receptacle

F Capsule: e.g. willowherb

- pericarp
- dehisced capsule
- hairy seed

Fig. 24.11 Diagrams of a selection of fruits showing how, in all cases, the fruit is formed from the ovary and the seeds from the ovules.

what?), middle **mesocarp** is fleshy and tasty, inner **endocarp** ('stone') protects the seed and resists digestion. Examples: plum, peach, sloe, cherry, apricot, elder, almond, walnut (partially fibrous mesocarp), coconut (completely fibrous mesocarp).

Cut open a representative drupe and examine its structure.

AGGREGATE OF DRUPES

Groups of drupes clumped together on the receptacle.
Examples: raspberry, loganberry, blackberry.

Examine the individual drupes of a representative berry. Remove the drupes and note the receptacle. Anything interesting about the receptacle?

BERRY

Similar to drupe but both the mesocarp and endocarp are soft, fleshy, and juicy, the epicarp forming a skin of variable thickness (Fig. 24.11C).

Seeds usually have hard seed coats which serve the same functions as the endocarp of drupe. They are generally formed by the fusion of two or more multi-seeded carpels.
Examples: tomato, gooseberry, black currant, red currant, grape, orange, lemon, grapefruit, banana, date (single seed).

Cut a representative berry transversely and note its internal structure. How many carpels does it have?

FALSE FLESHY FRUIT

In this case the fleshy, edible part of the fruit is formed from a **swollen receptacle**. *Examples*:
Strawberry (Fig. 24.11D): the receptacle bears on its surface numerous small dry fruits (achenes) each of which contains a single seed.

Cut a strawberry vertically and note the relationship between the achenes and the receptacle.
Apple (Fig. 24.11E): the pericarp of the true fruit is the 'core', which contains the seeds ('pips'). Since the ovary of the flower is inferior, the fruit comes to be enveloped by the swollen receptable. Pear is similar. Such fruits are known as **pomes**.

Cut an apple (or pear) vertically and horizontally through the centre. How many carpels are there? How many seeds in each carpel?
Hawthorn: here the pericarp is hard and woody and surrounded by a fleshy receptacle with purple skin.

Cut open a 'haw' and note the above features.
Rose: the fleshy receptacle does not completely surround the fruits (why?).

Cut open a 'hip' and identify the fruits each of which is an achene with a covering of hairs.

Self-dispersal Mechanisms

Many fruits disperse their seeds by splitting (**dehiscence**) or some kind of scattering mechanism.

PODS AND FOLLICLES

Fruits formed from a single carpel whose 'leathery' pericarp splits along one or both sides. The term 'pod' is normally reserved for the Leguminosae.
Examples: pea, bean, gorse, broom (all Leguminosae); larkspur, columbine, wallflower, shepherd's purse.

Examine representative pods and follicles before and after dehiscence.

CAPSULES

Fruits formed from several carpels joined together. The seeds are released from the capsule in various ways:
(a) **Pores** permitting release of seeds by a 'pepper-pot' mechanism, e.g. poppy.
(b) **Teeth** which bend outwards, e.g. campion.
(c) **Lid** which falls off, e.g. plantain.
(d) **Splitting** of capsule wall, e.g. iris, violet, geranium, willowherb (Fig. 24.11F).

Examine representative capsules in various stages of maturity and investigate the mechanism by which the seeds are released. In the case of toothed and splitting types place unopened capsules on a hotplate and observe the dispersal mechanism in action. What happens if you breathe on the open capsule, or place it in a drop of water? Explain.

Fruits and Seeds Dispersed by Wind

Such fruits (or seeds) are light and are usually equipped with wings or hairs to help them remain airborne.

WINGED FRUITS

The wall of the carpel is expanded to form one or two wings.
Examples: sycamore, maple (two wings); ash (one wing).

Examine representative winged fruits and note the twirling parachute effect when thrown into the air.

HAIRY FRUITS

The hairs provide the small, light fruit with a 'parachute'.

Examples:

Groundsel: crown of hairs derived from sepals located directly on top of fruit;

Dandelion: 'pappus' of hairs, derived from sepals, is borne on a long stalk.

Release some of the hairy fruits of groundsel or dandelion and note how long they take to reach the ground. What is the effect of a slight gust of air?

HAIRY SEEDS

Sometimes the seeds, rather than the fruit, bear a tuft of hairs to aid effective wind dispersal.

Example: willowherb.

Cause the release of hairy seeds of willowherb by placing a mature fruit on a hotplate. Test the ability of the seeds to float in air.

Fruits Dispersed by Water

Though many fruits may be spread incidentally by water, relatively few are specifically adapted for water dispersal. An example is provided by the coconut, a drupe whose **fibrous mesocarp** contains numerous **air spaces** to aid floating.

For Consideration

(1) How do lower plants (algae, fungi, mosses, and ferns) disperse their progeny and how do their mechanisms compare with fruit and seed dispersal?

(2) Dispersal is not only important to plants, it is important to animals too. How is dispersal achieved in animals generally and to what extent can animal mechanisms be compared with those of plants?

Requirements

Binocular microscope or hand lens
Hotplate
Needles
Small forceps
Razor blade
White tile

Plants bearing a range of different types of fruits (see list on pp. 261–4)

Investigation 24.8

Vegetative reproduction in plants

Often vegetative reproduction involves the formation of some sort of **storage organ** which lies dormant in the soil over the winter and develops into one or more new plants the following year. The storage organ may be a modified stem, root, leaves or bud, depending on the plant in question. But whatever structure the storage organ is developed from, the fundamental cycle of events is the same: food materials are translocated from the leaves of the plant to the developing storage organ, and the following year these food reserves are mobilized and moved to the growing regions of the new plant (Fig. 24.12).

Since these storage organs enable the plant to survive from one year to the next they are called **perennating organs**. Perennation and vegetative reproduction take place by different means depending on which plant is under consideration.

By means of a Swollen Taproot
(Fig. 24.13A)

The taproot expands and the above-ground parts of the plant die except for the axillary buds at the base of the stem from which new shoots develop the following year. Food is stored in the expanded taproot.

Examples: parsnip, turnip, beet, swede, mangel, radish.

Examine representative swollen taproots. How could you ascertain that this is a modified root and not some other part of the plant? Carry out tests to find out in what chemical form the food is stored (*see* p. 46).

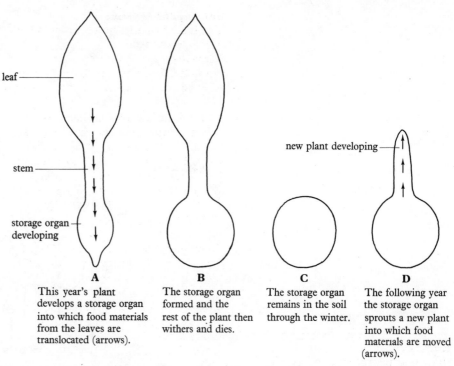

A	B	C	D
This year's plant develops a storage organ into which food materials from the leaves are translocated (arrows).	The storage organ formed and the rest of the plant then withers and dies.	The storage organ remains in the soil through the winter.	The following year the storage organ sprouts a new plant into which food materials are moved (arrows).

Fig. 24.12 Summary of the cycle of events involved in the formation and functioning of a perennating storage organ.

A Swollen taproot

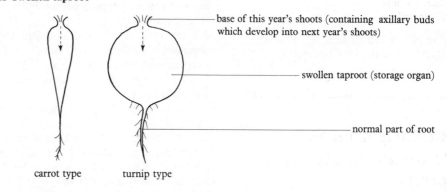

carrot type   turnip type

B Root tubers

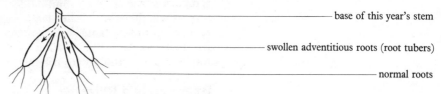

Fig. 24.13 Diagrams of perennating organs. Broken arrows indicate movement of food materials from the foliage leaves of this year's plant into the perennating organ. Fig. 24.13 is continued on pp. 266–269.

By means of Swollen Adventitious Roots (Root Tubers) (Fig. 24.13B).

Storage organs may be formed not from a tap root but from adventitious roots. The latter expand, store food and survive the winter, new plants being formed from axillary buds the following year.

Examples: dahlia, lesser celandine, sweet potato.

Examine plants with swollen adventitious roots and test the storage organs for their reserves.

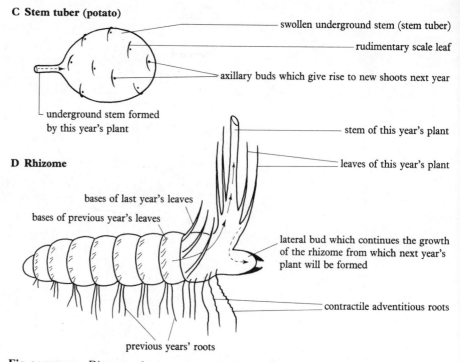

C Stem tuber (potato)

— swollen underground stem (stem tuber)

— rudimentary scale leaf

— axillary buds which give rise to new shoots next year

— underground stem formed by this year's plant

D Rhizome

— stem of this year's plant

— leaves of this year's plant

bases of last year's leaves

bases of previous year's leaves

— lateral bud which continues the growth of the rhizome from which next year's plant will be formed

— contractile adventitious roots

previous years' roots

Fig. 24.13 cont.   Diagrams of perennating organs. Solid arrows indicate movement of food materials from perennating organ to new plant or new perennating organ. Broken arrows indicate movement of food materials from the foliage leaves of this year's plant into the perennating organ.

By means of Stem Tubers
(Fig. 24.13C)

A stem tuber is a swollen underground stem which stores food, survives the winter, and gives rise to new plants from axillary buds. the following year. Example: potato.

Examine a potato plant showing new tubers and (if possible) the remains of the old tuber. Now look at a single tuber. Being a modified stem the tuber possesses axillary buds and leaves in the usual way. These are the so-called 'eyes' of the potato. The 'pupil' represents the axillary bud and the 'eyebrow' the tiny scale leaf (Fig. 24.13C). New plants are sprouted from the axillary buds. What is the food reserve in the potato tuber?

By means of a Rhizome
(Fig. 24.13D)

A rhizome is a horizontally growing underground stem which continues to live for many years. Each year the terminal bud at the end of the stem turns up and produces leaves and flowers above the ground, whilst contractile adventitious roots are formed below ground. The lateral bud closest to the terminal bud continues the growth of the rhizome, food materials for this being supplied by the aerial shoot. Other lateral buds may produce new rhizomes which branch off the parent stem. In some species the rhizome is short, thick, and slowly growing (Fig. 24.13D), whilst in others it is long, thin and quickly-growing. Examples: iris, water lily, Solomon's seal (short, thick type); fern, Michaelmas daisy, couchgrass, marram grass, mint, ground elder, thistle, bracken (long, thin type).

Examine a representative rhizome of the short-thick and/or long-thin type. Test it for food reserves. Cut transverse sections and stain in acidified phloroglucinol or FABIL. Conclusions?

By means of a Corm (Fig. 24.13E)

A corm is a short, swollen, vertically growing underground stem. It stores food and survives the winter, giving rise to new plants and corms the following year. The terminal bud gives rise to the new plant, axillary buds developing into either new plants or new corms. Contractile adventitious roots keep the corm anchored to the soil.
Examples: crocus, gladiolus.

Cut a corm in half vertically. Note the solid stem with a central vascular strand. Stain with acidified phloroglucinol. If available, examine corms at various stages of development.

E Corm

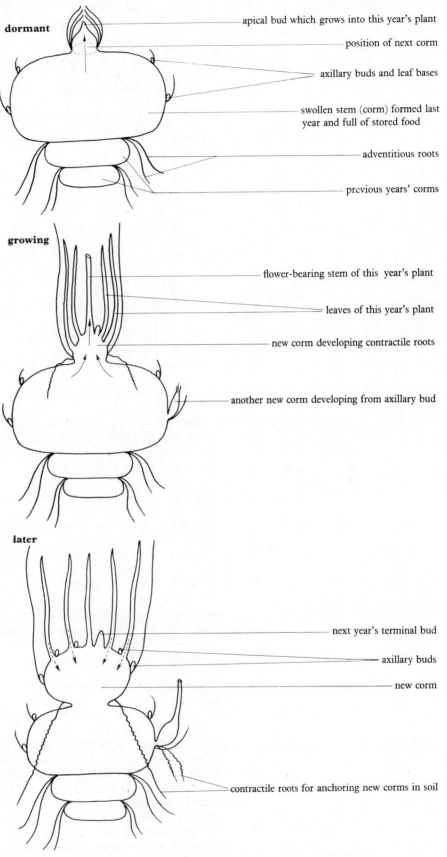

dormant

— apical bud which grows into this year's plant

— position of next corm

— axillary buds and leaf bases

— swollen stem (corm) formed last year and full of stored food

— adventitious roots

— previous years' corms

growing

— flower-bearing stem of this year's plant

— leaves of this year's plant

— new corm developing contractile roots

— another new corm developing from axillary bud

later

— next year's terminal bud

— axillary buds

— new corm

— contractile roots for anchoring new corms in soil

Fig. 24.13 cont.   Diagrams of perennating organs. Solid arrows indicate movement of food materials from perennating organ to new plant or new perennating organ. Broken arrows indicate movement of food materials from the foliage leaves of this year's plant into the perennating organ.

F Bulb

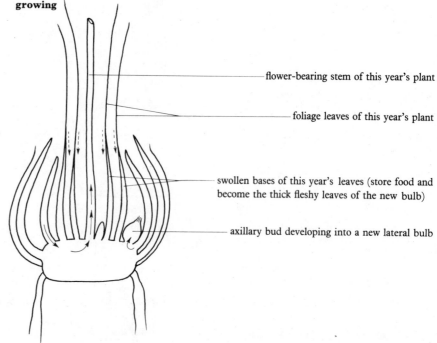

dormant

— apical bud which grows into this year's plant

— rudimentary leaves which will form this year's foliage leaves

— thick fleshy food-storing leaves formed from last year's swollen leaf bases

— scale leaf from the year before last

— axillary bud which will develop into new bulb

— next year's apical bud

— flattened stem

— adventitious roots

growing

— flower-bearing stem of this year's plant

— foliage leaves of this year's plant

— swollen bases of this year's leaves (store food and become the thick fleshy leaves of the new bulb)

— axillary bud developing into a new lateral bulb

Fig. 24.13 cont. Diagrams of perennating organs. Solid arrows indicate movement of food materials from perennating organ to new plant or new perennating organ. Broken arrows indicate movement of food materials from the foliage leaves of this year's plant into the perennating organ.

By means of a Bulb (Fig. 24.13F)
A bulb consists of a short vertical stem bearing adventitious roots, thick fleshy leaves (or leaf bases), and a variable number of axillary buds. In the centre is the terminal bud which develops into a new plant after the winter is over. The axillary buds develop into new bulbs.
Examples: onion, daffodil, tulip, snowdrop, lily, bluebell, hyacinth.

Cut a bulb in half vertically from below upwards. Identify the features noted above, particularly the terminal bud and the thick fleshy leaves. Is each fleshy structure a complete leaf or just the leaf base? How do the outermost leaves differ from those further in and what do they represent? Test a leaf for food reserves. If available examine bulbs at different stages of development.

By means of a Runner
(Fig. 24.13G)
Vegetative reproduction does not necessarily involve the formation of a storage organ. Some plants reproduce vegetatively by sending out side-branches which develop into new plants. Such is the case with runners. A runner is a horizontally growing,

G Runner

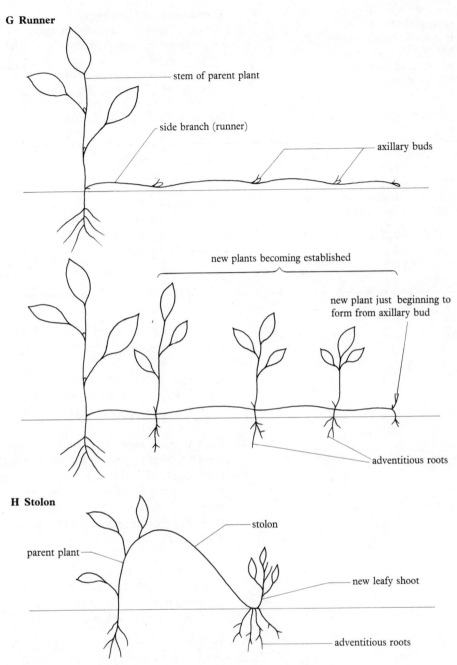

stem of parent plant

side branch (runner)

axillary buds

new plants becoming established

new plant just beginning to form from axillary bud

adventitious roots

H Stolon

stolon

parent plant

new leafy shoot

adventitious roots

Fig. 24.13 cont.   Vegetation reproduction by means of runner and stolon.

above-ground stem which grows from one of the lower axillary buds on the main stem. At regular intervals along the length of the runner are small axillary buds which give rise to new plants. Once the new plants are self-supporting the internodal sections of the runner have no further function. Examples: strawberry, creeping buttercup.

Examine a plant with one or more runners and note the above features. Sometimes the *main* stem grows horizontally along the surface of the soil, forming new plants at the nodes.

By means of a Stolon (Fig. 24.13H) Some stems grow so long that their weight causes them to bend over. Where a node touches the soil the axillary bud at that node develops into a new plant. Gardeners often encourage such plants to reproduce vegetatively by fixing the stolon to the soil with a staple, a procedure called layering.

Examples: blackberry, gooseberry, blackcurrant and redcurrant.

Requirements

Binocular microscope or hand lens
Microscope
Slides and coverslips
Watchglass
Needles
Small forceps
Razor blade
White tile
Apparatus for section cutting
Phloroglucinol and/or FABIL
Hydrochloric acid
Reagents and apparatus for testing
   for food reserves (*see* p. 46).

A range of perennating organs
   (*see* list on pp. 264–9)

By Means of Cuttings

Though not a natural method of vegetative reproduction, this is widely used by gardeners. A stem is cut and pushed into the soil. Adventitious roots grow out from the submerged part of the stem, particularly if the cut end is treated with a growth substance (*see* p. 306).

For Consideration

(1) What functions, other than reproduction, are performed by plant storage organs?

(2) Do animals show any processes that can be considered as equivalent to vegetative reproduction in plants?

Questions and Problems

1   Spermatozoa have been kept in a deep-frozen state for as much as ten years without losing their viability. Eggs, however, will only survive such treatment for a few hours at the most. Speculate on this difference.

2   What do you anticipate would be the effect on the reproductive future of a human female if, during the prepubertal period, the following took place:
(a) a course of FSH (follicle-stimulating hormone) was administered;
(b) One ovary was surgically removed;
(c) both ovaries were surgically removed;
(d) both fallopian tubes (oviducts) were ligatured?
Give reasons for your prognostications.   (*O and C*)

3   Give a concise account of the changes that take place in the course of the human oestrous cycle and the way the changes are regulated.
   Comment on the graph in Fig. 24.14 which shows how the body temperature of a human female changes in the course of the oestrous cycle.

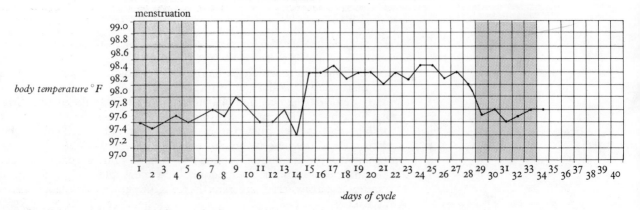

Fig. 24.14 Graph showing fluctuations in the basal body temperature of a human female in the course of the oestrous cycle. (*From* Demarest, R. J. and Sciarra, J. J., *Conception, Birth and Contraception*, Hodder and Stoughton, 1969).

B weight
+ T. All before root.
mating.

role of epididymis.

4 In each of the months January, March, July and October, five adult male specimens of a particular species of mammal were collected in the wild. Their bodies were weighed and after dissection, paired testis and paired epididymis weights were also obtained. The data were as follows:

B (kg)	T (g)	E (g)	B (kg)	T (g)	E (g)	B (kg)	T (g)	E (g)	B (kg)	T (g)	E (g)
5.2	12.4	2.4	6.5	4.7	1.5	6.8	2.6	0.8	5.4	4.7	1.4
6.4	13.8	2.5	5.7	6.0	1.8	6.3	1.8	1.1	6.6	4.3	1.3
6.9	16.6	3.6	6.8	5.0	1.3	6.3	2.5	0.8	7.4	3.9	1.0
7.5	12.0	2.9	7.5	5.6	1.9	5.2	1.6	0.3	6.3	6.6	1.5
8.0	10.7	2.6	8.5	5.7	1.5	8.4	2.5	1.0	6.8	9.5	1.8
January			**March**			**July**			**October**		

B = Body weight
T = Paired testis weight
E = Paired epididymis weight

(a) Present these data in graphical form in order to show weight variation of the testis in the months concerned.
(b) On graph paper plot testis weight against epididymis weight.
(c) Discuss the possible reproductive pattern of the species concerned.
(d) What additional evidence, if any, would you need to confirm your ideas on this pattern? *(O and C)*

5 Discuss the factors responsible for synchronizing sexual activity in animals.

6 In rabbits it has been found that ovulation occurs only after copulation has taken place. It has been shown that it is the act of copulation which is the stimulating factor rather than any chemical stimulus from semen or spermatozoa. Describe what must be the experimental evidence for this last statement.

7 'The fertilized egg contains a full set of maternal and paternal chromosomes.' Explain what this statement means and how it comes about. *(O and C)*

8 The data given below show the difference between the blood of a pregnant woman and that of the foetus developing in her uterus:

Partial pressure of oxygen (kN/m^2)	Percentage saturation of blood with oxygen	
	Mother	Foetus
1.3	8	10
2.7	20	30
3.9	40	60
5.3	65	77
6.6	77	85
8.0	84	90
9.3	90	92
10.6	92	92

Plot these results graphically and comment on the differences between the two curves.

9 Your biological interests have led you to a position where it is essential to establish the reproductive pattern of a particular species of mammal which is well represented and successful in the wild. You have been supplied with sufficient, healthy, sexually mature individuals of both sexes for normal breeding, but despite your having arranged optimum conditions for their captive existence, they have failed to breed. This failure is not because you have kept them for insufficient time. Assuming that you have adequate facilities and funds, how would you now proceed in your aim to discover the normal reproductive activity of this species? *(O and C)*

10 Comment on the following statements:
(a) Bacteria show sex in its simplest and most rudimentary form.
(b) The greater the degree of parental care, the fewer the offspring produced.
(c) Asexual reproduction is commonest amongst animals that also show marked powers of regeneration.
(d) In animals the sexes are usually in separate individuals whereas in plants they are usually on the same individual.
(e) Some mammals have young which are born at a more advanced stage than others.

11 Compare the mechanisms by which male and female gametes are brought together in the mammal and flowering plant.

12 What do you consider to have been the major evolutionary trends in the animal kingdom as far as reproduction is concerned? Illustrate your answer with examples. Are the same trends also seen in the plant kingdom?

13 'The evolution of flowering plants has been closely bound up with the evolution of insects.' Discuss.

14 'By means of various adaptions, living organisms have been able to exploit terrestrial environments.' Discuss this statement with reference to adaptation for reproduction. *(O and C)*

25 The Life Cycle

Background Summary

1 In all organisms with sexual reproduction **meiosis** (halving of the chromosome number) and **syngamy** (union of gametes) divide the life cycle into **haploid** and **diploid** phases.

2 In all animals and in most lower plants, the haploid phase is represented only by the gametes. However, in mosses, liverworts, ferns, and certain seaweeds the life cycle shows an alternation between a haploid gamete-producing **gametophyte** and a diploid **sporophyte**, which produces haploid **spores**. This is called **alternation of generations**.

3 In mosses the gametophyte is the more prominent generation, the sporophyte being attached to, and dependent upon, the gametophyte. For successful reproduction the gametophyte requires wet conditions, the sporophyte, dry.

4 In ferns the sporophyte is the more prominent generation, the gametophyte being variously reduced. As in mosses, wet conditions are essential for reproduction of the gametophyte and dry conditions for the sporophyte.

5 In certain ferns the sporophyte produces two kinds of spores (**heterospory**): **microspores** give rise to male sperm-producing gametophytes, whilst large **megaspores** give rise to female egg-producing gametophytes.

6 In some primitive plants, e.g. certain green algae, there is no sporophyte: the diploid zygote undergoes meiosis, giving rise to a haploid adult which produces gametes. In some cases motile haploid **zoospores** may be interpolated into the life cycle between the zygote and adult.

7 Conifers and flowering plants continue the tendency seen in ferns to reduce the gametophyte. In both groups the gametophyte is incorporated into the body of the sporophyte, being represented in the male by the protoplasmic contents of the **pollen grain** (microspore) and its derivative, the pollen tube; and in the female by the protoplasmic contents of the **embryosac** (megaspore).

8 The rise of the sporophyte, and decline of the gametophyte, can be seen as an adaptation to life on dry land. It is associated with the evolution of reproductive mechanisms involving transfer of pollen, and the seed habit, typical of higher plants.

9 Certain animals have an asexually reproducing stage in their life cycle, but, as far as is known, there is no true alternation of generations in the genetic sense in the animal kingdom.

Investigation 25.1
Life cycle of moss

Mosses show alternation of generations between a small leafy gametophyte and a spore-producing sporophyte which grows out of, and is dependent upon, the gametophyte.

Gametophyte

(1) Examine a whole gametophyte plant, noting simple **stem**, **leaves**, and **rhizoids** (Fig. 25.1A). Mount a leaf in water and examine. What can you say about its structure? Observe a rhizoid under the microscope. How does it compare with a true root? How capable do you think this plant is of surviving dry conditions?

(2) Examine transverse sections of the leaf and stem under high power.
Leaf: **lamina**, **midrib**;
Stem: **epidermis**, **cortex**, **conducting tissue**.

(3) Examine gametophytes under a lens or binocular microscope and look for **male rosettes** at the top of certain branches, and **female rosettes** lower down. Why are the male rosettes generally higher than the female rosettes? The male rosette consists of a group of **antheridia** enveloped by a 'cup' of leaves; the female rosette consists of a group of **archegonia** enveloped by a 'cup' of leaves.

Fig. 25.1 Structure of gametophyte of moss. These diagrams, and those in Fig. 25.2 are based on the moss *Mnium*.

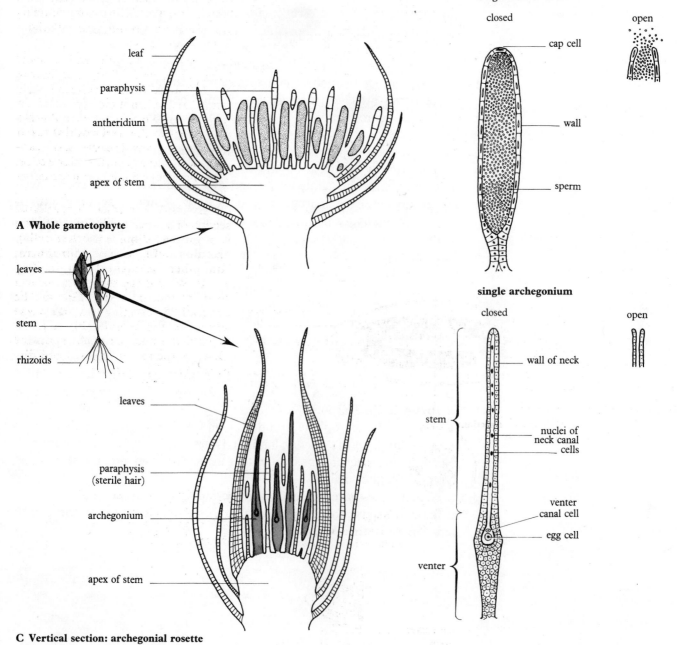

B Vertical section: antheridial rosette

leaf
paraphysis
antheridium
apex of stem

A Whole gametophyte

leaves
stem
rhizoids

leaves
paraphysis (sterile hair)
archegonium
apex of stem

C Vertical section: archegonial rosette

single antheridium

closed   open

cap cell
wall
sperm

single archegonium

closed   open

wall of neck
nuclei of neck canal cells
stem
venter canal cell
egg cell
venter

(4) Tease out the contents of a male rosette in a drop of water on a slide and examine under high power. Observe **antheridia**, **paraphyses** (sterile hairs) and, if possible, **sperm cells** (antherozooids). If sperm are present irrigate the slide with iodine or Noland's solution to see their **flagella**.

(5) Tease out the contents of a female rosette in a drop of water and observe under high power. Note **archegonia**, each with **egg cell** and **neck**.

(6) Examine a prepared vertical section of a male rosette under low power. Note enveloping **leaves**, **paraphyses**, **antheridia**, **sperm cells** (Fig. 25.1B). The paraphyses absorb water and are said to help hold water in the rosette. (Why should this be necessary?).

(7) Examine a prepared vertical section of a female rosette under low power. Note enveloping **leaves**, **paraphyses**, **archegonia** with **neck** and **egg-cell** (Fig. 25.1C). How is fertilization brought about?

Sporophyte

(1) After fertilization the zygote develops into the young sporophyte. This grows out of the female rosette carrying with it the upper part of the archegonium which eventually falls off.

Examine sporophytes in various stages of development. Note **spore capsule** located at the upper end of a stalk whose foot is embedded in a female rosette. The stalk contains vascular strands continuous with those in the gametophyte.

(2) Remove a capsule and a short length of its stalk. With a needle take off the **operculum**. Insert the stalk into a piece of plasticine on a slide in such a way that you can look down a microscope at the **peristomial teeth** (Fig. 25.2A). Now breathe on the capsule and observe down the microscope. What happens? Explain your observations.

(3) Examine a median longitudinal section of a capsule. Notice the **spore sac**, **spores** (or **spore mother cells**), **annulus cells**, **teeth**, **operculum**, and other features shown in Fig. 25.2B. What is the three-dimensional shape of the spore sac? This can be deduced by examining a transverse section of the capsule (Fig. 25.2C). What is the function of the annulus? How are the spores released and dispersed (*see* Fig. 25.2D)?

A End view of sporangium (spore capsule) after removal of operculum

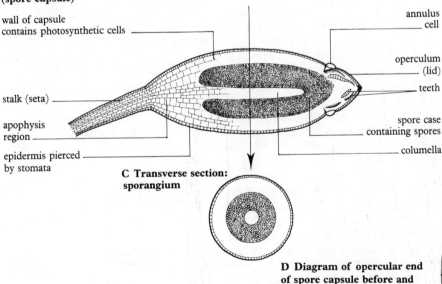

annulus

peristomial teeth

B Longitudinal section: sporangium (spore capsule)

wall of capsule contains photosynthetic cells

annulus cell

operculum (lid)

teeth

stalk (seta)

apophysis region

spore case containing spores

epidermis pierced by stomata

columella

C Transverse section: sporangium

D Diagram of opercular end of spore capsule before and during release of spores

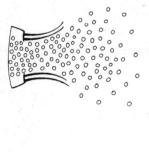

operculum

teeth

Fig. 25.2 Structure of sporophyte of moss.

Requirements
Microscope
Hand lens or binocular microscope
Slides and coverslips
Plasticine (modelling clay)
Mounted needle

Iodine solution
Noland's solution

Moss gametophytes and
  sporophytes (entire)
TS stem of moss
TS leaf of moss
VS male rosette (antheridial
  rosette) of moss
VS female rosette (archegonial
  rosette) of moss
LS and TS moss capsule
WM protonema

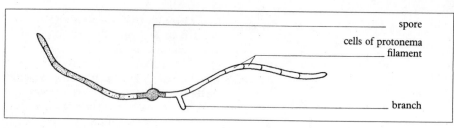

Fig. 25.3 Young protonema of moss.

Protonema
The spores do not develop into a new gametophyte direct, but first give rise to a filamentous alga-like protonema. This produces buds which, in turn, give rise to gametophytes.

Examine a protonema under high power (Fig. 25.3). Note branched filament whose cells have chloroplasts; also formative **spore**. **Buds** with **rhizoids** may be seen.

For Consideration
(1) Mosses have been described as the amphibians of the plant kingdom. Do you think this is justified?
(2) Which do you consider to be the dominant generation, the gametophyte or sporophyte?
(3) To what extent is the sporophyte (a) self-supporting; (b) dependent on the gametophyte?
(4) What function is fulfilled by the protonema stage in the life cycle? Do you think it is more of a liability than a help?

Investigation 25.2
Life cycle of fern

In ferns the sporophyte is the dominant generation and the gametophyte is reduced to a small alga-like **prothallus**. The two generations are independent of each other.

Sporophyte
(1) Examine an entire plant (Fig. 25.4) and note the horizontal **rhizome** (*see* p. 266) with old **leaf bases** and **roots**; **leaves** (**fronds**) subdivided into **pinnae** and **pinnules**.
(2) Examine prepared sections of rhizome, roots, leaves and a leaf stalk (rachis), noting the **vascular tissues**. How does the fern sporophyte compare with that of mosses? And how does it compare with the flowering plant?
(3) Certain pinnules, known as **sporophylls**, produce spores. The latter are formed in **sori** on the undersides of the sporophylls. In some species each sorus is protected by an umbrella-like **indusium** (Fig. 25.5A). With a needle remove an indusium from one of the sporophylls so as to reveal **sporangia** (spore capsules) beneath. Mount a mature sporangium in water and notice the **annulus cells** and other features shown in Fig. 25.5C.

(4) Place a few mature **sporangia** on a dry slide and either mount them in glycerine or leave them under a hot lamp to dry out. Examine under low power. Watch the capsule dehiscing. Whereabouts in the wall of the capsule does splitting occur? Compare with Fig. 25.5C. Can you explain the mechanism of dehiscence?

(5) Examine a transverse section of a sporophyll, noting the structures shown in Fig. 25.5B. In particular observe the sporangia cut in various planes.

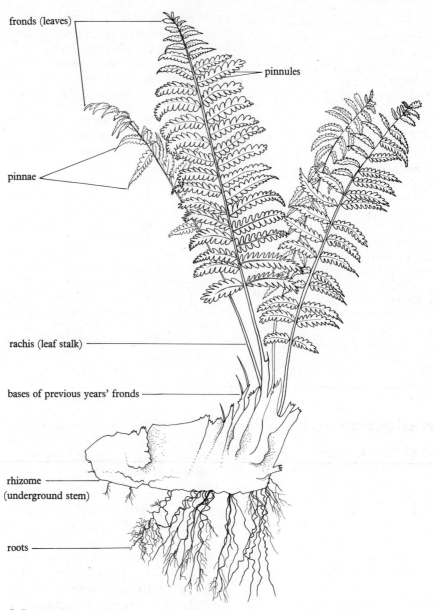

fronds (leaves)

pinnules

pinnae

rachis (leaf stalk)

bases of previous years' fronds

rhizome
(underground stem)

roots

Fig. 25.4 The sporophyte of the fern *Dryopteris* has a creeping underground rhizome. This represents the stem. The above-ground parts of the plant are the leaves (fronds) whose structure is illustrated here.

A Spore-bearing pinnule (sporophyll)

central vein of pinnule

sorus covered by indusium

B Transverse section: sporophyll

central vein of pinnule
small vascular bundle
palisade layer
spongy mesophyll
placenta
sporangium
indusium

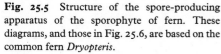

Fig. 25.5 Structure of the spore-producing apparatus of the sporophyte of fern. These diagrams, and those in Fig. 25.6, are based on the common fern *Dryopteris*.

C Sporangium

side view
(spore capsule)

a

stalk

annulus cells

spores

b

thick inner
thickish radial
thin outer

walls of annulus cells

view from a

side-wall cells

view from b

open

annulus cells now sprung back

spores released

line of weakness where splitting has occurred (stomium)

Gametophyte (Prothallus)

(1) Each spore is potentially capable of developing into a prothallus. This is small and reduced, living on the surface of damp soil.

Examine a mature prothallus, noting its simple structure and shape. Mount it in water, lower surface uppermost, and examine under low and high powers. In what ways is it adapted to lead an independent existence?

(2) Examine a prepared whole mount and/or horizontal section of a prothallus. Observe **antheridia** and **archegonia** in surface view. Both are located on the underside of the prothallus, as are the **rhizoids**. Archegonia are closer to the apical notch than the antheridia (Fig. 25.6A). In good preparations coiled **sperm** may

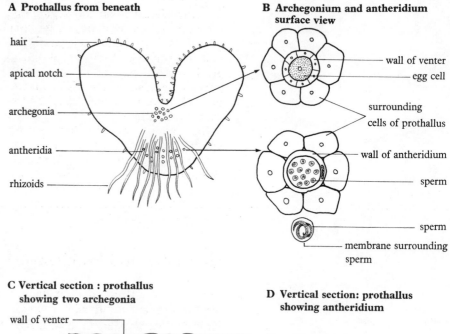

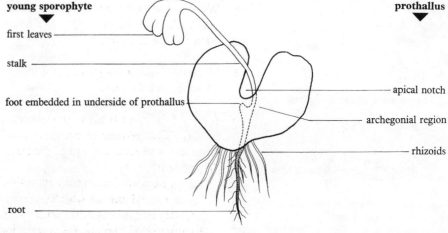

Fig. 25.6   Structure of the gametophyte of fern.

be visible in the antheridia, and an **egg cell** in some of the archegonia (Fig. 25.6B).

(3) Examine a vertical section of a prothallus to see antheridia and archegonia in side view (Fig. 25.6C,D). How are the sperms released from the antheridia and how do they enter the archegonia? How do the antheridia and archegonia of ferns compare with those of mosses?

(4) After fertilization the zygote develops into the sporophyte. At first this is small and rudimentary, with its foot embedded in the lower side of the prothallus. Later, when the sporophyte's roots and leaves develop, the prothallus—no longer required—withers and dies.

Examine a prothallus with **young sporophyte** attached and note the young **leaves**, **roots**, and **foot** embedded in the prothallus (Fig. 25.6E).

For Consideration

(1) To what extent are ferns adapted to life on land?

(2) What are the weak points in the life cycle?

(3) Assuming that ferns are derived from mosses in evolution, what major changes have occurred during the transition?

(4) Is the sporophyte entirely self-supporting?

(5) How does the mechanism of spore-dispersal compare with that of mosses?

Questions and Problems

1 Make a list of features of their anatomy and physiology which equip mosses and ferns to live on land. In what respects are they poorly equipped for life on land? How do your answers fit in with their distribution?

2 Compare the general plan of the human life cycle with that of insects and mosses.

3 Compare the life cycle of a named moss with that of a named fern.

4 It is generally agreed that flowering plants have evolved from (ultimately) ferns.
(a) What changes would have to take place in the structure and life cycle of the fern *Dryopteris* for it to evolve into a flowering plant?
(b) Do any present-day plants provide information as to the intermediate steps that may have occurred in the above transition?

5 What advantages are gained by mosses and ferns in having alternation of generations in the life cycle? Why do you think flowering plants have dispensed with such a life cycle?

6 'A flower is not an organ of sexual reproduction; it produces spores, not gametes.' How far is this true?

7 (a) (i) Distinguish between the diploid and haploid condition in cells.
(ii) What genetic advantages have diploid organisms over haploid organisms?
(b) Give a concise account of the roles of mitotic and meiotic divisions in the life histories of:
(i) a named plant which shows a dominant haplophase,
(ii) a named animal which shows a dominant diplophase,
(iii) any named organism which shows a regular alternation between a clearly distinguished haplophase and diplophase. (*JMB*)

8 'There is not a single case of true alternation of generation in the animal kingdom.' Discuss.

9 Write an essay on the colonization of dry land by plants.

10 Write an essay entitled 'Alternation of Generations'.

11 Explain the differences between:
   (a) haploid and diploid,
   (b) sporophyte and gametophyte,
   (c) microspore and megaspore,
   (d) sporophyll and sporangium,
   (e) spore and seed.

12 'Life cycles can generally be divided into diploid and haploid phases, but organisms differ in the relative emphasis given to each phase.' Discuss this statement.

26 Patterns of Growth and Development

Background summary

1 **Growth**, the permanent increase in size undergone by an organism, results from **cell division**, **assimilation**, and **cell expansion**.

2 Growth can be measured by estimating increase in a chosen linear dimension (e.g. height), or by estimating increase in volume, total weight or dry weight. Each has its snags and a source of inaccuracy common to them all is that growth may be **allometric**, different parts of the body growing at different rates.

3 Growth may be expressed in terms of a **growth curve** from which the growth rate and percentage growth can be derived. For many purposes this last is the most meaningful way of expressing growth.

4 In most organisms growth takes place in a smooth and regular pattern. An exception is provided by the arthropods in which growth is **intermittent**. This is due to arthropods possessing a hard cuticle which must be shed (**moulting, ecdysis**) before growth can take place.

5 Starting with fertilization, the development of form (**embryology**) can be divided into cleavage, gastrulation, and organogeny.

6 **Cleavage** may be equal or unequal, depending on the yolk content of the egg. It results in the formation of a hollow **blastula**.

7 At its simplest, **gastrulation** occurs by invagination of the blastula. It results in the formation of two layers of cells, the inner layer lining the **archenteron**. In most animals, particularly where much yolk is present, gastrulation involves cell migration as well as, or instead of, invagination.

8 Gastrulation in vertebrates is followed by development of the **notochord** and **neural tube** from the mid-dorsal region of the inner and outer layer of cells respectively.

9 Meanwhile the **mesoderm** develops either by evagination of the archenteron wall or by inward migration of cells from the lips of the blastopore. Either way, the mesoderm lies between the outer layer of cells (**ectoderm**) and lining of the gut (**endoderm**).

10 In most animals the mesoderm surrounds a cavity, the **coelom**. This expands, giving rise to the general body cavity.

11 In vertebrates the mesoderm becomes subdivided into **somitic, nephrogenic** and **lateral plate** mesoderm. The first two are metamerically segmented, the last unsegmented. The somites give rise to the axial skeleton and muscles, the nephrogenic mesoderm to the kidney, and the lateral plate mesoderm to the muscles of the gut wall and body wall and to the heart.

12 Various organs are now moulded out of the different parts of the ectoderm, endoderm, and mesoderm.

13 In **amniotes** (reptiles, birds, and mammals) the embryo is situated above a **yolk sac** from which nourishment may be gained. Also, extra-embryonic membranes are formed.

K

14 Developed as an adaptation to life on dry land where they are associated with the cleidoic ('closed') egg, or internal (uterine) development, the **extra-embryonic membranes** protect, provide a means of excretion for, and in mammals nourish, the embryo.

15 The extra-embryonic membranes are the **chorion** and **amnion** (enclosing the amniotic cavity), and the **allantois** (enclosing the allantoic cavity). Part of the allantois fuses with the chorion to form the highly vascularized **allanto-chorion**.

16 In all amniotes the chorion, amnion, and amniotic cavity are protective. In reptiles and birds the allantoic cavity is an excretory chamber, and the allanto-chorion, a respiratory surface. In eutherian mammals the allanto-chorion becomes the **placenta**, and the stalk of the allantoic cavity becomes the **umbilical cord**.

17 In mammals, at birth, fundamental changes occur in the respiratory and circulatory systems as a result of the respiratory function of the placenta being taken over by the lungs.

18 In some animals one or more **larval stages** are interpolated between egg and adult. In general, larvae are important in distribution, feeding and (in certain specialized cases) asexual reproduction.

19 Larvae develop into the adult by **metamorphosis** which may involve a total reorganization of larval structures. In insects and amphibians metamorphosis is brought about by a combination of external (environmental) and internal (hormonal) factors.

20 In insects a distinction is made between **hemimetabolous** insects with incomplete or gradual metamorphosis, and **holometabolous** insects with complete metamorphosis.

21 In flowering plants development starts with the growth of the zygote into the embryo within the **seed** (*see* p. 294). Development continues only when **germination** takes place.

22 Two types of germination are recognized: **hypogeal** and **epigeal**. In hypogeal germination the epicotyl elongates, in epigeal germination the hypocotyl elongates. These two types of germination are related to whether nourishment is provided by the cotyledons or endosperm tissue.

23 Conditions required for successful germination include water, correct illumination, suitable temperature, and presence of oxygen. Internally a rapid mobilization of food reserves takes place.

24 Growth and development of the shoot and root take place by cell division in apical (primary) meristems followed by cell expansion and differentiation. This **primary growth** results in the formation of **primary tissues**.

25 Increase in girth takes place by **secondary growth** which results in the laying down of **secondary tissues**, formed from secondary meristems (**cambium cells**).

26 There are two cambium layers in a typical woody perennial. An internal cambium gives rise annually to secondary xylem and phloem, an external **cork cambium** forms protective corky cell (**periderm**) and secondary cortex. **Lenticels** in the periderm permit gaseous exchange.

Investigation 26.1

Observations of live amphibian embryos

Amphibians are much used by embryologists because they are tolerably easy to handle and their development is relatively uncomplicated. Furthermore, they demonstrate the fundamental sequence of changes which occurs during the development of a typical chordate. True, the tadpole is unique to amphibians and the details of gastrulation are more complex than in primitive chordates, but the broad picture of development is typically chordate.

Observations can be carried out on any amphibian at the appropriate time of the year. A species much used by embryologists is *Xenopus laevis* the African clawed toad. The normal breeding season of *Xenopus* is July, but mating and spawning may be induced at any time of the year by injecting them with gonadotrophic hormone. So readily do they respond to this treatment that for many years they were used for testing pregnancy: if the urine of a woman, suspected of being pregnant, was injected into a female toad, subsequent spawning could be taken as unequivocal evidence that the woman was pregnant.

Procedure

With a wide pipette, transfer a recently laid egg to a watch glass. Cover with the same water that the egg came from. Examine it under a hand lens or binocular microscope. Watch for **cleavage**. In *Xenopus* the first cleavage division normally occurs one-to-two hours after laying. In what plane are the first three cleavage divisions? Note the many-celled **blastula**. This remains enclosed inside the transparent **vitelline membrane**.

Examine the developing embryos at intervals over the next few days. **Gastrulation** normally occurs about eight hours after laying: notice the yolk plug—what does it represent? **Neurulation** occurs about 24 hours after laying: notice the formation of the **neural folds** which eventually fuse to form the **neural tube**. Which end of the neural tube is going to form the **brain**?

The **larva** (**tadpole**) hatches (i.e. breaks out of the vitelline membrane) on about the third day. The exact rate of development depends on various factors including temperature, and also on the species. The figures given above are based on *Xenopus laevis* at normal room temperature.

For further development of the larva, *see* Investigation 26.3.

For Consideration

(1) How would you investigate the influence of environmental factors on the rate of development of an amphibian?

(2) How would you investigate the amphibian embryo's source of energy during cleavage?

(3) What factors might cause the egg of *Xenopus* to start cleaving?

(4) What factor or factors might control the plane of successive cleavage divisions?

(5) Does the development of an amphibian entail changes in bulk as well as form? Explain your observations on this point.

Requirements

Binocular microscope or hand lens
Watch glass
Wide pipette

Newly laid eggs of amphibian, e.g. *Xenopus laevis* (*see* p. 431)

Investigation 26.2

Internal changes during early chordate development

Examination of live embryos provides a dynamic picture of the external changes that occur during development. To study the internal changes requires examining sections of embryos cut in various planes. It is suggested that you should examine the following sections, all of which are of frog embryos unless otherwise stated. Preserved embryos should also be examined under a hand lens or binocular microscope for cross reference.

CLEAVAGE

(1) *Vertical section two-cell stage* (Fig. 26.1A): Note pigmented upper half of cleaved egg (**animal pole**), yolky bottom half (**vegetal pole**). Why are the polar ends of the egg described as 'animal' and 'vegetal'? Note vertical **cleavage furrow** and **vitelline membrane**.

(2) *Horizontal section four-cell stage* (Fig. 26.1B): Two vertical **cleavage furrows** are visible at right angles to each other. The four cells are equal in size.

A Vertical section: 2-cell stage

animal pole

pigment

vitelline membrane

cleavage furrow

vegetal pole

B Horizontal section: 4-cell stage

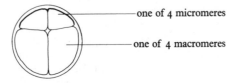

C Vertical section: 8-cell stage

one of 4 micromeres

one of 4 macromeres

D Vertical section of blastula

micromeres

blastocoel

macromeres

Fig. 26.1   Cleavage in the frog.

(3) *Vertical section eight-cell stage* (Fig. 26.1C): Why can only four cells be seen? Those towards the animal pole are smaller than those towards the vegetal pole—why the difference?

BLASTULA
(4) *Vertical section of completed blastula* (Fig. 26.1D): The blastula is essentially a hollow ball of cells. Take note of the large yolky cells (**macromeres**) towards vegetal pole, and the smaller pigmented cells (**micromeres**) towards animal pole; **blastocoel**.

GASTRULATION
The simplest form of gastrulation is seen in *Amphioxus* and it is advisable to look at this animal first before going on to the amphibian.
(5) *Blastula and gastrula of Amphioxus* (Fig. 26.2): Examine slides or photomicrographs of (a) blastula and (b) gastrula of *Amphioxus*. Gastrulation occurs by a simple process of **invagination**. In (b) notice the **archenteron**, **blastopore** and reduced **blastocoel**.

A Blastula

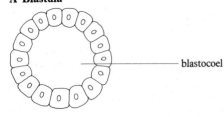

blastocoel

B Gastrula

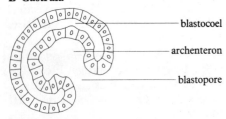

blastocoel

archenteron

blastopore

Fig. 26.2   Gastrulation in *Amphioxus*.

Owing to the presence of a greater quantity of yolk, gastrulation in amphibians is more complex than in *Amphioxus*, involving **cell proliferation** and **migration**, as well as invagination. Examine the following sections of amphibian gastrulae:

(6) *Vertical section of early gastrula* (Fig. 26.3A): Notice that the cells have started to migrate inwards on one side of the blastula, thus creating a small **archenteron** and slit-like **blastopore**. This is located just above the level of the yolky cells.

(7) *Vertical section of mid-gastrula* (Fig. 26.3B): Inward migration of cells has continued and the archenteron is expanding at the expense of the blastocoel. The blastopore, originally slit-like, has formed a complete circle, inward migration of cells occurring at all points round this circle. The yolky cells enclosed by the circle constitute the **yolk plug**. The cells turning inwards at the **dorsal lip** of the blastopore are future notochord cells, those turning inwards at the **ventral lip** are future mesoderm cells. What sort of cells migrate inwards at the **lateral lips** of the blastopore?

Also notice that on the ventral side of the gastrula the superficial layer of cells has grown over the surface of the yolky cells.

(8) *Vertical section of late gastrula* (Fig. 26.3C): The changes seen in (7) have continued further. Note the expanded **archenteron**, the still further diminished **blastocoel**, and the more extensive **mesoderm**. The yolk plug is now smaller—why?

(9) *Horizontal section of late gastrula* (plane a–a, Fig. 26.3C): Interpret this section yourself and try to identify the component structures. What you can see depends on the level of the section. Can you deduce the precise level at which your particular section has been cut relative to (8)?

(10) *Transverse section of late gastrula* (plane b–b, Fig. 26.3C): Can you interpret this section? At this stage the notochord and mesoderm cannot be distinguished from each other and together constitute the **chordamesoderm**. Notice the **neural plate** immediately above the chorda; also **ectoderm**, **endoderm** and **archenteron**.

A Vertical section of early gastrula

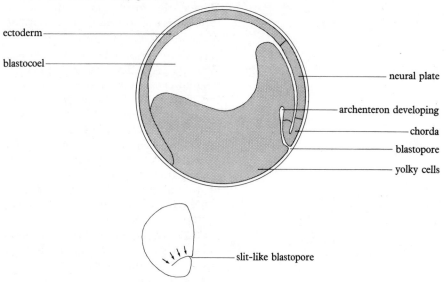

ectoderm

blastocoel

neural plate

archenteron developing

chorda

blastopore

yolky cells

slit-like blastopore

B Vertical section of mid-gastrula

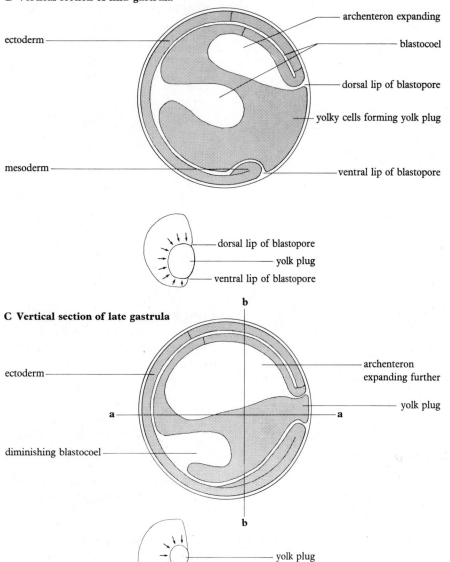

archenteron expanding

ectoderm

blastocoel

dorsal lip of blastopore

yolky cells forming yolk plug

mesoderm

ventral lip of blastopore

dorsal lip of blastopore

yolk plug

ventral lip of blastopore

C Vertical section of late gastrula

b

ectoderm

archenteron expanding further

yolk plug

a   a

diminishing blastocoel

b

yolk plug

Fig. 26.3 Gastrulation in frog, as seen in vertical sections. The small diagrams beneath each section are postero-dorsal views of the developing gastrula showing the blastopore.

NEURULATION

This is the formation of the neural tube and can be seen in transverse sections of frog embryos.

(11) *Transverse section of early neurula* (Fig. 26.4A): The **neural plate** is beginning to sink at the centre, and the **neural folds** to grow upwards. Note the **notochord** immediately beneath the neural plate.

(12) *Transverse section of mid-neurula*: You should have no difficulty interpreting this slide. The upward growth of the neural folds is now plainly evident. The neural folds flank the **neural groove**.

(13) *Transverse section of late-neurula* (Fig. 26.4B): The neural folds have fused in the mid-dorsal line to form the **neural tube**. As in all chordates the neural tube is dorsal and hollow

(note why) and lies immediately above the notochord. Can you see a **coelomic cavity** in the mesoderm on each side? Note also **neural crest cells** above, and to either side of, the neural tube. These cells are destined to give rise to the dorsal root ganglia of the spinal nerves.

(14) *Vertical longitudinal section of late neurula* (plane a–a, Fig. 26.4B): In this slide, which you should try to interpret for yourself, the structures seen in (13) can be observed in longitudinal section. Notice the **neural tube**, **notochord**, **pharynx**, not yet connected to the exterior by a mouth, **yolky endodermal cells**, forming floor of gut, and **blastopore**. At the stage reached in (13) and (14) the animal is ready to hatch from the egg membrane as the tadpole.

A Early neurula

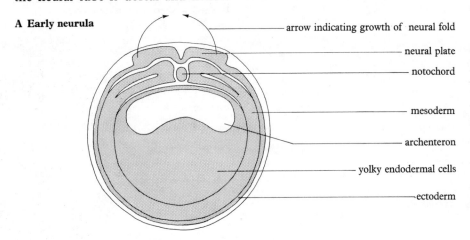

— arrow indicating growth of neural fold

— neural plate

— notochord

— mesoderm

— archenteron

— yolky endodermal cells

—ectoderm

B Late neurula

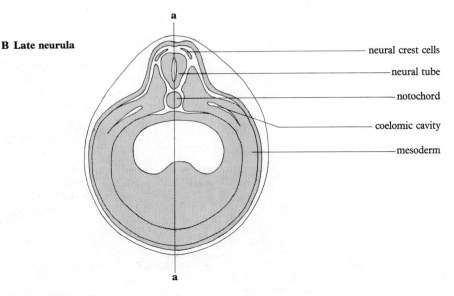

— neural crest cells

— neural tube

— notochord

— coelomic cavity

—mesoderm

Fig. 26.4 Neurulation in frog.

ORGAN FORMATION

Further development of the amphibian is continued in the tadpole. The fundamental cell layers have now been laid down and what remains is the formation of organs (**organogeny**). This takes place mainly, but not exclusively, by the differentiation of **mesoderm**.

(15) *Transverse section of early tadpole through trunk region* (Fig. 26.5A): Note in particular the subdivisions of the mesoderm: **somitic mesoderm (somites)** on either side of the neural tube and notochord; **nephrogenic mesoderm**; and **lateral plate mesoderm**. The somites become further subdivided into the **dermatome**, **sclerotome** and **myotome**. What do these subdivisions of the mesoderm give rise to?

(16) *Vertical longitudinal section of early tadpole* (Fig. 26.5B): In this slide some of the structures seen in (15) can be viewed in longitudinal section (plane a–a, Fig. 26.5A). Note the neural tube expanded anteriorly to form the **brain vesicles** (forebrain, midbrain, and hindbrain), the remainder of the neural tube forming the spinal cord; tubular **heart**; **pharynx**, still not connected to the exterior by a mouth; **anus**, immediately beneath blastopore, which has now closed. What you will see of the mesoderm depends on the level of the section: if the section is slightly off-centre **segmental somites** should be visible.

A Transverse section of embryo

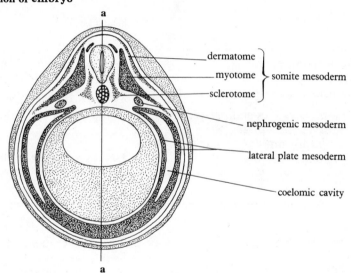

B Vertical longitudinal section of embryo (plane a–a)

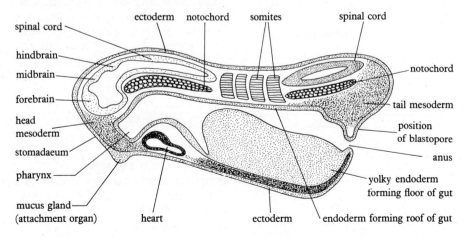

Fig. 26.5 Further development of the frog, including the differentiation of the mesoderm, as seen in sections of embryos prior to hatching. Semi-diagrammatic—sections vary according to level and exact age of specimen.

(17) *Transverse section of later tadpole through trunk region* (Fig. 26.5C): The **lateral plate mesoderm** remains unsegmented and develops a large **coelomic cavity**. The mesoderm itself becomes distinguishable into **somatic mesoderm** lining the body wall and **splanchnic mesoderm** lining the gut. How many of these features can you see in the section?

For Consideration
What are the disadvantages of studying embryology by means of prepared sections? How might the disadvantages be overcome?

C Transverse section of tadpole at later stage through trunk region

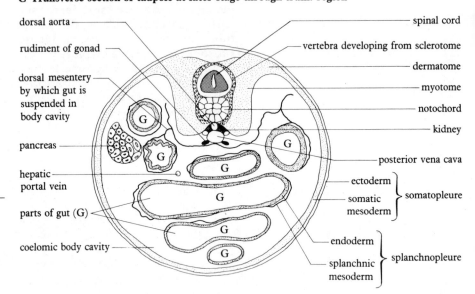

Fig. 26.5 cont. Further development of the frog, including the differentiation of the mesoderm, as seen in sections of embryos immediately after hatching.

Requirements
Microscope
Binocular microscope or hand lens

Amphibian slides and preserved embryos (*see* pp. 283–8)
Amphioxus slides or photomicrographs (*see* p. 284)

Investigation 26.3
Development of tadpole

The development of the tadpole can be followed by observing live tadpoles of *Xenopus*, or some other amphibian such as the common frog at the appropriate time of the year, and also by studying preserved specimens.

Live tadpoles can be observed semi-stranded in a watch glass under a binocular microscope or hand lens. They should be transferred to the watch glass with a wide pipette.

Observations of external features may be supplemented by an examination of transverse and/or longitudinal sections of the various stages.

Procedure

THE YOUNG TADPOLE
Examine a newly hatched tadpole. To what extent does it look like Fig. 26.6A which is based on the tadpole of the common frog?

In general tadpoles at this stage possess the following features:

(a) A short **tail** which gradually lengthens.

(b) On the ventral side of the head, a group of mucus glands, together forming a temporary **attachment organ**, by which the tadpole fixes itself to weeds, etc. The tadpole is, therefore, sessile at this stage and remains so until the tail is long enough to serve as an effective locomotory device.

(c) Three pairs of **external gills**, vascularized outgrowths of the skin, which are not connected with the pharynx. They are served by **arterial arches 3, 4 and 5**, the 6th arch being as yet undeveloped.

(d) A **straight gut** with, as yet, **no mouth**. The stomadaeum is visible externally but does not yet open to the pharynx (Fig. 26.5A). How does the tadpole gain nourishment at this stage?

A Immediately after hatching

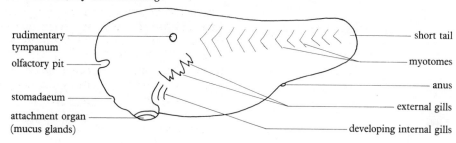

B Later

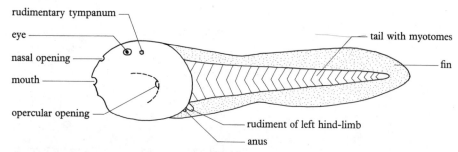

Fig. 26.6 Diagrams summarizing the major external changes that occur during the development of the frog tadpole. Based on the common frog *Rana temporaria*.

(e) An S-shaped tubular **heart** which, though simple at this stage, soon starts to sub-divide into atrium and ventricle.

(f) Four pairs of **pharyngeal pouches** which do not yet connect with the exterior. They are destined to develop into the internal gill pouches.

Observe the above features in live or preserved tadpoles and in prepared vertical and horizontal longitudinal sections.

FROM HATCHING TO
METAMORPHOSIS
During the ensuing three months the tadpole undergoes a series of changes the most important of which, in approximate sequence, are as follows:

(a) The **attachment organ degenerates** and the tail elongates so the tadpole ceases to be sessile and becomes motile. The tail acquires a fully functional series of segmental **myotomes** whose contractions permit lateral movements of the tail similar to those displayed by the tail of a fish.

(b) **Special sense organs develop** (nasal sacs, eyes and inner ears).

(c) **The gut becomes long and coiled**, suitable for a herbivorous diet.

(d) A **mouth develops**, the stomadaeum breaking through to the pharynx. The mouth develops horny jaws with small rasping teeth for feeding on algae, encrusing vegetation, and such like.

(e) The external gills are replaced by **internal gills** associated with four pairs of **gill pouches** leading from pharynx to exterior. These are associated with **arterial arches 3, 4, 5, and 6**, the sixth developing at this stage.

(f) An **operculum**, a ventro-lateral fold of skin, grows back to cover the gill openings. The resulting opercular chamber connects with the exterior by an opercular opening on the left hand side of the head (Fig. 26.6B).

(g) **Limb buds** appear which gradually develop into paired limbs. The hind limbs are fully visible in the later stages of development, but the forelimbs are concealed beneath the operculum (Fig. 26.6C).

C Immediately before metamorphosis

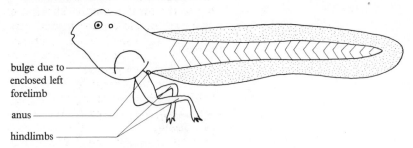

bulge due to enclosed left forelimb

anus

hindlimbs

D Immediately after metamorphosis

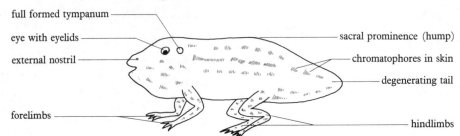

full formed tympanum

eye with eyelids

external nostril

forelimbs

sacral prominence (hump)

chromatophores in skin

degenerating tail

hindlimbs

Fig. 26.6 Diagrams summarizing the major external changes that occur during the development of the frog tadpole. Based on the common frog *Rana temporaria*.

(h) A pair of **lungs** develops as pouches from the ventral side of the pharynx, but the lungs are not yet functional.

Observe the above features in live or preserved tadpoles, and in prepared vertical and horizontal longitudinal sections. In particular note the gill pouches and gills.

METAMORPHOSIS

Towards the end of the third month the tadpole undergoes a comparatively sudden and dramatic **metamorphosis** into the adult, as a result of which it becomes terrestrial. Many of the adult structures are already present in rudimentary form, but at metamorphosis they have completed their development and become functional. Following is a summary of the important changes that occur, some of which can be seen in Fig. 26.6D.

(a) The **nasal sacs** acquire **internal openings** into the **bucco-pharynx**.

(b) **Eyelids** characteristic of the adult develop.

(c) The **middle ear** develops and a functional **tympanic membrane** forms.

(d) The **mouth** widens, the horny jaws and teeth being replaced by true jaws and teeth; jaw muscles and tongue develop.

(e) **Skull bones** develop, resulting in the head changing shape.

(f) The **hyoid pump** develops and becomes functional.

(g) The **limbs** increase in size: the left forelimb protrudes through the opercular opening, the right one breaks through the skin.

(h) The **pelvic girdle** develops resulting in the characteristic humpedback appearance of the adult.

(i) The **tail shortens**, its cells being resorbed by the action of phagocytes.

(j) The **colour pattern** characteristic of the adult develops.

(k) The **gut shortens**, this being associated with the change from a herbivorous to carnivorous diet.

(l) The gill pouches close up, the first pair forming the **Eustachian tubes**. The gill arches go to form the **laryngeal cartilages**.

(m) The **heart** completes its development, an **inter-atrial septum** developing.

(n) Changes occur in the pattern of **arterial arches**, a **double circulation** becoming established (*see* p. 291).

(o) The **lungs**, already present, become functional.

Metamorphosis is initiated and controlled by the release of thyroxine by the **thyroid gland**.

Premature metamorphosis can be induced by treating a tadpole with thyroxine. (See Investigation 27.3.)

CHANGES IN THE CIRCULATION

One of the most profound changes to occur at metamorphosis, and one with important evolutionary implications, concerns the heart and arterial arches. The changes are directly connected with the change in the method of respiration at metamorphosis: **gills** in the tadpole, **lungs** in the adult. The changes result in the establishment of a double circulation. Fig. 26.7 summarizes the situation.

For Consideration

(1) Can the adult frog be regarded as fully terrestrial? (Find out as much as you can about the adult).

(2) What are the natural hazards in the life cycle of amphibians? Plainly these hazards are not so acute as to have caused the group to become extinct. How have they managed to survive?

Tadpole with external gills

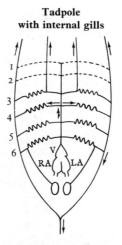

Arterial arches 1 and 2 never develop properly but traces of them may be seen. External gills are associated with arches 3, 4 and 5. Atrium (A) is undivided. Pulmonary artery starts to grow back from 6th arch.

Tadpole with internal gills

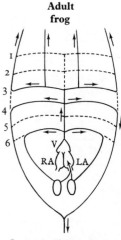

External gills replaced by internal gills which correspond to arterial arches 3, 4, 5 and 6. Lungs start to develop together with pulmonary arteries and veins. Inter-atrial septum starts to develop.

Adult frog

Internal gills lost. Lungs become functional. Vessel connecting arches 3 and 4 on each side closes to form ductus caroticus. Vessel connecting distal part of arch 6 on each side closes to become ductus arteriosus. 3 now becomes carotid arch to head; 4 becomes systemic arch to trunk etc. 6 becomes pulmonary arch to lungs. Atrium completely divided. Double circulation established.

Fig. 26.7 Changes in the heart and arterial arches of the frog before and during metamorphosis. RA, right atrium; LA, left atrium; V, ventricle.

Requirements

Binocular microscope or hand lens
Microscope
Watch glass
Wide pipette

Live tadpoles of common frog or
  Xenopus laevis
Preserved tadpoles (various stages)
VLS newly hatched tadpole
HLS tadpole with external gills
HLS tadpole with internal gills

Investigation 26.4

Development of chick

Hen's eggs hatch after three weeks' incubation following laying. During the first week it is possible to see organ-formation and the **extra-embryonic membranes**. What are the extra-embryonic membranes and which vertebrates possess them?

Examination of Live Embryo

Construct a plasticine cradle to hold an egg in the sideways position. Now take a three-to-five day egg from the incubator and, without rotating it or changing its orientation, place it in the cradle. Let the egg stand for several minutes so the embryo floats to the top of the yolk.

Now without removing the egg from the cradle, cut away the shell from the upper side of the egg. Use only the *tips* of a pair of scissors. When you have removed the shell you should see the embryo lying on top of the yolk. Draw off the albumen with a pipette until the surface of the yolk is uncovered.

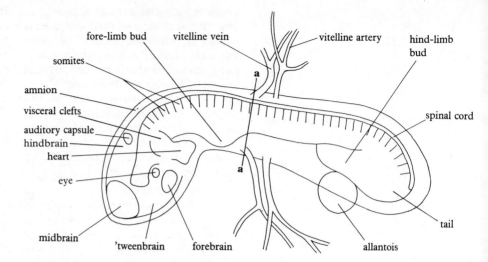

Fig. 26.8 Diagram of whole mount of chick embryo showing the various structures which are visible at the 3–4 day stage.

Examine the embryo under a binocular microscope or lens. Using Fig. 26.8 as a guide, note **brain** (**fore-, mid-,** and **hind-brain vesicles**), developing **eye**, **limb buds** and **tail**. Can the **heart** be seen, and is it beating?

Observe lining of the **yolk sac** containing the **vitelline arteries** and **veins**. Note blood corpuscles circulating outwards in the main vitelline artery, and back in the vein. What is the function of this circulation?

Next find the **allantois**: at this stage it is a small balloon-like sac, but later it will expand until it makes contact with most of the lining of the shell. What is the function of the allantois?

Now look for the **amnion** and **chorion**, the membranes immediately surrounding the embryo. What are their functions?

Using forceps and small scissors carefully remove these membranes from above the embryo. With a blunt needle gently deflect the embryo to one side and notice its connection with the yolk sac.

Observation of Live Embryo over Several Days

Take a three-day egg and mark out an 18 mm square on the shell in pencil immediately above the embryo. Carefully cut out the square with a hacksaw blade, taking care not to rupture the underlying white shell-membrane. Wear a sterile face mask from now on.

With a sterile blade and small forceps, remove the exposed shell membrane. Now place a 22 mm coverslip over the aperture and seal it in position with molten paraffin wax.

Mark the egg with your name and return it to the incubator. Observe the embryo at intervals during the next few days. Trace the development of organs, particularly the **eye**, **heart** and **limb buds**.

Making a Permanent Preparation of Embryo

The embryo from the first experiment can be used for this. Break open the yolk sac and carefully cut away the embryo plus the surrounding yolk sac wall containing the vitelline arteries and veins.

Float the embryo in a dish of warm sodium chloride solution (0.9 per cent). Flush away any yolk or albumen clinging to the embryo or to the yolk sac wall.

Using a wide pipette transfer the specimen to a small petri dish. Gently spread it out on the floor of the dish, so the embryo is lying in the centre of the yolk sac wall. Now cover the specimen with Bouin's fixative and leave for at least an hour.

Proceed as follows:

(1) After fixation, replace the Bouin's fluid with several changes of tap water to wash the specimen. This should be done for at least 30 minutes.
(2) Stain in borax carmine for up to 24 hours.
(3) Differentiate in acid ethanol until the specimen is semi-transparent. Observe in the petri dish under low power until internal structures become plainly visible. This may take as much as an hour.

(4) Dehydrate in 70 per cent ethanol (10 minutes), 90 per cent ethanol (10 minutes), and two lots of absolute ethanol (total of at least 30 minutes).
(5) Clear in xylene for at least 15 minutes.
(6) Mount in Canada balsam under a large coverslip supported round the edge by a cardboard frame.

N.B. The times for staining and differentiation are by no means absolute: both processes may be accelerated by warming on a hotplate.

Examine your preparation under low power noting: **neural tube** expanded anteriorly into **fore-, mid-** and **hind-brain vesicles, eye, heart, somites, vitelline blood vessels**. Are the **extra-embryonic membranes** visible? Compare with the structure in Fig. 26.8.

Transverse Sections of Chick Embryo

Examine transverse sections of two- and three-day embryos, and later stages if available. To what extent are the embryos similar to amphibian embryos at the same stage? In what way(s) do they differ? Identify as many of the structures shown in Fig. 26.9 as you can.

Notice the **amniotic folds** growing up above the embryo. What functions do they perform?

For Consideration

(1) What are the salient differences between the development of amphibians and birds?
(2) In what way is the embryo of birds adapted for terrestrial development? How does the bird embryo compare with that of other terrestrial vertebrates in this respect?
(3) From where does the developing heart of the chick embryo receive its oxygen and food supply? How does this compare with amphibians?
(4) How does the developing chick embryo dispose of its nitrogenous waste?

Requirements

Binocular microscope or hand lens
Sterile face mask
Hacksaw blade
Cardboard for coverslip support
Plasticine
Large petri dish
Small petri dish
22 mm coverslips (× 2)
Slide
Hot plate
Wide pipette

Beaker of molten paraffin wax with small paintbrush and spatula
0.9 per cent sodium chloride (NaCl) (25 cm^3)
Bouin's fluid (10 cm^3)
Acid ethanol (4 drops of strong HCl to 100 cm^3 of 70 per cent ethanol)
50 per cent ethanol (10 cm^3)
70 per cent ethanol (10 cm^3)
90 per cent ethanol (10 cm^3)
Absolute ethanol (20 cm^3)
Xylene (10 cm^3)
Canada balsam

Three-day fertilized eggs of chick (× 2) in incubator
TS two-day chick embryo
TS three-day chick embryo

Chick: transverse section of 2-day stage

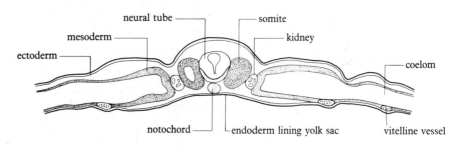

Chick: transverse section of 3-day stage

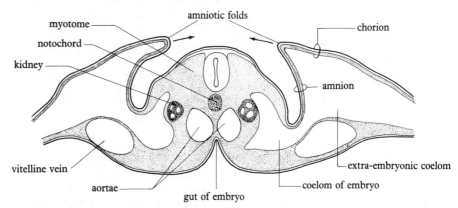

Fig. 26.9 Structure of chick and development of extra-embryonic membranes as seen in transverse sections through the trunk region of chick embryos. The lower diagram is the same stage as the whole mount in Fig. 26.8 and is cut at the level **a–a**.

Investigation 26.5
Development of embryo of flowering plant

After fertilization the following changes take place in the flower: the zygote develops into the **embryo**; the primary endosperm nucleus divides to form the **endosperm tissue**, which surrounds and nourishes the embryo; the ovule develops into the **seed**, the integuments forming the **seed coat**; and the ovary forms the **fruit**.

The developmental changes listed above can be seen in *Shepherdia canadensis*, shepherd's purse.

A Fruit soon after fertilization
(side wall of fruit removed)

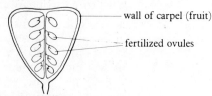

wall of carpel (fruit)

fertilized ovules

B Ovule soon after fertilization

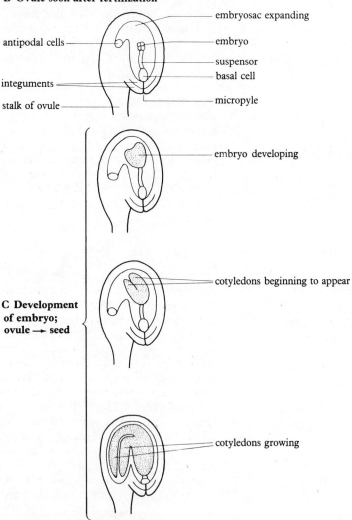

embryosac expanding

antipodal cells

embryo

suspensor

basal cell

integuments

stalk of ovule

micropyle

embryo developing

cotyledons beginning to appear

C Development of embryo; ovule → seed

cotyledons growing

D Fully formed seed

seed coat

plumule

central cylinder

cotyledons

radicle

} of embryo

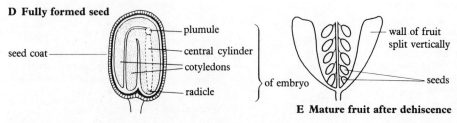

wall of fruit split vertically

seeds

E Mature fruit after dehiscence

Fig. 26.10 Diagrams illustrating development of the embryo and associated structures in shepherd's purse *Capsella bursa-pastoris*. The diagrams are not all drawn to the same scale: in fact the carpel and its contents grow steadily larger after fertilization.

Procedure

(1) First examine a whole plant of shepherd's purse. The youngest part of the plant is the apex where unopened **flower buds** may be seen; further back open **flowers** are visible, and further back still heart-shaped **fruits**, developed from the ovaries of post-fertilization flowers, will be seen. Trace the developmental sequence from apex to base.

(2) Open one of the youngest fruits (towards the apex) and note two rows of **ovules** (Fig. 26.10A). Remove a few of them and mount in chloral hydrate or acetocarmine. *Note*: **stalk** of ovule, **integuments**, **embryosac** containing **embryo** which is attached by a **suspensor** to the end of the sac nearest the micropyle (Fig. 26.10B).

(3) Mount ovules of different ages in chloral hydrate. Start with ovules from fruits towards the apex of the plant, and then work your way down the stem towards the base. In each case observe intact ovules first, then gently press the coverslip with a needle so as to burst the ovule and release the embryo. If the embryo is too transparent try mounting another one in acetocarmine.

Using Fig. 26.10 to help you, reconstruct the sequence of stages in the development of the embryo and the formation of the seed. In the fully formed embryo note the **plumule**, **radicle**, and **cotyledons**. What happens to the endosperm and suspensor as development proceeds?

(4) Supplement this investigation by examining prepared longitudinal sections of the fruit of shepherd's purse.

For Consideration

(1) What useful function is formed by the suspensor?

(2) During development the embryo bends over so the cotyledons point downwards. Significance?

(3) What is the function of the cotyledons?

(4) The final event in the formation of the seed is the drying out of the inside with the result that water content is reduced from approximately 80 to 10 per cent. How might this be achieved and what is its purpose?

Requirements

Microscope
Slides and coverslips
Mounted needle

Chloral hydrate
Acetocarmine

Shepherd's purse plant bearing flowers and fruits
LS fruit of shepherd's purse

Investigation 26.6
Structure and germination of seeds

In Investigation 26.5 we saw that the seed contains the embryonic plant. In most plants the embryo is surrounded by a variable amount of **endosperm tissue** which serves as a food supply during germination. In other seeds there is little or no endosperm tissue, food being supplied by the enlarged **cotyledons**. Flowering plants are divided into **monocotyledons** and **dicotyledons**: the former possess seeds with only one cotyledon, the latter have two cotyledons.

Whatever other factors are required for germination, one essential factor is water: the first clearly observable event in germination is the imbibing of water. As a result of this the embryonic tissues swell and rupture the seed coat.

Broad Bean

(1) Examine a dry broad-bean seed (i.e. the so-called 'bean') and notice the external **seed coat**, the **micropyle** and **hilum** (Fig. 26.11A). The hilum is the scar of the seed stalk, originally the stalk of the ovule. Split open a bean pod: the pod is the fruit—detach one of the seeds and

A End-on view

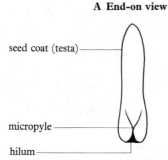

B Cut open

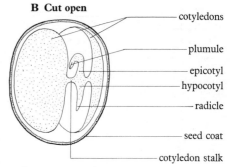

Fig. 26.11   Structure of broad-bean seed, a non-endospermous seed.

confirm that the hilum is the scar of the seed stalk.

(2) Compare the dry seed with one which has been soaked in water for 24–48 hours. Note the large increase in size. By weighing the two seeds, estimate the **percentage increase in weight**: does this represent the percentage increase in water content?

How, and by what route, might water be taken up by the seed? How could you investigate this by experiment?

(3) Remove the seed coat of the soaked seed and note that it consists of two layers, an external **testa** and inner **tegmen.** From what structures in the unfertilized ovule are these two layers derived?

(4) The broad bean is a non-endospermic dicotyledon. Remove one of the two cotyledons. Note **plumule, radicle, cotyledon** and **cotyledon stalk** (Fig. 26.11B). Note absence of endosperm tissue.

(5) Stain the whole of the embryo (including a cotyledon) in dilute iodine. Conclusion? Carry out tests for other food reserves.

Castor Oil

(1) Compare the dry seed with one that has been soaked for 48 hours. The seed stalk has an outgrowth (the **caruncle**) which covers the hilum and micropyle (Fig. 26.12A). The caruncle is said to play some part in imbibition of water. How could you test this?

(2) Slice the soaked seed horizontally down the centre of its flattened side and examine the cut surface. Note the very small **plumule** and **radicle** at one end of the seed. The embryo has two thin, delicate **cotyledons** flattened against the **endosperm** (Fig. 26.12B). The endosperm is plentiful.

Slice another soaked seed vertically, i.e. in a plane at right angles to the previous section.

(3) Treat the cut surface of the seed with Sudan III. Conclusions? Test for other food reserves.

Other Seeds

Using the same techniques as above, examine other seeds. In each case either remove the seed coat or split the seed lengthways in order to examine the embryo. Note the relative extents of the **cotyledons** and **endosperm** and test for food reserves.

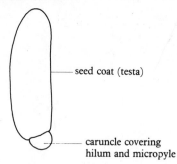

A End-on view

— seed coat (testa)

— caruncle covering hilum and micropyle

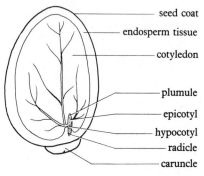

B Cut open

— seed coat
— endosperm tissue
— cotyledon

— plumule
— epicotyl
— hypocotyl
— radicle
— caruncle

Fig. 26.12 Structure of the seed of the castor oil plant, an endospermous seed.

For Consideration
Devise—and, if necessary, perform—a series of experiments to investigate the factors required for germination of seeds of a named plant.

Requirements
Balance
Dish
Razor blade

Dilute iodine
Sudan III
Reagents for other food tests
 (*see* p. 32)

Broad bean seed (dry)
Broad bean seed soaked for
 24–48 h
Castor oil seed (dry)
Castor oil seed soaked for 48 h
Other seeds as required

Investigation 26.7
Hypogeal and epigeal germination

These two types of germination are related to the different sources of nourishment available to the germinating seed. In **hypogeal germination** the epicotyl elongates, in **epigeal germination** the hypocotyl elongates.

Hypogeal Germination

Observe your broad bean embryo from Investigation 26.6. Imagine the epicotyl elongating rapidly: what will happen?

In hypogeal germination the epicotyl elongates, with the result that the plumule is thrust upwards through the soil, but the cotyledons remain below ground (hence *hypo*geal). Until the plumule develops its first green leaves nourishment comes from the large cotyledons which contain stored food reserves. There is little or no endosperm tissue.

Examine seedlings of a hypogeal plant (e.g. broad bean, wheat or maize) which have been caused to germinate at different times so the developmental sequence can be followed. Compare with Fig. 26.13A.

Epigeal Germination

Observe the sliced castor oil seed from Investigation 26.6. Imagine the hypocotyl elongating rapidly. What will happen?

In epigeal germination the hypocotyl elongates with the result that the delicate cotyledon(s), protected by the ruptured seed coat, are thrust up above the soil surface where, exposed to light, they turn green and photo-

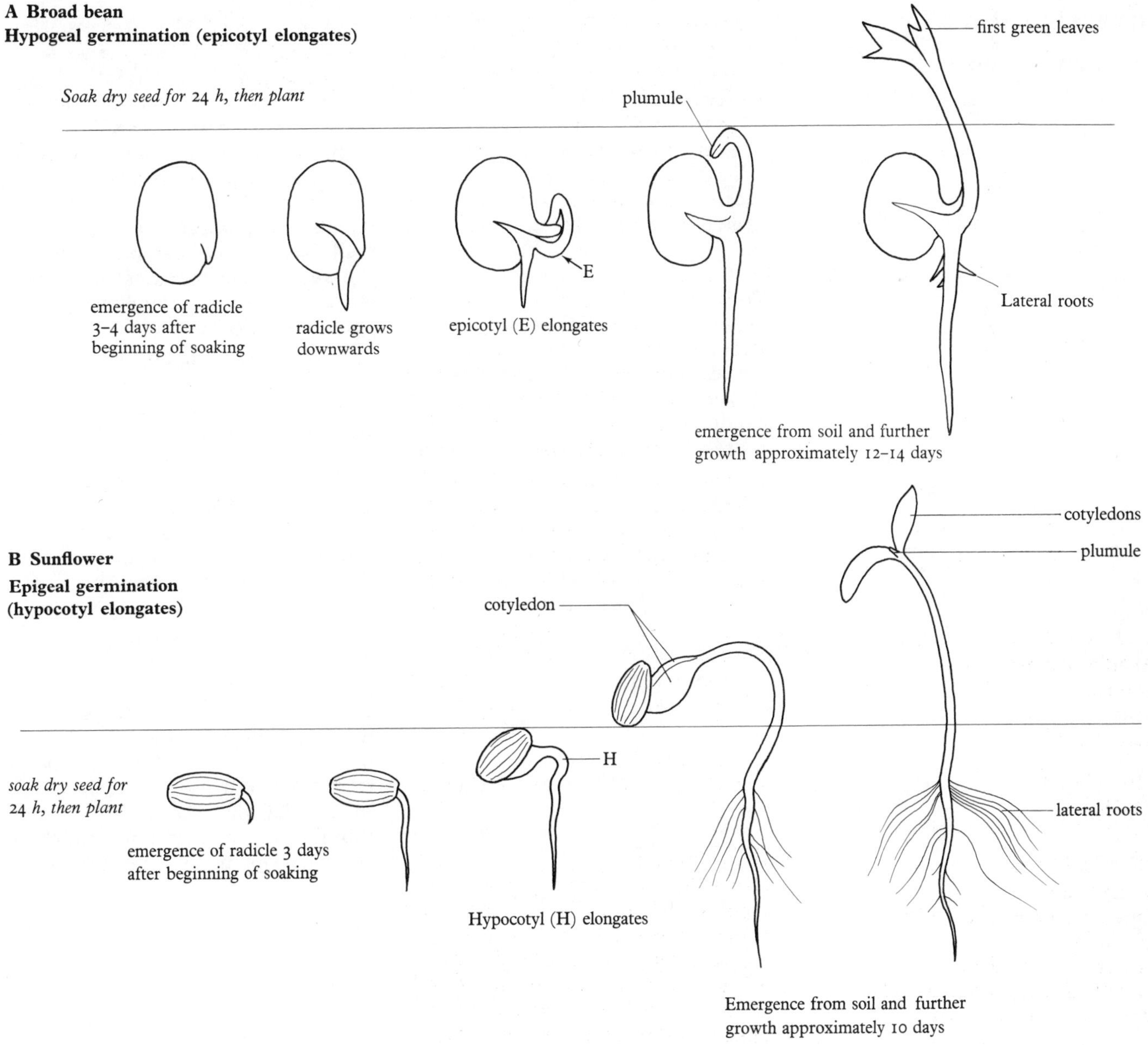

Fig. 26.13 Stages in germination of (**A**) broad bean, and (**B**) sunflower to illustrate hypogeal and epigeal germination respectively. The times, which are given as the number of days after the beginning of soakings, are very approximate and assume good growing conditions.

A Broad bean
Hypogeal germination (epicotyl elongates)

Soak dry seed for 24 h, then plant

first green leaves

plumule

emergence of radicle 3–4 days after beginning of soaking

radicle grows downwards

epicotyl (E) elongates

Lateral roots

emergence from soil and further growth approximately 12–14 days

B Sunflower
Epigeal germination (hypocotyl elongates)

cotyledons

plumule

cotyledon

soak dry seed for 24 h, then plant

emergence of radicle 3 days after beginning of soaking

H

lateral roots

Hypocotyl (H) elongates

Emergence from soil and further growth approximately 10 days synthesize. During germination nourishment comes from the endosperm tissue which is usually, but not invariably, extensive.

Examine a series of seedlings of an epigeal plant (e.g. castor oil, sunflower, or marrow) resulting from staggered germination, and note the developmental sequence. Compare with the sequence in Fig. 26.13B.

Requirements
Dish
Razor blade

Broad bean seeds } from
Castor oil seeds } Investigation 26.6

Staggered seedlings showing hypogeal germination (e.g. bean)
Staggered seedlings showing epigeal germination (e.g. sunflower)

For Consideration
(1) Compare hypogeal and epigeal germination from a structural and physiological point of view.
(2) The seeds of epigeal plants usually contain abundant endosperm tissue—why is this?
(3) The seeds of hypogeal plants usually contain large, fleshy cotyledons—why?
(4) Although the seeds of epigeal plants usually contain abundant endosperm tissue, there are some that do not, e.g. marrow and sycamore. How do you suggest such plants obtain nourishment during and immediately after germination?

Investigation 26.8
Primary growth of flowering plant

Requirements
Microscope
Slides and coverslips
Razor blade
Watch glasses
Pith
Chloral hydrate
Phloroglucin
Concentrated HCl
FABIL
Dilute glycerol

Germinating maize seed (Fig. 26.14)
Germinating pea or bean with radicle 10–20 mm long (Fig. 26.16)
LS root tip of onion (*Allium* sp.)
LS stem apex of lilac

Primary growth takes place at the apex of the stem and root. It involves division of primary meristematic cells (**apical meristem**) and it results in the formation of **primary tissues** (*see* Investigations 12.3 and 12.4).

To investigate primary growth it is necessary to study sections and/or cleared whole mounts of young stem and root tips.

Root Apex
(1) Obtain a germinating seed of maize or some other comparable species (Fig. 26.14). Instead of having a main root the seedling develops a bundle of **adventitious (fibrous) roots**, typical of grasses. The plumule is protected by a sheath-like **coleoptile**.

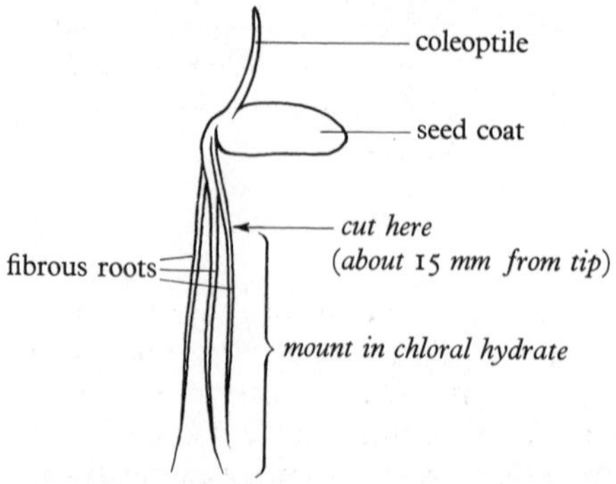

— coleoptile
— seed coat
fibrous roots —
cut here (about 15 mm from tip)
mount in chloral hydrate

Fig. 26.14 Preparation of young maize roots for viewing under the microscope.

Cut one of the roots about 15 mm from the tip and mount in chloral hydrate which clears the tissue. Examine under medium power, reducing the illumination as much as possible. Start with the extreme tip, where the cells are youngest, and work back to the older

parts of the root. Note successively:
(a) **Root cap cells** (protective)
(b) **Zone of cell division** (cells square-shaped, dividing mitotically)
(c) **Zone of cell expansion** (cells progressively more elongated as one passes back along the root).
(d) **Zone of differentiation** (cells acquire features, particularly of their walls, characteristic of specific tissues, e.g. xylem, phloem, etc. Xylem vessels with annular or spiral thickening should be evident).

(2) Take a germinating pea or bean with a radicle 10–20 mm long and examine it under a hand lens or binocular microscope. With a razor blade cut thin longitudinal sections and mount them in water or dilute glycerine. Examine under microscope. Note **zones of cell division, expansion** and **differentiation**. How do these relate to the **root hair zone**?

(3) Examine a prepared longitudinal section of the root tip of, e.g. onion (*Allium* sp.) under low and high powers (Fig. 26.15). Start at the tip and work back noting in particular **mitotic figures** in the **zone of cell division**. The **root cap** and **zone of expansion** will also be clearly visible and possibly the **zone of differentiation** showing spiral or annular thickening in developing xylem elements.

B Detail of zone of cell division

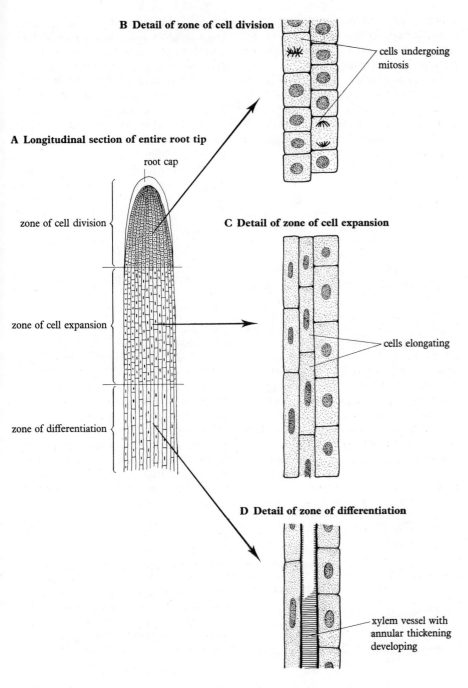

cells undergoing mitosis

A Longitudinal section of entire root tip

root cap

zone of cell division

zone of cell expansion

zone of differentiation

C Detail of zone of cell expansion

cells elongating

D Detail of zone of differentiation

xylem vessel with annular thickening developing

Fig. 26.15 Microscopic structure of the root tip of *Allium* sp.

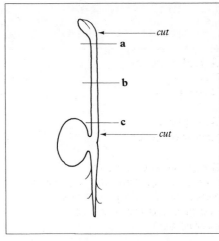

Fig. 26.16 Young seedling of broad bean. Cut the stem at the two points indicated. Section at levels **a**, **b**, and **c**.

(2) Examine a longitudinal section through the stem apex of, e.g. lilac (Fig. 26.17). Start at the tip and work back. How does the section compare with that of the root apex. In addition to the zones of cell division, expansion and differentiation, note the developing **vascular strands**, **leaf primordia** with vascular tissue going to them, and **axillary buds**.

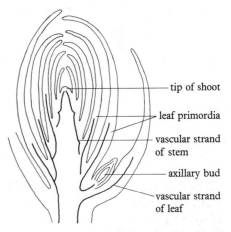

tip of shoot

leaf primordia

vascular strand of stem

axillary bud

vascular strand of leaf

Fig. 26.17 Longitudinal section of the stem apex of lilac.

Stem Apex

(1) Remove the tip of the plumule of a germinating broad bean, and cut it near to its attachment to the seed. Holding the isolated stem in a piece of moistened pith (*see* p. 123), cut transverse sections at levels **a**, **b**, and **c** in Fig. 26.16 and place them in water in separate watch glasses.

Stain a thin section from each watch glass in either acidified phloroglucin or FABIL, then mount in dilute glycerol.

Compare your sections. What conclusions can you draw regarding the development of primary tissues in the stem?

For Consideration

Compare the structure of a developing stem and root. How would you explain the differences between them?

Investigation 26.9
Secondary growth of flowering plant

Primary growth is mainly concerned with growth of the plant in a longitudinal direction. In contrast, secondary growth is responsible for increasing the girth of the stem and root. It occurs by division of **secondary meristematic cells** (**cambium**) within the primary tissues, and it results in the formation of **secondary tissues**.

Procedure

(1) Cut transverse sections of a woody twig of lime tree (*Tilia* sp.), dated Autumn of second year. From your knowledge of how secondary tissues are formed, what would you *expect* the appearance of such a section to be?

Stain some sections in iodine, others in acidified phloroglucin, then mount in dilute glycerine. Note the general appearance of your sections under low power. Are your predictions confirmed?

Now examine the detailed distribution of tissues (Fig. 26.18). Note **primary** and **secondary xylem**, **cambium**, **secondary phloem**. The secondary phloem contains blocks of lignified sclerenchyma in between the sieve tubes, etc., which—though unusual—makes it easier to see. Little or no **primary phloem** will be visible (why?).

Observe the **primary** and **secondary medullary rays**: what are their functions? Distinguish between first- and second-year wood and between spring and autumn wood. What is the functional significance of the difference between spring and autumn wood?

At the periphery note **corky tissue.** What functions are performed by this tissue?

(2) What would you expect a lime twig to look like in transverse section in the autumn of its first year? Cut sections and stain. Are your predictions correct?

(3) What would you expect a lime twig to look like in the spring of its first year? Test your predictions by cutting sections, staining them and examining.

(4) Cut transverse sections of woody twigs of other plants, e.g. beech, oak, etc. Stain and mount. Report on differences and similarities between them.

(5) The main structure formed in secondary growth is **wood** which, more than any other tissue, is responsible for increasing the girth of the stem, with, of course, important functional consequences.

Reconstruct the three-dimensional anatomy of wood by cutting transverse, radial, and tangential sections of a woody twig (Fig. 26.19). Stain some sections in iodine, and others in acidified phloroglucin.

Note the relationship between lignified (woody) **xylem elements**, constituting the secondary xylem tissue, and non-lignified **medullary rays**. Do the medullary ray cells always contain starch?

Describe the different types of element that make up the wood. What functions do you think they perform?

Supplement the information gained from your own sections by examining prepared sections.

(6) Examine the cells towards the surface of a secondarily thickened stem. A layer of **corky cells**, of variable thickness, is formed beneath the original epidermis from a **cork cambium** which also forms a limited amount of **secondary cortex** (Fig. 26.20A). Periodically the layer of corky cells is interrupted by a mass of loosely packed cells which constitute a **lenticel**. Examine a prepared slide showing the detailed structure of the corky cells and lenticel (Fig. 26.20B).

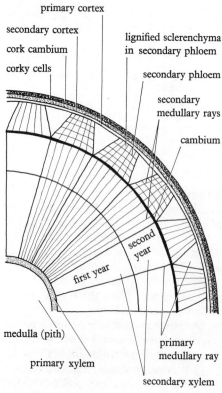

primary cortex
secondary cortex
cork cambium
corky cells
lignified sclerenchyma in secondary phloem
secondary phloem
secondary medullary rays
cambium
second year
first year
medulla (pith)
primary xylem
primary medullary ray
secondary xylem

Fig. 26.18 Transverse section of lime twig *Tilia* sp., autumn of the second year.

Transverse section

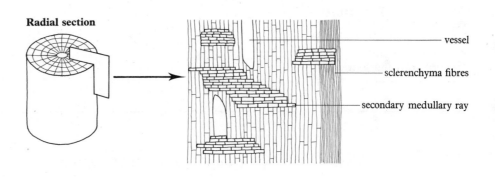

vessels

primary medullary ray

secondary medullary ray

xylem parenchyma and sclerenchyma

Radial section

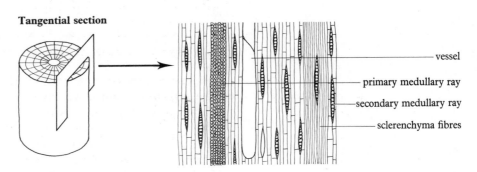

vessel

sclerenchyma fibres

secondary medullary ray

Tangential section

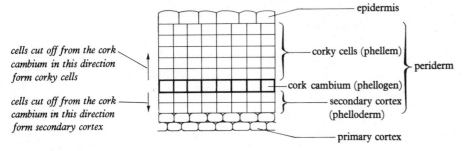

vessel

primary medullary ray

secondary medullary ray

sclerenchyma fibres

Fig. 26.19 Section of secondary xylem (wood).

A. Diagram showing the cork cambium and its derivatives

epidermis

corky cells (phellem)

periderm

cells cut off from the cork cambium in this direction form corky cells

cork cambium (phellogen)

secondary cortex (phelloderm)

cells cut off from the cork cambium in this direction form secondary cortex

primary cortex

Fig. 26.20 Diagrams showing the corky cells and associated structures found at the surface of a secondarily thickened stem.

B Cork and lenticel (based on elder)

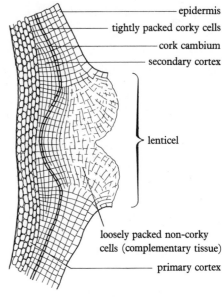

epidermis

tightly packed corky cells

cork cambium

secondary cortex

lenticel

loosely packed non-corky cells (complementary tissue)

primary cortex

(7) Examine transverse sections of secondarily thickened roots of different ages, noting the form and distribution of the **primary** and **secondary tissues**. How do the sections differ from those of stem? How would you explain the differences?

Requirements
Microscope
Slides and coverslips
Watch glasses
Razor blade

Iodine solution
Phloroglucin
Concentrated hydrochloric acid
Dilute glycerol

Twigs of lime (*Tilia* sp.): spring
  and autumn of 1st year, and
  autumn of 2nd year
Twigs of other woody perennials,
  e.g. beech, oak, etc.
Transverse, radial, and tangential
  sections of wood
TS cork showing lenticel

For Consideration
(1) Make a sketch showing the distribution of tissues in a two-year-old stem and root. Explain to yourself (or, better, to someone else!) how this pattern is arrived at during development.
(2) To what extent can the properties of wood be explained by the structure of its cellular elements?

Questions and Problems

1 Write a short essay on the problem of measuring growth.

2 The female sheep, the ewe, may have one, two or three lambs per pregnancy. At first the lambs feed only on milk, but after about five weeks they begin to eat solid food.

A research worker was interested in the effect of the number of lambs per ewe on the growth of lambs. To investigate this problem he weighed lambs of different classes every four weeks. The classes were as follows:
Singles (single lambs reared by their own mother);
Twins (twins reared by their own mother);
Triplets (triplets reared by their own mother);
Twins reared as singles (lambs born as twins, but shortly after birth one was transferred to another ewe, and both lambs were reared as singles).

His data are summarized in the Table below where each observation is the mean for a large number of lambs:

Age (weeks)	**Weight** (kg)			
	Singles	Twins	Triplets	Twins reared as singles
2	10.0	7.2	6.5	7.9
6	19.5	13.7	13.1	16.8
10	28.5	20.9	20.2	25.0
14	37.0	28.6	27.0	33.6
18	43.7	35.1	33.1	41.0
22	48.5	40.0	37.9	46.1
26	51.8	44.3	42.4	49.3

(a) For each period of four weeks, calculate the growth rate (as kg per four-week period) for each class and tabulate your results.
(b) Present your calculated data in the form of a graph of growth rate against age of lamb.

(In the answers to parts (c) and (d) below you should refer to the appropriate parts of the graph and give quantitative data to amplify your statements.)

(c) Comment upon the differences in growth rate of the different classes.

(d) Do you consider that food supply is the only factor which affects the growth rate of the different classes? *(CL)*

3 The following data are the results of making daily growth measurements on a locust over a period of 24 days during its development:

Day	Total body weight (mg)	Width of head (mm)	Length of hind femur (mm)
1	100	3.0	7.0
2	100	3.5	7.5
3	100	4.0	8.0
4	140	4.0	8.0
5	160	4.0	8.0
6	200	4.0	8.0
7	230	4.0	8.0
8	230	4.4	9.2
9	230	4.7	10.5
10	230	5.0	12.0
11	280	5.0	12.0
12	350	5.0	12.0
13	410	5.0	12.0
14	470	5.0	12.0
15	530	5.0	12.0
16	600	5.0	12.0
17	600	5.6	13.3
18	600	6.4	14.8
19	600	7.0	16.4
20	600	7.6	18.0
21	760	7.6	18.0
22	900	7.6	18.0
23	1050	7.6	18.0
24	1200	7.6	18.0

Plot these three sets of data on a single sheet of graph paper so that they can be readily compared.

Interpret each of the graphs and account for any differences between them.

Draw the form of graph you would expect to obtain if you plotted, for the human, body weight against time, from birth to maturity. Compare and contrast this with the results obtained for the locust.

4 The tip of a bean root was marked with Indian ink. Each mark was 2 mm from the next. After several days it was noticed that some of the marks were no longer the same distance apart. Here are two hypotheses to account for this result:

(a) Cells have divided in the regions between adjacent marks; more divisions have occurred in the region where the marks have moved furthest apart.

(b) Cells between certain marks have elongated more than others. How would you investigate which hypothesis is correct?

5 Write short explanatory notes on cleavage, gastrulation, neurulation and organogeny.

6 Compare the structure and functions of the extra-embryonic membranes in mammals and birds.

7   The following results were obtained from a study of the germination and early growth of a cereal. The grains were sown in soil in a greenhouse and at two-day intervals samples were taken and separated into the two components, endosperm and embryo (or young plant), which were then oven-dried and weighed:

Time after sowing (days)	Total dry weight (g)	Dry weight of endosperm (g)	Dry weight of embryo (g)
0	0.045	0.043	0.002
2	0.043	0.041	0.002
4	0.040	0.032	0.008
6	0.036	0.020	0.016
8	0.035	0.009	0.024
10	0.040	0.006	0.034

Plot the results graphically and describe in words what they show.   (O and C)

8   A crop of oats was sown on 14th April. At intervals after this estimates were made of the amount of minerals in the crop, as well as of the amount of the dry matter (i.e. the weight after drying plants at 100°C for 24 hours). The results obtained, expressed in arbitrary units, were as follows:

Date	Dry matter	Potassium	Nitrogen	Phosphorus	Other mineral elements
1 May	30	0.6	0.6	0.2	0.2
1 June	200	17	13	3	9
1 July	2400	53	45	8	17
1 Aug	4200	56	46	10	18
14 Aug (harvest)	4400	55	46	10	18

No fertilizers were added during the growth of the crop, and the soil was not found to be exhausted of nutrients at harvest time; the results are to be regarded as typical of the growth of oats in this country. Comment on any features of interest shown by these figures.   (OCJE)

9   The dry weight of potato tubers and the shoots they produced on sprouting were determined at weekly intervals over a period of 8 weeks with the results given below:

Organ		Week and dry weight (g)							
		1	2	3	4	5	6	7	8
Tuber		14.1	13.0	10.6	8.2	7.5	6.2	5.3	4.5
Shoots	stalks	0	0	0.2	0.5	0.8	2.6	6.0	10.2
	leaves	0	0	0.1	0.2	0.7	3.2	6.8	11.0

Comment on the weekly variations and suggest an explanation of them. How, in your opinion, were these results obtained?   (O and C)

10   Germination starts with the uptake of water by the seed. (a) What factors may trigger this process to start taking place? (b) Make a list of the subsequent events which this makes possible.

11   Summarize the changes which occur in the circulation of the human foetus at, or soon after, birth.

12 Compare the growth of a flowering plant with that of a vertebrate animal.

13 Describe in detail the ontogeny (development) of (a) a sieve tube, and (b) a xylem vessel. (From your knowledge of the finished product in each case try to *work out* the sequence of events rather than look them up in a book).

14 In one of the Californian National Parks there is a redwood tree (*Sequoia* sp.) which, having leaned over to one side, has produced a wedge to support itself as shown in Fig. 26.21.
What would you expect a transverse section cut at level A to look like?
Explain how this situation may have come about.

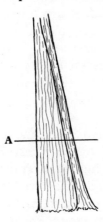

Fig. 26.21 The redwood tree (*Sequoia* sp.).

27 The Control of Growth

Background Summary

1 Growth is influenced by a combination of external and internal factors. The internal factors are affected by the external ones.

2 In flowering plants experiments indicate that a hormone (**auxin, indole-acetic acid, IAA**) promotes growth of the shoot, and (except in very low concentration) inhibits it in the root. Auxin achieves its effect by influencing cell elongation. A variety of synthetic **growth substances** have the same effect.

3 Plants often respond to directional stimuli by means of growth movements, (**tropisms**). Experiments suggest that the tropic responses of shoots and roots to gravity and light (**geotropism** and **phototropism**, respectively) can be explained by a differential distribution of auxin within the plant.

4 Further experiments indicate that a shoot's positive phototropic response results from auxin moving to the dark side of the shoot, rather than being destroyed on the illuminated side.

5 Other growth responses shown by plants include **hydrotropism**, **chemotropism** and **thigmotropism**. Plants also respond to diffuse stimuli that do not come from a particular direction (**nastic responses**).

6 Although it normally exerts its effects on cell elongation, auxin can also promote cell division. Its other functions include maintenance of **apical dominance** by suppression of lateral buds; growth of adventitious roots; initiation of secondary growth; ripening of fruit; abscission of leaves, flowers, and fruits; and dormancy of buds. Synthetic auxins are of considerable commercial importance.

7 A second category of plant growth substances are **gibberellins**. These are responsible for bolting, promoting internodal growth, and side-branching of stems, inhibiting root growth, and bringing about parthenocarpy. They achieve their growth-promoting effects by initiating both cell division and elongation.

8 **Cytokinins** (**kinins**) are a further group of plant growth substances. They promote cell division and, in conjunction with other growth substances, including auxins, they control cell differentiation and break dormancy.

9 Of all external factors, **light** and **temperature** have the most pronounced effects on plant growth and development. Each influences a variety of important developmental processes. Lack of light results in **etiolation**.

10 Experiments on Grand Rapids lettuce seeds indicate that germination is promoted by red light and inhibited by far-red, a phenomenon with important ecological implications.

11 Further experiments have shown that light is perceived by a photo-chemical substance, **phytochrome**, which exists in two inter-convertible forms, one sensitive to red light, the other to far-red. A plant's response to light is determined by the balance between these two forms of phytochrome.

12 In addition to germination the phytochrome system is involved in stem elongation, leaf expansion, growth of side-roots, and flowering.

13 Phytochrome probably exerts its action through the intermediacy of a hormone. There is particularly strong evidence for this in the photoperiodic control of flowering.

14 **Photoperiodism** is the response of an organism to day-length. Photoperiodic responses are shown by animals as well as plants, and may involve behaviour and other responses as well as growth.

15 In photoperiodic control of flowering, plants can be divided into **long-day**, **short-day**, and **day-neutral plants**. The category to which a plant belongs is related to its ecological situation. As in other phenomena involving the phytochrome system, the plant's response depends on a balance between the two forms of phytochrome. The link between the perception of the stimulus and the flowering response appears to be hormonal.

16 The influence of temperature on plant growth is most obvious in the case of flowering. Sometimes flowering will only take place if the seeds or plants have been subjected to a period of chilling (**vernalization**).

17 Growth in animals is also controlled by hormones, e.g. the **pituitary growth hormone** of mammals, and the **moulting** and **juvenile hormones** of insects. The latter are involved in the control of metamorphosis, (in which respect they are equivalent to) **thyroxine** in amphibians.

18 There is mounting evidence that in both plants and animals hormones controlling growth exert their action by activating the appropriate genes.

19 Growth and development may be temporarily interrupted by **dormancy**. Important in distribution and survival over unfavourable periods, dormancy is found in seeds, buds, spores (including zygospores), eggs, perennating organs, and adult organisms.

20 There is much variation in the length of time a dormant structure can survive, and the severity of the conditions which it can endure.

21 Dormancy is initiated and terminated by a combination of external and internal factors. The external factors include day-length (photoperiodism) and temperature. The dormant seeds of some plants require an obligatory period of chilling before germination can take place (**stratification**).

22 Internally chemical substances appear to be involved in dormancy. This is established in the case of seeds and buds, and also in the initiation of leaf-fall, a prerequisite to winter dormancy in deciduous trees. In buds dormancy is controlled by the relative amounts of **abscisic acid** (**dormin**) and **auxin** present.

23 In insects **diapause** is comparable to dormancy in plants. **Hibernation** and **aestivation** are also types of dormancy found in certain animal groups.

Investigation 27.1

Effect of indoleacetic acid on the growth of coleoptiles and roots

Requirements
Incubator at 25°C
Ruler with mm divisions
Petri dishes with lids (× 5)
Wax pencil
Razor blade
White tile

2 per cent sucrose solution ⎱ *see*
Indole acetic acid solutions ⎰ below
  (IAA) in sucrose

Germinating wheat seeds grown
in the dark at 25°C for at least
five days. (The coleoptiles must
be approximately 15 mm long)

IAA solutions made up as follows:
Make up 1 dm³ of stock IAA solution: dissolve
200 mg IAA in 2 cm³ ethanol (absolute). Add
900 cm³ distilled water and heat for 5 minutes at
80°C to evaporate the ethanol. Cool, and make up
to 1.0 dm³ with distilled water. Store in refrigerator.
  Prepare the five solutions thus:
(a) **100 p.p.m. IAA + 2 per cent sucrose:**
to 100 cm³ of stock IAA add 100 cm³ of 4 per cent
sucrose solution in a stoppered flask.
(b) **1.0 p.p.m. IAA + 2 per cent sucrose:**
take 10 cm³ of stock IAA and make up to 1.0 dm³
with water; to 100 cm³ of this solution add 100
cm³ of 4 per cent sucrose.
(c) **10⁻³ p.p.m. IAA + 2 per cent sucrose:**
take 1.0 cm³ of (b) and make up to 1.0 dm³ with
water. To 100 cm³ of this solution add 100 cm³ of
4 per cent sucrose.
(d) **10⁻⁵ p.p.m. IAA + 2 per cent sucrose:**
take 10 cm³ of (c) and make up to 1.0 dm³ with
water. To 100 cm³ of this solution add 100 cm³
of 4 per cent sucrose.
(e) **2 per cent sucrose:** to 100 cm³ of 4 per cent
sucrose solution add 100 cm³ of water.

It has been known for many years that
indoleacetic acid (IAA) affects the
growth of shoots and roots. In this
experiment these effects are investigated in relation to specific concentrations of IAA.

Procedure
In this experiment it is convenient to
work in groups of five.

(1) Prepare five labelled petri dishes,
with lids, containing the following:
(a) 2 per cent sucrose plus 100 parts
per million IAA.
(b) 2 per cent sucrose plus 1.0 parts
per million IAA.
(c) 2 per cent sucrose plus 10^{-3} parts
per million IAA.
(d) 2 per cent sucrose plus 10^{-5} parts
per million IAA.
(e) 2 per cent sucrose alone.
  Other dilutions can also be used if
desired.

(2) Take 25 germinating wheat seeds
which have been kept in the dark.
Work in as dark a place as possible, or
in red light in a darkroom.
  With a razor blade cut off the distal
2.0 mm of the coleoptiles. Then sever
each coleoptile exactly 10.0 mm further
back, thereby isolating a 10.0 mm
length of coleoptile. Do this for all 25
seeds, so you have 25 lengths of
coleoptile, all 10.0 mm in length. Place
five coleoptiles in each petri dish.

(3) Repeat with the radicles, placing
five in each petri dish.

(4) Put lids on the five petri dishes
and place them in darkness in an
incubator at 25°C.

(5) After approximately 48 hours,
remove the petri dishes and measure
the length of (a) the coleoptiles, (b) the
portion of leaf which was enclosed
within each coleoptile when it was cut,
and (c) the radicle. Record your
measurements and for each treatment
calculate the *average* lengths of the
coleoptile, leaf, and radicle.

(6) Express your results as change in
length of coleoptile, leaf, and radicle
in mm. Plot changes in length for all
three on a single sheet of graph paper:
length on the vertical axis, auxin concentrations equally spaced on the
horizontal axis.

For Consideration
(1) Why was it necessary to remove
the tips of the coleoptiles and radicles
before performing this experiment?
(2) For what reason were the auxin
solutions made up with sucrose?
(3) What general conclusions would
you draw from your results?
(4) How do your results relate to the
normal growth of the coleoptile, leaves
and radicle?

Investigation 27.2

Measurement of the phototropic response of coleoptiles

It is well known that stems normally grow towards light, i.e. they are **positively phototropic**. In this experiment the phototropic response of a coleoptile is observed and measured.

Procedure

For this experiment you require a microscope whose eyepiece has been fitted with a micrometer scale (*see* p. 424).

(1) Bend the microscope downwards towards you, so the tube is horizontal and the stage vertical. Darken the room and arrange for a source of light to strike the microscope stage from one side.

(2) Take a germinating oat, barley, or wheat seed, which has a straight coleoptile. Mount the seed in a moist chamber constructed from a plastic syringe (Fig. 27.1). The coleoptile should protrude from the chamber as shown.

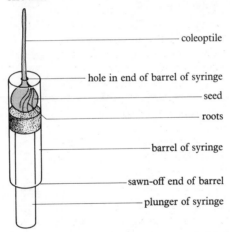

coleoptile

hole in end of barrel of syringe

seed

roots

barrel of syringe

sawn-off end of barrel

plunger of syringe

Fig. 27.1 Moist chamber, constructed from plastic syringe, for mounting a germinating seed for observations on phototropism.

(3) With an elastic band, attach the moist chamber to a slide. Clip the slide to the microscope stage so the coleoptile is pointing directly upwards. Position it so that the tip of the coleoptile, when viewed down the microscope, corresponds exactly with the mid-point of the micrometer scale.

(4) With the coleoptile illuminated from one side, observe the tip of the coleoptile for 5–10 minutes until it starts moving in a definite direction. Then rotate the moist chamber through 180° on its vertical axis so the other side of the coleoptile now receives the strongest illumination. Re-adjust the position of the moist chamber so the coleoptile tip corresponds to the centre of the micrometer scale, as before.

(5) Take readings of the position of the coleoptile tip at regular intervals for up to an hour.

(6) Plot position of coleoptile tip (vertical axis) against time. What conclusions can you draw?

For Consideration

(1) Have you obtained any inconsistent or anomalous results in the course of this experiment? If so, how would you attempt to explain them?

(2) What was the reason for rotating the coleoptile through 180° at step (4)?

(3) If you were to perform the same experiment on the radicle, what would you predict the results to be?

(4) What is the relevance of your results to the normal life of the plant?

Requirements

Microscope
Eyepiece with micrometer scale
Moist chamber (*see* Fig. 27.1)
Slide
Elastic band

Germinating seed of oat, barley or wheat grown in the dark at 25°C

Investigation 27.3

The control of metamorphosis in amphibians

There is abundant evidence that in animals, as in plants, growth and development are controlled by hormones. In vertebrates the hormone most directly involved is **thyroxine**, secreted by the **thyroid gland**. Thyroxine is a complex compound which contains iodine derived from the diet.

A developmental event in which thyroxine has been implicated is **metamorphosis** in amphibians. This is the development of the larva (tadpole) into the adult (*see* Investigation 26.3, p. 288). In this investigation certain aspects of this event are explored.

The African clawed toad, *Xenopus laevis*, is suitable for this investigation. Instructions on the rearing of *Xenopus* tadpoles are given on p. 431. Alternatively tadpoles of the common frog or toad may be used.

In the experimental part of this investigation it is convenient to work in pairs, one student investigating thyroxine, the other iodine.

Requirements

Microscope

Dishes: 10 cm diameter and at
  least 6 cm high (× 7)

Wax pencil

Pond water or Holtfreter's solution

Thyroxine solutions $\Big\}$ *see* below
Iodine solutions

Pre-metamorphosis tadpoles of,
  e.g. *Xenopus laevis*

TS pre-hind limb tadpole

TS post-hind limb tadpole

Thyroxine solutions made up as follows:
Make up stock solution of thyroxine: dissolve
0.1 g thyroxine in 10 cm³ of 0.1 M sodium
hydroxide. Add 90 cm³ of water. Add 1.0 cm³ of
water. Add 1.0 cm³ of this solution to 1.0 dm³ of
pond water or Holtfreter's solution. This gives a
solution of 1 p.p.m. thyroxine.

Prepare the three solutions thus:

(a) **0.1 p.p.m. thyroxine:** add 100 cm³ of stock
thyroxine to 1.0 dm³ of pond water or Holtfreter's
solution.

(b) **0.05 p.p.m. thyroxine:** add 50 cm³ of stock
thyroxine to 1.0 dm³ of pond water or Holtfreter's solution.

(c) **0.01 p.p.m. thyroxine:** add 10 cm³ of stock
thyroxine to 1.0 dm³ of pond water or Holtfreter's
solution.

Iodine solutions made up as follows:
Make up stock solution of iodine: dissolve 0.1 g
of iodine crystals in 10 cm³ of ethanol. Add 90
cm³ of water. Add 1.0 cm³ of this solution to
1.0 dm³ of pond water or Holtfreter's solution.

Prepare the three solutions thus:

(a) **0.1 p.p.m. iodine:** add 100 cm³ of stock
iodine to 1.0 dm³ of pond water or Holtfreter's
solution.

(b) **0.05 p.p.m. iodine:** add 50 cm³ of stock
iodine to 1.0 dm³ of pond water or Holtfreter's
solution.

(c) **0.01 p.p.m. iodine:** add 10 cm³ of stock
iodine to 1.0 dm³ of pond water or Holtfreter's
solution.

Holtfreter's solution made up as follows:
To 1.0 dm³ of distilled water add 3.5 g sodium
chloride, 0.05 g potassium chloride, 0.10 g
calcium chloride, 0.02 g sodium bicarbonate, and
dissolve.

Further Information
Nuffield Advanced Science, *The Developing
Organism* (Penguin, 1970), p. 78.

Effect of Thyroxine on Metamorphosis

Obtain seven dishes of 10 cm diameter.
Label them **A** to **G**. In **A** place natural
pond water, or Holtfreter's solution,
to a depth of 5 cm. In **B** place a solution
containing 0.1 parts per million
(p.p.m.) of thyroxine; into **C** a solution containing 0.05 p.p.m. thyroxine;
into **D** a solution containing 0.01
p.p.m. thyroxine; into **E** a solution
containing 0.1 p.p.m. iodine; into **F**
a solution containing 0.05 p.p.m.
iodine and into **G** a solution containing
0.01 p.p.m. iodine.

Into each dish place at least five, and
not more than ten, pre-metamorphosis
tadpoles. There should be the same
number of tadpoles in each dish.

Examine the tadpoles at regular
intervals during the course of their
subsequent development. Record their
relative sizes and state of maturity.
Note the times at which metamorphosis takes place.

What conclusions would you draw
from your results?

Changes in the Thyroid Gland during Development

First be sure that you can recognize the
thyroid gland when you see it in a
microscopic section, and that you
understand its histology. If you have
not already done so, examine a section
of mammalian thyroid gland (*see* Investigation 18.7, p. 188).

Now examine a transverse section
of a tadpole whose hind limbs have
developed (**post-hind limb tadpole**).
The thyroid gland will be seen immediately beneath the floor of the
pharynx. Can you recognize it? If
necessary refer to Fig. 27.2 to help you
locate it.

Now examine a transverse section
of a tadpole whose hind limbs have not
yet developed (**pre-hind limb tadpole**). Can you find the thyroid
gland? How does its appearance compare with what you observed in the
previous section? Conclusions?

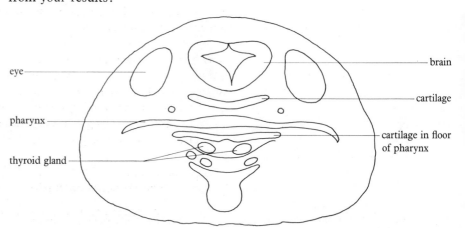

Fig. 27.2 Diagram of transverse section of post-hind limb tadpole (*Rana*) to show position of the
thyroid gland beneath the pharynx.

For Consideration

(1) In the control of amphibian metamorphosis thyroxine is an *internal*
factor. What *external* factors might be
involved in controlling metamorphosis?

(2) How could you investigate whether
or not each of the external factors listed
in your answer to (1) *actually* controls
metamorphosis?

(3) What are the consequences if (a) a
child and (b) an adult human fail to
receive sufficient iodine in their diets?

(4) Thiourea is said to block the in-
corporation of iodine into thyroxine
in the thyroid gland. How would you
investigate the effect of this thyroid
inhibitor on metamorphosis?

Questions and Problems

1   The graphs in Fig. 27.3 show the rate of growth in three populations of the
    fairy shrimp *Chirocephalus*, an aquatic crustacean, over a period of 30 days under
    different temperature conditions. Population **A** is kept at 25°C, population **B**
    at 15°C and population **C** at 5°C. Suggest an explanation for the different
    growth rates and discuss the implications of the data.   (*VUS*)

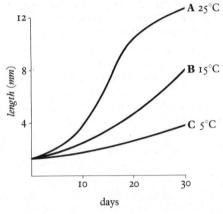

Fig. 27.3   Graphs showing the growth rate in
three populations of the fairy shrimp *Chiro-
cephalus* at three different temperatures.

2   Devise a series of experiments to find out what factors are necessary for the
    germination of seeds.

3   The effects of the application of the same quantity of various concentrations of
    auxin on the rate of elongation of the roots and shoots of oat seedlings was
    determined. The results (given below) were expressed as percentage stimulation
    of growth compared with untreated controls; thus a minus sign indicates
    relative inhibition. Suggest explanations for the results and discuss their
    relevance in relation to the opposite geotropic responses of typical roots and
    shoots:

Percentage stimulation of growth								
Concentration of applied auxin (parts per million)								
Organ	10^{-6}	10^{-5}	10^{-4}	10^{-3}	10^{-2}	10^{-1}	1.0	10.0	100.0
Root	+5	+20	+20	+10	−18	−55	−90	−95	−98
Shoot	0	0	0	+5	+20	+100	+180	+150	+10

(*O and C*)

4   What evidence is there to support the view that a stem is (a) negatively geotropic,
    (b) positively phototropic?
       Suggest physiological explanations for the response of stems and roots to the
    force of gravity. To what extent are such responses comparable to those of
    animals?
       Describe how you would show the presence of a growth substance in a stem
    or coleoptile.   (*AEB modified*)

5   Define and illustrate 'apical dominance'. Explain briefly how you believe the
    dominance of the apex is maintained.   (*O and C*)

6 Three similar groups of shoots were treated in the dark as follows:
Group **A**: the shoots were laid horizontal (*horizontal throughout*).
Group **B**: the shoots were laid horizontal for 1 hour only and then returned to the vertical position (*horizontal 1 hour only*).
Group **C**: the shoots were laid horizontal but their apices were removed (*decapitated horizontal throughout*).

One-half, one, two, three and four hours after the beginning of the experiment the curvature of the shoots was measured. A positive (+) curvature indicates upward bending and a negative (−) curvature downward bending. The results of the experiment were:

Time (hours)	Group A Horizontal throughout	Group B Horizontal 1 hour only	Group C Decapitated horizontal throughout
$\frac{1}{2}$	−5	−5	−4
1	0	0	−2
2	+25	+23	+4
3	+40	+16	+5
4	+70	+3	+7

All results are given in degrees.
Plot the results and outline the main conclusions you can draw from the experiment. *(O and C)*

7 How would you expect young, green, terrestrial higher plants to respond:
(a) if given extra auxin
(b) if given extra gibberellin
(c) if auxin action were totally inhibited?
Give reasons for your answers. *(O and C)*

8 In a carefully designed and executed experiment, groups of pea plants were subjected to the following treatments:
A Apical bud removed
B Apical bud removed and auxin placed on cut stump
C Apical bud removed and gibberellic acid placed on cut stump
D Apical bud removed and kinetin placed on cut stump
E Plants left intact. *Inhibited by auxin from apical bud*
At intervals after treatment the lengths of the axillary shoots were determined and the results are given below.

Days after start of treatment	A	B	C	D	E
2	3	3	3	3	3
4	10	4	12	9	3
6	30	4	45	32	3
8	50	5	90	47	3
10	78	6	116	80	3
13	118	30	150	119	3

Mean total axillary shoot lengths per plant (mm)

N.B. Assume that the environmental conditions were the same for all treatments throughout.
(a) Plot the results graphically.
(b) Discuss the results. *(O and C)*

9   Three species of the genus *Lolium* and a hybrid between two of them were tested for their vernalization requirements. All the strains were grown under controlled conditions (23°C day temperature; 17°C night temperature; long photoperiods). Sample plants of each strain were then subjected to different periods of time at 4°C before being returned to the original conditions. In the table below are recorded the number of days which elapsed between the end of the cold treatment and the onset of flowering:

	Number of days between end of cold treatment and onset of flowering			
	Strains of *Lolium*			
Weeks at 4°C	**A**	**B**	**C**	**A × B** (hybrid)
0	★	40	25	75
1	160	38	25	62
2	110	36	25	50
4	90	34	25	40
8	35	32	25	32
16	24	28	25	24
	★ did not flower			

(a) Plot the results graphically.

(b) Discuss the results.   *(O and C)*

10   The relation between moisture content and respiration of wheat grains (milligrammes of CO_2 produced per 100 grammes dry weight per day) was determined with the following results:

Moisture content %	13.0	14.0	15.0	16.0	16.5	17.0
Respiration rate	0.6	0.7	1.0	2.9	6.0	11.0

Suggest possible explanations for these results. What bearing do they have on the problems of seed dormancy and the storage of grain in large quantities by man?   *(O and C)*

11   To what extent is amphibian metamorphosis comparable with the breaking of dormancy in plants? What general principles do the similarities illustrate?

12   Describe an experiment (not necessarily on planarians) which shows the involvement of nervous tissue in regeneration. State as precisely as possible the information which your experiment provides on the role of the nervous tissue in this process. What further investigations on this matter could you recommend?

13 Fig. 27.4 summarizes the sizes of the testes and length of the antlers of roebuck deer in Australia in relation to the seasons of the year. What does this information suggest to you about the way the growth of the antlers is controlled? (SA)

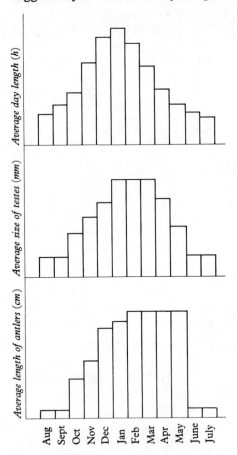

Fig. 27.4 Graphs showing the seasonal growth of the antlers in roebuck deer in relation to testis size and day length.

14 Planarians (flatworms) show marked powers of regeneration and have been much used in research on this phenomenon.

Discuss the conclusions to be drawn from each of the following experiments, which are illustrated in Fig. 27.5:

(a) A planarian is transected at three levels; the four sections regenerate but the anterior sections are larger and have better heads than the posterior sections (Fig. 27.5A).

(b) Two *very short* sections are cut out of a planarian towards the anterior and posterior ends. The anterior section develops a head at each surface, whereas the posterior section develops a 'tail' at each surface (Fig. 27.5B).

(c) The anterior third of a planarian is transected by three median longitudinal cuts. Each of the four sections regenerates a complete new head (Fig. 27.5C).

(d) A piece of the head is excised from one planarian and implanted into the skin at the side of another worm. The result is that a head develops on the side of the recipient's body. The recipient's original head is now removed. Result: a tail is formed at the anterior end of the recipient's body (Fig. 27.5D).

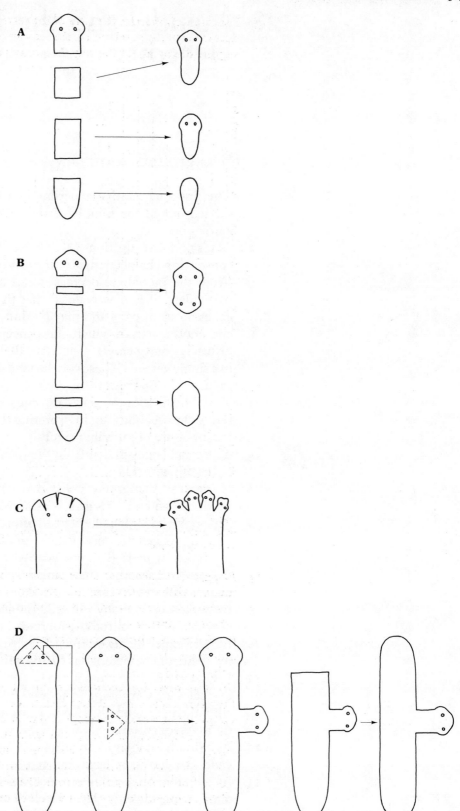

Fig. 27.5 Experiments on regeneration in planarian flatworms.

28 Mendel and the Laws of Heredity

Background Summary

1. The first quantitative experiments on heredity of any significance were carried out in the middle of the nineteenth century by Mendel on the garden pea.

2. In his first investigations Mendel studied the inheritance of a single pair of contrasting characteristics (**monohybrid inheritance**).

3. From the monohybrid cross it can be concluded that (i) inheritance is particulate (it is now realized that the 'particles' are the genes), (ii) that genes occur in pairs and may be **dominant** or **recessive** with respect to one another (from which the concept of **alleles, heterozygosity** and **homozygosity** emerge), and (iii) that only one such gene may be carried in a single gamete. These conclusions are embodied in Mendel's First Law, the **Law of Segregation**.

4. The transmission of genes depends on chance and can be expressed in terms of probability. In interpreting the results of genetic experiments one must beware of sampling error.

5. The conclusions drawn from the monohybrid cross can be interpreted in terms of **meiosis**.

6. If identical homozygotes are crossed, nothing but homozygotes will be produced and a **pure line** is established. This is called **true breeding**.

7. An organism's characteristics (**phenotype**) are determined, at least in part, by its genetic constitution (**genotype**). A single phenotype may be produced by more than one genotype.

8. To establish whether an organism is homozygous or heterozygous a **back cross** (**test cross**) can be carried out.

9. Examples of monohybrid inheritance in man include albinism, cystic fibrosis, and chondrodystrophic dwarfism (achondroplasia).

10. In later experiments on the garden pea Mendel studied the inheritance of two pairs of characteristics (**dihybrid inheritance**).

11. From the dihybrid cross it can be concluded that each member of one pair of alleles may combine randomly with either of the other pair. This is embodied in Mendel's Second Law, which is called the **Law of Independent Assortment**.

12. As with monohybrid inheritance, the transmission and assortment of genes depends on chance and can be expressed in terms of probability.

13. As with monohybrid inheritance, a single phenotype may be produced by more than one genotype. A **back cross** (**test cross**) can be carried out to establish an organism's genotype.

14. The conclusion drawn from the dihybrid cross can be interpreted in terms of **meiosis**.

15. Although it forms the foundation of modern genetics, Mendel's work was carried out without any knowledge of chromosomes, genes, or meiosis. Its significance was not realized until 1900.

Investigation 28.1

Mendelian inheritance in corn (maize) and tobacco seedlings

Requirements

Trays of soil approx. 40 × 20 cm

Petri dish containing fine damp soil or 1.5 per cent agar

Corn (maize) seeds*

Seeds of tobacco plant (*Nicotiana tabacum*)*

*Available from supplier.

Procedure

(1) You are provided with samples of corn (maize) seeds. All the seeds were obtained from one plant.

Plant out 60 seeds in a tray of soil that measures approximately 40 × 20 cm. Keep in a warm, light place and water frequently. When the seedlings have grown sufficiently (about 28 days after sowing), examine each tray for observable differences. Are all the seedlings alike, or are there visible differences? If there are differences, count each type. What is the ratio between them? What conclusions can you draw?

(2) Sprinkle tobacco seeds on a petri dish of fine damp soil or agar jelly (1.5 per cent). To ensure that the distribution of seeds is not too dense, first mix the tobacco seeds with fine sand in the proportion of about two-thirds sand to one-third seeds.

When the seedlings have grown sufficiently (10–14 days after sowing), examine them for any observable differences. Count the different types, and calculate the ratio between them. Conclusions?

For Consideration

(1) How close is your ratio to the expected ratio? If they are not very close what should you do next?

(2) What genetic principle emerges from your results?

Investigation 28.2

Mendelian inheritance in corn (maize) seeds

In corn (maize) a single ear or cob is covered with several hundred kernels which represent the seeds. Each seed contains an embryo formed as a result of a single fertilization. It is, therefore, diploid. The fertilizations are independent of each other: each seed originated as a diploid cell which, prior to fertilization, underwent meiosis independently of all the others. In human terms what is the relationship between two embryos belonging to the same cob?

The seeds display a number of easily recognized characteristics such as colour and shape: thus they may be purple or yellow, smooth or shrunken, etc.

Procedure

(1) Examine cob **A**. This was produced by self-pollinating a flower of the parent plant. Count the purple and yellow seeds and determine the ratio between them. Do you think it is sufficient to count the seeds in one (or several) rows, or should all the seeds in the cob be counted?

(2) Which is the genetically dominant seed colour, purple or yellow?

(3) Letting **P** be the symbol for purple seed, and **p** for yellow, what can you say about the genetic constitution (genotype) of the purple and yellow seeds?

(4) What was the genotype of the plant on which this cob developed?

(5) Construct a genetic diagram to illustrate the cross involved in the production of cob **A**.

(6) Examine cob **B**. Determine the ratio of purple to yellow seeds.

(7) Construct a genetic diagram to illustrate the cross involved in the production of this cob.

(8) How would you determine the genotype of one of the purple seeds in cob **A**?

(9) Examine cobs **C** and **D**. The seeds differ from each other with respect to *two* pairs of characteristics; each is purple or yellow, and smooth or wrinkled.

Estimate the number of seeds that are:
  (a) purple and smooth;
  (b) purple and wrinkled;
  (c) yellow and smooth;
  (d) yellow and wrinkled.

(10) Construct a genetic diagram illustrating the cross involved in the production of cobs **C** and **D**.

For Consideration
Can you state Mendel's first and second laws? How precisely do the results of your analysis of corn support these two laws?

Requirements
Corn cobs showing segregation for purple and yellow seeds (**A** and **B**)*

Corn cobs showing segregation for purple and yellow, and smooth and wrinkled seeds (**C** and **D**)

*Available from supplier.

Investigation 28.3

The genetic constitution of maize pollen

It follows from the result of monohybrid crosses that the gametes formed by a heterozygous organism differ in their genetic constitution: approximately half contain the dominant gene, the other half the recessive gene. Although this is an inevitable theoretical conclusion, it is difficult to prove because genetic differences seldom reveal themselves in the gametes. Usually gametes look identical. However, an exception to this is provided by maize.

In maize a distinction can be made between **starchy** and **non-starchy** plants. Starchy plants have cells containing normal starch which stains blue-black with iodine; non-starchy plants contain an abnormal form of starch, which gives a reddish-brown reaction with iodine. The same distinction is seen in the seeds (kernels). Starchy is dominant to non-starchy.

Consider a heterozygous maize plant resulting from a cross between starchy and non-starchy parents. Since starchy is dominant to non-starchy, the heterozygous plant will be starchy. Now the genes responsible for this condition determine the presence or absence of starch in the pollen grains as well as in the adult. So pollen grains produced by this plant will be expected to be starchy and non-starchy in approximately equal numbers. In this investigation we shall test this prediction.

Procedure
You are provided with a floret of a heterozygous maize plant. With needles, tear off the enveloping bracts and remove one of the six anthers. Place the anther in a drop of iodine on a microscope slide. With needles break the anther open and tease out the pollen grains. Put on a coverslip and wait for the stain to take effect.

View the pollen grains under medium power. If they appear uniform, alter the illumination by changing the angle of the substage mirror. Count the number of dark and light ones in a total of up to 100.

Do your results confirm the prediction that the heterozygous plant produces two types of gametes in approximately equal numbers?

For Consideration
Which of Mendel's laws is given considerable support by the results of this experiment? Explain fully.

Requirements
Microscope
Slide and coverslip
Needles
Iodine solution

Maize floret (available from supplier)

Investigation 28.4

Handling *Drosophila*

Drosophila, the fruit fly, is an ideal animal for experimental genetics. It can be kept easily in the laboratory, and at 25°C the complete life cycle takes only 10–14 days. Moreover, a single female lays between 80 and 200 eggs, so it does not take long to produce a large population.

Fruit flies used in experimental work are descendants of wild species that feed on yeasts and plant sugars such as are found on damaged fruits. In the laboratory they can be cultured in bottles containing a nutrient medium. This consists of syrup, raisins, banana, yeast, etc., set into a jelly with agar. The life cycle is summarized in Fig. 28.1. Following impregnation, the female stores sperm in her spermotheca from which a large number of eggs are fertilized. Once laid, the eggs develop into **larvae** which burrow into the medium. After two moults the larvae leave the medium and crawl into drier parts of the bottle, usually up a roll of filter paper provided for the purpose. They then **pupate** and after a few days the **adults** emerge, mate, and the cycle starts again.

Before you can do genetic experiments with *Drosophila* you must learn to handle the flies, recognize the different strains, and identify the sexes.

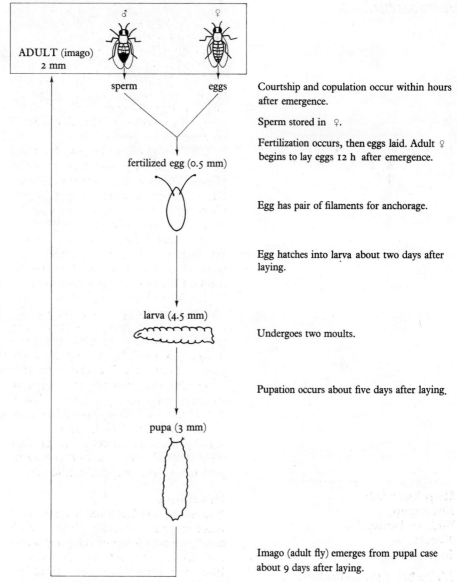

Courtship and copulation occur within hours after emergence.

Sperm stored in ♀.

Fertilization occurs, then eggs laid. Adult ♀ begins to lay eggs 12 h after emergence.

Egg has pair of filaments for anchorage.

Egg hatches into larva about two days after laying.

Undergoes two moults.

Pupation occurs about five days after laying.

Imago (adult fly) emerges from pupal case about 9 days after laying.

Fig. 28.1 Life cycle of the fruit fly, *Drosophila melanogaster*. The times apply to fruit flies kept at 25°C. (*Based on* Haskell).

Examination of Live Flies

Examine a **culture bottle** containing a **pure line** (what's that?). Adult flies of both sexes should be present, also larvae burrowing in the culture medium and, possibly, pupae on the glass or filter paper.

Now examine several pure-line flies from the bottle marked '**wild-type**' Being active animals they must be anaesthetized first. This is done as follows (Fig. 28.2):

(1) Pipette a few drops of ether on the cotton wool of an **etherizer**. Return unwanted ether to the bottle and replace the stopper immediately: ether fumes are dangerous.

(2) Tap the culture bottle so the flies are dislodged from the mouth of the bottle and the cotton-wool bung.

(3) Quickly remove the cottonwool bung and place the funnel of the etherizer over the open mouth of the culture bottle. Turn the etherizer and culture bottle the other way up and tap the latter gently until at least ten flies have entered the etherizer. Then replace the stopper of the culture bottle.

(4) As soon as the flies in the etherizer stop moving, remove the funnel of the etherizer and tip the flies onto a white tile. Examine them under a lens or binocular microscope.

Anaesthetized flies usually remain unconscious for about ten minutes. Avoid giving them too much ether; a fly with arched abdomen, legs bunched together and wings deflected upwards, is probably dead.

Recognizing the Wild Type

The **wild type** is the normal (non-mutant) fly. It is designated by the symbol +. Its characteristics include the following: eyes round and red, body grey, wings approximately the same length as (or slightly longer than) the abdomen. A fly may of course be normal for some characteristics but mutant for others.

In examining wild type flies notice the above features. The wild type bottle contains both males and females (*see* below). Young flies are pale coloured with incompletely expanded wings.

Recognizing Sex Differences

With a paintbrush separate the wild type flies into **males** and **females**. Male and female specimens of *Drosophila* can be distinguished as shown in Fig. 28.3 and in the accompanying table on the opposite page.

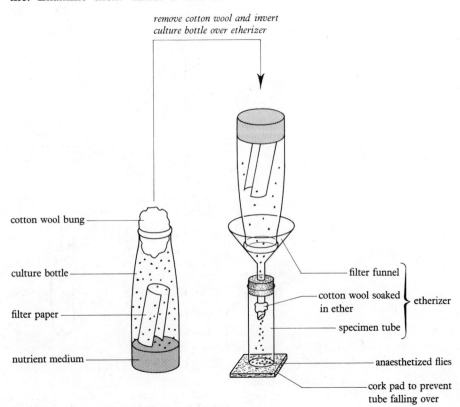

remove cotton wool and invert culture bottle over etherizer

cotton wool bung

culture bottle

filter paper

nutrient medium

filter funnel

cotton wool soaked in ether

specimen tube

etherizer

anaesthetized flies

cork pad to prevent tube falling over

Fig. 28.2 Method of anaesthetizing *Drosophila*.

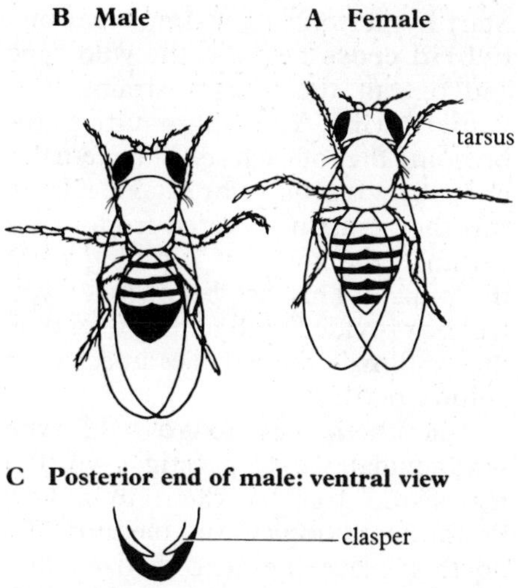

B Male   A Female

C Posterior end of male: ventral view

—clasper

Fig. 28.3 *Drosophila melanogaster*, wild-type male and female.

	Male	Female
1	Rounded abdomen	Pointed abdomen
2	Black transverse stripes at posterior end of abdomen so close together they appear as a single dark blob	Black transverse stripes at posterior end of abdomen narrow and clearly separated
3	Pair of chitinous claspers on ventral side of abdomen at posterior end	Claspers absent
4	Row of small bristles (sex comb) on first tarsal joint of forelegs	Sex combs absent

Recognizing the Mutants

Having examined the wild type it is now necessary to identify some of the **mutant strains** of *Drosophila*.

Examine one or more mutants from the culture bottles provided. Use a hand lens or binocular microscope as required.

Table 28.1 gives a list of five of the more easily recognized mutants which may be available to you.

For Consideration

Before proceeding to Investigations 28.5 and 28.6, plan in detail a series of breeding experiments which you yourself would carry out to investigate heredity in *Drosophila*. Base your programme on the flies examined in this laboratory session.

Requirements

Hand lens or binocular microscope
White tile
Paintbrush
Etherizer (Fig. 28.2)

Wild-type flies
Mutant flies (*see* Table 28.1)

All flies should be pure-breeding (homozygous) in labelled culture bottles.

Table 28.1 Summary of the five mutants of *Drosophila melanogaster* **featuring in the investigations outlined in Chapters 28 and 29.**

Name of mutant	Symbol	No. of chromosome on which gene is located	Phenotypic Characteristics
White	w	1 (X chromosome)	White eyes
Yellow	y	1 (X chromosome)	Yellow body
Brown	bw	2	Brown eyes
Vestigial	vg	2	Reduced wings
Ebony	e	3	Black body

Investigation 28.5
Monohybrid inheritance in *Drosophila*

Start by performing a simple **monohybrid cross** between the wild type and one of the mutant strains, e.g. vestigial wing. You will require pure-breeding flies of each sex. The females must be **virgins** (why?). To ensure this they must be isolated from males within eight hours of hatching from the pupae. This has been done: you will be provided with female wild type flies and male mutant flies in separate culture bottles.

Anaesthetize up to 10 wild type virgin females and 15 vestigial-winged males and transfer them to a new bottle containing culture medium. To do this, place the anaesthetized flies on a white tile, then with a paintbrush push them gently one by one into the new culture bottle.

Leave the bottle on its side until the flies recover so they do not stick to the medium. Use only mature adults for crossing, and make sure you do not over-anaesthetize them (*see* p. 320).

Label your bottle with your name, indicating the cross which you have set up, as follows:

P   ♀+ × ♂vg (date:)

When you are sure that at least one female and two male flies have recovered, put your bottle in an incubator at 25°C.

One week later, when the eggs have hatched into larvae, remove the parent flies. (Why?). The **F₁** flies should emerge over a period of several days approximately 10–14 days after the cross was set up.

Do you think you should set up the reciprocal cross, i.e. ♂+ × ♀vg? If you think it does not matter, explain your reasoning.

Setting up F₁ Crosses
Anaesthetize all the **F₁** flies from your first cross and tip them onto a white tile. How would you describe them phenotypically? What is the ratio of males to females? Explain your observations.

Now cross **F₁** flies amongst themselves. The females need not be virgins. (Note why.) Label the bottle as follows:

P   ♀+ × ♂vg (date:)
F₁   all +
F₁   ? × ? (date:)

Incubate at 25°C.

Backcross **F₁** flies with vestigial wing (double recessive). In this case the females must be virgins. (Why?). Incubate at 25°C.

After one week remove the **F₁** parents from the culture bottle. Allow a further week to elapse and then examine the **F₂** offspring.

Examination of F₂ Flies
Ideally, analysis of the **F₂** flies should be started the day after the first flies emerge and continued daily until all the flies have emerged. Counted flies should be killed each time. Alternatively, wait until sufficient flies have emerged to provide an adequate sample.

Anaesthetize in the usual way and tip all the **F₂** flies onto a white tile. With a paintbrush separate them into groups according to wing-length. Then subdivide each of these groups into males and females.

Count the numbers in each group. How many of each type? What is the ratio between them? Within each group how many are males and how many females?

Other Monohybrid Crosses
Other possible crosses involving the inheritance of clear-cut characteristics are as follows:

Wild type with brown eye: + × bw

Wild type with ebony body: + × e

Backcrosses between the **F₁** or **F₂** wild types and the mutant flies can also be carried out.

Results
Write a full account of your experiments and the results obtained.

For Consideration
You have probably suggested that your **F₂** flies relate to each other in a particular ratio. Look up Appendix 19 and make sure that your conclusions are valid.

Requirements
Hand lens or binocular microscope
White tile
Paintbrush
Etherizer
Culture bottles

Virgin ♀ wild type flies
♂ vestigial-winged flies
(Other types of fly if required)

All flies should be pure-breeding (homozygous) in labelled culture bottles.

Investigation 28.6

Dihybrid inheritance in *Drosophila*

Requirements

Hand lens or binocular microscope
White tile
Paintbrush
Etherizer
Culture bottles

Virgin ♀ wild type flies
♂ ebony vestigial flies
(Other types of fly if required)

All flies should be pure-breeding (homozygous) in labelled culture bottles.

Using exactly the same technique as that described in Investigation 28.5, set up a **dihybrid cross** between wild type virgin females and ebony vestigial males, i.e. ♀+ × ♂e vg, and/or a reciprocal cross.

Label the bottle. Remove the parent flies one week after you have set up the cross. After two weeks examine the F_1 offspring.

Now cross F_1 flies amongst themselves and carry the investigation through to the F_2. After one week remove the adults in the usual way.

In analysing the F_2 flies record the numbers of each type, i.e. grey-bodied, normal wing; grey-bodied, vestigial wing; ebony-bodied, normal wing; and ebony-bodied, vestigial wing. Calculate the ratio between them. Explain your results.

Dihybrid inheritance involving clear-cut features can also be investigated by crossing wild-type virgin females with ebony bodied, brown-eyed males (♀+ × ♂e bw). Backcrosses can be carried out between F_1 and F_2 wild type flies and the double recessive.

Results

Write a full account of your experiments and the results obtained.

For Consideration

As with the monohybrid cross, you have probably suggested that the F_2 flies relate to one another in a particular ratio. Consult Appendix 19 and make sure that your conclusions are valid.

Questions and Problems

1 In *Drosophila* straight wing is dominant over curved wing. What would be the result in the F_1 generation of crossing a homozygous straight-winged fly with a curved-winged fly? What would be the result in the F_2 generation of crossing two of the F_1 flies? How would you determine the genotype of one of the F_2 straight-winged flies?

2 In *Drosophila*, black body colour is recessive to grey body colour. A geneticist had three pairs of flies with grey bodies designated **A**, **B**, and **C**. He crossed **A** × **B** and obtained 109 grey-bodied flies; **A** × **C** gave 80 grey-bodied and 28 black-bodied; whilst **B** × **C** gave 76 grey-bodied flies. Explain, giving reasons, the expected genotypic and phenotypic ratio when the flies **A**, **B**, and **C** are crossed with flies having black bodies. (*UL*)

3 Fibrocystic disease of the pancreas is a fatal condition caused by a single recessive gene. Heterozygous individuals are normal. A pair of normal parents produce a defective child, their first. What is the probability that their next child will be defective? These particular parents know no biology, but they are intelligent and want to know the reasoning on which your conclusion is based. Explain it to them simply, but thoroughly.

4 Albinism in man is caused by a recessive gene which is transmitted in a normal Mendelian fashion. A phenotypically normal (non-albino) couple have four children: the first three are normal, the fourth is an albino:
(a) What can you say about the genotypes of the parents?
(b) What is the probability that their next child will be an albino?
(c) One of the normal children marries a normal woman. What predictions can be made concerning their first child?
(d) The albino child marries a normal woman. What predictions can be made regarding their first child?
Where there are several possibilities, state them. Show your reasoning.

5 Four flowers were chosen on a tall sweet-pea plant. The unripe anthers were carefully removed, and the receptive stigmas dusted with pollen from a short sweet-pea plant. The pollinated flowers were covered with paper bags, and at the end of the season 26 seeds were collected from the four pods. Sixteen of these seeds grew into tall plants and ten into short plants. Suggest explanations of this result and state what further experiments you might carry out to test the truth of your suggestions. *(O and C)*

6 In human beings the ability to taste phenylthiourea depends upon the presence of a dominant gene.
(a) Construct diagrams showing
  (i) the genotype and phenotype of the parental generation and the F_1 generation, and the gametes of the cross between a homozygous 'taster' and a homozygous 'non-taster';
  (ii) the genotype and phenotype of the progeny of a cross between two genotypes of the F_1 type in (i).
(b) Study the diagram below. State, with reasons, the possible genotype of persons **A**, **B**, **C**, **D**, and **E**. *(CL modified)*

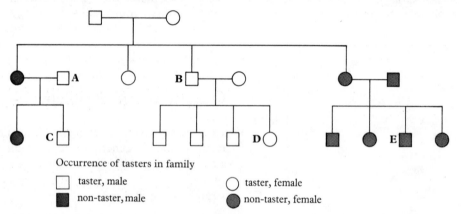

Occurrence of tasters in family

☐ taster, male   ◯ taster, female
■ non-taster, male   ⬤ non-taster, female

7 Tongue-rolling is due to a dominant gene. If a man, whose parents are both non-rollers, marries a rolling girl, whose mother and grandparents are rollers and whose father and sister are non-rollers, what are the chances that their first child will be a roller? Show your reasoning.

In fact their first child, a boy, turns out to be a roller. He eventually marries another roller and between them they produce two sons, neither of whom can roll, and three daughters, two of whom can roll, whilst the third cannot. Express the transmission of tongue-rolling in this family, through all four generations, in the form of a pedigree chart like the one given in Question **6**.

8 Discuss the genetic implications of the following experiment:
(a) An inbred plant of barley, susceptible to mildew, was pollinated by hand using pollen from an inbred plant showing resistance to mildew, precautions being taken to prevent self-pollination.
(b) Eleven F_1 plants, which were all resistant to mildew, were grown and were allowed to self-pollinate naturally. All the seeds were collected and later sown to produce the F_2 generation.
(c) Five hundred and twelve F_2 plants were grown, but as there was no attack by mildew no observations could be made. The plants were allowed to self-pollinate and all the seeds were collected.
(d) A sample of the seed was sown and the F_3 plants were found to consist of 586 which were infected with mildew and 1,014 which were resistant to mildew. *(O and C)*

9   The bread mold *Neurospora* belongs to a class of fungi known as the Ascomycetes. The life cycle of *Sordaria*, a related genus, is given in Fig. 29.1 (p. 329).

    In a laboratory experiment an albino strain of *Neurospora* was crossed with the normal pink wild-type strain. Ascospores formed by the offspring were isolated at random and allowed to germinate. The resulting cultures were 2,018 albino and 1,996 pink. Explain this result.

10   Tall, cut-leaved tomato plants are crossed with dwarf, potato-leaved plants giving in the F_1 generation nothing but tall, cut-leaved plants. These are allowed to cross with each other producing in the F_2 generation 926 tall, cut-leaved; 288 tall, potato-leaved; 293 dwarf, cut-leaved; and 104 dwarf, potato-leaved. What explanation can be offered for these results?   (*O and C*)

11   In the summer squash plant white fruit is dominant over yellow, and the disc fruit shape is dominant over sphere. A cross between a plant with white disc fruits and one with yellow sphere fruits gave:

25 plants with white disc fruits; 26 plants with white sphere fruits; 24 plants with yellow disc fruits; 25 plants with yellow sphere fruits.

If the white disc parent is self-fertilized, what proportion of its offspring will have yellow sphere fruits? Show how you arrive at your answer.   (*O and C*)

12   After pollinating a plant with rough stems and yellow flowers with pollen from another plant of the same species having rough stems and white flowers, adequate precautions being taken to prevent pollination occurring in any other way, 80 seeds were obtained. These grew into plants showing four combinations of character:

    26 with rough stems and yellow flowers;
    12 with smooth stems and yellow flowers;
    33 with rough stems and white flowers;
     9 with smooth stems and white flowers.

What deductions would you make concerning the genetic constitution of the parent plant if it is known from other experiments that 'yellow flower' is dominant to 'white flower'? Set out your answer clearly to show the steps leading to your final conclusion.   (*O and C*)

13   In the fruit fly *Drosophila* the wild type (normal) is grey in colour with wings that extend beyond the tip of the abdomen. Among the mutants of *Drosophila* are two which are respectively distinguished by dark body colour (ebony) and a vestigial condition of the wings (vestigial). A fruit fly with vestigial wings and ebony body colour is crossed to the wild type. The F_1 flies are back-crossed to the double recessive (ebony-vestigial) and the result is:

    35 wild type; 32 ebony, normal wing; 33 vestigial, grey body colour; 34 ebony, vestigial.

Discuss this result with the aid of a diagram, commenting on the relationship between phenotypic appearance and genetic make-up. What would be the result of mating the wild types of this last experiment among themselves? (*O and C*)

14   In guinea pigs, black coat colour is dominant to brown and short hair is dominant to long hair. These characters are not linked. A breeder has only stocks of pure breeding, long-haired brown and pure breeding, short-haired black guinea pigs. Explain clearly the breeding programme to be followed to obtain pure breeding, long-haired black guinea pigs.   (*UL*)

15   State Mendel's first and second laws (a) as he originally stated them, and (b) using modern terminology. How would you explain these two laws?

    Do you think it is justified to call them *laws*?

16 In garden peas, tall (**T**) is dominant over dwarf (**t**), green pods (**G**) over yellow (**g**), and round seeds (**R**) over wrinkled (**r**).

If a homozygous dwarf, green, wrinkled pea plant is crossed to a homozygous tall, yellow, round one, what will be the appearance of the F_1 plants? What will be the appearance of the offspring resulting from:

(a) self-pollination of one of the F_1 plants;

(b) cross-pollinating an F_1 plant with its dwarf, green, wrinkled parent;

(c) cross-pollinating an F_1 plant with its tall, yellow, round parent?

17 The seeds from a maize plant were sown and 452 green and 148 white (non-photosynthetic) seedlings emerged. How do you account for this? What progeny would you expect in the next two generations? (*OCJE*)

18 Normal (wild-type) strains of the fruit fly *Drosophila melanogaster* have greyish-brown bodies if developed on food media free of silver salts, but have yellow bodies if certain silver salts are added to the food on which the larvae develop. Strains homozygous for the recessive allele yellow(**y**) have yellow body-colour regardless of whether their food contains silver salts of not.

You have a single living yellow fruit fly, and the food on which it has developed is unknown; you also have a stock of normal (wild-type) fruit flies. Describe how you would proceed to find out whether or not the yellow body-colour of the single yellow individual was genetically or environmentally produced.

(*HK*)

29 Chromosomes and Genes

Background Summary

1   In dihybrid crosses genes sometimes fail to assort freely. This can be explained by postulating that such genes are carried on the same chromosome (**linkage**).

2   Genes linked together on the same chromosome constitute a **linkage group**. In *Drosophila* the linkage groups correspond in size and number to the chromosomes.

3   Homologous chromosomes are normally identical in appearance (**autosomes**). An exception is provided by the **sex chromosomes** which are **heterosomes**. Sex is determined by a pair of alleles carried on the **X** and **Y** chromosomes. In man the female is **homogametic** (**XX**), the male **heterogametic** (**XY**). In certain other organisms the reverse is the case.

4   Experiments and observations indicate that other genes, in addition to those determining sex, are carried on the **X** chromosome (**sex linkage**). Though open to question, it is generally thought that no such genes are carried on the **Y** chromosome.

5   In man sex linkage is seen in the inheritance of colour-blindness and haemophilia, the latter being of historic interest in the royal families of Europe.

6   Usually crosses involving linked genes produce a small proportion of offspring with new combinations (**recombinations**) in addition to the **parental combinations**. This can be explained by **chiasmata-formation** and **crossing-over** during meiosis.

7   Crossing-over enables geneticists to work out the relative positions of genes on a chromosome. In general the distance between two genes is proportional to their **cross-over value** (**cross-over frequency**). In this way **chromosome maps** can be established.

8   **Gene loci** in chromosome maps of *Drosophila* can be correlated with the **bands** seen in the **giant chromosomes** of the larval salivary glands. This, and much other information, suggests that genes are located in a linear sequence on the chromosomes.

9   Certain characteristics, e.g. blood groups, are controlled by **multiple alleles**. In the case of the **ABO** blood-group system, three alleles are involved, only two of which can be present in any one individual. The blood group alleles are inherited in a normal Mendelian manner. Knowledge of the inheritance of blood groups is sometimes used in legal questions of paternity.

10   The inheritance of blood groups, together with data from a wide range of breeding experiments, indicate that between the two extremes of complete dominance and no dominance there are all degrees of **partial dominance**.

11   A medically important example of partial dominance is provided by the inheritance of **sickle-cell anaemia**. Studies on the cause of this condition have also thrown light on the way genes express themselves.

12 Many situations are known where a particular combination of genes is lethal (**lethal genes**). Sometimes a lethal gene is rendered relatively, but not completely, harmless in the heterozygous state, another example of partial dominance.

13 A particular characteristic is sometimes controlled by two or more pairs of alleles interacting with one another, e.g. the comb condition of poultry.

Investigation 29.1

Relationship between genes and chromosomes in *Drosophila*

Requirements

Hand lens or binocular microscope
White tile
Paintbrush
Etherizer
Culture bottles

Virgin ♀ wild-type flies
♂ brown-eyed, vestigial-winged flies
♂ wild-type flies
Virgin ♀ white-eyed, yellow-bodied flies
♂ white-eyed, yellow-bodied flies

All flies should be pure-breeding (homozygous) in labelled culture bottles.

Using the same technique as that described in Investigation 28.5 (p. 322), set up the following crosses:

(1) Wild-type virgin females with brown-eyed, vestigial-winged male flies, i.e. ♀ + × ♂ bw vg. Label the bottle. After one week remove the parent flies, and after two weeks examine the F_1 offspring.

Now cross F_1 flies amongst themselves and carry the investigation through to the F_2. Do not forget to remove the F_1 flies.

Analyse the F_2 flies, recording the numbers of individuals showing each combination of characteristics. Calculate the ratio between them.

Explain your results. What do they tell you about the relationship between the genes responsible for the brown eye and vestigial wing conditions, and the chromosomes that carry them?

(2) Wild-type male flies with white-eyed, yellow-bodied virgin females, i.e. ♂ + × ♀ w y.

Label the bottle. After one week remove the parent flies and after two weeks examine the F_1 offspring. Analyse the F_1 flies, recording the numbers and sexes of each type. Explain your results. What results would you expect to get if you crossed the F_1 flies amongst themselves.

Now cross F_1 flies amongst themselves and carry the investigation through to the F_2. As before, do not forget to remove the F_1 flies. Analyse the F_2 flies, recording the numbers and sexes of each type. Are your predictions confirmed? What conclusions would you draw regarding the relationship between the genes controlling white eye, yellow body, and sex?

(3) Wild-type virgin females with white-eyed, yellow-bodied males, i.e. ♀ + × ♂ w y. Proceed exactly as in (2) above. Explain the results obtained in both the F_1 and F_2 generations.

Results

Write a full account of your experiments and the results obtained.

For Consideration

What conclusions would you draw from Investigations 28.6 and 29.1 regarding the relationship between genes and chromosomes in *Drosophila*? Are your conclusions supported by the information in Table 28.1?

Investigation 29.2

Genetic analysis of the products of meiosis in *Sordaria fimicola*

Sordaria is a fungus belonging to the Ascomycetes. It consists of branched filaments (**hyphae**) which are haploid. The life cycle is summarized in Fig. 29.1. Sexual reproduction takes place between neighbouring hyphae. Nuclei migrate from one hypha into the other and fuse in pairs. Each diploid nucleus resulting from this fusion gives rise to a **perithecium** consisting of a group of spore-producing structures called **asci** (singular **ascus**).

Initially each ascus has a single diploid nucleus. This undergoes three successive divisions: the first two are meiotic (meiosis I and II respectively) and the third is mitotic. The result is that each ascus contains eight haploid **spores** (known as **ascospores**). Within each ascus the eight ascospores are arranged in a linear sequence corresponding to the order in which they are formed. In other words the spores containing the two nuclei resulting from any particular nuclear division are next door to one another.

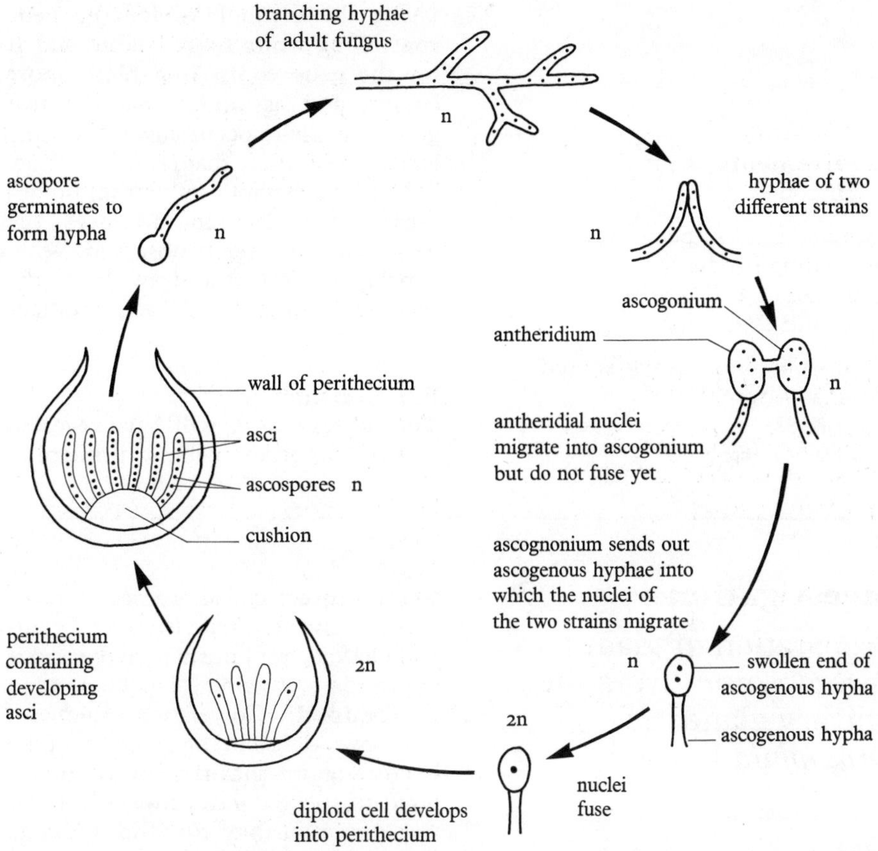

branching hyphae
of adult fungus

n

hyphae of two
different strains

n

ascopore
germinates to
form hypha   n

n

ascogonium

antheridium

antheridial nuclei
migrate into ascogonium
but do not fuse yet

wall of perithecium

asci

ascospores n

cushion

ascognonium sends out
ascogenous hyphae into
which the nuclei of
the two strains migrate

n

swollen end of
ascogenous hypha

ascogenous hypha

perithecium
containing
developing
asci

2n

2n

nuclei
fuse

diploid cell develops
into perithecium

Fig. 29.1 Simplified life cycle of an ascomycete fungus of the *Sordaria* type. The eight ascopores inside each ascus are formed by three successive nuclear divisions, meiosis I and II followed by mitosis. The ascospores are therefore haploid, as are the hyphae to which they give rise. **n**, haploid; **2n**, diploid.

The colour of the ascospores varies: some are white, others black. The colour of a particular spore is determined by its genetic constitution. White spores contain a gene which we can designate **W**, black spores contain a gene which can be designated **B**.

Procedure

With a sterile needle inoculate a plate of cornmeal agar with about 1 mm³ of white-strain and 1 mm³ of black-strain *Sordaria*, separated by a distance of approximately 30 mm. Incubate at 25°C in darkness for ten days.

Ten days later small dark **perithecia** will be observed resting on the agar. With a needle pick off several of the largest ones from the area where the two cultures have met, and also from the area close to where you inoculated.

Mount each perithecium separately in water on a slide. Put on a coverslip and tap it *gently* to squash the perithecium and spread out the asci. Do not tap too hard or the asci may burst.

Examine perithecia under high power and look for **three types of asci**: those with only black spores, those with only white, and those with both. Does an individual perithecium contain only one type, or more than one? Explain.

Now carefully examine asci containing black and white spores. How many of each are there within a single ascus? Is the number fixed or variable? Explain.

Make diagrams illustrating all the different arrangements of black and white spores which you can observe in your asci. Do they agree with Fig. 29.2? Which arrangements are most common?

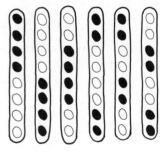

Fig. 29.2 Diagram showing the possible arrangements of spores in the asci of *Sordaria*.

Requirements
Microscope

Cornmeal agar in petri dish
Inoculating needle
Incubator at 25°C
Slides and coverslips

White-spored and black-spored
 strains of *Sordaria**

*Available from supplier.

Using the symbol **W** for the gene
controlling white spore colour, and **B**
for the gene controlling black spore
colour, and assuming that the two
genes are alleles occurring at the same
locus, construct diagrams to show
how the behaviour of chromosomes
during cell division produces the
various arrangements observed. What
conclusions do you draw from the
relative frequency of each arrangement?

For Consideration
Your observations will have revealed
certain important genetic principles.

What are these principles and what is
their general significance?

Investigation 29.3

Preparation of giant chromosomes from the salivary glands of *Drosophila* larvae

Certain insect chromosomes, notably
those in the salivary glands of *Drosophila* larvae, are unusually large and if
appropriately stained can be seen to
be **banded**. Correlations between
cross-over data and the positions of the
bands suggest that the bands correspond to specific **gene loci**, including
those responsible for the mutants
listed in Table 28.1. **Giant chromosomes** can be prepared for observations under the light microscope as
follows:

Temporary Preparation
(1) Obtain a *Drosophila* larva from a
culture bottle and place it in a drop of
sodium chloride (0.7 per cent) in
hydrochloric acid (M) on a microscope slide. Examine the larva under
a binocular microscope (a hand lens
is less satisfactory) and identify the
structures shown in Fig. 29.3A. Be
sure you can tell the anterior from
the posterior end.

(2) Hold the posterior end down with
a needle or grasp it with fine forceps;
place a needle on the head and pull the
head from the body as shown in
Fig. 29.3B. Identify the two elongated
transparent **salivary glands** on either
side of the gut. Detach them, and
remove everything else from the slide.

(3) Cover the glands with a drop of
acetic orcein stain. Leave for 15–20
minutes.

(4) Put on a coverslip. Press down on
the coverslip and tap it with the handle
of your mounted needle. Remove
surplus stain with filter paper. The
cells should rupture and the chromosomes spread out. Locate **giant**

chromosomes under low power and
examine them in detail under high
power.

A Intact larva

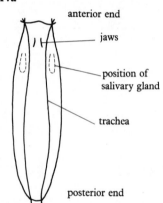

B After decapitation

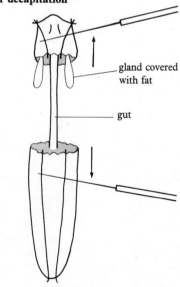

Fig. 29.3 Technique for exposing the salivary
glands of *Drosophila* larva.

Requirements
Microscope
Binocular microscope or hand lens
Slide
Coverslip
Fine forceps
Mounted needles
Filter paper
Acetic orcein stain
0.7 per cent sodium chloride
 (NaCl) in M HCl

Drosophila larvae

Additional Requirements for Permanent Preparation
Siliconized coverslip
Albumen
Hotplate
Ethanol (96 per cent)
Canada balsam

To siliconize coverslip, dip coverslip quickly into silicon fluid.

Permanent Preparation
(1) Dissect out the salivary glands as for the temporary preparation.

(2) Place a small drop of acetic orcein onto a thick **siliconized coverslip** and transfer the glands to it. Leave for 15–20 minutes.

(3) At the end of the 15–20 minute period, remove most of the stain from the coverslip with the edge of a piece of filter paper so the glands stick to the coverslip.

(4) Cover a slide with a thin layer of albumen by placing a very small drop of albumen in the centre of the slide and rubbing it all over the surface with a *clean* finger.

(5) Place a drop of stain onto the **albuminized slide** and invert your coverslip, plus the glands, onto it.

(6) Press down on the coverslip and tap it. Remove surplus stain with filter paper.

(7) Dip your slide in a jar of ethanol (96 per cent) and wait until the coverslip falls off. The stained squashed chrosomes should remain spread out on the slide.

(8) Remove the slide from the ethanol, allow it to dry, and then mount it in Canada balsam under a thin coverslip. Leave on hotplate to dry.

For Consideration
(1) Why is it necessary in the making of a permanent preparation for the coverslip to be siliconized and the slide albuminized?

(2) Can you suggest why the chromosomes in the salivary glands of *Drosophila* larvae should be abnormally large?

(3) How would you investigate what the bands in the giant chromosomes represent?

Questions and Problems

1 In *Drosophila* red eye is dominant over brown eye and yellow body is dominant over dark body. In an experiment a red eyed, yellow-bodied fly is crossed to a brown eyed, dark-bodied fly. The F_1 flies are all red eyed with yellow bodies. Two of the F_1 flies are crossed with the following results in the F_2:

Red eyed, yellow body	126
Brown eyed, dark body	39
Red eyed, dark body	5
Brown eyed, yellow body	3

Offer a full explanation of these results. (*O and C*)

2 In the fruit fly *Drosophila* normal antennae and grey body are linked and are dominant to twisted antennae and black body.

When a normal fly was crossed with one carrying the recessive alleles the offspring were all of the normal type. One of these latter was then crossed with a fly homozygous for the recessive alleles and the following numbers of offspring were obtained:

Normal antennae and grey body	90
Normal antennae and black body	9
Twisted antennae and grey body	11
Twisted antennae and black body	86

Explain this result and make an annotated diagram of the bivalent which produced the recombinant classes. (*AEB*)

3 Explain the term 'cross-over frequency'. In a certain organism, genes **A**, **B**, **C**, and **D** were studied and it was found that the cross-over frequency between **A** and **B** was 20 per cent, **A** and **C** 5 per cent, **B** and **D** 5 per cent, and **C** and **D** 30 per cent. What is the probable sequence of these genes on the chromosome and what would you expect the cross-over frequency between genes **A** and **D** to be? Discuss the significance of crossing over. (*O and C*)

4 The meiotic products in many ascomycetes form a linear tetrad and, in the majority, meiosis is followed by a mitotic division. The mitotic products also remain in the positions in which they were formed. Consequently, eight spores in single file are eventually produced in each ascus.

Fig. 29.2 (p. 329) shows the types of asci produced by a hybrid between a white spored and black spored strain of the ascomycete *Sordaria fimicola*. How would you account for the various spore arrangements? What would you expect the frequency of each type to be? (*O and C modified*)

5 Using some means to distinguish maternal and paternal chromosomes, draw two bivalents each with one chiasma at a stage prior to the disruption of the nuclear membrane (late prophase). Locate one pair of alleles (**A**/**a**) on one of the bivalents in such a way that the chiasma lies between the gene and the centromere. Locate a second pair of alleles (**B**/**b**) on the other bivalent in the same way. Assuming that these are two bivalents in the same cell, illustrate the various ways in which they can behave subsequently and show clearly the spore tetrad you expect from each type of behaviour. What products would you expect if the chiasma in each bivalent had formed outside the segment between the gene and the centromere? (*O and C*)

6 **A**, **a** and **B**, **b** are two pairs of *alleles* located in different *linkage groups*, and showing *independent assortment*. They are, therefore, due to genes located on different chromosomes. In fact the genes for **A**, **a** are on a pair of *autosomes*, and those for **B**, **b** on the *sex-chromosomes*. The male is the *heterogametic sex*, and the **Y** chromosome is genetically empty.

A female *homozygous* for the *dominant* alleles, **A** and **B**, is mated to a male which *phenotypically* shows both in the recessive condition:
(a) What is his *genotypic constitution* in respect of these factors, and what will be the constitution of the F$_1$ generation?
(b) Explain the meanings of the words in italics. (*O and C*)

7 Light Sussex fowls have mostly white plumage, and Rhode Island fowls have mostly red. The character white-feathered (**R**) is dominant to the character red (**r**). Explain why, on crossing Rhode Island cockerals with Light Sussex hens, all male offspring have white plumage and all females red. (In birds, males are the homogametic sex (**XX**), females are heterogametic (**XY**).) (*O and C*)

8 In the following human pedigree circles indicate females, squares males. Filled-in circles and squares indicate red–green colour-blind individuals; open circles and squares indicate normal vision:

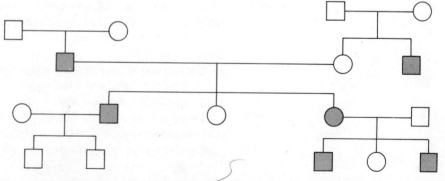

State, as far as possible, the genotypes of each individual in the pedigree. What conclusions can you draw as to the inheritance of colour-blindness?

9 Imagine you are a genetic counsellor. You are visited by a boy called John and his sister Susan. Both are normal, as are their parents, but they have a haemophiliac brother. John and Susan are worried that their children, when they have them, might turn out to be haemophiliacs. What would your advice be to each?

10 A farmer had two sons. The first, born when he was young, grew into a handsome, healthy youth in whom he took great pride. The second, born much later, was always a sickly child, and neighbours' 'talk' induced the farmer to bring his wife to court disputing its paternity. The grounds of the dispute were that the farmer, having produced so fit a first son, would not be the father of the weakling. The blood groups were:

	Group	Type
Farmer	O	M
Mother	AB	N
First son	A	N
Second son	B	MN

What advice would you give the court[1]?

11 After eight years of married life, during which time she had failed to become pregnant, Mrs X met and fell in love with Mr Y. During the ensuing five years three children were born. In the meantime the persons involved had tried to come to an understanding and wished to determine which of the two men was the father of each child. The blood types of those involved were determined, with the following results:

	Group	Type	
Mr X	O	MN	
Mrs X	O	MN	
Mr Y	A	N	
First child	O	MN	
Second child	O	M	Mr X
Third child	A	N	Mr Y

What do you conclude as to the paternity of each child? Give your reasoning. (Case history from A. S. Wiener, *Blood Groups and Transfusion*, Thomas, 1943)

12 A mother in a maternity ward claimed that the baby (a) allocated to her was not her own. There was only one other baby (b) in the ward at the time. The mother is blood group O and MN and cannot taste phenylthiocarbamide (PTC). Baby (a) is group A, M and can taste PTC. Baby (b) is O, MN and cannot taste PTC. The husband is dead, but she has three other children:
  (i) A MN taster
  (ii) B N taster
  (iii) A MN non-taster.
Which baby do you think belongs to her? Give your reasoning.

13 In man there is a gene which, when homozygous, causes a condition called sickle-cell anaemia, so named because the red blood cells are abnormal and assume a sickle shape. Death usually occurs before adulthood. Heterozygous persons generally appear normal, but under low-oxygen concentrations their red blood cells may also become sickle shaped.

A young woman, about to be married, had a brother who died of sickle-cell anaemia. Worried about the possible prospect of the condition occurring in her own children, she seeks medical advice. When samples of her blood are taken and placed in a low oxygen concentration, her red cells become sickled. However, her prospective husband's blood remains normal. What could you tell her about her children? What can you say about her parents?

[1] The blood groups A, B, AB and O are controlled by three alleles, A, B, and O, any two of which may be combined together in a heterozygous individual. The blood types M, N and MN are controlled by two alleles, M and N, the heterozygote being MN.

14 Thalassaemia is a type of human anaemia rather common in the Mediterranean population, but relatively rare in other populations. The disease occurs in two forms, minor and major; the latter is much more severe, in fact usually lethal. The severely affected individuals are homozygous for a defective dominant gene whereas the mildly affected are heterozygous for the same gene. Persons free of the anaemia are homozygous for the normal allele. It can be assumed that the disease is inherited in a normal Mendelian manner.

(a) Letting **T** be the allele for thalassaemia, and **t** the normal allele: a man with minor thalassaemia marries a normal woman. What types of children with respect to thalassaemia may they expect and in what proportions? Diagram the germ cell unions producing children in this marriage.

(b) In another family both father and mother have thalassaemia minor. What are the chances that their children will be affected? Diagram the unions as before.

(c) A child has thalassaemia major. From the information given so far, what might you expect the genotypes of the parents to be? Diagram the possible genotypes and germ cells as before.   (*OCJE*)

15 A recessive mutant of the gene which is responsible for chlorophyll synthesis in the tomato plant causes the plant to be colourless when present in the homozygous condition. Such a plant dies as a seedling after it has used up its supplies of food. In the heterozygous state the mutant produces a pale plant but one that does survive.

A normal green tomato plant was crossed with a pale heterozygote and the seeds formed from the cross were collected. These seeds were subsequently planted and a number of tomato plants reared which were self-pollinated. Once again the seeds were collected and a further generation of plants grown.

The ratio of normal green to pale plants in this generation was found to be 5:2. How can these results be explained?   (*AEB*)

16 In shorthorn cattle the coat colour can be red (gene **W**), white (gene **w**), or in the heterozygous condition, roan. In addition, the polled (without horns) condition is dominant to the horned condition. What will the genotypes and phenotypes of the F_1 and F_2 generations be if homozygous polled red cattle are crossed with white horned cattle? How would you establish a pure breeding strain of red polled shorthorns from your F_2 generation?   (*O and C*)

17 A broad leaved, red flowered snapdragon was crossed with a narrow leaved, white-flowered plant and the offspring were all broad leaved and had pink flowers. One of these F_1 plants was crossed with a broad leaved, white-flowered plant of unknown breeding behaviour, and the following offspring were obtained:

   100 broad leaved, pink
    95 broad leaved, white
    35 narrow leaved, pink
    30 narrow leaved, white.

How would you explain these results?   (*CCJE*)

18 In poultry the genes for rose comb (**R**) and pea comb (**P**), if present together, produce walnut comb. The recessive alleles of both, when present together in the homozygous condition, produce single-comb. What are the possible results that might be obtained by crossing a rose-combed fowl with pea comb? Explain your reasoning in full. What general principle is demonstrated here?

19 Explain why brothers never look exactly alike. Why are 'identical twins' so similar and 'non-identical twins' dissimilar? Would you expect 'identical twins' to be *exactly* alike?

20 Four plants **A, B, C,** and **D** each true breeding for white flowers were crossed in all possible ways. The phenotypes of the progenies are given below:

	A	B	C	D
A	W	W	M	M
B	W	W	M	R
C	M	M	W	W
D	M	R	W	W

W = white flowers
R = red flowers
M = mauve flowers

Explain these results and suggest further breeding experiments to test your hypothesis. Write down all the steps of your argument. The way you argue is just as important as a correct solution. (*CCJE*)

21 The following data (*from* Sinnott, Dunn, and Dobzhansky) show the results of a large number of matings carried out with a yellow variety of house mouse:

Parents	Offspring	
	Yellow	Non-yellow
yellow × non-yellow	2,378	2,398
yellow × yellow	2,396	1,235

How would you explain these data?

22 Summarize the evidence—direct and indirect, ancient and modern—which leads to the conclusion that genes are located on the chromosomes.

30 The Nature of the Gene

Background Summary

1   Experiments first carried out by Avery on *Pneumococcus*, and since repeated on other bacteria, point to **nucleic acid** as being the carrier of genetic information.

2   Two principal types of nucleic acid exist in cells: **deoxyribonucleic acid (DNA)** and **ribonucleic acid (RNA)**. Both consist of chains of **nucleotides**, each nucleotide consisting of a **pentose sugar, phosphate group** and one of five **organic bases** (**A, G, C, T**, and **U**). DNA contains **A, G, C**, and **T**; in RNA **T** is replaced by **U**.

3   According to the **Watson-Crick hypothesis**, based on X-ray crystallography and chemical analysis, DNA is a double helix consisting of two coiled chains of alternating phosphate and sugar groups, the latter being interconnected by pairs of bases linked in a specific way: **A** with **T** and **C** with **G**.

4   Evidence suggests that DNA undergoes accurate **replication**, the mechanism being **semi-conservative**. In the intact cell this occurs during interphase, prior to cell division.

5   There is good reason to believe that DNA exerts its influence by controlling **protein synthesis**. A **gene** can be looked upon as a segment of the DNA chain; in general a single gene is responsible for the synthesis of one protein, or part of it (the **one gene–one protein hypothesis**).

6   Protein synthesis takes place on the **ribosomes** in the cytoplasm. DNA, confined to the nucleus, controls protein synthesis by determining the order in which amino acids are linked together on the ribosomes. Each amino acid is coded for by a triplet of bases in the DNA (**codon**). The sequence of amino acids in the protein is determined by the sequence of base triplets in the DNA.

7   In controlling protein synthesis the DNA is first transcribed into **messenger RNA** which is then, through the intermediacy of **transfer RNA**, translated into protein structure.

8   Ribosomes occur in groups or chains (**polyribosomes, polysomes**) and it is thought that during protein synthesis they move in convoy along the messenger RNA strand, thereby speeding up the assembly of amino acids.

9   It has been found that more than one triplet of bases can code for a single amino acid. The code is, therefore, **degenerate**. Triplets corresponding to all the known amino acids have now been characterized. Other triplets are involved in stopping and starting the message.

10   The potency of DNA in controlling protein synthesis can be seen in viruses, notably bacteriophage, whose nucleic acid has been shown to take over the metabolic machinery of the host cell.

Investigation 30.1

Making models of DNA, RNA, and protein synthesis

Requirements
Eight different colours of
  plasticine (modelling clay)
Match sticks

Using match sticks and plasticine (modelling clay) of different colours, construct flat models (i.e. models which lie flat on a table) illustrating the following:

Procedure

(1) **The molecular structure of DNA**. Show the relationship between the sugar, phosphate and organic bases, the molecular basis of the helical configuration and the complimentary relationship between the bases.

(2) **How DNA replicates**. Build a short length of DNA, made up of, say, five pairs of nucleotides. Construct ten further nucleotides of the same type as those in your DNA molecule. Now make your DNA replicate.

(3) **Formation of messenger RNA**. Show how DNA is transcribed into messenger RNA, one of the two strands of the DNA serving as the template for the synthesis of the RNA.

(4) **How messenger RNA controls the assembly of a protein**. Make models of amino acids, transfer RNA molecules and a short length of messenger RNA. Show how the sequence of bases in the messenger RNA is translated into protein structure.

(5) **The action of polyribosomes**. Make models of two ribosomes (more if you wish). Move the ribosomes in convoy along a strand of messenger RNA and show how each one results in the formation of a polypeptide chain.

For Consideration
What is the evidence for each of the structures and processes which you have represented as a model?

Investigation 30.2

Distribution of DNA and RNA in root-tip cells

Requirements
Microscope
Oven at 60°C
Slides and coverslips
Watch glasses
Razor blade
Filter paper

M HCl (or 50 per cent HCl)
Feulgen solution
Acetocarmine
Methyl green–pyronin stain
Distilled water
Ethanol: 70, 90 per cent and
  absolute
Xylene
Canada balsam

Bean roots fixed in acetic ethanol
Bean roots fixed in absolute ethanol

One piece of evidence supporting the DNA–RNA theory is that RNA is found in the cytoplasm as well as in the nucleus, but DNA is confined to the nucleus. In this experiment the distribution of DNA and RNA in undifferentiated cells will be investigated by staining with chemicals specific for each nucleic acid. Two techniques will be tried: the **Feulgen technique** stains only the DNA; the **methyl green–pyronin technique**, however, stains the RNA and DNA with different colours. Root-tip cells of bean provide suitable material.

Testing for DNA with Feulgen Stain
The technique involves hydrolysing the DNA with acid. This liberates aldehydes which restores the red colour of bleached Feulgen solution. The roots should be fixed in acetic ethanol for two hours before you begin.

(1) Transfer the fixed material to *either* M HCl at 60°C for six minutes in an oven, *or* 50 per cent HCl at room temperature for fifteen minutes. This treatment hydrolyses the DNA and macerates the tissue.

(2) Transfer the root to a watch glass of colourless Feulgen solution for one to two hours. If time is short the

reaction can be accelerated by placing the watch glass on a warm surface.

(3) Cut off the terminal 3 mm of the root and transfer it to a microscope slide, discarding the rest. Add acetocarmine which intensifies the Feulgen stain.

(4) Tease out the stained root tip and put on a coverslip. Place a piece of filter paper on the coverslip and press gently so as to spread out the tissue and soak up surplus stain.

(5) Examine under low and high power. What is the distribution of DNA in the cells?

Testing for DNA and RNA with Methyl Green–Pyronin Stain
The stain is a mixture of methyl green and pyronin. DNA takes up the methyl green, and RNA the pyronin. This enables the two types of nucleic acid to be distinguished, at least in good preparations. The roots should be fixed in absolute ethanol for 30 minutes before you begin.

(1) With a clean sharp razor blade cut thin longitudinal sections of the terminal 3 mm of a root. Place the sections on a slide and cover with aqueous methyl green–pyronin stain for 30 minutes.

To obtain roots

Germinate broad bean seeds 10 days previously. When radicle is 1.5 cm long, cut off tip to stimulate growth of lateral roots. Fix lateral roots in acetic ethanol for Feulgen, and absolute ethanol for methyl green–pyronin.

Feulgen's reagent prepared as follows:

Dissolve 1.0 g basic fuchsin in 200 mm³ boiling distilled water. Filter. Add 30 mm³ M hydrochloric acid and 3.0 g potassium metabisulphite to the filtrate. Allow to bleach for 24 h in the dark. If solution is still coloured the residual colour should be adsorbed on carbon. Filter. Store in tightly stoppered bottle in dark.

Methyl green–pyronin stain prepared as follows:

Methyl green 0.15, pyronin 0.25, ethanol 2.5, glycerine 20 vols. Make up to 100 vols. with 0.5 per cent carbolic acid.

(2) Draw off the stain with a pipette and replace with distilled water. Change the water several times so as to thoroughly wash the sections.

(3) Mount the sections in distilled water and view under low and high powers. DNA should be stained blue-green, RNA red.

If your preparation is successful and you wish to make it permanent, continue as follows:

(4) Remove the coverslip and cover the sections with successively 70 per cent, 90 per cent and two changes of absolute ethanol, spending one to two minutes in each. Do not spend too long in the ethanol or you will wash out the red stain.

(5) Cover with xylene for one minute.

(6) Mount in Canada balsam.

For Consideration

From your observations in this investigation what predictions can you make about the way the nucleus communicates with the cytoplasm in controlling the development of the cell?

Investigation 30.3
Identification of bases in DNA and RNA

DNA consists of two strands of alternating sugar and phosphate groups linked by pairs of organic bases. Two of the bases, **adenine** and **guanine**, are **purines**; the other two, **thymine** and **cytosine**, are **pyrimidines**. What evidence suggests that from the genetic point of view the bases are the most important components of the molecule?

To identify the four bases they must first be released from the DNA molecule. This can be done by hydrolysing DNA with a reagent which splits it into its chemical subunits. The bases in the resulting mixture can then be separated by **paper chromatography**.

Procedure

HYDROLYSIS OF THE DNA

(1) Weigh out 5 mg of DNA. Dissolve it in 0.4 cm³ of distilled water and 0.1 cm³ of 6 M hydrochloric acid or 72 per cent perchloric acid in a small test tube.

(2) Maintain in a water bath at 100°C for 1 hour.

(3) Cool by placing the test tube in a beaker of ice chippings. Add a few drops of 50 per cent potassium hydroxide solution to make the contents of the tube less acidic.

(4) If necessary centrifuge so as to throw down the insoluble perchlorates. The supernatent liquid (hydrolysate) contains the soluble products of hydrolysis of the DNA, including its four bases.

SEPARATING THE BASES OF DNA

Separation of the bases should be carried out in a chromatography tank, but a gas jar or large beaker will do. The tank should have a cover and some means of suspending the chromatography paper.

Proceed as follows:

(1) Cut a strip of chromatography paper of the same length as the height of the chromatography tank. Draw a pencil line across one side of the strip about 3 cm from the bottom. In the centre of the line make a pencil dot. With a fine capillary tube, apply a tiny drop of the hydrolysate to the dot. Allow the drop to dry, and then apply another drop. Repeat this process of loading with hydrolysate at least 20 times. The smaller and more concentrated the spot, the better.

(2) If it has not already been done for you, pour the solvent (**propan-2-ol–hydrochloric acid**) into the chromatography tank to a depth of 3 cm. Suspend your strip of chromatography paper in the tank so the bottom edge of the paper dips into the solvent to a depth of about 1 cm. Cover the top of the tank and leave for not less than eight hours and not more than 24.

Requirements

Small test tubes
Water bath at 100°C
Small beaker of ice chippings
(Centrifuge)
Dropping pipette
Fine capillary tube
Chromatography tank with cover
 (gas jar or 2–3 dm³ beaker will do)
Chromatography paper (Whatman
 No. 1 in 5-cm width roll)
Ultra-violet lamp emitting light
 of wavelength 260 mμ
(Hair drier, or oven at 100°C)
DNA (5 mg per student)*
72 per cent perchloric acid or 6 M HCl
 (0.1 cm³ per student)
Potassium hydroxide solution
Propan-2-ol–hydrochloric
 acid (solvent)

For two-way chromatography
 As above plus:
 Chromatography paper in 24 cm width
 roll
 Butan-1-ol-ammonia (second solvent)

If standard solution of bases are required :

Adenine Guanine	dissolve 0.25 g in 5 cm³ of 6M HCl, warming if necessary
Thymine Cytosine Uracil	dissolve 0.5 g in 5 cm³ of 6M HCl, warming if necessary

If RNA is to be analysed
 RNA (0.2 g per student)
 6M HCl

*Solvents made up as follows (volumes):
Propan-2-ol–hydrochloric acid: propan-2-ol
(170), conc. HCl (41), distilled water (39). Butan-
1-ol-ammonia: butan-1-ol (12), ammonia (3),
distilled water (5).

---

(3) *After 8–24 hours*: Remove your strip of chromatography paper. Draw a line across the strip to mark the highest point reached by the solvent. Dry the strip at room temperature. A heater can be used to accelerate evaporation of the solvent. If you are short of time the strip can be dried by putting it in an oven at 100°C for ten minutes. You have now completed your **chromatogram**.

IDENTIFYING THE BASES
The spots on your chromatogram are invisible in ordinary light, but they become visible if viewed in ultra-violet light of wavelength 260 mμ.

Light of wavelengths less than 310 mμ can damage the eyes: when viewing your chromatogram wear safety goggles or hold a glass plate in front of your eyes.

Mark the position of the bases by drawing a ring round each spot. The lowest base to appear in the chromatogram is purple (guanine); the other spots are brown or white.

If the spots are too close for identification, replace the strip in the chromatography tank and run for a further six to eight hours.

Assuming that they have separated sufficiently, work out the R_f value for each base:

$$R_f = \frac{\text{distance moved by spot}}{\text{distance moved by solvent}}$$

Compare your figures with the following published R_f values:

Purines
Adenine   0.32
Guanine   0.22 (purple spot)
Pyrimidines
Thymine   0.76
Cytosine   0.44
(Uracil   0.66)

Although it is not a component of DNA, uracil is included in the above list because, during the hydrolysis of DNA, some of the cytosine is usually degraded to uracil.

Further Experiments

A MORE THOROUGH SEPARATION
OF THE BASES IN DNA
A snag that may have been encountered is that the bases have not completely separated, the spots touching or merging together.

This difficulty can be overcome by means of **two-way chromatography**. In this case a concentrated spot of DNA hydrolysate is placed near the bottom right-hand corner of a *sheet* of chromatography paper. The paper is suspended for eight hours in a tank containing the solvent **propan-2-ol–hydrochloric acid**. The paper is then dried and rotated through 90° so the spots which have been separated by the solvent lie along the bottom. Thus orientated, the paper is suspended in a second solvent, **butan-1-ol-ammonia**. The two solvents in succession, working at right angles to one another, achieve a more complete separation of the bases than is possible with a single solvent.

COMPARISON OF BASES IN DNA
WITH STANDARD BASES
For identification of the bases in DNA it is obviously helpful if your chromatogram contains spots corresponding to standard solution of the bases.

This can be achieved as follows. Along the bottom of a sheet of chromatography paper, spaced evenly, place concentrated spots of the DNA hydrolysate, and standard solutions of adenine, guanine, thymine, cytosine and uracil. On viewing the chromatogram, spots of the standard solutions of the bases should occur at the same level as corresponding bases in the hydrolysate.

SEPARATION OF THE BASES IN RNA
RNA is similar to DNA in its component bases except that thymine is replaced by uracil. If you wish to separate the bases of RNA you can hydrolyse RNA by heating it with 6M HCl in a boiling water bath for 15 minutes. After hydrolysis proceed as for DNA.

For Consideration

(1) Why do you think the bases are visible in ultra-violet light but not in ordinary light?
(2) What are the other chemical constituents of DNA besides the bases, and why don't they appear in your chromatogram?
(3) What information relating to the genetic function of DNA might be provided by this experiment, or an extension of it?
(4) Imagine nothing is known about the chemical sub-units of the DNA molecule. How could the techniques used in this investigation be extended to include a chemical analysis of the bases?

Questions and Problems

1 Trace, as concisely as you can, what happens to the DNA in a piece of meat from the moment it enters a person's mouth to the moment it becomes incorporated into his or her genes.

2 Briefly, and without resorting to chemical formulae, explain the structure of DNA. How does it differ from RNA?

The following is the sequence of bases in one of the two strands of part of a DNA molecule:

CAGGTACTG

What will be the sequence of bases in the other strand?
What evidence supports the sequence you suggest?

3 Bacterial cells were fed with labelled nitrogen-containing food. On division the DNA strands of the daughter cells were found to contain labelled nitrogen. These daughter cells were then fed on normal food and when they, in turn, divided, it was found that the cells produced contained DNA in which only half the strand contained labelled nitrogen atoms. How can this be explained?

(AEB modified)

4 Let the amount of DNA present in the nucleus of a cell at the beginning of interphase be x. What will be the amount present at (a) the end of interphase, (b) the end of mitosis, (c) the end of a first meiotic division, and (d) the end of a second meiotic division? Give reasons for your answers.

5 Francis Crick put forward the hypothesis that each amino acid is coded for by a triplet of bases in DNA. Why a triplet?

6 The following sequence of bases in DNA codes for the formation of a polypeptide consisting of ten amino acids:

GTTAACCGAACGGTTAGATGTACATTTAAG

Give the initial letters of the messenger RNA bases responsible for transcribing the above sequence. What will be the sequence of amino acids in the resulting polypeptide?

7 Briefly explain how it is believed that the DNA in the nucleus of a cell controls the synthesis of proteins in the cytoplasm. Mention some of the more important evidence supporting your suggestions.

8 What do you understand by the one gene–one enzyme hypothesis? What evidence supports it? This hypothesis was first postulated in the 1940s. Do you consider that it should be modified in the light of more recent research?

9 In the bread mould *Neurospora crassa*, a biosynthetic pathway can be interrupted by three mutations. Mutant **X** will grow if provided with cystathionine, homocystine, or cysteine; mutant **R** will grow only on homocysteine but accumulates cystathionine; mutant **W** will grow if provided with homocysteine or cystathionine, but not on cysteine.

Sketch the sequence of biosynthesis of these three chemicals and indicate the positions of the metabolic blocks imposed by mutants **X**, **R**, and **W**.

What do you predict would be the result of crossing the mutants in pairs **X** with **R**, **X** with **W**, and **W** with **R**, and what conclusion would you draw if your predictions turned out to be correct?

(SA modified)

10   The following metabolic pathway was discovered by A. E. Garrod:

dietary and/or tissue protein
↓
tyrosine
↓
hydroxyphenylpyruvic acid
↓
homogentisic acid
↓
fumaric and acetoacetic acids
↓
Krebs' citric acid cycle

This pathway ensures that in normal circumstances the amino acid tyrosine, derived from excess protein, is disposed of metabolically. However, in certain individuals large quantities of homogentisic acid are excreted in the urine, which consequently turns black on standing. This condition is known as alkaptonuria and Garrod showed that, though rare, it is familial, i.e. it runs in families. Suggest explanations of this condition in terms of modern genetic theory.

11   What experiments would you perform to test whether DNA is the heriditary material of organisms?   (*CCJE*)

12   'The relation between DNA and protein is rather like that between the hen and the egg in the well-known question of which came first.' Discuss this statement.
                                                                                      (*CCJE*)

13   In what ways has research on (a) bacterial genetics, (b) nutrition of Fungi, and (c) reproduction of viruses helped scientists to elucidate the genetic code?

14   'A gene performs its function of influencing the phenotype of an organism by controlling the production of an enzyme.' Discuss this statement.   (*O and C*)

31 Genes and Development

Background Summary

1 Development involves a highly ordered sequence of events, elaborately controlled in space and time.

2 Experiments on unicellular organisms indicate that the nucleus is not only necessary for development to proceed, but that it also determines what sort of structure the cell develops into.

3 It is also apparent that the cytoplasm can play an important part in controlling development. In some multicellular organisms the fate of a particular cell is determined by which part of the original egg cytoplasm it comes to contain.

4 In chordate embryos the origin of certain structures can be traced back to specific **presumptive areas** in the blastula.

5 Although the destiny of most tissues is determined as early as the blastula stage, some tissues show considerable plasticity in what they can develop into. The fate of such tissues is determined by the inductive influence of neighbouring tissues (**organizers**).

6 An example of a powerful organizer is **chorda-mesoderm** whose inductive capacity has been demonstrated by transplantation experiments in amphibian embryos.

7 Development is controlled by the action of a *sequence* of organizers. This can be seen in the development of the vertebrate eye which has been analysed in some detail, but little is known about the chemical nature of the organizers involved.

8 Returning to the role of the nucleus, does the DNA content of the nucleus of a particular cell change during the course of development, or does it remain unaltered? Evidence, principally from nuclear-transfer experiments, suggests that the latter is the case.

9 If this is so, how does cell differentiation take place? Evidence suggests that different parts of the genetic code operate at different times during development. It would seem that at any given moment a proportion of the genes are in some way masked or 'switched off'. This concept can be illustrated by the formation of different types of haemoglobin during development and is supported by work on the synthesis of sugar-splitting enzymes in the colon bacillus *Escherichia coli*.

10 A variety of external environmental influences can affect development. Generally these tend to be cruder, more all-or-nothing, than the subtler internal influences.

Investigation 31.1

Investigation into the organizing ability of the dorsal lip of the amphibian gastrula

Requirements

Operating instruments all of which must be sterilized by flaming or ethanol before use:

Fine forceps (2 pairs)
Glass needle (*see* Fig. 31.1)
Needle knife (*see* Fig. 31.1)
Hair loop (*see* Fig. 31.1)
Sterile pipette (*see* Fig. 31.5)
Binocular microscope
Operating dish (solid watchglass containing 2 per cent agar)
Culture dish (5 cm petri dish containing 2 per cent agar)

Niu-Twitty solution (*see* below)

Early gastrulae of *Xenopus*

Niu-Twitty solution made up as follows:
A
3.4 g sodium chloride
50 mg potassium chloride
80 mg hydrated calcium nitrate
100 mg manganese sulphate
500 cm³ distilled water

B
110 mg sodium hydrogen phosphate
20 mg potassium hydrogen phosphate
250 cm³ distilled water

C
200 mg sodium hydrogen carbonate
250 cm³ distilled water

Dissolve **A**, **B**, and **C** in warm water, then mix. To prevent bacterial growth add 1.0 g sulphadiazene sodium per 500 cm³.

During development certain cells exert a powerful influence over other cells. This was originally demonstrated by the classical experiments of Spemann and Mangold, who demonstrated that in amphibians a secondary embryo could be produced by transplanting a piece of dorsal lip of the blastopore from one embryo to the ventral surface of another. You can test this for yourself by a slightly different method. It has been found that secondary embryonic tissues can be produced by implanting a small piece of the dorsal lip into the blastocoel of an early gastrula. If the implantation is carried out on the dorsal side of the gastrula, subsequent movements of the endodermal cells during gastrulation will carry the implant round to the ventral side where, lodged between the ectoderm and endoderm it should, with luck, induce the formation of a secondary embryo.

Equipment

This experiment involves carrying out an operation under a binocular microscope. Micro-dissection instruments should be used: **fine forceps** for decapsulation of embryo, a **hair loop** for holding the embryo in position, and a **needle-knife** and **glass needle** for embryonic surgery. These are illustrated in Fig. 31.1.

You are also provided with **operating dishes** (solid watchglasses containing agar), and **culture dishes** (petri dishes containing agar). The purpose of the agar is to provide a soft bed for the embryo. **Niu-Twitty solution** is also provided: this has been found to be an ideal medium for amphibian development.

Precaution

As you will see below, the first step in the experiment is to decapsulate the gastrulae. Once they are decapsulated be careful to ensure that in their dishes and pipettes the *naked embryos do not come into contact with the air–liquid interface*. If they do they will burst.

Procedure

(1) Decapsulate two gastrulae using fine forceps. Do this in a small petri dish containing Niu-Twitty solution. First remove the jelly coat (quite easy) and then remove the vitelline membrane (difficult).

(2) Transfer the gastrulae with a sterile pipette to an operating dish containing Niu-Twitty solution.

(3) Excise a small piece of the **dorsal lip** of the blastopore from one of the gastrulae (the **donor**). Use the hair loop to hold the gastrula steady and use a needle-knife for cutting. Do not apply too much pressure or you will burst the embryo.

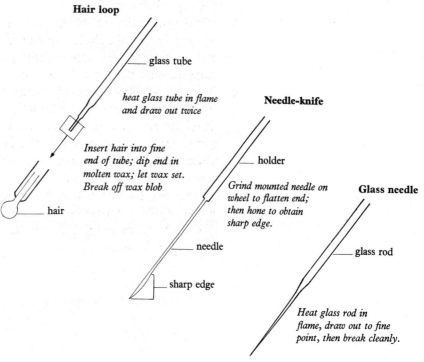

Fig. 31.1   Micro-dissection instruments for embryonic surgery.

(4) Make a slit in the roof of the blastocoel of the other gastrula (the **recipient**) and push in the piece of dorsal lip using the glass needle as shown in Fig. 31.2.

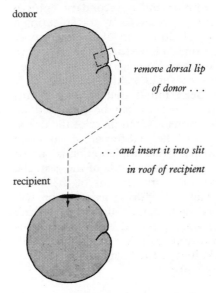

donor

remove dorsal lip

of donor . . .

. . . and insert it into slit

in roof of recipient

recipient

Fig. 31.2 Procedure for transplanting the dorsal lip of amphibian gastrulae.

(5) Leave the recipient for 30–60 minutes in the operating dish for healing, and then transfer it to a separate culture dish containing Niu-Twitty solution. Label the lid with a wax pencil.

(6) Set up a control: dacapsulate a gastrula, make a slit in the roof of the blastocoel but do not implant anything into it. Treat as (5).

(7) Observe at regular intervals during the next week. If your embryos survive make comparative drawings of the recipient and control.

For Consideration

(1) Build up an hypothesis to explain the mechanism by which the dorsal lip causes the production of a complete secondary embryo on the ventral side of the recipient.

(2) Why do you think the recipient does not reject the dorsal lip of the donor by an immune response?

Investigation 31.2

Experiments to test the self-differentiating capacity of embryonic structures

The **allanto-chorionic membrane** (ACM) of the chick is a vascular sheet which will support the life of cells placed in contact with it. Viruses, bacteria, and adult or embryonic cells of mammals and birds will multiply, grow and differentiate after being grafted onto it. In this experiment we will graft structures from one chick embryo (the donor) onto the ACM of another (the recipient) to investigate the extent to which a structure will continue to differentiate after it has been isolated from the rest of the embryo.

Precautions

Technically allanto-chorionic grafting is not unduly difficult given a reasonably steady hand and some patience. Failure is most often due to infection leading to death of the donor tissue. Minimize the risk of infection by taking the following precautions:

(1) Wear a sterile face mask throughout the experiment.

(2) Touch living materials *only* with sterile instruments (*see* p. 432). Forceps and scissors must be either flamed or dipped in ethanol before use. Pipettes must be kept in a sterile condition and should be returned to a sterile test tube when not in use.

(3) Work as quickly as possible. Decide what you are going to do and then do it as quickly as you can.

Procedure

You will be given two fertilized eggs, one of five days incubation and the other of about twelve days incubation. The former is the **donor egg** (marked with a **D**) and the latter is the **recipient** (marked **R**). Prepare the donor first, and then the recipient, as follows:

PREPARATION OF DONOR

(1) Swab the shell with ethanol over its whole surface and allow it to dry.

(2) Open the shell with sterile forceps and scissors. Using your sterile pipette, remove the embryo to a large petri dish containing sterile saline. The embryo can be separated from the yolk sac by sucking it into the pipette, but it may be necessary to cut the umbilical cord with scissors.

(3) Dissect out one or more of the following structures for grafting: eye, limb bud, tail, wing bud (Fig. 31.3).

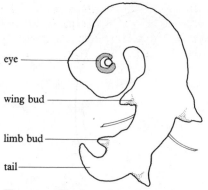

Fig. 31.3 Diagram of five-day chick embryo to show four of the organs which can be grafted to the allanto-chorion of a recipient.

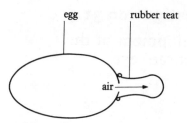

Fig. 31.4 Technique for getting the chorio-allantoic membrane to fall away from the shell in a hen's egg.

Requirements

Incubator at 37°C
Sterile face mask
Sterile dissecting instruments
Wide-mouthed sterile pipette (*see* Fig. 31.5)
Cotton wool
Small petri dish
Hack-saw blade
Plasticine (modelling clay)
Rubber teat of pipette
Sellotape (Scotch tape)
Tray

Ethanol for sterilization
Chick saline (*see* below)

Fertilized hen's eggs (*see* p. 432)
5-day marked **D**
12-day marked **R**

Chick saline (Locke's solution) made up as follows:
sodium chloride (0.9 g)
potassium chloride (0.4 g)
anhydrous calcium chloride (0.024 g)
sodium hydrogen carbonate (0.02 g)
distilled water (100 cm³)
Sterilize by boiling, then seal.

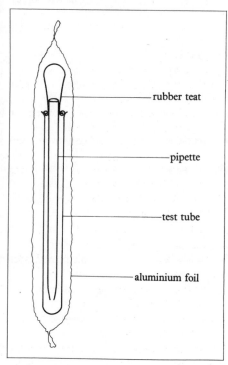

Fig. 31.5 Sterile pipette for use in experimental embryology. The whole assemblage should be autoclaved before use.

(4) Transfer the isolated structure to a small petri dish of sterile saline. Put the lid on and leave until the recipient is ready.

PREPARATION OF RECIPIENT
(1) Make a hole through the shell into the air space at the blunt end of the egg. Do this by sawing through the hard part of the shell with a hack-saw blade and then piercing the soft shell membrane with a needle. Be careful not to pierce too deep.

(2) Lay the egg on its side in a cradle of plasticine. Swab the upper surface lightly with ethanol.

(3) Using a hack-saw blade, etch a square of 1 cm side on the upper surface of the shell. Saw only through the hard part of the shell; do *not* penetrate the shell membranes.

(4) At one point of exposure of the shell membranes pierce them carefully with the tip of a sterile scalpel. Be careful not to damage the underlying ACM which is immediately beneath.

(5) The object of this next step is to get the allanto-chorion to fall away from the shell beneath your etched square. To do this suck air out of the air space by placing the opening of a large deflated pipette teat against the hole at the blunt end of the egg, and then allowing it to inflate (Fig. 31.4). Do this several times. With luck the suction should cause the allanto-chorion to fall away from the shell, creating a new air space beneath the etched square.

(6) Remove the etched square by cutting the shell membranes with a sterile scalpel. Place the piece of shell in a covered petri dish with the outer surface on the sticky side of a piece of sellotape. Orientate the piece of shell for subsequent replacement.

MAKING THE GRAFT
(1) Use your sterile pipette to transfer the piece of donor tissue in a small volume of sterile saline to the *middle* of the exposed allanto-chorionic membrane. Be careful that it does not slide down the side of the egg.

(2) Replace the piece of shell, and sellotape it in position. Ensure that the sellotape seal round the edge is air-tight.

(3) Take care not to rock the egg from this time onwards. Keep it correctly orientated as you place it, together with the plasticine cradle, on a tray. Incubate at 37°C for a week, after which you should examine the graft.

For Consideration
What can be learned from this experiment about the properties and potentialities of (a) the organs of the donor, and (b) the allanto-chorionic membrane of the recipient?

Investigation 31.3
Development of the vertebrate eye

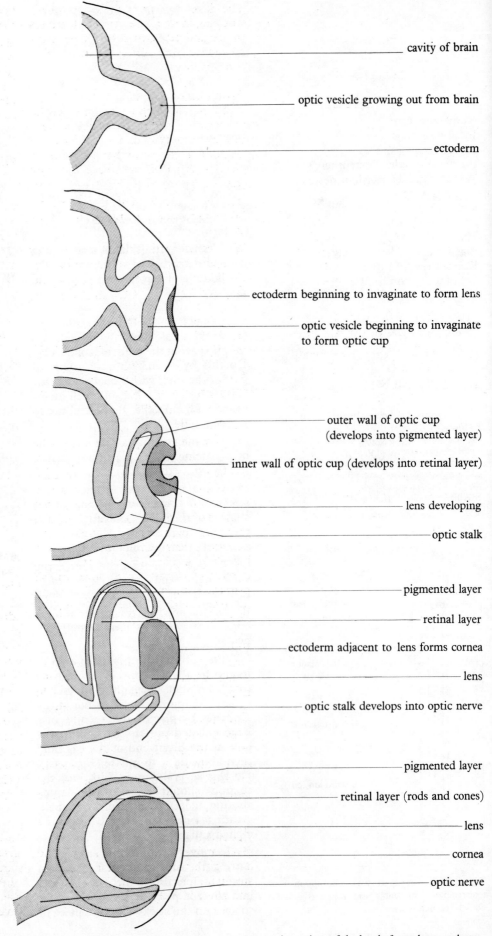

- cavity of brain
- optic vesicle growing out from brain
- ectoderm
- ectoderm beginning to invaginate to form lens
- optic vesicle beginning to invaginate to form optic cup
- outer wall of optic cup (develops into pigmented layer)
- inner wall of optic cup (develops into retinal layer)
- lens developing
- optic stalk
- pigmented layer
- retinal layer
- ectoderm adjacent to lens forms cornea
- lens
- optic stalk develops into optic nerve
- pigmented layer
- retinal layer (rods and cones)
- lens
- cornea
- optic nerve

Fig. 31.6 Development of the vertebrate eye as seen in sections of the head of vertebrate embryos

The development of the eye is worth studying in detail because it illustrates the organized sequence of events involved in the differentiation of a complex organ.

Procedure

Using prepared slides reconstruct the sequence of changes that takes place as the eye develops. You will need to examine a large number of slides to piece together all the stages. The key stages are as follows (Fig. 31.6):

(1) The **neural tube** is formed by invagination of the mid-dorsal ectoderm.

(2) The anterior end of the neural tube swells up to form the **brain**.

(3) An **optic vesicle** is formed on each side of the head as an outgrowth from the side of the brain.

(4) A **lens** is formed by thickening and invagination of the ectoderm adjacent to the optic vesicle.

(5) The optic vesicle invaginates to form an **optic cup**.

(6) The inner wall of the double-walled optic cup gives rise to the sensory and nervous layers of the **retina**; the outer wall gives rise to the pigment layer.

(7) The ectoderm adjacent to the lens forms the **cornea**.

For Consideration

(1) Suggest what induces the formation of the neural tube, optic vesicle, lens, optic cup, and cornea. How would you test your suggestions experimentally?

(2) The vertebrate eye is described as having an **inverted retina**. What does this mean and what causes it? Octopuses and squids have an eye which is almost identical to the vertebrate eye but it is *not* inverted. Explain.

Requirements
TS head of embryo, amphibian, or chick through eyes at various stages of development

---

Investigation 31.4

To investigate the ability of isolated cells to grow and differentiate *in vitro*

The growth and differentiation of cells is controlled by a complex system of interacting factors, some derived from the cells themselves and some from the environment. In recent years much information about these factors has come from **tissue culture** experiments in which isolated groups of cells obtained from embryos and adults have been grown in a nutrient medium outside the body, i.e. *in vitro*. In the present experiment you will attempt to grow isolated heart cells of the chick in tissue culture.

Precaution

One of the main reasons for failure in tissue culture work is infection of the tissue. For this reason you should wear a sterile face mask throughout the procedure and you should touch living material only with sterile instruments. Forceps and scissors must be absolutely clean and either flamed or dipped in ethanol before use. Sterile pipettes should be returned to a sterile test tube when not in use.

Procedure

(1) Obtain a fertilized hen's egg (8–10 days old). Support it in a cradle of plasticine. Wipe the egg with ethanol and allow it to dry.

(2) Crack the shell over the air space (blunt end). Remove the shell at the blunt end with forceps and then the white opaque membrane beneath it.

(3) Using sterilized forceps remove the embryo and place it in sterile saline in a petri dish.

(4) Dissect out the heart and transfer it to another petri dish of saline. Cut the heart into small pieces of 0.5 mm diameter using a sterilized blade.

(5) Sterilize a large coverslip in ethanol and dry it by rubbing it between two sheets of sterile filter paper. With a 1 cm³ disposable syringe without needle place a piece of the chopped tissue in the centre of the coverslip. This is now known as the **explant**.

(6) Blot around the explant with sterile filter paper till almost dry. Then with a 1 cm³ syringe pipette on one drop of nutrient medium. Spread out the medium round the tissue to cover a circular area not more than $\frac{3}{4}$ of the diameter of the depression of a large cavity slide or watch glass.

(7) Sterilize a large cavity slide or watch glass in ethanol and dry it with sterile filter paper. Spot it with four spots of vaseline at the corners of where the coverslip will lie. Now invert it and bring it down onto the top of the coverslip.

Requirements

Microscope (phase-contrast if
  possible)
Incubator at 37°C
Sterile face mask
Dissecting instruments
Large cavity slide, or watch glass
Large coverslip
Sterile 1-cm^3 syringe without
  needle
Petri dishes ($\times$ 2)
Glass rods ($\times$ 2)
Paintbrush
Sterile filter paper
Cotton wool
Plasticine (modelling clay)
Vaseline
Molten wax–vaseline mixture

Ethanol for sterilization
Chick saline
Nutrient medium

Fertilized hen's egg 8–10 days (*see*
p. **432**)

Nutrient medium prepared as follows:
A medium 199
B chick serum
C chick embryo extract
  (All obtainable from supplier)
Mix immediately before laboratory work commences in the proportions: **A**, 80 per cent; **B**, 10 per cent; **C**, 10 per cent. Each of the above ingredients can be stored indefinitely, the medium 199 in the ordinary part of a refrigerator, the serum and embryo extract in the freeze compartment.

(8) While still inverted seal around the edge of the coverslip with molten wax-vaseline mixture using a paintbrush. Make sure that the seal is good.

(9) Incubate upside down on glass rods at 37°C for two days (Fig. 31.7).(N.B. the slide should be upside down and the coverslip beneath it).

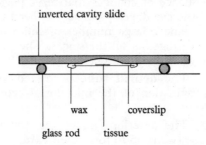

Fig. 31.7 Set-up for tissue culture. A watch glass can be used in place of the cavity slide.

TWO DAYS LATER
(10) Turn the slide and coverslip over so that the coverslip is now on top of the slide and the explant is enclosed in a **hanging drop** of culture solution.

(11) Observe under the high-power of, preferably, a phase-contrast microscope. You should be able to see fibroblasts and possibly epitheliocytes and amoebocytes growing out from the original explant.

For Consideration

(1) To what possible uses might tissue culture be put?

(2) To what extent are the cells which have grown out from your explant similar to, and different from, those of the original heart tissue?

(3) What light does rearing tissues *in vitro* throw on the normal development of tissues in the intact embryo, i.e. *in vivo*?

For Further Information: Paul, J. *Cells and Tissues in Culture* (Churchill-Livingstone).

Questions and Problems

1 Explain what is meant by the term *organizer*, illustrating your answer by reference to the development of the vertebrate eye. Describe experiments which you would carry out (a) to demonstrate the validity of the organizer concept, and (b) to discover the chemical nature of the organizer.

2 The haemoglobin synthesized by an adult human consists of an iron-containing core (called haem) surrounded by two types of polypeptide chains. The two types of polypeptide are called *alpha* and *beta* respectively. Each consists of a long chain of amino acids in the usual way.

  The kind of haemoglobin synthesized in the foetus differs from that produced in the adult. The haem part is the same but *one* of the two types of polypeptide is different. In fact the haem part of the molecule is surrounded by *alpha* and *gamma* polypeptide chains.

  It is known that the synthesis of the two kinds of haemoglobin, foetal and adult, is under genetic control. Explain, as precisely as you can, the *replacement* of foetal by adult haemoglobin at birth in terms of the action of genes.

  Why is it necessary for the foetus and adult to have different kinds of haemoglobin?

3 Fig. 31.8 shows the system of organizers which is believed to control cell differentiation in the newt. Examine the diagram carefully and then answer the following questions:

(a) What do the second line of structures (head endoderm, head mesoderm, etc.) *develop* into?

(b) What is the spatial relationship between an organizer and the structure it induces?

(c) What cell layers in a late gastrula appear to play little or no part as organizers?

(d) What do you understand by the primary organizer in Holtfreter's scheme?

(e) What sort of experiments do you think the scheme is based on?

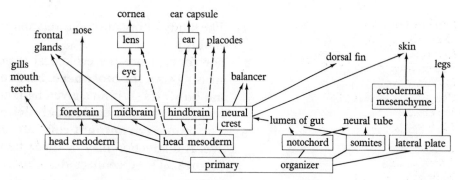

Fig. 31.8 Diagram of secondary organizers in the newt (*after* Holtfreter, from *Cell Biology*, A. J. Ambrose and D. M. Easty, Nelson, 1970).

4 Describe experiments which have been carried out to investigate the role of the nucleus in cell differentiation. What conclusions would you draw from the experiments you describe?

5 Consider the following information:

(a) When a small piece of tissue from the secondary phloem of a carrot tap root is cultured in a liquid medium containing salts, sugar and vitamins, the tissue grows slowly mainly by cell division.

(b) In contrast, when similar tissues are exposed to the same medium supplemented by coconut milk (the liquid endosperm of the coconut) the tissues grow rapidly by cell division and turn green.

(c) Single cells which break off from the dividing carrot tissue in (b) divide actively, and some develop roots, leaves and eventually flowers.

(d) Single cells removed from a carrot *embryo* and grown in a nutrient medium with coconut milk pass through a sequence of developmental stages similar to those gone through by a fertilized egg when it develops into an embryo. They eventually develop into carrot plants.

Now answer the following questions:

(i) What can you infer from these experiments about the genetic content of mature carrot cells?

(ii) Speculate on the role of coconut milk.

(iii) Put forward reasons for the different results in (b) compared with (c), and in (c) compared with (d).

(iv) Suggest ways in which these results could be put to practical use.

6 Several instances are known (see question **5**, for example) where differentiated plant cells from an established tissue can be shown to be totipotent, i.e. capable of developing into a complete new organism. In contrast, attempts to demonstrate the totipotency of differentiated animal cells have only been successful when the nucleus of the differentiated cell is transferred to the cytoplasm of an enucleated fertilized egg. Can you explain this difference?

7 Write an essay on nucleo-cytoplasmic interactions.

(*CCJE*)

32 The Organism and its Environment

Background Summary

1 The study of organisms in relation to their environment is the science of **ecology**.

2 For convenience the ecological system can be divided into units of decreasing size: the **biosphere, biogeographical regions, biomes, zones** and **habitats**. Within a given habitat, or range of habitats, each organism occupies a specific **ecological niche**.

3 The term **environment** embraces all the conditions in which organisms live. A distinction may be made between **biotic** and **abiotic** (**physical**) **factors**. Those connected exclusively with the soil are called **edaphic factors**.

4 **Abiotic factors** include temperature, water, light, humidity, wind and air currents, pH, mineral salts and trace elements, water currents, salinity, wave action, topography, and background.

5 **Biotic factors** include predation, competition, and other interrelationships between organisms including one using another as a habitat or for other purposes.

6 Organisms generally live in **communities** which together make up an **ecosystem**. The study of communities and ecosystems is known as **synecology**.

7 Communities do not happen suddenly; they grow gradually by a natural process in which one species is succeeded or replaced by another. This **succession** results eventually in the establishment of a stable **climax community**.

8 An ecosystem is composed of **producers, consumers**, and **decomposers** whose activities result in matter being cycled. The **carbon** and **nitrogen cycles** are applications of this general principle.

9 The nutritional relationships between producers and consumers can be seen in **food chains** and **food webs**. Whilst matter is cycled, energy is transferred through the system, a proportion being lost as heat at each level.

10 Estimations can be made of the numbers and total weight of individuals at each level of a food chain. The latter is known as the **biomass**. The drop in numbers and biomass at each level gives the **pyramid of numbers** and **pyramid of biomass** respectively. The pyramid of biomass has important economic implications for man.

11 The study of the individual species in a community is called **autecology**. Autecological studies aim to analyse the relationship between the organism and its environment in the broadest sense.

12 One of the most powerful influences in the environment is man. Pollution, the indiscriminate use of pesticides and herbicides, radioactive contamination, and urbanization, aggravated by the population explosion, are some of the more obvious ways in which man has affected his environment.

Investigation 32.1

Investigating the distribution of organisms

The ultimate aim of most autecological studies is to assess why a given organism should abound in one locality and be sparse in another. Obviously to answer this question it is necessary to investigate the distribution of the organism as accurately as possible.

Method

As it is rarely possible to count all the individuals present in a given locality, one resorts to **sampling**. Sampling methods take many different forms, but one commonly used involves the use of a **quadrat frame**. A metal or wooden frame of suitable area is laid on the ground and the number of individuals within it are counted. Quadrats are placed, either systematically or randomly, over as large an area within the locality as is feasible, so an accurate assessment can be made of the numbers of individuals per unit area.

Plainly there are certain organisms for which this technique is unsuitable, but for many it is a useful and successful method. Which particular organisms you choose to investigate must depend on where you live. The following examples illustrate the principles:

EARTHWORMS

Assess the relative population densities of earthworms in the soil in two different undisturbed localities. A simple, though not entirely satisfactory, method is to extract the worms by chemical means.

Choose a day when the ground is neither frozen nor dry. Lay a series of one-metre quadrat frames systematically on the surface of the ground in each of two localities. Alternatively mark out a series of one-metre squares with string.

Using a watering can with a rose, spray each square metre with one gallon of water containing potassium permanganate (7–9 g). Within 30 minutes most or all of the worms in the top metre or so should come to the surface. Some may not quite reach the surface but they can be exposed by gentle digging. Count and record the number of worms in each square metre.

Alternative method: Choose a day which is relatively warm and wet. Before nightfall lay a series of one-metre quadrat frames on the ground, or mark out with string as above. After nightfall examine the ground with a flash-lamp. Look for earthworms lying on the surface. Tread softly, for the worms will jerk back into their burrows if the ground is vibrated. Count and record the number of worms in each square. Do you think this is an adequate method of estimating the worm population in the soil?

PLANARIANS

Planarians are flatworms which live under stones in streams. Choose two different streams, or two stretches of the same stream, both of which contain stones of approximately the same size. Examine the underside of one or more stones and identify any planarians present. Initially, they will appear as small blobs 2–3 mm across, varying in colour from grey to black. When you have learned to identify planarians, lay a *metal* quadrat frame on the floor of the stream and examine the underside of all the stones within the frame. Count and record the number of planarians in one square metre. Repeat with the quadrat in a different position. Take as many samples as possible in the first locality, and then go on to the second.

Other animals that live under stones which might also be investigated include the freshwater shrimp, *Gammarus*, leeches, limpets and other molluscs, caddis fly larvae.

PLANTAIN

Plantain is a common inhabitant of fields and lawns. Using a one-metre

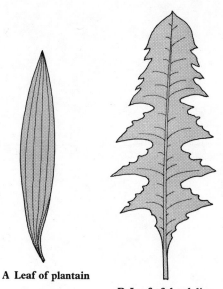

A Leaf of plantain

B Leaf of dandelion

Fig. 32.1 Leaves of plantain and dandelion.

quadrat frame estimate the population of individual specimens of plantain in two different localities.

Dandelions can also be conveniently estimated by the same method. If not in flower they can be distinguished from plantains by the structure of their leaves (Fig. 32.1). In both cases a single plant consists of a main flower-bearing stem surrounded by a ring of six to twelve leaves.

Results
You now have a series of population samples from two different localities. Calculate the average population of organisms per square metre in each locality. This is the **arithmetical mean**.

Is the arithmetical mean the same for both localities? By how much do they differ? Do you think the difference is simply the result of chance. What do you understand by the term **sampling error**? Do you think this may be responsible for the difference?

Is it sufficient only to compare the arithmetical means of the two populations? What further information do you need to make a more meaningful comparison of the two populations?

Now consult **Appendix** 19 (p. 439). Calculate the **standard deviation** for each of the two localities, and then work out the **standard error**. Do you consider the difference between the two populations to be significant?

If you do consider the difference to be significant, suggest hypotheses to explain the difference. How would you test your hypotheses?

For Consideration
(1) Do you have any criticisms of the methods used in this investigation for estimating populations? Can you suggest any improvements.
(2) Were the statistical tests necessary? If you feel they were unnecessary justify your reasoning.
(3) If you have obtained a significant difference between two populations, attempt to explain why the difference exists.
(4) What further investigations should be carried out?

Requirements
Quadrat frame (1 m^2)
Watering can with rose
Spade or trowel
Flash-lamp

Potassium permanganate

For counting aquatic animals the quadrat frame should be made of metal.

Investigation 32.2
Measuring the rate of water loss from willow and beech

When a population study reveals a significant disparity in the distribution of an organism, it is natural to ask *why* the organism should occur in one place and not another.

Why, for example, do willow trees thrive on river banks, whereas beech trees thrive on drier chalky soil? A number of factors might be instrumental in bringing about this difference in distribution: what are they? In this investigation we will test one possibility, namely that willows lose water more readily than beeches.

Procedure
Two methods are suggested:

WEIGHT METHOD
Take two or three leafy shoots of willow, and weigh them together. Aim to have a total weight of about 5 g (about 20 leaves will be necessary).

Note the time, then hang your shoots in such a way that they do not touch one another. Re-weigh at 15 minute intervals at least four times during the course of the session.

Do the same for beech shoots, hanging them in the same conditions as the willow.

Calculate the percentage decrease in weight at each interval for both plants:

Percentage decrease in weight =
$$\frac{a - b}{a} \times 100$$

Where a = initial weight at start of experiment
b = new weight.

Construct a graph comparing the weight loss of the two plants: put weight on the vertical axis, time on the horizontal axis.

POTOMETER METHOD
Assemble a potometer with a fresh willow shoot (*see* p. 118). Place the set-up in a part of the laboratory where conditions are unlikely to change in the course of the experiment. After one or more trial runs, when the plant has settled down to a steady rate of transpiration, take regular readings of the water uptake over a period of at

least 20 minutes. Express the rate of water uptake in distance moved by the air bubble per unit time (e.g. cm/min).

When consistent results have been obtained, remove all the leaves from the shoot and weigh them together.

Express your final result in, e.g. cm/min/g.

Using the same potometer, repeat the experiment on a beech shoot. The conditions should be identical with those given to the willow. Express the results in the same units.

Requirements
Balance
Potometer (*see* p. 118)
Beaker

Leafy shoot of willow
Leafy shoot of beech

For Consideration
(1) What are the merits and demerits of these two methods of measuring water loss from a plant?
(2) What ecological conclusions do you draw from your results? (Beware of concluding more than is justified).
(3) What further investigations would need to be made in order for firmer conclusions to be drawn?
(4) What structural and/or physiological differences between the two plants might explain their respective abilities to control water loss?
(5) What factors, other than water loss from the leaves, might be responsible for the observed distribution of willow and beech trees?
(6) What are the limitations of this kind of investigation as a means of studying the ecology of organisms?

Investigation 32.3

Comparison of water loss of arthropods in dry and humid air

The ecological distribution of animals, like plants, is closely related to their ability to control water loss. Some animals, for example mammals and insects, have marked powers of water retention and are, therefore, able to exploit habitats which are denied to animals whose water retaining capabilities are not so good.

In this investigation the water-retaining ability of the mealworm, an insect, is compared with that of the woodlouse, a terrestrial crustacean. Other small animals can be compared by the same method. Water loss is estimated by measuring changes in weight.

Procedure

(1) Obtain four small plastic boxes with lids. The diameter of the box should be about 5 cm and a large hole of about 4 cm diameter should be cut in the lid. Label the boxes 1–4.

(2) Place ten woodlice in each of boxes 1 and 2 and a similar number of mealworms in boxes 3 and 4. Cover the hole in each box with a single layer of nylon gauze trapped under the lid (Fig. 32.2). Then invert the box so the animals are resting on the gauze.

(3) Weigh each box and its contents. Record the weights.

(4) Place boxes 1 and 3 over a watch glass containing anhydrous calcium chloride; place boxes 2 and 4 over watch glasses containing water. In all cases make sure that the contents of

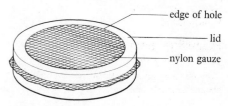

Fig. 32.2 Box in which to place small arthropods for estimating weight changes.

the watch glass do not touch the gauze. Note the time.

(5) Re-weigh the boxes and their contents at 15-minute intervals at least four times during the laboratory session. Then return the animals to their stock jars.

Results

Calculate the percentage decrease in weight of the animals in each of the four boxes at each interval of time (*see* p. 352).

Plot the results for all four boxes on one sheet of graph paper: weight loss on the vertical axis, time on the horizontal axis.

For Consideration

(1) What assumption have we made in this investigation? Is this assumption justified?
(2) What ecological conclusions do you draw from your results? (As with Investigation 32.2 beware of concluding more than is justified.) Are your conclusions supported by the occurrence of woodlice and mealworms in their natural habitats?

Requirements
Balance
Plastic boxes (× 4) with hole in
  lid (Fig. 32.2)
Nylon gauze
Watch glasses (× 4)

Anhydrous calcium chloride

Woodlice (× 20)
Mealworms (× 20)

(3) Is one of the two species investigated better at controlling water loss than the other? What structural and/or physiological features might explain the differences observed?
(4) To what extent do the results of this experiment relate to the *behaviour* of woodlice and mealworms?

Investigation 32.4
Analysis of a simple ecosystem

In the preceding investigation we were interested in the ecology of one specific organism. Such studies fall under the general heading of **autecology**. In contrast, **synecology** is the study of communities. Individual organisms within a community interact to produce an **ecosystem**. Ecosystems are generally very complex and it is as well to start with a comparatively simple system. Such is illustrated by a community of freshwater organisms in a laboratory aquarium tank. If there is a large number of students the organisms can be conveniently divided into a series of glass jars.

In studying any ecosystem a number of fundamental questions need to be answered. These include the following:
(1) What organisms are present?
(2) Where do they live?
(3) How do they relate to each other nutritionally, etc.?
(4) How do they relate to each other in other ways, e.g. does one organism provide protection for another, etc.?
(5) What physical factors are important in maintaining the ecosystem?

Procedure
Present the results of your observations in a table with the following headings:

Name of organism	Location	Food	Eaten by	Other relationships with other organisms	Abiotic factors on which dependent

NAME OF ORGANISM
Using a key (*see* p. 2), identify all the organisms, both animal and plant, present in your aquarium tank or jar. Include microscopic as well as macroscopic organisms. In the case of microscopic organisms, pipette a drop of water from a series of different levels (surface, middle, bottom) onto a slide, put on a coverslip and examine under the microscope.

LOCATION
Where does the organism occur within the habitat? Does it move from place to place, and if so, how? Or is it sessile? Draw a diagrammatic vertical section of the aquarium showing the positions of the various organisms.

FOOD
If possible observe each organism feeding, and/or examine its feeding apparatus and draw such conclusions as you can about the nature of its food. If necessary, examine the contents of its gut. With small animals like the water flea, *Daphnia*, this can be done by placing the specimen in the cavity of a depression slide and viewing it as a transparent object under the low power of the microscope. Which animals are predators?

EATEN BY
What organisms, if any, is each animal and plant consumed by?

OTHER RELATIONSHIPS WITH OTHER ORGANISMS
The following are the kinds of questions to answer here:
(a) What is the spatial relationship between members of the same species, and different species? Do the organisms live temporarily or permanently in groups? If so, why?
(b) Does the organism, either individually or collectively, provide protection for another organism? This might be achieved by its providing cover from predators or physical factors, camouflage, etc.
(c) Does the organism reside temporarily or permanently on the body of another organism? An animal which lives attached to another animal with-

out being parasitic on it is known as an **epizoite**. A plant which lives attached to another plant without being parasitic on it is known as an **epiphyte**.

(d) Is the association between animals or plants which live in or on one another **parasitic**, **symbiotic** or **commensalitic**?

ABIOTIC FACTORS ON WHICH DEPENDENT

Abiotic factors are those which are not directly related to the presence of other organisms. The most obvious abiotic factor is **water**, but are all the organisms completely dependent on it? Can any of them survive on land? What is the relevance of this to a real pond?

Other abiotic factors to consider include **light**, **temperature**, **oxygen content**, **water movement**. Can you think of others?

To what extent do each of the organisms appear to be dependent on these factors? Is there evidence that their distribution within the aquarium is influenced by any of these factors? How would you test your suggestions experimentally?

Requirements

Binocular microscope or hand lens
Microscope
Slides and coverslips
Cavity slide
Pipette capable of reaching bottom of aquarium
Watchglasses

Freshwater aquarium divided, if necessary, into a series of glass jars.

For Consideration

(1) Construct a food web to show the nutritional inter-relationships between the various organisms in your aquarium.

(2) An aquatic ecosystem, such as a pond or river, usually contains one or more animals which can be called **top carnivores**. What does this term imply? What do you predict would happen in your aquarium ecosystem if a top carnivore was introduced? Suggest examples of such top carnivores.

(3) Draw a diagram to show how carbon circulates in the aquarium. Do not simply produce a generalized carbon cycle, but show how it occurs in *your* particular ecosystem.

(4) Do the same for the nitrogen cycle.

Investigation 32.5

Relationship between predator and prey

One way in which animals relate to one another is in their nutrition. For example, in a natural community one particular species of animal may prey upon another. The feeding relationship between these two species will determine the relative abundance of each, and may be instrumental in moulding its evolutionary future.

In this investigation the predator-prey relationship between two common pond animals is analysed. The two animals are the damsel fly nymph (predator) and the water flea, *Daphnia* (prey).

Procedure

(1) Obtain six jars, all of approximately the same shape and capacity, and place an equal volume of pond water in each. Label them **A–F**. They should not be more than about two-thirds full.

(2) Place one damsel fly nymph in each jar, providing it with a short twig on which to cling. Leave the larva for about ten minutes to settle down.

(3) With a pipette transfer a specific number of *Daphnia* specimens from a rich culture to each jar. To jar **A** add five *Daphnia*, to **B** add ten *Daphnia*, to **C** add 15 *Daphnia*, to **D** add 20 *Daphnia*, to **E** add 30 *Daphnia*, and to **F** add 50 *Daphnia*.

(4) Leave each jar for 40 minutes after you have added the *Daphnia*.

(5) At the end of the 40 minutes for each jar, remove the damsel fly nymph, and count the number of *Daphnia* still remaining.

(6) Record your results, and pool them with those of the rest of the class. From the class results, calculate the average number of *Daphnia* remaining in each jar at the end of the 40 minutes.

(7) Plot the average number of *Daphnia* eaten (vertical axis) against the number of *Daphnia* in each jar before predation started (horizontal axis).

For Consideration

(1) What can you say about the relationship between the rate of predation and prey-density?

(2) What conclusions can be drawn regarding the control of an animal population by predation?

Requirements

Jars of approximately equal shape and capacity (× 6)
Pipette

Rich culture of *Daphnia*
Damsel fly nymphs (× 6)
Short twigs (× 6)

Questions and Problems

1 How do plants survive periods of low temperature? How would you expect plants living in a Mediterranean climate, where the main limiting factor is summer drought, to differ from those in Northern Europe, where winter cold is critical?

Commercial varieties of the perennial grass, *Dactylis glomerata* (cocksfoot), from different geographical areas were studied with regard to growth and survival at low temperatures with the results given below:

Source of variety	Relative increase in leaf area at 5°C	% survival after freezing at −5°C
Israel	27.9	0
Portugal	24.2	0
Denmark	16.4	14
Norway	9.3	88

Comment on these results. (*OCJE*)

2 Phytoplankton was collected in summer from the Sargasso Sea, from the surface **A**, and from depths to which ten per cent **B**, and one per cent **C** of the surface light penetrated. The effect of light intensity on the rate of photosynthesis in these samples was found to be as follows:

	Photosynthesis per cell (arbitrary units)					
	Light intensity (kilolux)					
Sample	5	10	20	40	60	80
A	28	50	72	91	95	92
B	34	68	90	95	81	58
C	74	95	58	38	30	27

Suggest possible explanations for these results and comment on their ecological significance. (*OCJE*)

3 A quadrat frame with 25 divisions, each of 10 cm side, was thrown at random ten times in each of two near-by areas of chalk grassland, one heavily, the other lightly, trampled. A species was awarded one point for each of the smaller divisions of the frame in which it occurred, irrespective of the number of individuals by which it was represented in those divisions. The results obtained for five species are given below. Comment on the results and describe how you would investigate further the differences observed. (*O and C*)

	Trampling	
Species	Heavy	Light
Sheep's fescue (*Festuca ovina*)	205	241
Mouse-ear hawkweed (*Hieraceum pilosella*)	2	60
Bulbous buttercup (*Ranunculus bulbosus*)	10	25
Hoary plantain (*Plantago media*)	55	10
Ribwort (*Plantago lanceolata*)	27	29

4 Show, by describing named examples, how:
   (a) plants may affect the distribution of other plants;
   (b) plants may affect the distribution of animals;
   (c) animals may affect the distribution of plants;
   (d) animals may affect the distribution of other animals.   (*JMB*)

5 What kind of features characterize an organism's biotic environment?
    Fig. 32.3A shows the normal black and yellow form of the African swallowtail butterfly, *Papilio dardanus*. This species produces females (Fig. 32.2B) whose colour and markings are similar to another species of butterfly, *Amauris albimaculata* (Fig. 32.3C), which is distasteful to predators. *Papilio dardanus* is not itself distasteful. Discuss this phenomenon.

A

B

C

Fig. 32.3   African butterflies. **A**, *Papilio dardanus*, normal black and yellow form; B, *planemoides* form which resembles **C**, *Amauris albimaculata* (*From* Wigglesworth, V. B., *The Life of Insects*, Weidenfeld and Nicolson).

6 The tropical jungle has been described as a 'vegetative frenzy'. Discuss the factors that permit the prolific growth of plants typical of the jungle.

7 Suggest explanations of the following and in each case indicate how you might test your hypotheses:
(a) Mosses are usually more abundant on north-facing than south-facing walls.
(b) Certain species of lichen are totally absent from most industrial areas, though common elsewhere.
(c) The ground flora is comparatively sparse in a beech wood.
(d) Rosebay willowherb (*Epilobium* sp.) is generally one of the first plants to become established in a desolate area such as a bomb site.
(e) In a freshwater stream planarian flatworms are found only under stones.

8 Explain with examples what you understand by the terms habitat and environment and make clear the differences between them.

9 Investigations have been made of competition between two species of flour beetles of the genus *Tribolium*. These small beetles can live out their entire life cycles in a jar of flour, provided fresh flour is added from time to time.

If individuals of the species *Tribolium castaneum* and *Tribolium confusum* are introduced into the same jar of flour, one species is always eventually eliminated while the other continues to thrive. Sometimes one species 'wins' the competition, sometimes the other. The relative numbers of individuals of each species originally introduced into the jar do not seem to affect the outcome.

One series of investigations involved the effect of temperature and relative humidity on the outcome of the competition between the two species. The results of these investigations are shown in the table below. Twenty to thirty different jars of flour containing individuals of both species were maintained at each of the six different 'climates' and the eventual 'winner' of the competition in each jar was noted. The percentage of 'wins' for each species at each climate is shown in the table. (Each species can survive alone in any of the climates.)

Climate	Temperature (°C)	Relative humidity (%)	Results of competition	
			% wins for *T. castaneum*	% wins for *T. confusum*
Hot–wet	34	70	100	0
Hot–dry	34	30	10	90
Warm–wet	29	70	86	14
Warm–dry	29	30	13	87
Cool–wet	24	70	31	69
Cool–dry	24	30	0	100

Comment on the above data.   (*VUS*)

10 What do you understand by the term desert? What kind of deserts are there? Write an account of animal and plant life in a desert known to you.

11 What do you understand by the term ecological niche?
The competitive exclusion principle states that co-existence is impossible between two species that are in total and complete competition within one another, i.e. which both occupy exactly the same ecological niche. Discuss.

12 Find out as much as you can about the history of your own habitat, say within a one-mile radius of your home, school or college. How has it changed as a result of man's activities and what effects have the changes had on the ecology of the area?

13 Fig. 32.4 indicates the relationships between certain plants and animals, and between one animal and another, in a wood. Using this information, and other examples known to you, explain what is meant by (a) a food chain, (b) a food web, (c) an ecosystem. *(AEB)*

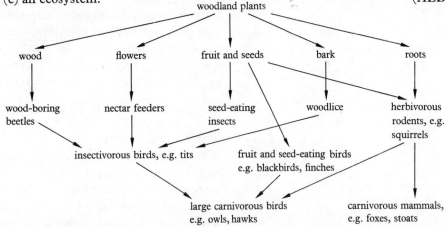

Fig. 32.4 Diagram showing the relationship between organisms in a wood.

14 The fish population of a river in the industrial part of England was sampled using a fine-mesh net in June 1950. A standard sample was shown to comprise fish species **A** to **G** as shown in the table below, Column **1**. Species **H** to **J** are known to occur elsewhere in the river but were not represented in the sample.

In 1955 a power station was constructed nearby and heated water was discharged into the river just above the sampling site. Samples taken in June 1957 showed the species composition to be as shown in Column **2** of the table.

In 1959 it was discovered that the sewage works of a town three miles upstream was discharging raw sewage into the river from a cracked pipe. The sampled fish population in June 1960 is shown in Column **3**.

In late 1960 a new river board inspector was appointed. More stringent control of effluents was enforced; the sewage pipe was repaired and the water from the power station was allowed to cool before entering the river. During the years 1962–1965 samples were taken and the results are shown in Columns **4–7**.

Discuss these results. What can be said about the changes in the fish populations that have occurred? What are the factors which are likely to have been important in producing the changes? How would you design laboratory experiments to find out more about the effects of pollution on fish?

Further industrial and urban developments are now planned close to the river. What advice would you give to the planning authorities for keeping the river clean and the fish population in balance?

Species composition samples from the same region of a river

Species	1 1950	2 1957	3 1960	4 1962	5 1963	6 1964	7 1965
A	5,000	100	10	200	560	3,100	6,500
B	100	2	0	3	18	200	300
C	20	150	0	0	0	1	0
D	1,100	65	2	0	0	5	25
E	2	0	1	0	0	0	0
F	3	0	0	0	0	0	0
G	85	560	150	480	560	120	25
H	0	30	45	83	120	200	350
I	0	2	0	1	0	3	0
J	0	26	38	40	20	3	2

(OCJE)

15 The seasonal changes in the concentration of inorganic nitrogen in the water of an overgrown canal were determined at monthly intervals over a period of one year. Comment on the following results:

Month	Inorganic nitrogen (parts per million)
Jan.	0.101
Feb.	0.105
Mar.	0.115
Apr.	0.153
May	0.172
June	0.128
July	0.103
Aug.	0.090
Sept.	0.085
Oct.	0.094
Nov.	0.097
Dec.	0.098

(*O and C*)

16 The crown of thorns starfish, *Acanthaster planci*, inhabits tropical waters where it feeds on coral polyps. Observations by underwater divers have shown that in certain coral areas large numbers of starfish gather together to form large aggregations of hundreds, even thousands, of individuals. When the food supply in one area has been used up the starfish move to another neighbouring reef.

(a) Suggest hypotheses to explain what causes the starfish to congregate together.

(b) Describe experiments you might carry out to test your hypotheses.

(c) What are the possible advantages and disadvantages of *Acanthaster*'s aggregation behaviour?

Off the north–east coast of Australia, *Acanthaster* has caused very severe damage to the coral of the Great Barrier Reef. Before 1959 the starfish population was kept in check by the giant triton, a large carnivorous mollusc; but since then the number of tritons has decreased with a corresponding increase in the starfish population.

(d) Suggest hypotheses to explain why the triton population has declined since 1959.

(e) Put forward suggestions as to how the starfish might be kept under control by man. Consider the relative merits and demerits of each method.

17 During the 1939–45 war a large quantity of fertilizer was emptied into a Scottish loch (enclosed sea-water lake) in the expectation that a more rapid growth of its fish population would occur. What in fact happened was a great increase in the seaweed population fringing the loch. The sequence of events is shown in Fig. 32.5. (*AEB*)

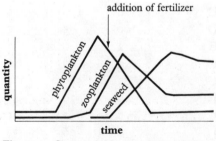

Fig. 32.5 Changes in the population of phytoplankton (**P**), zooplankton (**Z**) and seaweed (**S**) in a Scottish loch.

Explain, as you might to an intelligent layman, what you think was the original scientific basis of the project, and why it went wrong. (*AEB*)

18 Domestic sewage is rich in saprophytic bacteria and in organic material which is readily decomposed. When such sewage is discharged into a river it causes a number of chemical and biotic changes for some distance downstream from the 'outfall', i.e. the point at which it enters the river.

Typical changes have been summarized by Hynes (1960) and are shown graphically in Fig. 32.6.

Fig. 32.6A deals with chemical changes, indicating the amounts of various materials present in the river above the outfall and at various distances down-stream from the outfall.

Fig. 32.6B shows the changes in abundance of some of the biotic components over the same stretch of river as that covered by Fig. 32.6A.

Give reasoned explanations for the changes expressed by the curves, relating them to each other where you are able. (*JMB*)

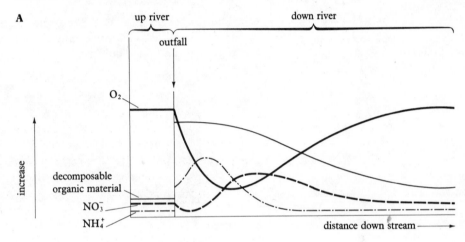

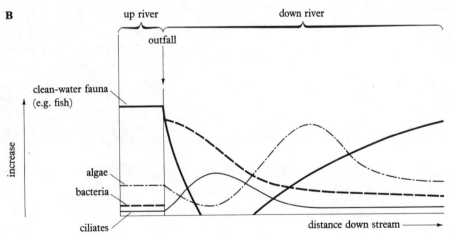

Fig. 32.6 (**A**) Chemical and (**B**) biotic changes in a river into which sewage flows. (Data from Hynes, 1960).

19 (a) Faced with an unexplored and undescribed habitat of your own choice, how would you set about investigating its ecology? State the particular habitat which you would choose and indicate the programme of investigations which you would plan to carry out.

(b) Give an account of an ecological project which you yourself have carried out either on your own or with other students.

20 Some of the energy derived from sunlight may ultimately be passed on to all species in a community. Discuss the physiological and ecological basis of this statement.

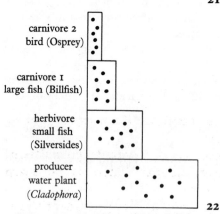

Fig. 32.7 Biomass and amounts of DDT in a food chain.

21 One of the problems man has created for himself is pollution of the environment with injurious chemicals such as DDT. The data shown in Fig. 32.7 have been collected in the U.S.A. The area of each rectangle represents the total mass of organisms of the species involved in the food chain. This mass is called the biomass. The number of dots in each rectangle indicates the amount of DDT at each level of the chain,

(a) What conclusions would you draw from the data?

(b) Attempt to explain the data in physiological terms.

(c) In parts of the U.S.A. it has been found that the death rate from DDT poisoning is higher among carnivorous birds than other organisms. Why do you think this is?

(d) Suggest ways of overcoming the problem of DDT pollution.   (VUS)

22 The graphs in Fig. 32.8 show the changes in populations of two species of *Paramecium* (*P. caudatum* and *P. aurelia*) grown in culture either singly or together in a mixed culture.

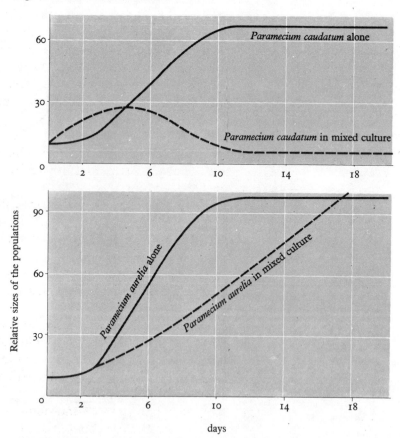

Fig. 32.8 Graphs showing changes in populations of two species of *Paramecium* grown in culture either singly or together in mixed culture.

Discuss these results.

(JMB)

33 Associations between Organisms

Background Summary

1 Associations between organisms may be **intraspecific** or **interspecific**. Both are widespread and often extensive.

2 Three types of interspecific association are recognized: **parasitism**, **commensalism** and **symbiosis** (**mutualism**). There is no sharp distinction between them.

3 In parasitic associations the parasite lives temporarily or permanently in or on the host, deriving benefit from it and causing harm to it. To qualify as a *bona fide* parasite an organism must fulfil the criteria embodied in this definition.

4 Parasites fall into two groups: **ectoparasites** and **endoparasites**. Endoparasites in turn may be either **intercellular** or **intracellular**.

5 A so-called parasite need not be totally dependent on parasitism for its existence. Certain parasites can feed on dead as well as living material. Others depend on parasitism for some, but not all, constituents of their diet.

6 Parasites generally have clearly observable effects on their hosts. The harm caused varies with the parasite. Those that kill their hosts are considered to be poorly adjusted and less highly evolved than those which cause relatively little harm.

7 Parasites show a number of **adaptations** to the parasitic way of life. These adaptations relate to their structure, physiology and life cycle.

8 A full knowledge of the parasite's life cycle can enable effective **control measures** to be devised.

9 In commensalism the commensal benefits but the host neither loses nor gains. This rather loose relationship can be considered as midway between parasitism and symbiosis.

10 In symbiotic associations both organisms benefit. The association may be obligatory, neither partner being able to survive without the other, or facultative in which either or both can, if necessary, exist alone. In some cases the association is so intimate that the two form what may be regarded as a single organism.

11 Nothing is known for certain about the evolution of these associations, and comparatively little about their physiology.

12 Intraspecific associations can be seen in **social organization**, which is particularly well developed in insects and mammals.

13 Amongst insects social organization, though efficient, is, in the main, rigid and stereotyped. This contrasts with primates, particularly man, where flexibility is a noticeable feature.

14 In social species, whether mammalian or insect, **communication** between individuals is of paramount importance. Only in this way can the integrity and efficiency of the colony or society be maintained.

Investigation 33.1

Examination of the liver fluke *Fasciola hepatica*

The liver fluke *Fasciola hepatica* inhabits the liver and bile passages of sheep where it causes much damage culminating in 'liver rot'. In its adult structure and life cycle the fluke shows many adaptations for parasitism. Pay special attention to these adaptive features as you examine it.

Adult

(1) Examine a whole mount of the adult, first under a hand lens or binocular microscope, then under the low power of an ordinary microscope. Note in particular those features which enable it to live in the host's liver.

Observe its size and shape, anterior and ventral **suckers**, **mouth**, **pharynx**, **gut** with two main branches and numerous side branches. Why is the gut branched?

Using Fig. 33.1 to help you, identify the various components of the **genital system**. Flukes are hermaphroditic. The genital system is geared to producing large numbers of eggs. During copulation the penis of one individual protrudes through the **genital pore** and is inserted into either the **genital atrium** or **Laurer's canal** of the other fluke. Fertilization takes place at x in the diagram, the point where

Fig. 33.1 Reproductive system of the liver fluke *Fasciola hepatica*. The reproductive system has two openings to the exterior: the genital pore (ventral) and Laurer's canal (dorsal).

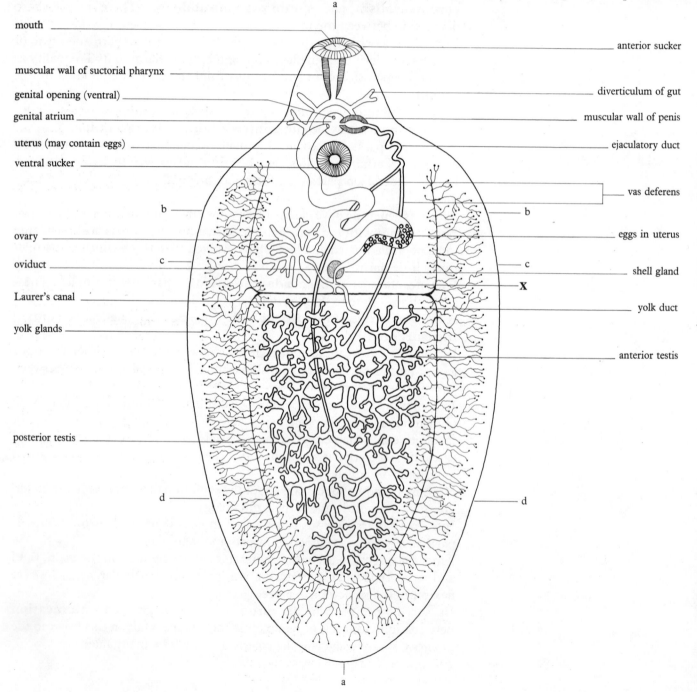

mouth

muscular wall of suctorial pharynx

genital opening (ventral)

genital atrium

uterus (may contain eggs)

ventral sucker

b

ovary

oviduct

c

Laurer's canal

yolk glands

posterior testis

d

a

anterior sucker

diverticulum of gut

muscular wall of penis

ejaculatory duct

vas deferens

b

eggs in uterus

c

shell gland

X

yolk duct

anterior testis

d

a

the **oviduct, yolk ducts** and **Laurer's canal** converge upon the **uterus**. After fertilization the egg, surrounded by a few yolk cells, is enclosed in a capsule hardened by the **shell gland** and then stored in the uterus. As well as serving as a tube for the entry of sperm, Laurer's canal is said to allow surplus eggs to escape.

In ripe specimens note numerous eggs in the uterus. Almost certainly self-fertilization occurs on a fairly large scale. Does this matter?

(2) Next examine sections of *Fasciola* to obtain a full picture of the spatial relationship between the various structures seen in the whole mount. First determine the level of the body at which the section was cut and decide whether it is a transverse, longitudinal or oblique section: what you will see depends on the level and plane of the section. Looking at Fig. 33.1, what structures would you *expect* to see in sections cut in the planes a–a, b–b, c–c, and d–d?

In addition to the structures listed for the whole mount, observe thick **cuticle** with spines; **mesenchyme**; circular, longitudinal and dorso-ventral **muscles**. What do the muscles achieve?

Make a list of the ways in which the adult fluke is adapted to a parasitic mode of life.

Larval Stages

Sexual reproduction takes place in the liver of the host. The fertilized eggs eventually get into the host's small intestine (how?) and leave with the faeces. In the course of its life-cycle the liver fluke passes through three larval stages (Fig. 33.2). From the egg capsule a ciliated **miracidium larva** emerges. If water is present the miracidium swims to, and bores into, the fleshy foot of the snail *Limnaea truncatula*, the intermediate host.

In the tissues of the snail the miracidium develops into a **sporocyst**. Internal propagation inside the sporocyst gives rise to **rediae larvae** which congregate in the snail's digestive gland. Within each redia internal propagation gives rise to further larvae which escape via the birth pore. These larvae are either more rediae or **cercaria larvae**.

Cercariae migrate from the snail's digestive gland to the mantle cavity and thence to the exterior via the pulmonary aperture. Attached to blades of grass, the cercariae encyst, developing no further unless eaten by a sheep, the final host. In the sheep's gut the cyst is digested, releasing the young fluke which bores through the gut wall and invades the liver.

(1) Examine a prepared slide of a miracidium larva under high power (Fig. 32.2A). Note the **cilia, anterior glands, eye spots** (if visible), and **propagatory cells** which will give rise asexually to the next generation of larvae. Longitudinal and circular muscles are also present in the wall of the miracidium.

In what respects does the miracidium differ from the adult? How would you explain the differences in terms of the roles of the miracidium larva and adult in the life cycle?

(2) Obtain live specimens of the snail *Limnaea truncatula* or closely related species. Species of snail other than *L. truncatula* harbour liver flukes similar to *Fasciola*. With a glass rod crush the shell in a watchglass with a little water. Examine the contents of the watchglass under low power for rediae and/or cercariae larvae (Fig. 33.2 B and C).

Study the behaviour of **live cercaria larvae** in the watchglass. How do they move? Test their response to various kinds of stimulation.

(3) Mount cercariae in water on a slide. Immobilize them by irrigating with ten per cent ethanol. Observe the **muscular tail** and general resemblance to adult. The flickering of **flame cells** may also be seen.

(4) If **rediae** are available note the **posterior processes, muscular collar** and **birth pore**. Can you see cercariae or further rediae larvae developing inside them?

(5) Augment your study of rediae and cercariae larvae by examining prepared whole mounts.

What are the most obvious observable differences between the redia and cercaria? How would you explain the differences?

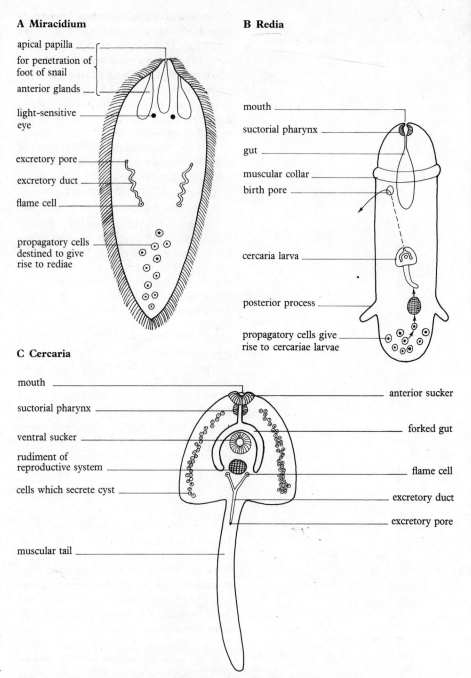

A Miracidium

apical papilla
for penetration of
foot of snail

anterior glands

light-sensitive
eye

excretory pore

excretory duct

flame cell

propagatory cells
destined to give
rise to rediae

B Redia

mouth

suctorial pharynx

gut

muscular collar

birth pore

cercaria larva

posterior process

propagatory cells give
rise to cercariae larvae

C Cercaria

mouth

suctorial pharynx

ventral sucker

rudiment of
reproductive system

cells which secrete cyst

muscular tail

anterior sucker

forked gut

flame cell

excretory duct

excretory pore

Fig. 33.2 The three larval stages of the liver fluke *Fasciola hepatica*.

Requirements
Binocular microscope or hand lens
Microscope
Slide and coverslip
Watchglass
Glass rod

Ethanol (10 per cent)

Limnaea truncatula (live specimens)
WM *Fasciola*
TS and LS *Fasciola*
WM miracidium
WM redia
WM cercaria

For Consideration
(1) The purpose of this investigation has been to observe the liver fluke's adaptations for a parasitic mode of life. Make a list of all the adaptations you can think of? How many of them have you yourself been able to observe?
(2) What are the natural hazards in the liver fluke's life cycle and how are they overcome?
(3) From your knowledge of the life cycle suggest methods by which the liver fluke might be controlled.

(4) The Chinese liver fluke, *Clonorchis sinensis*, a parasite of man, has a life cycle similar to that of *Fasciola hepatica* except that there are two intermediate hosts, an aquatic snail and then a fish to which the cercaria swims after leaving the snail. The cercaria encysts in the muscles of the fish, further development being suspended until the fish is eaten by man. Would you say that this life cycle is more, or less, hazardous than that of *Fasciola*?

Investigation 33.2

Examination of three parasitic fungi

A fungus characteristically consists of a branching system of thread-like **hyphae**. The whole network is known as a **mycelium**. There is no chlorophyll present, feeding taking place by saprophytic or parasitic means. Reproduction generally takes place sexually and asexually.

Parasitic fungi are of considerable economic importance. They include the downy and powdery mildews, potato blight fungus, and the rusts of wheat and other cereals. In general, hyphae enter the leaves and feed on the contents of the cells, destroying them in the process. Parasitic fungi display many adaptations for parasitism which can be well seen in the following cases.

Downy Mildew

Peronospora is parasitic on various plants including onion, beet, clover, cabbage, wallflower, and shepherd's purse. Its principal method of reproduction is by means of spores (conidia). When a conidium lands on the surface of a leaf, if moisture is present it germinates to produce a **hypha** which enters the leaf through a stoma and then branches into an extensive mycelium. Short side branches called **haustoria** penetrate into, and absorb the contents of, the host's cells. Eventually hyphae grow out through the stomata and form tree-like **condiophores** which produce **conidia** at the tips of their branches. On being released, the conidia are dispersed by wind to other host plants. Although asexual condidia-formation is the more prolific method of reproduction, the fungus can also reproduce sexually within the host plant.

(1) Observe the fungus on the leaves of a plant. Examine under a hand lens or binocular microscope. Note **conidiophores** bearing **conidia**.

(2) With needles tease out part of the mycelium, mount it in lactophenol on a slide and examine under high power. How would you describe its appearance? The hyphae are multinucleate and aseptate, i.e. the mycelium is a **coenocyte**.

(3) Examine the mycelium in a prepared section of an infected plant. Notice the hyphae *between* the host's cells, and **haustoria** devouring the contents. Stages in sexual reproduction may also be seen: **antheridia** and **oogonia** in various stages of development; **zygotes** and **zygospores**.

Potato-Blight Fungus

The life cycle of *Phytophthora infestans* is shown in Fig. 33.3. This important fungus is similar to *Peronospora* but differs from it in the following respects:

(1) Hyphae may penetrate through the epidermis as an alternative to entering the leaf via the stomata.

(2) The mycelium eventually grows through the entire plant including the tubers.

(3) Asexual reproduction takes place by the formation of conidia like *Peronospora*, but also by **sporangia** which, when they land on a wet leaf, give rise to motile **zoospores**.

If available, examine an infected potato plant and compare it with a healthy one. What are the symptoms of the disease?

Examine the mycelium in a section of an infected leaf and note its general similarity to *Peronospora*.

Damping-off Fungus

Pythium attacks seedlings, particularly of mustard, cress, cucumber and grasses. The stems of infected plants lose their rigidity.

In its structure and life cycle *Pythium* is basically similar to *Phytophthora*. The principal difference is that, instead of growing *between* the cells, the hyphae grow *through* them, feeding on their contents as they do so. The host is, therefore, damaged quickly and extensively. After the host has died the fungus continues feeding on it saprophytically.

If available, examine, e.g. cress seedlings infected with *Pythium*. How do they differ from healthy plants? What is the cause of the symptoms?

Examine a section of infected plant and observe the hyphae in the cells.

For Consideration

(1) For successful growth these three fungi depend on being able to penetrate into the host's cells. How do you think they achieve this?

(2) Make a list of the ways these three fungi are adapted to a parasitic mode of life. Which particular adaptations do they share with the liver fluke?

(3) Now that you have studied the life cycle of *Phytophthora* suggest methods by which it can be controlled.

(4) Which of these three parasites do you consider to be least well adjusted to its host, and why?

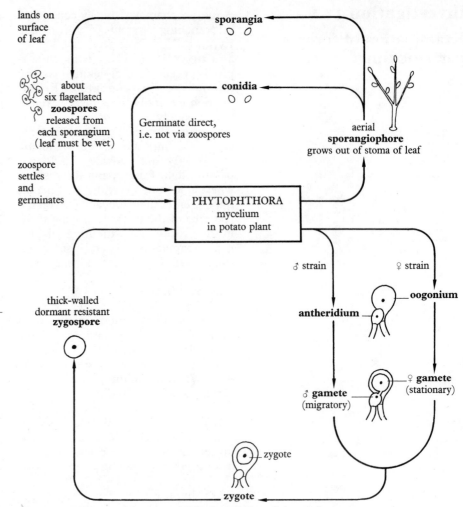

Fig. 33.3 Life cycle of the potato-blight fungus *Phytophthora infestans*.

Requirements
Microscope
Binocular microscope or hand lens
Slide and coverslip

Lactophenol

Section of plant infected with
  Peronospora
Section of plant infected with
  Phytophthora
Section of plant infected with
  Pythium
Live plants infected with fungi

Investigation 33.3
Observations of various interspecific associations

Interspecific associations show all gradations from situations where one species benefits and the other is demonstrably harmed (**parasitism**), through cases where one gains and the other neither loses nor gains (**commensalism**), to cases where both species derive benefit (**symbiosis**, mutualism).

Symbiotic partnerships may be **facultative**, each of the two species being capable of leading an independent existence if necessary, or **obligatory**, where one or other partner cannot survive without the other. In extreme cases a symbiotic association may be so intimate that the two partners form what can be regarded as a single organism.

There are also varying degrees of spatial intimacy between the species. Least intimate are associations where one species lives on the surface of the other. Such associations include the epiphytic and epizoic ones referred to on p. 355. At the other extreme are associations where one of the two species may live inside the cells of the other.

In what other respects do interspecific associations show gradations from one extreme to another?

This investigation provides you with an opportunity of examining a selection of associations. Try to decide into which category each one fits.

Green Hydra and Chlorella
Examine a green hydra (*Chlorohydra viridissima*) under low and high powers. Observe the unicellular green alga *Chlorella* (in this case called *Zoochlorella*) in its endodermal cells. This well known association is a case of symbiosis. What precisely does each organism gain? Could each survive without the other? How would you test this?

Nematodes and Earthworm

Examine an active nephridium of an earthworm in saline under low power and observe roundworms (*Rhabditis* sp). Where exactly are they in relation to the nephridium? What do they appear to feed on?

Examine a small piece of tissue from a decaying earthworm under low power. Are live roundworms visible? Conclusions?

Aphid and Green Plant

Examine a leaf of, e.g. sycamore or rose, under a hand lens or binocular microscope. Observe green flies (aphids). Can you see them feeding? What can you say about this association? How does the association between aphid and plant compare with that between mosquito and man?

Malarial Parasite and Man

Examine a human blood smear showing the trophozoite (amoeboid) stage of the malarial parasite. Note that some of the red blood corpuscles contain the parasite (darkly stained). What kind of parasite are we dealing with here?

Trypanosome and Mammal

Examine a mammalian blood smear showing trypanosomes, the causative agent of African sleeping sickness. The parasites are small worm-like organisms. Where are they in relation to the red blood corpuscles?

Gall Wasp and Flowering Plant

Gall wasps lay their eggs in the stems or branches of various plants. The larvae induce the surrounding plant cells to form a characteristic wart-like protuberance or **gall**.

Examine a gall and cut it open. What can you observe? What is the nature of the relationship between the plant and the wasp?

Mosquito and Mammal

Examine a mosquito under low power, noting in particular the piercing and sucking mouthparts (*see* p. 72). How would you describe the relationship between the mosquito and the mammal whose blood it sucks? How does this relationship differ from that between the mosquito and the malarial parasite?

Nitrogen-fixing Bacteria and Flowering Plant

Certain plants, particularly those belonging to the family Leguminosae (peas, beans, etc.) harbour nitrogen-fixing bacteria in their roots. The bacteria cause the cortical cells to proliferate forming a **root nodule**. Examine a transverse section of a root showing a root nodule and compare it with an uninfected root. Notice bacteria in the nodule cells. What do the bacteria and plant gain in this well-known symbiotic association? How does the presence of the bacteria affect the ecological distribution of the host plant?

Hydra and Kerona

Examine a specimen of *Hydra* under a binocular microscope. Often a small ciliate, *Kerona*, can be observed gliding over the surface of the hydra's body. Observe it in more detail under an ordinary microscope. What kind of relationship is this? What is the potential danger in *Kerona*'s way of life, and how does this seem to be overcome?

Flea and Dog

The dog flea *Ctenocephalides canis* is adapted to live amongst the hairs close to the host's skin. If available, examine a live specimen in a closed tube. Examine a whole mount under the low power. In what respects is it adapted to its particular mode of life?

Mycorrhiza

Many plants, particularly those that grow in nitrogen-deficient soil, show an association between their roots and a fungus. Such an association is termed a mycorrhiza. Examine a transverse section of the root of, e.g. pine, and notice the fungal network enveloping the root. Hyphal branches of the network penetrate between the cortical cells. Compare it with an uninfected root. Speculate on the nature of this relationship: what do the fungus and the host gain from it? How could you test your suggestions?

In the bird's-nest orchid (*Neottia nidus-avis*) the host is devoid of chlorophyll and the mycorrhizal fungus penetrates *into* the cortical cells. What are the benefits in this case?

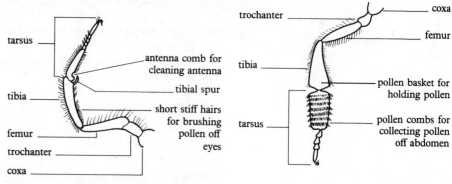

Fig. 33.4 The fore-leg and hind-leg of the worker of the honey bee showing adaptations for gathering pollen.

Requirements
Microscope
Binocular microscope
Hand lens
Slides and coverslips
Watchglass

Earthworm saline

Green hydra (*Chlorohydra viridissima*)
Live earthworm (or nephridium thereof)
Decaying earthworm
Aphids on green leaf
Blood smear with trophozoite stage of malarial parasite
Blood smear with trypanosomes
Plant galls
WM mosquito
Plant with root nodules
TS root with nodule
Hydra with *Kerona*
Live fleas in specimen tube
WM flea
TS root showing ectotrophic mycorrhiza
TS root without mycorrhiza
Live hermit crab with *Calliactis parasitica*
Honey bee (worker) entire
WM fore- and hind-limbs of honey bee
Bee-pollinated flower
Freshwater shrimp *Gammarus* with *Vorticella*
Lichen on bark
VS lichen (e.g. *Xanthoria*)

Hermit Crab and the Sea Anemone Calliactis parasitica
Observe a hermit crab in a whelk shell and notice the sea anemone *Calliactis parasitica* attached to the shell. What do you make of this association? Do you consider the sea anemone to be appropriately named?

Honey Bee and Flowering Plant
Examine a worker bee (*Apis mellifera*) and a flower that is normally pollinated by it (*see* p. 259). In what ways is the bee adapted for collecting pollen? Observe the fore- and hind-legs under the microscope and notice the pollen-collecting adaptations (Fig. 33.4).

Does the flower show any special adaptations which facilitate its being pollinated by the bee?

How would you describe the association between the bee and the plant?

Freshwater Shrimp and Vorticella
Observe the head and/or anal regions of the shrimp *Gammarus* under low power and look for the stalked ciliate *Vorticella* attached to the surface of the shrimp's body. Is *Vorticella* found elsewhere? How would you classify this association?

Lichens
Examine a sample of lichen attached to a piece of bark. Look at a vertical section of lichen (e.g. *Xanthoria*) under low and high powers. Notice the algal and fungal components of the organism. The algal cells are small and round and will be seen amongst the fungal hyphae towards the centre of the section. This represents the ultimate in symbiosis. What does each gain? What effect has this association on the ecology of lichens?

How would you describe the relationship between the lichen and the tree on whose bark it grows?

For Consideration
(1) How clear-cut do you consider the distinction to be between the various associations considered in this investigation?
(2) In the course of your observations you have speculated on the nature of these associations. How would you test the validity of any hypothesis which you have formulated?

Questions and Problems

1 'Big fleas have little fleas
Upon their backs to bite 'em;
And little fleas have lesser fleas,
And so ad infinitum.'
Comment.

2 Define the term 'parasite'. Discuss the problem of deciding whether a given association is parasitic or not.

3 Give a *general* account of how parasites are adapted for a parasitic existence.

4 How do parasites and their hosts influence one another?

5 The following organisms, or organisms and tissues, may be found associated together:
(a) *Rhizobium leguminosarum* in the root of a leguminous plant.
(b) Active tubercle bacteria in alveolar tissues of the lung.
(c) *Zoochlorella* in the endoderm cells of *Chlorohydra*.
(d) Non-pathogenic species of bacteria in the rumen of a cow.
In each case describe the substance, or substances, which may pass from one organism to the other and comment on the possible significance of these exchanges. *(AEB)*

6 Show how a knowledge of the life cycle and physiology of a *named* parasite and its host can lead to the development of effective methods of controlling the parasite.

7 Suggest explanations for the following:
(a) There is never more than one tapeworm (*Taenia* sp.) in a single host gut at any time.
(b) Certain termites can digest wood.
(c) Some species of flowering plant are devoid of all traces of chlorophyll and yet they survive.
(d) Many intercellular parasites flourish in the mammalian body and yet foreign tissue is rejected.

8 Compare the mode of life of a named endoparasite within its host with that of a mammalian foetus in the uterus. *(UL modified)*

9 Describe the methods by which parasites may be transmitted from one human being to another. How may such parasites be controlled? *(UL)*

10 Speculate on the possible evolutionary relationships between parasites, commensals and symbionts, and between different kinds of parasites.

11 The polynoid worm *Arconoë* has two species **A** and **B**, each of which lives as a commensal on the surface of a particular species of echinoderm. **A** lives on the starfish *Evasterias troschelii*, whilst **B** lives on the holothurian (sea cucumber) *Stichopus californicus*. The two associations are highly specific, each species of worm living exclusively on its own particular host. If the worm is removed from its host it will climb back onto it.
(a) How would you confirm, beyond all reasonable doubt, that the association is commensalitic and not parasitic or symbiotic?
(b) Describe experiments which you would carry out to investigate the mechanism by which the worm associates with its own particular species of host.

34 Evolution in Evidence

Background Summary

1 Charles Darwin was the first person to put forward a theory of evolution based on firm scientific evidence.

2 Darwin's contribution was twofold: he put forward evidence supporting the fact of evolution and he formulated a plausible hypothesis explaining the mechanism of evolution.

3 Evolution reveals itself in the geographical distribution of animals and plants, comparative anatomy, taxonomy, embryology, cell biology, and palaeontology. Studies in all these fields furnish evidence for evolution.

4 Evolution is seen in **continental distribution**. For example, the distribution of mammals in the southern continents of the world, together with fossil evidence, suggests that their ancestors migrated there from the northern hemisphere, became isolated and then evolved independently.

5 Particularly cogent evidence for isolation followed by independent evolution is provided by the Australian fauna where the marsupials occupy the ecological niches filled elsewhere by eutherian mammals.

6 Further evidence for isolation and independent evolution is provided by the distribution of organisms on **oceanic islands** such as the Galapagos archipelago. On the Galapagos islands, which were visited by Darwin during his voyage on the *Beagle*, the finches furnish particularly striking evidence for evolution.

7 The process of isolation followed by independent evolution depends on the establishment of natural barriers. These include mountain ranges, water, and various climatic factors.

8 From **comparative anatomy** the principle of **homology** emerges. This can be explained in terms of **divergent evolution** and **adaptive radiation** and provides evidence for evolution. A good example is the vertebrate pentadactyl limb.

9 In certain situations comparative anatomy can provide information which enables **evolutionary pathways** to be reconstructed. This is seen, for example, in the vertebrate heart and arterial arches. However, to gain an accurate picture of the evolutionary history of animals, the comparative anatomy of modern forms should be accompanied by a study of the fossils.

10 On occasions phylogenetically unrelated groups may appear to be similar as a result of **convergent evolution**. Superficially similar structures shared by such groups are called **analogous structures**.

11 **Taxonomy**, the classification of organisms, is not so much evidence for evolution as the inevitable result of it. The purpose of a natural classification is to reflect the degree of evolutionary affinity between different groups. A good example is seen in the classification of the chordates.

12 Sometimes the **embryology** of two seemingly unrelated groups reveals a phylogenetic connection which is not evident from studying the adult forms. Thus **comparative embryology** can provide important evidence for evolutionary affinities.

13 **Cell biology** also provides evidence for evolution. The basic similarity in the structure and functioning of cells, and the ubiquitous occurrence of many biochemical molecules, suggests a common ancestry for all animal and plant cells. **Biochemical homology** and **serological tests** provide further evidence for evolutionary affinities between certain groups.

14 The most direct evidence for evolution derives from **palaeontology**, the study of fossils. Fossils take the form of petrified remains, moulds, impressions and preservation in amber, asphalt, or ice.

15 Fossils can be dated by sedimentation data or, more accurately, by estimating the decline in radioactivity as, for example, in **carbon dating**.

16 From the fossil record evolutionary pathways can be reconstructed in detail as in the case of the evolution of horses, and the vertebrate ear ossicles.

Investigation 34.1

The vertebrate pentadactyl limb: an exercise in homology

Homologous structures are those that, though they may serve different functions, are fundamentally similar to one another and are believed to share a common ancestry. An example of homology is provided by the **pentadactyl limb** of vertebrates. Animals possessing the pentadactyl limb include amphibians, reptiles, birds, and mammals.

The pentadactyl limb is so called because typically it terminates in the form of **five digits**. However, in the course of evolution it appears to have undergone considerable modification in different groups. These modifications have involved enlargement, fusion, degeneration or, in some cases, total loss of certain components.

Procedure

A generalized pentadactyl limb is shown in Fig. 34.1. We shall examine the limbs of various tetrapods and observe the extent to which they conform to, or depart from, the idealized pattern shown in the diagram. In each case try to correlate the structure of the limb with the functions it performs.

Fig. 34.1 A generalized pentadactyl limb. This kind of limb is possessed by all terrestrial groups of vertebrates. The preaxial border (i.e. the edge of the limb which generally points towards the main axis of the body) is to the left, the postaxial border to the right. The fore- and hindlimbs both conform to the pattern illustrated: the nomenclature used for each is shown to the left and right of the diagram respectively. (*After* Grove and Newell, *Animal Biology*, University Tutorial Press).

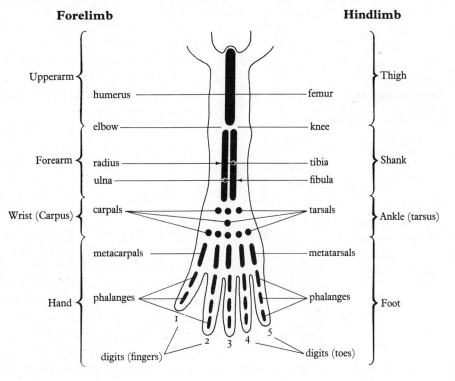

Forelimb   Hindlimb

Upperarm — humerus, elbow — Thigh, femur, knee

Forearm — radius, ulna — tibia, fibula — Shank

Wrist (Carpus) — carpals — tarsals — Ankle (tarsus)

Hand — metacarpals, phalanges — metatarsals, phalanges — Foot

digits (fingers)   1 2 3 4 5   digits (toes)

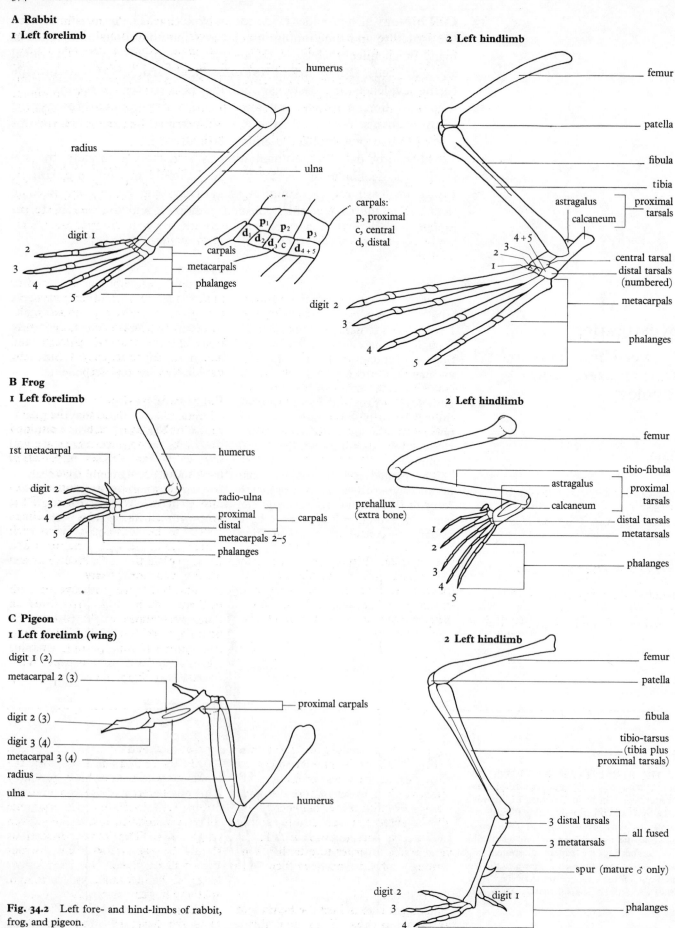

A Rabbit
1 Left forelimb

humerus

radius

ulna

carpals:
p, proximal
c, central
d, distal

digit 1

2

3

4

5

carpals

metacarpals

phalanges

2 Left hindlimb

femur

patella

fibula

tibia

astragalus

calcaneum

proximal tarsals

4 + 5

2 3

1

central tarsal

distal tarsals (numbered)

metacarpals

phalanges

digit 2

3

4

5

B Frog
1 Left forelimb

1st metacarpal

digit 2

3

4

5

humerus

radio-ulna

proximal

distal

carpals

metacarpals 2–5

phalanges

2 Left hindlimb

femur

tibio-fibula

astragalus

calcaneum

proximal tarsals

prehallux (extra bone)

distal tarsals

metatarsals

phalanges

1

2

3

4

5

C Pigeon
1 Left forelimb (wing)

digit 1 (2)

metacarpal 2 (3)

digit 2 (3)

digit 3 (4)

metacarpal 3 (4)

radius

ulna

proximal carpals

humerus

2 Left hindlimb

femur

patella

fibula

tibio-tarsus (tibia plus proximal tarsals)

3 distal tarsals

3 metatarsals

all fused

spur (mature ♂ only)

digit 2

3

4

digit 1

phalanges

Fig. 34.2 Left fore- and hind-limbs of rabbit, frog, and pigeon.

RABBIT

Examine the fore- and hind-limb bones of a rabbit. Identify the component parts, regarding the carpus and tarsus as single units for the moment. To what extent does each limb depart from the ideal pentadactyl pattern? Why the differences?

Now examine the **carpus** and **tarsus** in detail. Are all nine component bones present? If not, what do you think has happened to them? Try to answer this for yourself *before* you look at Fig. 34.2A. What functional explanation would you suggest for the modifications seen in the carpus and tarsus?

FROG

Examine the fore- and hind-limb bones of a frog. Again regarding the carpus and tarsus as single units, identify the component parts of each limb. To what extent do the fore- and hind-limbs depart from the ideal pentadactyl pattern, and how do they differ from each other? How would you explain the differences?

Now examine the **carpus** and **tarsus** in detail. In each case try to account for the nine component bones *before* looking at Fig. 34.2B. What do you think is the evidence for the statements made in Fig. 34.2B?

BIRD

Examine the hind-limb of a bird, e.g. pigeon. This is more modified than either the rabbit or frog. Try to decide for yourself what has happened, and *then* look at Fig. 34.2C. What are the functional reasons for the modifications in this case?

Now examine the fore-limb. This is even more modified. It is obvious that great reduction has taken place in the **hand**, but there is some controversy as to which particular components have been lost. Two different interpretations are given in Fig. 34.2C. What do you think is the evidence for such views?

To explain the fore-limb skeleton in functional terms it is useful to examine an intact wing with the skin and feathers in position. Do this and try to interpret in functional terms the fore-limb's departure from the pentadactyl pattern.

OTHER TETRAPODS

If available, examine the fore- and/or hind-limb bones of other tetrapods, e.g. man, monkey, pig, horse, mole, bat, etc. In each case note its departure from the pentadactyl pattern and interpret, in functional terms, the modifications you observe.

For Consideration

(1) Are you convinced that the pentadactyl limbs which you have examined share a common ancestry, or are you sceptical? What further information do you think you should have?

(2) If available, examine entire skeletons of rabbit, frog, and bird. What other structures besides the limbs appear to be homologous with each other? Compare such structures and try to explain the differences between them in functional terms.

(3) The bird's wing is covered with feathers. With what structures in other vertebrates might feathers be homologous? How might you test your suggestion? If time permits examine a feather under the microscope. What functions do feathers perform?

Requirements

Hand lens

Skeleton of fore- and hind-limb of rabbit
Skeleton of fore- and hind-limb of frog
Skeleton of fore- and hind-limb of bird (e.g. pigeon)
Intact wing of bird
Mounted skeletons of rabbit, frog, and bird
Other limb skeletons as available

Investigation 34.2

Comparative dissections of the vertebrate heart and arterial arches

An important purpose of dissection is to throw light on functional processes. However, another reason for dissecting animals is to investigate the similarities and differences between them. The object of this exercise is to explore the internal anatomy of four different vertebrates in order to establish evolutionary affinities between them.

Procedure

The object is to dissect the **heart** and **arterial arches** of an amphibian, reptile, bird, and mammal, so they can be readily compared. Start with the mammal which you have probably dissected before and then go on to the others. In dissecting the mammal follow the detailed instructions given on p. 104. The other dissections should be exploratory: few instructions will be given. Use your knowledge of the mammal, and vertebrate anatomy in general, to guide you.

The guiding principle is this: in all cases the heart lies underneath the

sternum, the mid-ventral component of the pectoral (shoulder) girdle.

The sternum must be removed to expose the heart. The heart is enveloped by a **pericardial membrane** which must also be removed. The arteries will be seen on the ventral side of the atria. They should be identified and traced outwards.

MAMMAL
Dissect a rat to show its heart and arterial arches (*see* p. 105). Observe the **heart**; **pulmonary arteries** to lungs; **systemic arch (aorta)** *on left side* giving off **carotid arteries** to the head. Note that there is no systemic arch on the right side. The **ductus arteriosus** linking the systemic and pulmonary arteries may be visible. What does it represent? Do your observations agree with Fig. 34.3A?

BIRD
Dissect a bird (e.g. pigeon) to show its heart and arterial arches. Proceed as follows:

First pluck the feathers from the ventral side ('breast'), then remove the skin from this region. Next remove the thick pectoral muscles (flight muscles) from the sternum. Note the deep **keel** of the much enlarged **sternum** (function?).

Carefully remove the sternum by cutting through the ribs and coracoid to which it is attached at its outer edge. In lifting it away be careful not to damage the heart which lies immediately beneath it.

Observe the **heart** covered by the **pericardium**. Remove the pericardium.

Observe the arteries lying on the ventral side of the heart. Trace them outwards sufficiently to see the following: **pulmonary artery** to lungs; **systemic arch (aorta)** *on right side*; **carotid arteries** to head. Note that in this case there is no systemic arch on the left side. Check against Fig. 34.3B.

In what other respects, apart from the systemic-arch condition, does the bird system differ from that of the mammal?

REPTILE
Dissect a reptile (e.g. lizard) to show its heart and arterial arches. Proceed as follows:

Make a mid-ventral cut through the skin and body wall and deflect them on either side.

Next cut through the pectoral girdle in the mid-ventral line and pin back the two halves of the girdle on either side.

Observe the **heart** within the **pericardium**. Remove the pericardium and identify the arteries lying on the ventral side of the atria.

Trace the arteries forwards and outwards, noting; **pulmonary arteries** to lungs; **systemic arches** *on both sides*; **carotid arteries** to head (Fig. 34.3C).

Ascertain that the left and right systemic arches unite on the dorsal side of the viscera in the mid-line to form the **dorsal aorta**, and that the carotid arteries arise from the *right* systemic arch.

Towards the edge of the animal on either side an artery may be seen connecting the carotid and systemic arches. This is the **ductus caroticus**. What does it represent in evolutionary terms?

How does the reptile system compare with the mammal and bird? What would have to happen in order for the reptilian system to be turned into those of the bird and mammal?

AMPHIBIAN
Dissect an amphibian (e.g. frog or salamander) to show the heart and arterial arches. Proceed as for the lizard: open up the body cavity, cut through the pectoral girdle in the mid-ventral line and pin back the two halves; remove the pericardium, identify the arteries and trace them outwards.

Note that, as in the reptile, there is a **systemic arch** on both sides. However, in this case, the pulmonary, systemic, and carotid arches arise from a common vessel, the **conus arteriosus**, which lies on the ventral side of the atria (Fig. 34.3D).

Ascertain that the left and right systemic arches unite in the mid-line on the dorsal side of the viscera to form the **dorsal aorta**.

In what respects is the amphibian system similar to, and different from, the reptilian one? How does it compare with the bird and mammal? What changes would have to take place in order to turn the amphibian system into those of the other three animals?

A Mammal (rat)

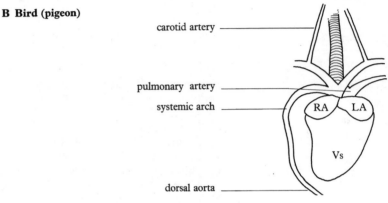

carotid arteries

systemic arch

ductus arteriosus

pulmonary artery

dorsal aorta

RA

LA

Vs

B Bird (pigeon)

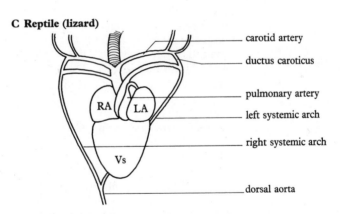

carotid artery

pulmonary artery

systemic arch

RA   LA

Vs

dorsal aorta

C Reptile (lizard)

carotid artery

ductus caroticus

pulmonary artery

left systemic arch

right systemic arch

dorsal aorta

RA   LA

Vs

D Amphibian (frog)

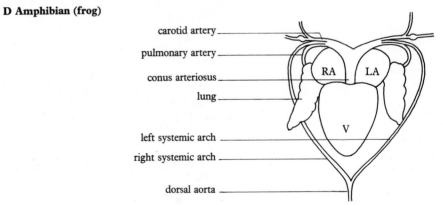

carotid artery

pulmonary artery

conus arteriosus

lung

left systemic arch

right systemic arch

dorsal aorta

RA   LA

V

Fig. 34.3 The vertebrate heart and arterial arches as seen in dissections of rat, pigeon, lizard and frog. These are all ventral views so the right side of the heart is on the observer's left, and the left side of the heart is on the observer's right. RA, right atrium; LA, left atrium; V, ventricle.

Requirements
Dissecting instruments

Mammal (e.g. rat)
Bird (e.g. pigeon)
Reptile (e.g. lizard)
Amphibian (e.g. frog)
 (these are all for dissection)

For Consideration
(1) On the basis of the arterial arches *only*, draw up a scheme suggesting how amphibians, reptiles, birds and mammals might be related to each other in evolution.
(2) What other information needs to be taken into account in order to gain a true picture of their evolutionary relationships?

(3) While changes were taking place in the system of arterial arches, developments were also occurring in the internal structure of the heart. What were these changes?
(4) Put forward a *functional* explanation of the changes that have occurred in the heart and arterial arches during the evolution of vertebrates.
(5) What general principles are illustrated by this exercise?

Investigation 34.3
Examination of the chordate characters

Evolutionary relationships can be seen in classifications. Provided it is based on homologous structures, a classification of animals and plants into groups reflects the evolutionary affinities between them.

A good example of this is provided by the **chordates**. The phylum Chordata includes all the vertebrates (fishes, amphibians, reptiles, birds, and mammals) together with several invertebrate groups (protochordates) which include *Amphioxus*, the lancelet. This wide range of animals are believed to be related phylogenetically on the grounds that they all share the six **chordate characters**. These are:
(1) A dorsally suited strengthening rod, the **notochord**, present at some stage during development.
(2) A **hollow, tubular CNS** on the dorsal side of the body immediately above the notochord.
(3) **Pharyngeal clefts** connecting the cavity of the pharynx with the exterior on either side of the body.
(4) A series of segmental **muscle blocks** (**myotomes**) on either side of the body.
(5) The anus, instead of being terminal, is situated some way in front of the posterior end of the body, thus leaving a **post-anal tail** which extends beyond the anus.
(6) **Blood flows** forwards in a ventral vessel and backwards in the dorsal vessel.

These features can be readily observed in, e.g. a fish, amphibian, and *Amphioxus*.

FISH
(1) Examine the external features of, e.g. a dogfish or bony fish, and locate the cloaca (common vestibule receiving the anus, excretory, and genital openings) and **post-anal tail**. Function of the latter?

(2) Strip off a piece of skin from the side of the body shortly behind the cloaca. Note the ≤ -shaped segmental **muscle blocks** (**myotomes**) separated by tough sheets of connective tissue (**myocommata**). How are the muscle fibres of the myotomes orientated? Function?

(3) Examine a transverse section of a dogfish embryo through the pharyngeal region. How many chordate characters can you detect? Observe the **notochord**, dorsal **nerve tube** with small cavity, **pharyngeal clefts** lined by gills, myotomes in dorso-lateral regions, dorsal and ventral aortae.

How would you describe the microscopic structure of the notochord? How does its structure relate to its function as a strong, but flexible, rod?

What function is served by the pharyngeal clefts in the dogfish?

(4) Examine a transverse section through the tail region of a dogfish embryo and compare with the previous section. How does this section compare with the previous one with respect to the chordate characters?

(5) Examine a transverse hand section through the tail region of an adult dogfish. How does it differ from the embryo section? Explain the differences.

AMPHIBIAN
(1) First compare the external features of a newt or salamander with a frog or toad. How many chordate characters can you observe in each case?

(2) Examine a late tadpole of frog or toad (*see* p. 290) and notice that it possesses certain observable chordate characters which are absent in the adult. What principle is illustrated by the differences?

(3) Examine transverse sections of a tadpole through the head, pharyngeal,

abdominal, and tail regions. Observe the occurrence (or otherwise) of the chordate characters in each section.

AMPHIOXUS

(1) Examine a whole mount of *Amphioxus* noting the chordate characters, particularly the **post-anal tail** and < shaped **myotomes** (Fig. 34.4A).

(2) Examine a transverse section of *Amphioxus* through the pharyngeal region. Most of the chordate characters will be clearly detectable. The **pharyngeal clefts** will appear as a series of perforations in the wall of the pharynx which is enclosed by a flap of the body wall called the **atrial fold** (Fig. 34.4B).

How do you think the atrial fold is formed? The cavity created by it is called the **atrium**. Into this the pharyngeal clefts open. At the back of the pharyngeal region the atrium opens to the exterior by a pore, the **atriopore**.

Amphioxus is a **filter feeder**: water containing food particles in suspension is drawn by cilia into the pharynx and expelled through the pharyngeal clefts. As the water goes through the clefts food particles get caught up in mucus secreted by the **endostyle organ** and carried by cilia on the **pharyngeal bars** to the **dorsal groove**. Cilia lining the dorsal groove draw the mucus and food particles backwards to the intestine. Water leaves the atrial cavity via the **atriopore**.

For Consideration

(1) From your observations of the chordate characters are you, personally, convinced that *Amphioxus* should be placed in the same phylum as the dogfish and frog?

(2) This phylum also includes *you*. What has happened to your chordate characters?

(3) What principle is illustrated by the fact that the pharyngeal clefts are

Fig. 34.4 Structure of *Amphioxus* as seen under the microscope. The chordate characters are marked with an asterisk.

A Side-view of whole mount

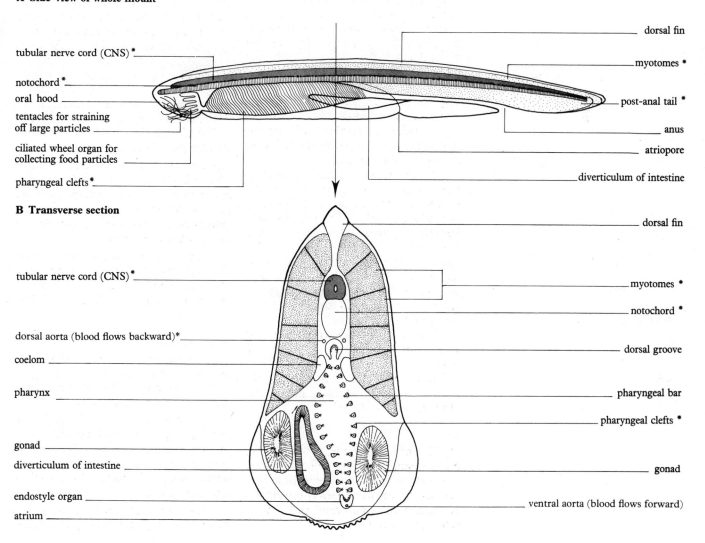

B Transverse section

Requirements
Microscope
Scalpel and forceps

Dogfish or bony fish (entire)
Hand section of adult dogfish (tail)
Frog or toad
Salamander or newt
Tadpole internal gill stage

TS dogfish embryo (pharyngeal region)
TS dogfish embryo (tail region)
TS tadpole at various levels
WM *Amphioxus*
TS *Amphioxus* pharyngeal region

used for respiration in fishes, but for filter-feeding in *Amphioxus*?
(4) *Balanoglossus* (the acorn worm) is a marine worm-like animal with a muscular proboscis which it uses for burrowing into sand and mud. It has pharyngeal clefts but the dorsal CNS is, for much of its length, an open groove rather than a closed tube; the anus is terminal; the body wall muscles appear to be unsegmented; the blood system is rudimentary, and the notochord is absent though a short structure with a similar microscopic appearance is present in the proboscis.
  Do you think it is justified to include this animal in the chordates?

(5) What other information, besides that which you have obtained in this investigation, would you need to have in order to establish whether or not a given animal is a member of the chordates?

Questions and Problems

1 What kind of differences would you expect to find between the fauna and flora of an oceanic island and a comparably sized island on a continental shelf? Explain your answer. *(JMB)*

2 Dinosaurs belonging to the same genus have been found in North America and East Africa, and others in South America and India. However, the mammals found in these areas belong to different genera. How would you explain these facts?

3 'It is useless and futile to try to classify animals and plants into groups and is of no more scientific value than classifying men according to their height or place of birth.' Critically discuss this statement supporting your views with examples drawn from the animal or plant kingdoms.

4 In an attempt to investigate the evolutionary affinity between the domestic rabbit and various other mammals, Moody, Cochran, and Drugg carried out serological tests. Their results are summarized in Fig. 34.5. The length of the bars represents the amount of precipitation in the tests.

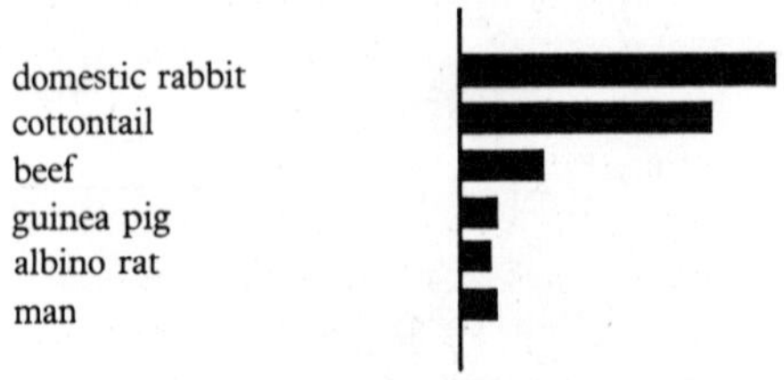

domestic rabbit
cottontail
beef
guinea pig
albino rat
man

Fig. 34.5 Results of serological tests on the domestic rabbit and other mammals. (*Based on* Moody, Cochran and Drugg).

(a) Describe the procedure by which you think these results were obtained.
(b) What do the results suggest to you?
(c) Consult a classification of mammals in a textbook of zoology.
Do the results of the serological tests fit in with the classification?
(d) What further information might help to clarify the evolutionary relationships of these animals?

5 'Thorough comparison of living things with each other gives the biologist crucial information which can be gained in no other way.' Discuss this statement using *named* examples wherever possible.

(*O and C*)

6 Given below is information from a biologist's notebook about twelve animals (**A–L**) in the phylum Mollusca, the group which includes snails, oysters, and octopuses:

A Has head; fleshy foot; one pair of tentacles; shell single and coiled; one pair of simple eyes; single gill on left side; lives between low-tide mark and 180 metres; active and carnivorous; separate sexes; internal fertilization.

B Body compressed from side to side; shell in two halves, slightly coiled and held together by pair of adductor muscles; head rudimentary; no tentacles or eyes; one pair of sheet-like gills; common between high and low-tide marks in 'beds'; reduced foot secretes sticky threads for attachment to rocks; filter-feeder; separate sexes; external fertilization.

C Shell single, internal and reduced; well developed head into which the foot is incorporated; mouth surrounded by ten tentacles with suckers; one pair of gills; large CNS; well developed eyes; marine; active swimmer; carnivorous; separate sexes; internal fertilization.

D Mouth surrounded by eight prehensile tentacles; no shell; well developed head with eyes and large brain; foot incorporated into head; marine; active carnivore; separate sexes; internal fertilization.

E Body flattened from side to side; rudimentary head with no tentacles or eyes; bivalve shell with one adductor muscle; left valve of shell larger than right and 'cemented' to rock; no foot; filter-feeder; separate sexes; some species hermaphroditic; external fertilization; marine.

F Lives on land; herbivorous; head with two pairs of tentacles, the posterior pair bearing eyes; single coiled shell; no gills but has chamber which functions as lung; flat fleshy foot; hermaphrodite; internal fertilization.

G Tubular shell open at both ends; no gills; head with numerous prehensile tentacles; no eyes; foot reduced and used for burrowing; found in sand and mud just below low-tide mark.

H Shell single, flattened and very slightly coiled and pierced by a row of holes; moderately well developed head with three pairs of tentacles, the posterior pair bearing eyes at their tips; large fleshy foot for attachment to rocks; numerous short tentacles (tactile) round edge of body; one pair of gills displaced to left by large shell muscle; occur from low-tide mark to depths of over 20 fathoms; feed on encrusting plant material; separate sexes; fertilization external.

I Small head with no tentacles or eyes; numerous small gills on either side of body; large flat foot for attachment to rocks; shell consists of eight calcareous plates which, being separate, allow the animal to roll up like a woodlouse; lives under stones and rocks near low-tide mark; separate sexes; external fertilization.

J Shell single, internal and reduced; head with two pairs of eyeless tentacles; pair of eyes just in front of posterior tentacles; hermaphroditic; internal fertilization; single gill; elongated foot with flattened outgrowths used for swimming; marine.

K Much reduced foot; shell laterally compressed in two halves, one half flatter than the other; adductor muscle with striated fibres; no head but tentacles and simple eyes round edge of shell; can swim by repeatedly opening and closing the shell; filter-feeder; hermaphroditic; external fertilization; one pair of gills.

L Occurs in fresh water; flattened body; shell in two halves; well developed muscular foot for burrowing; one sheet-like gill on either side of foot; much reduced head without tentacles or eyes; slow moving, filter-feeder; separate sexes; internal fertilization.

(a) On the basis of the information given, construct an evolutionary tree depicting the relationship between these animals.

(b) In constructing your evolutionary tree, which features have you found (i) most useful, and (ii) least useful? How would you justify your use of certain features and not others?

(c) Make a natural classification of the animals, dividing them into groups and sub-groups. Invent a name for each group, reflecting, if possible, the particular feature which distinguishes that group from others.

(d) Construct a key (see p. 2) enabling each animal to be quickly identified by its code letter.

7 In animals phosphate groups for the synthesis of ATP from ADP are derived from compounds called phosphagens. Two principal phosphagens are found in the animal kingdom: phosphocreatine and phosphoarginine whose chemical formulae are shown below:

$$HN=C\begin{cases} NH-H_2PO_3 \\ \\ N-CH_2-COOH \\ | \\ CH_3 \end{cases} \qquad HN=C\begin{cases} NH-H_2PO_3 \\ \\ N-CH_2-CH_2-CH_2-CH-COOH \\ | \qquad\qquad\qquad\qquad\qquad | \\ H \qquad\qquad\qquad\qquad\qquad NH_2 \end{cases}$$

phosphocreatine   phosphoarginine

Table 34.1 summarizes the distribution of these two phosphagens in the main groups of the animal kingdom.

Group	phospho-arginine	phospho-creatine
Coelenterates	+	−
Annelids	−	+
Molluscs	+	−
Arthropods	+	−
Echinoderms		
Asteroids (starfish)	+	−
Holothurians (sea cucumbers)	+	−
Echinoids (sea urchins)		
Some species	+	−
Other species	+	+
Ophiuroids (brittle stars)	−	+
Protochordates		
Hemichordates (acorn worms)	+	+
Tunicates (sea squirts)		
Ascidia	+	−
Other species	−	+
Cephalochordates (Amphioxus)	−	+
Vertebrates	−	+

Table 34.1 Occurrence of the phosphagens, phosphoarginine and phosphocreatine, in the principal animal groups. (After Prosser and Brown 'Comparative Animal Physiology', W. B. Saunders Company).

Although many annelids possess phosphocreatine as indicated in the table, one genus of fanworms (Spirographis) has been found to contain phospho-arginine. Annelids are also unusual in possessing other phosphagens. These include phosphoglycocyamine found in the ragworm Nereis, and phospholombricine found in the earthworm Lumbricus. The formulae of these annelid phosphagens are as follows:

$$
\begin{array}{c}
\text{NH—H}_2\text{PO}_3 \\
\diagup \\
\text{HN=C} \\
\diagdown \\
\text{N—CH}_2\text{—COOH} \\
| \\
\text{H}
\end{array}
$$

phosphoglycocyamine

$$
\begin{array}{c}
\text{NH—H}_2\text{PO}_3 \\
\diagup \qquad\qquad \text{OH} \\
\text{HN=C} \qquad\qquad\quad | \\
\diagdown \\
\text{N—CH}_2\text{—CH}_2\text{—O—P—O—CH}_2\text{—CH—COOH} \\
| \qquad\qquad\quad || \qquad\qquad\quad | \\
\text{H} \qquad\qquad\quad \text{O} \qquad\qquad \text{NH}_2
\end{array}
$$

phospholombricine

Discuss these findings. Comment on their possible evolutionary significance and draw attention to any difficulties of interpretation.

8 What do you understand by the term 'adaptive radiation'? Illustrate your answer by reference to either the Australian fauna or to that of the Galapagos Islands.

9 The haemoglobin molecule of mammals contains four polypeptide chains: two α and two β chains. Each α chain consists of 141 amino acids, and each β chain consists of 146 amino acids.

  Analyses of the sequence of amino acids in the polypeptide chains of haemoglobin have revealed that gorilla haemoglobin differs from human haemoglobin with respect to two amino acids (one in the α chain and one in the β chain). However, horse haemoglobin differs from human haemoglobin with respect to 17 amino acids.

  Comment on these observations.

10 From fossils occurring in the Rocky Mountains, U.S.A., a biologist recorded the frequency of rodents and multituberculates. The animals belonging to these two groups are similar in size and have gnawing teeth, but rodents have only two front teeth whereas multituberculates have many.

  The table below shows his data:

Millions of years ago	Epoch		Rodents		Multi-tuberculates	
			Genera	Species	Genera	Species
45		Late	13	31	0	0
50	Eocene	Middle	9	19	0	0
55		Early	4	12	3	5
60		Late	1	1	7	11
65	Palaeocene	Middle	0	0	6	17
70		Early	0	0	5	7

(a) Why were the Rocky Mountains a particularly favourable place for studying such fossils?

(b) How do you think the ages of the fossils were estimated?

(c) What conclusions would you draw from the data?

(d) Suggest explanations for the particular change in fauna which the data indicate may have occurred during the Palaeocene and Eocene epochs. (SA)

11 Discuss the usefulness (or otherwise) of the following features in assessing the evolutionary affinities of a plant:
    (a) Its height.
    (b) The absorption spectrum of its photosynthetic pigment.
    (c) The number of petals in its corolla.
    (d) The colour of its flowers.
    (e) The degree of elaboration of its gametophyte generation.

12 The similarities between the fauna of different continents are cited as evidence for evolution, but so are the *differences* between them. How would you reconcile this apparent contradiction?

13 How would you convince a sceptic of the truth of the theory of evolution?

14 Make a list of the principal lines of evidence which support the theory of evolution. Which ones do you find the least convincing and the most convincing? Go on to discuss in detail *one* piece of evidence which you personally find most convincing.

35 The Mechanism of Evolution

Background Summary

1 According to the Darwinian theory, evolution occurs by the **natural selection** of **chance variations**. In contrast, Lamarck's theory suggests that evolution occurs by the **transmission of acquired characters**.

2 Of the two theories Darwin's has stood the test of time. The Darwinian theory, updated by the findings of modern genetics, is referred to as **neodarwinism**.

3 Darwin's theory depends on the existence of chance variation within species. Such variations may be **continuous** or **discontinuous**.

4 Continuous variation may be seen in **frequency distribution histograms** and generally approximates to a **normal** (**Gaussian**) **curve**.

5 Such variation may be attributed to reshuffling of genes resulting from the behaviour of chromosomes during meiosis, and new combinations established by fertilization. Variation is accentuated by the fact that characteristics are often controlled by the cumulative effects of numerous genes (a **polygenic complex**) whose expression may be controlled by **modifier genes**. The effects of the latter may themselves be altered as a result of the behaviour of chromosomes during meiosis.

6 Discontinuous variation can be attributed to **mutations**. These may be caused by chromosomal abnormalities (**chromosome mutations**) or changes in the structure of the genes (**gene mutations**).

7 Mutation frequencies are comparatively low. However, the mutation rate can be increased by various **mutagenic agents**.

8 Mutations arise spontaneously and are relatively persistent. Usually they are harmful but sometimes they may confer beneficial characteristics on the individual.

9 Mutations are generally recessive at first, but they may become dominant in the course of time.

10 Types of chromosome mutation include **deletion, inversion, translocation, duplication**, and **addition**. The last may involve addition of one or more whole chromosomes as a result of **non-disjunction** during meiosis.

11 In extreme cases non-disjunction results in an organism containing one or more complete sets of extra chromosomes. This is known as **polyploidy** and in certain situations it is associated with beneficial characteristics.

12 Beneficial characteristics may also result from **hybridization** in which crosses occur between closely related species.

13 Gene mutations involve changes in the sequence of nucleotide bases in DNA. These may involve **substitution, insertion, deletion**, or **inversion** of one or more bases. Seemingly trivial changes may have far-reaching effects.

14 The conflict between organisms and their environment has been described as the **struggle for existence**. This results from **environmental resistance** as population numbers increase.

15 Factors contributing to environmental resistance include lack of food, water, oxygen, light, and shelter; predation, disease, accumulation of toxic substances; and psychological factors.

16 Fluctuating about a set-point, populations are generally held reasonably constant by a homeostatic mechanism involving negative feedback (*see* Chapter 13).

17 However, sudden changes may occur in a population if some factor of the environment is suddenly changed. This can be seen in the growth of the human population over the last 200 years.

18 The relevance of population growth to evolution is that variation is at its greatest during periods of increase and the mortality which results from over-population is differential (**differential mortality**). This is the basis of **natural selection**.

19 Natural selection is instrumental in holding species constant (**stabilizing selection**), but, if the environment changes, it favours the emergence of new forms (**progressive** or **directional selection**).

20 Natural selection may be seen in action in various rapidly breeding organisms such as micro-organisms and insects.

21 Some species possess two or more different types of individuals (**polymorphism**). Where the differences are genetic (**genetic polymorphism**), the different morphs in a given population occur in more or less constant proportions, this constancy being maintained by selection.

22 The genetic constitution of a population is known as its **gene pool**. The Hardy–Weinberg principle states that the **gene frequencies** in a given gene pool are generally held constant. This **genetic equilibrium** is maintained by stabilizing selection.

23 However, genetic equilibrium may be upset as a result of mutation, environmental change and selection, and the loss or gain of genes. Changes in genetic equilibrium are the basis of directional selection.

24 For new species to arise (**speciation**) some degree of **isolation** is necessary. Isolation need not necessarily be geographic, though this is often the first step in speciation. It may also be ecological, reproductive, or genetic. Once two separated populations become reproductively or genetically isolated they may re-unite geographically without losing their identity.

25 The principles of genetics and evolution are employed (sometimes unwittingly) by man in animal and plant breeding, in which natural selection is replaced by **artificial selection**.

Investigation 35.1

Variation in a population of animals or plants

Variation provides the 'raw material' for natural selection. One way of investigating if evolution is taking place in a species is to study the occurrence of a variable feature in a population over a period of time, or in several different populations. Mathematical techniques can then be used to determine whether or not there are significant differences between the populations.

Studying the 'occurrence' of a particular feature means, in practice, determining the numbers of individuals within a given population showing the different variations of the feature. These numbers give us the **frequencies**. Plants lend themselves particularly well to this sort of investigation. Features suitable for analysis include: length of stem or flower stalk, length of petals, length of leaves, number of petals in a plant where this is variable (e.g. daisy), number of flowers per inflorescence, etc.

Obviously to be of any value, investigations of this kind need time. In a short session one can do little more than learn the basic techniques.

Procedure

First select a feature which you would like to investigate in a particular population of plants. Specimens may be either measured in the field or collected and measured in the laboratory.

Score the frequencies on squared paper as shown in the hypothetical example in Fig. 35.1. This gives you a **frequency distribution histogram**.

If you are investigating variations in a *number* (e.g. number of petals), then each number should be written along the horizontal axis of your frequency diagram. On the other hand, if you are investigating variations in, for example, a *length*, then it may be more appropriate to group the various lengths into classes (e.g. 10–15, 16–20, 21–25, etc.). The classes should be placed along the horizontal axis in your frequency diagram.

Obtain measurements from as large a sample of the population as possible. If time permits, do the same for a second population in a different locality.

Analysis of Results

Refer to Appendix 19, p. 439.

Construct a curve by joining the tops of the vertical bars in your frequency distribution histogram. Your curve should pass through the centre of the top of each bar.

Does the curve approximate to a curve of **normal distribution**? If

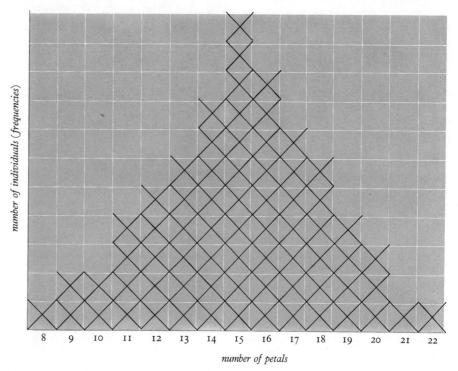

Fig. 35.1 An example of a frequency diagram. Starting from the base, a cross is placed in the appropriate box each time a flower with a particular number of petals is encountered in the population. In this particular population seventy individuals were examined and recorded.

not, suggest reasons for its failure to do so.

Calculate the **arithmetic mean** for your population. Is this the same as the **mode** in your histogram? If not, why not?

Now calculate the **standard deviation**. What percentage of the population departs from the mean by more than twice the standard deviation? What does this tell us about the population?

If you have obtained frequency data for two separate populations, calculate the **standard error** of the difference. Is the difference between the arithmetic means of the two populations less than, or greater than, twice the standard error? Conclusions?

Requirements

Hand lens or binocular microscope
  if required
Squared paper

For Consideration

(1) Suggest reasons why the 'class' represented by the mode in your histogram is the most common situation.

(2) If you were to continue with this investigation, what further observations and experiments would you carry out?

(3) Suppose a significant difference was found between the frequency data obtained in a given area in one year, and the frequency data obtained in the same area ten years later. How might this be explained?

(4) Suppose a significant difference was found between the frequency data obtained in two different areas. How might that be explained?

Investigation 35.2

Population growth: multiplication of yeast cells

The growth of populations is of great importance in ecology and evolution. The principles involved in population growth and its checks can be seen clearly in the growth of unicellular organisms such as yeast. Such organisms have a comparatively high rate of reproduction and, therefore, lend themselves well to studies over a short period of time. Yeast cells reproduce rapidly by **budding**.

In this investigation a small group of yeast cells are allowed to multiply in a nutrient medium for about a week, and the concentration of yeast cells present is estimated at regular intervals.

Method

A **haemocytometer** can be used for counting yeast cells. This is described on p. 102.

Procedure

(1) Using a clean graduated cylinder, measure 50 cm^3 of nutrient solution (e.g. cider) and pour it into a 250 cm^3 conical flask.

(2) Swirl the flask containing a suspension of yeast cells and, once they are well mixed, add one drop to the flask of nutrient medium.

(3) Plug the latter flask with cotton wool and place it in a warm cupboard.

(4) Using a haemocytometer, take periodic cell counts twice daily over a minimum of three days. Swirl the flask before withdrawing a sample onto the haemocytometer slide. There is no need to dilute the sample. It is essential that the cells should be evenly distributed before they are counted. On each occasion carry out at least two independent cell counts keeping the results separate. Yeast cells multiply by budding: to ensure comparability, count as a cell every bud you can clearly see, even if the bud is distinctly smaller than the parent cell.

Procedure for counting

Choose an appropriately sized square on your haemocytometer grid. Count in as many squares as possible and record the numbers systematically. From this information calculate the number of cells per cubic centimetre.

Results

Plot your results on linear (arithmetical) graph paper, time on the horizontal axis, number of cells per cubic centimetre on the vertical axis.

For convenience it is also advisable to plot the results with time on the horizontal axis and the *logarithm* of the number of cells on the vertical axis. The quickest way of doing this is to use semi-logarithmic graph paper. In this kind of graph paper the marks on the vertical axis have already been calibrated on a logarithmic scale, but labelled on a direct scale. In other words the figure 5, for instance, is located at the logarithm of 5.

Compare the semi-logarithmic and linear plots. Allowing for the difference in scale, is there any difference in the appearance of the two curves? Explain the difference.

For Consideration

(1) How would you describe the shape of the curve in your linear graph? Explain the reason for each phase in the curve. The period during which the population is growing at its fastest is the exponential phase. Explain this term.

(2) From your semi-logarithmic plot determine the doubling time during the exponential phase, that is the time it takes for the population to double. From this work out the division rate, that is the number of cell divisions per unit time.

(3) Assuming that *Paramecium* is capable of living in the culture medium, what do you predict may have happened to the yeast population if a few *Paramecia* had been added soon after the beginning of the exponential phase. Consider the long-term future of the yeast and the *Paramecium* populations in such a mixed culture. Consider *Paramecium*'s maximum division rate to be once every twelve hours.)

(4) You have made predictions concerning the behaviour of yeast and *Paramecium* populations in mixed culture. What sudden evolutionary changes in either the yeast or *Paramecium* might substantially alter the course of events?

(5) Below are given figures for the world human population since 1800:

Year	Population size
1800	8.1×10^8
1850	1.1×10^9
1900	1.6×10^9
1950	2.5×10^9
1960	2.9×10^9
1970	3.5×10^9

Plot these data on semi-logarithmic graph paper, putting population on the vertical (semi-logarithmic) axis and years on the horizontal axis.

How does the curve compare with that for yeast? What factors may affect the world population in the future?

Requirements

Microscope
Haemocytometer (*see* p. 102)
Graduated cylinder (50 cm^3)
Conical flask (250 cm^3)
Cotton wool
Pipette
Linear (arithmetical) graph paper
Semi-logarithmic graph paper
Nutrient medium for yeast (e.g. cider or 2 per cent sucrose solution)

Suspension of brewer's yeast

Questions and Problems

1 Variation is a very obvious characteristic of living things. What are the causes of variation? Discuss its role in evolution. (*O and C*)

2 Do laws prohibiting marriages between brothers and sisters make biological sense?

3 The following are two examples of spelling errors, the first in a newspaper report of a speech delivered by the mayor of a certain industrial city (which will remain nameless), the second in a letter sent by a small boy at a boarding school to his parents:

IN CONSIDERING THE FUTURE OF THIS CITY, ONE MUST REGARD IT AS A HOLE

ON SUNDAY WE HAD ROAST BEEF FOR LUNCH AND THE HEADMASTER'S WIFE CALVED.

These errors can be taken as analogies to illustrate two types of gene mutation. Using hypothetical, or actual, examples, explain precisely the nature of each type of mutation. (Describe the chemical structure of genes in sufficient detail to make your answer clear.)

4 The sex chromosome constitution of human males is usually **XY**, but some individuals are abnormal in this respect. The data given below refer to three samples:

A Random samples of males.

B Mentally subnormal males confined to a state hospital under conditions of special security owing to persistent violent behaviour, and

C Males from the same hospital in a wing for patients suffering from various psychoses, some of whom were also mentally subnormal.

Sample	No. in sample	Chromosome abnormality and incidence			
		XYY	XXY	XXXY	Others
A	2,000	1	0	0	0
B	197	7	1	1	3
C	314	9	2	1	4

Comment on these findings and discuss their significance. *(CCJE)*

5 Two populations of *Drosophila* initially containing an equal number of wild type and of individuals having the gene ebony (dark body colour) were kept in two boxes and under different conditions of temperature.

The graph shown in Fig. 35.2 shows the changes with time of the relative proportions of ebony flies. Comment on these results.

(N.S.W.H.S.C. modified)

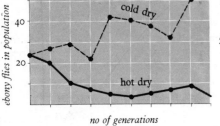

Fig. 35.2 Graph showing the proportion of ebony flies in two populations of *Drosophila* kept under different temperature conditions.

6 Discuss the biological basis and implications of the newspaper report shown in Fig. 35.3.

Weirdo, the rooster that kills cats

By IAN BALL in New York

A 17-YEAR-OLD Californian boy who has developed a flock of super-hens led by a 22lb rooster, named Weirdo, said yesterday he has been besieged with offers since stories about Weirdo gained international circulation.

Grant Sullens, of West Point, California, has received offers from governments, companies, chicken breeders and cockfight organisers.

He is hoping to increase his flock to more than 300 by the autumn so he can offer his breed of heavyweight hens to the world. By crossing and re-crossing the bigger members of various breeds he produced hens that grew rapidly, could survive the cold nights of the Sierra Mountains and "laid eggs like hell."

The only problem is that the super-roosters are unusually aggressive. Weirdo is so big and rough that he has killed two cats, crippled a dog and ripped through a wire fence to attack and kill one of his progeny, an 18lb rooster.

Flock of 40

Grant's superbreed now numbers 40—hens and roosters produced by Weirdo, many of them almost as big as he is. "I was lucky," he says. "Hybrids aren't supposed to be able to reproduce—but Weirdo could."

The youth's chicken breeding began seven years ago when his father won a lorry-load of chickens in a dice game at a bar he used to own.

Grant began selling eggs, and all was going well until the first snows of the winter began to fall on the Sierras. He lost more than 100 hens and roosters.

"Then somebody told me that Rhode Island Reds could stand plenty of cold," he recalled. "I threw some in and out I came with grey ones that laid eggs like hell."

Five thousands chickens later he had Weirdo. "I named him Weirdo because what else could you call something like that?"

Grant now has to decide who will look after his superchickens while he is away at university studying business administration.

Fig. 35.3 The newspaper report referred to in Question 6. (Reproduced by kind permission of the *Daily Telegraph*, London).

7 In nature living things compete with each other.
   (a) For what do they compete?
   (b) Giving examples, try to show some of the characteristics of organisms which may give advantage in competition.   (*O and C*)

8 Comment on this statement: 'The use of a new anti-bacterial drug may result in the development of new types of bacteria which are resistant to the drug.'

9 As part of a study of population growth an investigator introduced an equal number of flour beetles (*Tribolium confusum*) into two environments which were equal in every respect except that the first environment contained only 16 grammes of flour (food) and the second 64 grammes of flour. At intervals of time after the introduction the number of individual beetles was counted in each environment. The table below gives the results:

Days after introduction of Tribolium	Approximate number of individuals present in	
	16 g environment	64 g environment
0	20	20
7	20	20
40	200	300
60	550	800
80	560	1,300
100	650	1,750
120	640	1,600
135	650	1,900
150	645	1,500

Plot these results on a graph and discuss them.   (*O and C*)

10 Pheasants were introduced on to two islands, one a hundred square miles in area and the other two square miles in area. Population counts gave the following results:

	Large Island		Small Island	
	Spring	Autumn	Spring	Autumn
1938	30	100	30	68
1939	81	426	59	299
1940	282	844	101	307
1941	705	1,540	21	30
1942	1,325	1,898	18	24
1943	1,129	1,402	12	17
1944	1,037	1,463	6	7
1945	1,089	1,377	—	—

Comment on these figures.   (*CCJE*)

11 It has been suggested that senility has survival value because it leads to the death of old individuals which have become redundant. Comment.   (*OUE*)

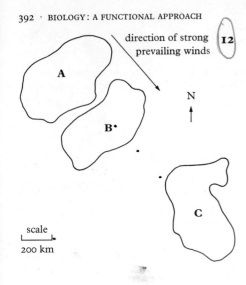

direction of strong
prevailing winds

N

scale
200 km

12 Three populations of birds **A**, **B**, and **C**, live separately on three isolated oceanic islands. The birds all eat both insects and nectar, but have slightly different beaks and plumage as shown in Fig. 35.4. The islands are swept throughout the year by strong prevailing winds from the north-west as shown in the figure.

Fig. 35.4 The birds **A**, **B**, and **C** which live on the three oceanic islands.

Populations **A** and **B** can interbreed and produce fertile offspring. Population **B** can mate with population **C**, but the young are sterile. Matings do not occur at all, even under laboratory conditions, between populations **A** and **C**.

From your knowledge of the way evolution works, formulate an hypothesis to explain how successful breeding occurs between **A** and **B**, but not between **A** and **C**, or **B** and **C**. (NSW)

13 A species may be defined as a group of individuals, recognizably different from other species, which are able to breed amongst themselves, but not with individuals of another species. Sub-species are groups within a species, recognizably different from one another, but capable of interbreeding.

In Australia four species of the rosella parrot (*Platycercus*) are distributed as shown in Fig. 35.5. The four species differ in the following respects:
P. icterotis: red head, yellow cheeks, underparts red.
P. eximus: red head, white and blue cheeks, red breast, bright-yellow belly.
P. adscitus: pale yellow head, white and blue cheeks, underparts pale blue.
P. venustus: black head, white and blue cheeks, pale yellow underparts.
Within each of the four species there is variation from one locality to another. In some cases the variations are sufficiently clear-cut for the variants to be regarded as sub-species.

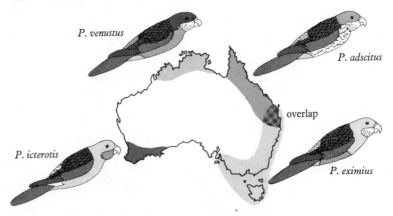

Fig. 35.5 Distribution of the four species of rosella parrots in Australia.

Discuss how the situation described above may have arisen. What difficulties do you think are involved in trying to define the terms species and sub-species?

14 *Inbreeding* is the crossing of individuals which are related to one another by descent. *Outbreeding* is the crossing of individuals which are unrelated.

Comment on the statement that 'inbreeding often results in lessened vigour, reduced size, and diminished fecundity and ultimately in inability of the stock to survive'. (Sinnott, Dunn, and Dobzhansky, 1952)

Answering questions **15** and **16** involves making use of the Hardy–Weinberg formula: if alleles **A** and **a** are present in a population with the frequencies of p and q, the proportion of individuals homozygous for the dominant allele (**AA**) will be p^2, the proportion of heterozygotes (**Aa**) will be $2pq$, and the proportion of double recessives (**aa**) will be q^2, where $p + q = 1$. (1 represents 100 per cent of the population).

15 Tongue-rolling (the ability to roll the tongue longitudinally into a U-shape) is caused by a dominant allele. Tongue-rollers are either homozygous for this allele, or heterozygous. Non-rollers are homozygous recessive.

In a survey carried out in Baltimore it was found that out of 5,000 people, 3,200 could roll their tongues, where 1,800 could not.

(a) What percentage of persons in the Baltimore sample are homozygous recessive for tongue-rolling?

(b) Calculate the frequency of the recessive allele in the population. In other words, what proportion of the alleles for the tongue-rolling trait in the population of Baltimore are recessive?

(c) Calculate the frequency of the dominant allele in the population.

(d) What percentage of persons in the Baltimore sample are homozygous dominant for tongue-rolling?

(e) What percentage of persons are heterozygous for tongue-rolling?

(f) What would you expect the answers to questions (a)–(e) to be in the next generation?

(g) What factor or factors might cause you to put forward a different answer to question (f)?

(h) What is the relevance of all this to the mechanism of evolution?

(BSCS modified)

16 Phenylketonuria (PKU) is a disease which may result in deterioration of mental ability and ultimately death. It is determined by a single pair of autosomal alleles in which normal is dominant to PKU.

(a) If the incidence of PKU in the population was one in every 10,000 people, what proportion of people are carriers of the disease? (Assume that the population is in equilibrium with respect to these alleles.) Show details of your calculations.

(b) If the incidence of mutation to the recessive allele was 1 mutation per 10^4 cells, would this markedly affect the frequency of the PKU gene in the population over a period of time? Explain.

(c) Would complete selection against homozygous PKU markedly affect the frequency of this gene in the population? Explain.

(d) It has been suggested that individuals exhibiting the PKU condition should not be permitted to reproduce. If all affected individuals in the population were sterilized, give three ways in which more affected individuals could appear in subsequent generations. *(SA)*

17 In populations of white clover, allele **Ac** controls the production of a cyanide-forming substrate while allele **Li** controls the formation of an enzyme releasing the cyanide. The recessive alleles **ac** and **li** indicate that either substrate or enzyme is lacking.

Clover that contains both dominant alleles gives off cyanide when the leaves are crushed and may also evolve cyanide spontaneously at low temperatures.

Individuals which include **Ac** but not **Li** in their genotypes give off cyanide slowly when the leaves are crushed. Where **Ac** is not present in the genotype no cyanide can be evolved on crushing.

From the above information suggest the likely genotypes of plants found (a) in a low-altitude field grazed by cattle, (b) in a hedgerow verge seldom grazed and (c) from the highest part of a mountain where the species can survive.

Explain how natural selection would tend to retain a mixture of genotypes within the species as a whole. *(AEB)*

18 Thalassaemia is a disease similar to sickle-cell anaemia in which persons homozygous for a recessive allele show a severe, and often lethal, effect (*see* p. 327).

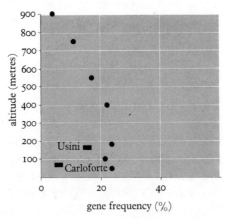

altitude (metres)

gene frequency (%)

Fig. 35.9 Diagram showing the frequency of the thallasaemia gene at different altitudes in Sardinia.

Fig. 35.9 illustrates the correlation between altitude and the gene frequency for thalassaemia in a group of villages in Sardinia. Each dot shows the average gene frequency for thalassaemia in these villages at the altitudes indicated.

The gene frequencies for the towns of Carloforte (altitude 70 m) and Usini (altitude 180 m) are indicated in the same diagram by black rectangles. Usini is a small village in which the local dialect shows definite influences of the Catalon language (of Spanish origin). Carloforte was only founded in 1700 by fishermen from Genoa (also of Spanish origin).

In Sardinia malaria is endemic at low coastal altitudes.

(a) Why might the gene frequency of thalassaemia be so high in villages (other than Carloforte and Usini) below 800 m?

(b) Suggest explanations for the low gene frequency of thalassaemia in Carloforte and Usini. (*SA*)

19 The following extracts are taken from *The Origin of Species* by Charles Darwin. Read the extracts carefully and then answer the questions concisely:

(**a**) Variation under domestication.

'Not one of our domestic animals can be named which has not in some country drooping ears; and the view which has been suggested that the drooping is due to the disuse of the muscles of the ear, from the animals being seldom much alarmed, seems probable.'

Comment briefly on this extract in the light of present-day knowledge.

(**b**) Struggle for existence.

'The causes which check the natural tendency of each species to increase are most obscure.'

Using observations you have made in your field studies, comment briefly on the view expressed by Darwin.

(**c**) On the imperfection of the geological record.

'The number of intermediate forms which have formerly existed must be truly enormous. Why then is not every geological formation and every stratum full of intermediate links?'

(i) What do you think Darwin means by 'intermediate links'?

(ii) How do you explain the imperfection of the fossil record? (*CL*)

20 Write a letter to Charles Darwin bringing him up to date on the theory of evolution, or to Gregor Mendel bringing him up to date on genetics. (*OCJE*)

36 Some Major Steps in Evolution

Background Summary

1 Pasteur, in his famous flask experiment, disproved, once and for all, the **theory of spontaneous generation**.

2 The **heterotroph theory** of the origin of life proposes that the first organisms to exist were heterotrophic. This is considered more likely than that the first organisms were autotrophs.

3 It is thought that the first step in the formation of living organisms was the synthesis of organic molecules from simple gases. This has been repeated in the laboratory under primitive earth conditions.

4 The first organism must have been able to obtain energy and reproduce. Circumstantial evidence suggests that, at first, energy was obtained **anaerobically**. Present knowledge of cell structure and function provides a possible insight into how they may have fed and reproduced.

5 Further evidence suggests that in the course of time autotrophic nutrition developed, paving the way to the development of **aerobic respiration**. It is envisaged that primitive autotrophs gave rise to the plant kingdom, heterotrophs to the animal kingdom.

6 The present day **phytoflagellates**, a group of primitive protists, provide some evidence as to how the split into the animal and plant kingdoms may have occurred.

7 There are various theories as to how **multicellularity** may have arisen. In general multicellular organisms may have arisen by failure of cells to separate after cell division, or from a colonial or multinucleate intermediate.

8 It is possible that certain present-day animals may have evolved from sexually mature larvae of ancestral forms. The gaining of sexual maturity by a larva (**neoteny**) is known to occur in certain animals.

9 A major development in the evolution of both animals and plants has been the exploitation of dry land. For this to take place various anatomical and physiological changes were necessary.

10 On grounds of comparative anatomy, embryology and biochemistry, the animal kingdom can be divided into the **annelid–mollusc–arthropod stock** and the **echinoderm–chordate stock**.

11 On the plant side, an early trend led to the separation of **reproductive** and **somatic cells**, and a later trend to the distinction between **sporophyte** and **gametophyte**. The great development of the sporophyte in flowering plants, with suppression of the gametophyte, is associated with adaptation to life on land.

12 In both animals and plants there have been trends towards larger size. These have not always been successful, as witnessed by the dinosaurs.

13 The evolution of man has been marked by the development of three salient features: an increased **mental ability**, manipulative **hands**, and **speech**.

14 Man has now entered upon a new phase of evolution, which may be called **psycho-social evolution**.

Investigation 36.1

The most primitive living organism?

Requirements

Microscope
Slides and coverslips
Filter paper
Lens paper
Mounted needle

Vaseline
Noland's solution
Methyl green in acetic acid
Acetocarmine
Methyl cellulose
Stained yeast suspension (*see* p. 21)
Silver nitrate: 2 per cent aqueous solution
Canada balsam

Euglena
Chlamydomonas
Vorticella
Amoeba
Paramecium

One way of gaining an insight into what the earliest organisms may have been like is to look at the simplest organisms living to-day. These include the protists. Can you think of any others? In doing this one must bear in mind that, although it is tempting to regard these organisms as primitive, they are, themselves, the products of many millions of years of evolution and may be markedly different from their ancestors.

Procedure

Protists are a large group of microscopic animals and plants. Most biologists regard them **unicellular**, i.e. single cells; however, some biologists prefer to think of them as **acellular**, i.e. organisms where the body is not subdivided into cells. The argument is academic.

Examine the following protists: *Euglena*, *Chlamydomonas*, *Vorticella* and *Amoeba* (*see* Investigation 2.2, p. 10), and *Paramecium* (*see* Investigation 3.3, p. 20). In what respects do they differ structurally? Determine how each moves and feeds.

Which do you think is the most primitive of these five protists, and why?

If available, examine other protists from mixed cultures. Do any strike you as being more primitive than the five which you have examined in detail?

For Consideration

(1) From your knowledge of protists in general, do you think there are more primitive ones than those looked at here?

(2) What do you regard as the most important criteria to take into account in deciding which are the most primitive protists?

(3) Structurally more simple than the protists studied here are bacteria and viruses. Would you regard these latter as the most primitive living or organisms?

(4) Critically evaluate the meaning of the word 'primitive'.

Investigation 36.2

Evolution of the multicellular state

The development of multicellularity was an important step in the evolution of both animals and plants. How did it arise? As in the previous investigation we may seek an answer to this question by looking at present-day forms. We can look, for example, at unicellular (or acellular) organisms to see if there is any tendency for them to become multinucleate or multicellular. Three common organisms—*Microstomum*, *Opalina*, and *Pleurococcus*—show interesting tendencies which may help to shed light on this question. However, remember that when it comes to interpreting evolutionary history on the basis of modern forms, we can *never* say what actually happened; at best, we can only say what *may* have happened.

Procedure

(1) Examine live specimens of the unicellular ciliate *Microstomum* under a microscope. How does it compare in size with other protists?

Irrigate the slide with methyl green in acetic acid. This stains the nucleus. How would you describe the shape of the **nucleus**? Suggest explanations for its shape.

(2) Examine the unicellular flagellate *Opalina* under a microscope. This lives in the rectum of the frog and toad. It may be seen by removing the contents of the rectum of a fresh-killed host onto a slide in a drop of amphibian Ringer's solution. Irrigate with methyl green in acetic acid, or acetocarmine, to see the nuclei.

Examine a prepared slide of *Opalina*, noting the **numerous nuclei**.

What light does the nuclear condition shed on the question of multicellularity? Why is it unlikely that *Opalina*, itself, represents the ancestor of multicellular animals?

(3) *Pleurococcus* is a simple green alga, found commonly on damp tree trunks, etc. Examine *Pleurococcus* under the microscope. Note that it occurs as a single cell, or as small **groups of cells**. What is the structural relationship between the cells where they occur in groups, i.e. are they stuck together or intimately united? What is the maximum number of cells in a group? Do the cells of a given group remain together permanently or do they tend to separate? How do you think the group condition arises?

(4) Compare *Pleurococcus* with (a) the permanently unicellular alga *Chlorella*, (b) permanently multicellular algae such as *Spirogyra* (*see* Investigation 2.2, p. 10), *Chaetophora*, and sea lettuce *Ulva*, and (c) colonial algae such as *Volvox* and its less elaborate relatives *Gonium*, *Pandorina*, and *Eudorina*. The cells in these colonial algae are similar to the unicellular alga *Chlamydomonas* (*see* Investigation 2.2, p. 10).

Examine these various algae alive under low and high powers.

For Consideration
From your observations in this investigation summarize the possible ways the multicellular state may have arisen in the animal and plant kingdoms generally.

Requirements
Microscope
Slides and coverslips

Methyl green in acetic acid
Acetocarmine
Amphibian Ringer's solution

Microstomum
Contents of rectum of frog or toad (for *Opalina*)
Pleurococcus
Chlorella
Spirogyra, *Chaetophora* and/or *Ulva*
Volvox and other colonial algae
Permanent slides of the above organisms

Questions and Problems

1 In 1953 Stanley Miller succeeded in synthesizing certain biologically important compounds under primitive earth conditions. A certain newspaper headlined this achievement as '*Life created in a test tube*'.

Write a letter to the newspaper putting the record straight.

2 Summarize the main steps in the heterotroph hypothesis for the origin of life, considering briefly the evidence for each of its propositions.

3 'Surely plants evolved before animals—they're so much simpler.' Discuss.

4 Giving due attention to caution, discuss the possible evolutionary significance of the following:
(a) The bodies of certain flatworms are incompletely cellularized.
(b) The larva of a certain amphibian becomes sexually mature and does not normally undergo metamorphosis into the adult.
(c) Flagellated collar cells are found in both the protists and the sponges.
(d) In flowering plants it is difficult to identify a gametophyte.

5 Why is it sometimes difficult to classify an organism as plant or animal? Use named examples to illustrate your answer wherever possible.   (*O and C*)

6 'The bigger the better.' Discuss.   (*CCJE*)

7 Find out the name of one species of animal or plant which has become extinct during recorded history. What were, or might have been, the causes of its extinction?

Can you think of any animals or plants which might, by now, have become extinct but for the direct or indirect intervention of man? Why do you think they were heading for extinction, and how has man saved them?

8 Speculate on the possible causes of the extinction of the dinosaurs.

9 Summarize the structural and physiological changes required for an aquatic animal to colonize dry land. What changes would be necessary for a plant to achieve the same thing? (*CCJE*)

10 What do we mean when we say that one group of organisms is more *primitive* than another? Illustrate your answer with examples.

11 By what criteria can the success of a group of animals or plants be assessed? What factors have contributed to the success of *either* insects, *or* mammals, *or* man?

12 'Dinosaurs became extinct when their great size changed from being an advantage to a disadvantage. Man has reached a comparable point: he is too clever by half, and faces the prospect of extinction'. Discuss. (*CCJE*)

Appendix 1

Who's Who in the Animal Kingdom

✓Phylum **Protozoa** ('first animals')

Main features
(1) Single celled (**unicellular**), unless one regards them as acellular.
(2) Usually contains a single nucleus, but sometimes multinucleate.

Sub-groups

FLAGELLATA (MASTIGOPHORA)★

One or more **flagella** which are generally used for locomotion.

Euglena: contains chlorophyll, looks green, occurs in freshwater ponds; photosynthetic.
Trypanosomes: no chlorophyll, colourless; blood parasite causing African sleeping sickness.
Collar flagellates: stalked, sessile; single flagellum creates water current for feeding; solitary or colonial.

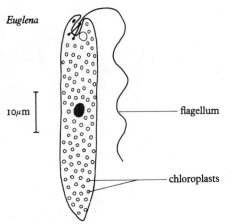

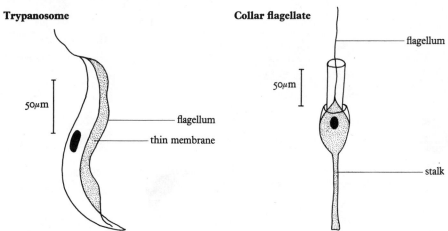

SARCODINA

Move by formation of **pseudopodia** ('false feet').

Amoeba: inhabits ponds; marine forms also known.
Entamoeba: different species parasitic in mouth cavity and intestine.
Foraminiferans and Radiolarians: shelled amoebas important in formation of calcareous and siliceous rocks.

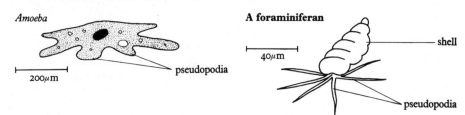

★ This group includes unicellular forms which are also considered members of the Algae in the plant kingdom (*see* footnote p. 412).

SLIME MOLDS: semi-terrestrial colonies of amoeboid cells which can move as a unit and reproduce by spores. Systematic position controversial.

CILIOPHORA

Possess covering of cilia, at least in juvenile stage; generally have two nuclei, large meganucleus and small micronucleus.

Paramecium: free-swimming, cilia present throughout life.
Stylonichia: groups of cilia fused to form leg-like cirri for locomotion.
Vorticella: stalked, sessile; cilia used for feeding.
Suctorians: cilia present during development only; adult has one or more suctorial tentacles for feeding.

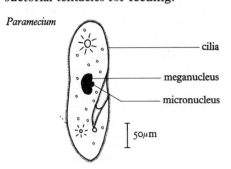

Paramecium

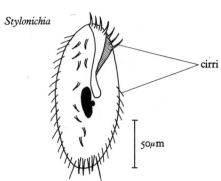

Stylonichia

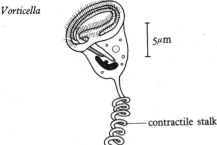

Vorticella

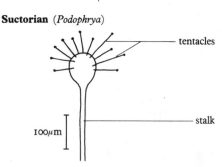

Suctorian (*Podophrya*)

Sporozoan (*Monocystis*)

SPOROZOA

No external locomotory devices; **spore-forming**; parasitic.

Plasmodium: blood parasite, causes malaria.
Monocystis: parasite in seminal vesicles of earthworm.

All animals from this point onwards are multicellular.

Phylum **Porifera** ('porous' sponges)

Main features

(1) Body contains cavity, or system of cavities, connected to exterior by **pores** (hence name of phylum).
(2) Cavity lined by flagellated **collar cells** for creating water current. (Collar cells closely resemble collar flagellates—*see* Protozoa).
(3) Skeleton of calcareous, siliceous or horny **spicules**.
(4) Incapable of locomotion, always **sessile**.
(5) Little inter-cellular integration and co-ordination; body virtually a colony of single cells.

Ascon: simple sponge with uniform body wall.
Sycon: more complex sponge with folded body wall.

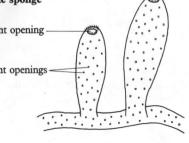

Simple sponge

exhalent opening

inhalent openings

Phylum **Coelenterata** ('gut-sac')

Hydra

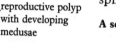

Main features

(1) Sac-like body cavity also serves as gut cavity (**enteron**).

(2) **Radially symmetrical**, i.e. section passing through any diameter gives two equal and opposite halves.

(3) Body wall contains two layers of cells (**diploblastic**): outer **ectoderm** and inner **endoderm** separated by non-cellular fibrous **mesogloea**.

(4) Exist in two forms, **polyp** and **medusa**, which, in a typical coelenterate life cycle, alternate with each other.

(5) Tentacles bear explosive cells (**nematoblasts**), certain of which can pierce and poison prey.

Sub-groups

HYDROZOA 'Hydra-like animals'

Hydra: solitary, semi-sessile polyp; no medusa.
Obelia: colonial, sessile polyp stage and free-swimming medusoid stage.

SCYPHOZOA 'Cup-like animals'

Jelly fish; complex medusa with a **gastrovascular system** of water-filled canals. Polyp stage reduced.

Aurelia: this is a complex medusa; polyp is a much reduced juvenile stage which splits horizontally to form medusae.

A schyphozoan jelly fish

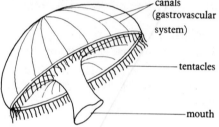

Obelia

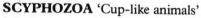

ANTHOZOA 'Flower-animals'

In the life cycle the medusa has been lost.

SEA ANEMONES

These are polyps; enteron is divided up by radial partitions (mesenteries).

CORALS

Mainly colonial, sea-anemone-like animals which secrete a calcareous skeleton.

Sea anemone   **A coral**

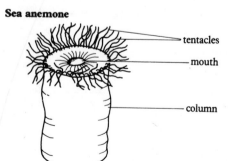

Animals from this point onwards show the following characteristics:
(1) **Triploblastic**: body composed of three layers of cells, outer ectoderm, inner endoderm and—between these two—the **mesoderm**.
(2) **Bilateral symmetry**: the body may be sectioned in only one plane to give two equal and opposite halves. (Note: the echinoderms are an exception).
(3) **Anterior end** ('head') and **posterior end** ('tail'); **dorsal side** ('back') and **ventral side** ('belly'). (Again the echinoderms are an exception).

Phylum **Platyhelminthes** ('flat worms')

Main features
(1) **Body flattened** dorso-ventrally (hence name of phylum).
(2) Possess a mouth but **no anus**.
(3) Gut generally has numerous blindly-ending branches.
(4) **Flame cells** for excretion and osmo-regulation.
(5) Phylum contains many important parasites.
(6) **Hermaphroditic**, often with elaborate precautions for minimizing self-fertilization.

Sub-groups

TURBELLARIA
Free-living flatworms, usually ciliated on underside.
Planaria: common flatworm found under stones in freshwater streams.

Planarian

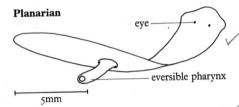

eye
eversible pharynx
5mm

TREMATODA
Ecto- and endoparasitic flukes; no cilia; suckers etc. for attachment to host.
Sphyranura: ectoparasite attached to skin of newts.
Fasciola hepatica: liver-fluke, endoparasite inhabiting liver and bile passages of sheep.

Sphyranura

anterior sucker
mouth

1mm

posterior sucker
hooks
anchor

Liver fluke

anterior sucker
mouth
ventral sucker

natural size

Tapeworm
(*Taenia*)

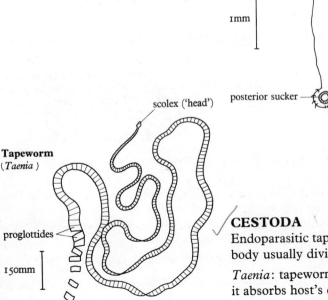

scolex ('head')
proglottides
150mm

CESTODA
Endoparasitic tapeworms; no cilia; no gut; suckers etc. for attachment to host; body usually divided into sections (proglottides).

Taenia: tapeworm in gut of pig (*Taenia solium*) and cattle (*T. saginata*) where it absorbs host's digested food; may exceed four metres in length.

Animals from this point onwards have an anus as well as a mouth.

Phylum **Nematoda**★ ('thread worms', 'round worms')

Main features
(1) **Narrow** body, **pointed** at both ends.
(2) In contrast to platyhelminthes, **rounded** in cross-section.
(3) Possess thick **elastic cuticle**.
(4) Longitudinal muscles in four quadrants, but no circular muscles.
(5) Phylum contains many important **parasites**.

Rhabditis: very common, mainly free-living; microscopic.
Ascaris: large worm, up to 20 cm long, parasitic in gut of man and pig.

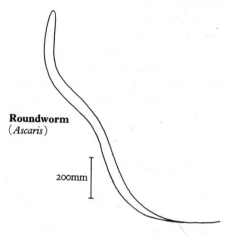

Roundworm
(*Ascaris*)

200mm

Phylum **Rotifera**★ ('wheel animalcules')

Main features
(1) Microscopic multicellular animals, amongst the smallest known.
(2) Circle of cilia on head (**trochal disc**) for locomotion and collecting food.
(3) This is a comparatively minor group but its members are very common inhabitants of freshwater ponds, etc.

Rotifer

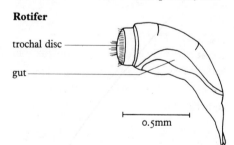

trochal disc

gut

0.5mm

★ In some classifications the nematodes and rotifers, together with certain other groups which do not feature in this survey, are combined into a single phylum, the **Aschelminthes**.

Animals from this point onwards generally display the following characteristics:
(1) **Coelom**, a fluid-containing cavity located between the body wall and gut and surrounded by mesoderm.
(2) **Metameric segmentation**, the serial repetition of some, but not all, of the structures and organ systems along the length of the body.
(3) **Circulatory system** containing an oxygen-carrying pigment such as haemoglobin.

Phylum **Annelida** ('ringed worms')

Main features

(1) **Metameric segmentation** is exhibited by more structures and organ systems than in virtually any other phylum.

(2) Externally, metamerism shows itself in constrictions (rings) between successive segments—hence the name of the phylum.

(3) Segments separated from each other by sheet-like **septa**.

(4) In most annelids each segment bears bristle-like **chaetae**.

(5) Segmental **nephridia** for excretion and osmo-regulation.

(6) Typically, there is a **trochophore larva** during development.

Sub-groups

POLYCHAETA

Numerous chaetae borne on lateral extensions of body wall called parapodia; separate sexes; marine.

Nereis: free-swimming, carnivorous ragworm with paddle-like parapodia and well developed head.

Arenicola: burrowing lugworm with reduced parapodia and head.

Sabella: tube-living fanworm with crown of ciliated tentacles for filter feeding.

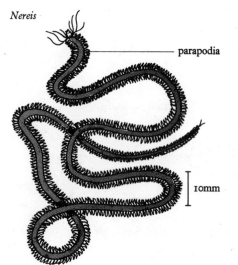

Nereis

parapodia

10mm

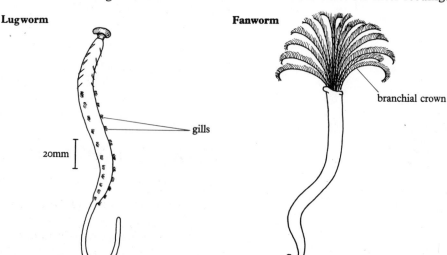

Lugworm

20mm

gills

Fanworm

branchial crown

OLIGOCHAETA

Chaetae few and not borne on parapodia; hermaphroditic; freshwater and terrestrial.

Lumbricus: burrowing earthworm with four pairs of chaetae per segment, and reduced head.

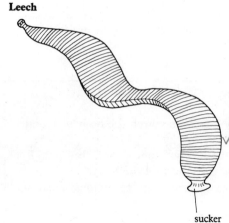

Leech

sucker

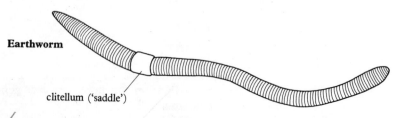

Earthworm

clitellum ('saddle')

HIRUDINEA

Leeches; **no chaetae** or parapodia; possess suckers at anterior and posterior ends; some forms ectoparasitic.

Hirudo: the medicinal leech; anterior sucker and sharp teeth for sucking blood.

Glossiphonia: common freshwater leech; carnivorous, feeds on small aquatic snails.

Phylum **Mollusca** ('soft-bodied')

Main features

(1) Ventral side of body typically has a soft muscular **foot**, hence name of phylum.
(2) On the dorsal side is a **visceral hump** containing the main digestive organs.
(3) Visceral hump generally protected by a **shell**.
(4) Most molluscs have a rasping tongue-like **radula** for feeding.
(5) **Gills (ctenidia)** for respiration and, in some cases, filter-feeding, located in chamber called **mantle cavity**.
(6) Majority have lost all traces of metameric segmentation.
(7) Typically there is a **trochophore larva** during development, thus linking them with annelids.

Sub-groups

MONOPLACOPHORA

The only group of molluscs showing metameric segmentation; single conical shell.

Neopilina: deep sea form with conical shell.

Neopilina — conical shell — foot (natural size)

AMPHINEURA

Limpet-like, with shell divided transversely into many units; marine.

Chiton: shell made up of 8 calcareous plates; numerous spicules project from sides of body.

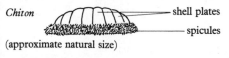

Chiton — shell plates — spicules (approximate natural size)

GASTROPODA

Large flat foot; visceral hump rotates (torsion) during development resulting in coiling of shell.

Buccinum: whelk; marine snail-like creature with coiled shell.
Helix: common garden snail; gills lost, mantle cavity becoming a 'lung'.
Testacella: slug; similar to snail but with greatly reduced shell.

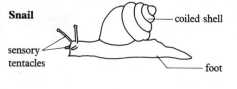

Snail — coiled shell — sensory tentacles — foot

LAMELLIBRANCHIATA

Laterally compressed; shell divided into two halves; sheet-like gills (hence name of group); head and foot generally reduced.

Anodonta: freshwater mussel, foot used for burrowing.
Mytilus: similar to *Anodonta* but marine and sessile, foot greatly reduced.

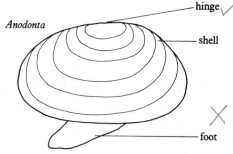

hinge — *Anodonta* — shell — foot

CEPHALOPODA

Foot incorporated into head (hence name of group); 8 or 10 suckers bearing tentacles; shell internal and reduced or absent.

Sepia: squid; 10 tentacles, reduced shell.
Octopus: basically similar to squid but no shell and 8 tentacles instead of 10.

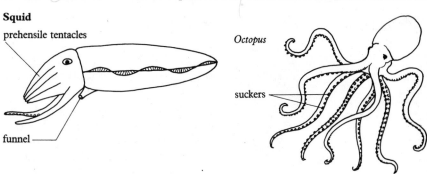

Squid — prehensile tentacles — funnel. *Octopus* — suckers

Phylum **Arthropoda** ('jointed limbs')

Main features
(1) **Chitinous cuticle**, hardened to varying degrees.
(2) **Jointed appendages** (hence name of phylum).
(3) Muscles attached to cuticle which therefore functions as an **exoskeleton**.
(4) Cuticle shed periodically (**moulting, ecdysis**) to allow for growth.
(5) Body cavity is a blood-filled **haemocoel**, derived from the blastocoel.
(6) Coelom much reduced.
(7) Metameric segmentation as in annelids but segments not separated from each other by septa, definite in number, and showing varying degrees of specialization.

Sub-groups

ONYCHOPHORA

Possess a mixture of **part-annelid** and **part-arthropod** features, thus providing a link between these two phyla.

Peripatus: caterpillar-like, confined to damp places in warm parts of the world.

Peripatus

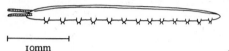

10mm

CRUSTACEA

Wide range of mainly aquatic arthropods held together by the fact that the second and third segments bear **antennae** ('feelers'), and the fourth segment bears **mandibles** ('jaws'). Dorsal side of body usually protected by a shield-like **carapace**.

Daphnia: 'water-flea', laterally compressed carapace; antennae used for swimming, limbs for filter-feeding.
Barnacles: sessile, head attached to rocks etc., limbs used for filter-feeding.
Gammarus: shrimp; common in streams and ponds; marine and freshwater; body compressed laterally.
Astacus: crayfish; as elaborate as any crustacean can be; extensive carapace; flexible abdomen. Lobster and crabs are close relatives of *Astacus*.
Armadillidium: woodlouse, the only fully terrestrial crustacean with several aquatic or semi-aquatic relatives; body compressed dorso-ventrally.

Crayfish

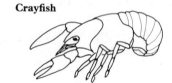

Woodlouse

Daphnia

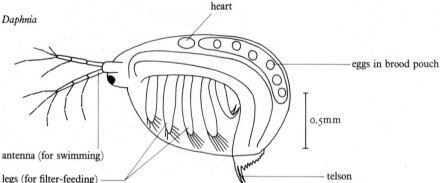

heart
eggs in brood pouch
0.5mm
antenna (for swimming)
legs (for filter-feeding)
telson

Centipede

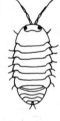

MYRIAPODA

Body elongated with numerous legs; internal features similar to insects; terrestrial.

CENTIPEDES: body flattened dorso-ventrally; carnivorous with pair of poison claws; 30 segments on average, each bearing one pair of legs.
MILLIPEDES: body cylindrical; herbivorous, no poison claws; 70 segments on average, each bearing two pairs of legs.

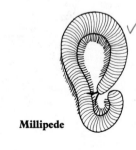

Millipede

Horseshoe (king) crab
Limulus

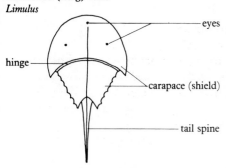

- eyes
- hinge
- carapace (shield)
- tail spine

Scorpion

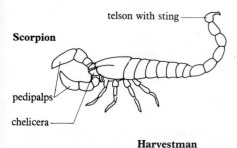

- telson with sting
- pedipalps
- chelicera

Harvestman

ARACHNIDA

Body divided into cephalothorax (head plus thorax) consisting of 6 segments, and abdomen consisting of 13 segments. Cephalothoracic segments bear, in the following order, a pair of jaw-like **chelicerae**, a pair of prehensile or sensory **pedipalps**, and 4 pairs of legs.

Limulus: misleadingly called the 'king crab' or 'horseshoe crab'; primitive marine form with large protective 'carapace' but recognizably an arachnid if one looks hard enough.

SCORPIONS: large pedipalps, flattened body with recurved stinging tail.

SPIDERS: abdomen secondarily unsegmented and separated from cephalothorax by a waist-like constriction; chelicerae contain poison glands.

Arachnida also includes ticks, mites and harvestmen.

Spider

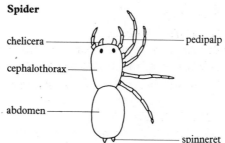

- chelicera
- pedipalp
- cephalothorax
- abdomen
- spinneret

Tick

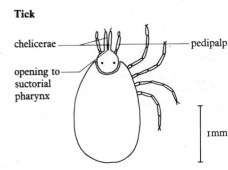

- chelicerae
- pedipalp
- opening to suctorial pharynx
- 1mm

INSECTA

Body divided into **head** (six segments fused); **thorax** (three segments bearing legs, second and third may have wings); and **abdomen** (11 segments). Single pair of long antennae. Well developed compound eyes for vision; tracheal system for respiration; and malpighian tubules for excretion.

APTERYGOTA Wingless insects

Lepisma: 'silver fish', common inhabitant of bathrooms and kitchens.

PTERYGOTA Winged insects

HEMIMETABOLA (EXOPTERYGOTA) **metamorphosis gradual**
(egg → nymphs → adult);
Wing rudiments during development external.
Cockroach; locust; grasshopper (orthopterans); dragonflies; earwigs; stick and leaf insects; stone flies; bugs; lice.

HOLOMETABOLA (ENDOPTERYGOTA) **metamorphosis complete**
(egg → larva → pupa → adult);
Wing rudiments during development internal.
Beetles; butterflies and moths (lepidopterans); bees, wasps, ants (hymenopterans); flies and mosquitoes (dipterans); fleas.

Lepisma

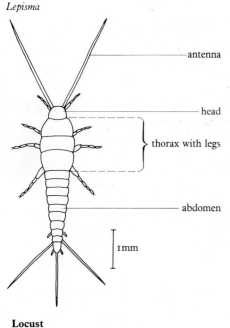

- antenna
- head
- thorax with legs
- abdomen
- 1mm

Locust

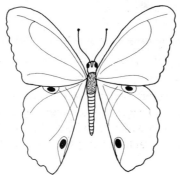

Butterfly

Beetle

Phylum **Echinodermata** ('spiny-skinned animals')

Starfish

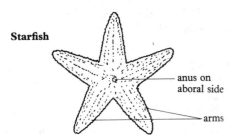

anus on aboral side

arms

Brittle star

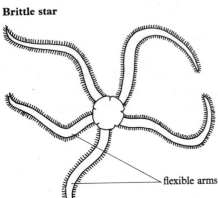

flexible arms

Main features

(1) Skin contains calcareous **ossicles** and **spines** (hence name of phylum).
(2) Exclusively marine.
(3) **Penta-radiate** in adult stage, but larva is bilaterally symmetrical.
(4) Mouth generally on lower (**oral**) side; anus on upper (**aboral**) side.
(5) Move slowly by concerted action of numerous suctorial **tube feet**.

Asterias: Common starfish of rock pools; five short arms each with tube feet on lower side.

Echinus: common sea urchin, very spiny; spherical, with five rows of tube feet and rasping jaw apparatus ('Aristotle's lantern').

Ophiothrix: brittle star; slender arms can move horizontally fairly rapidly, permitting crawling and swimming movements; tube feet reduced.

Cucumaria: sea cucumber; imagine a sea urchin, elongated in the oral–aboral axis, lying on its side: that's a sea cucumber; severe reduction in calcareous plates permits muscular worm-like movements.

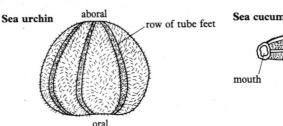

Sea urchin

aboral

row of tube feet

oral

Sea cucumber

mouth

(5-15cm long)

Phylum **Chordata**

Sea squirt (*Ciona*)

inhalant opening

exhalant opening

(natural size)

Main features

(1) **Notochord**, dorsally situated strengthening rod.
(2) Hollow, dorsal **nerve tube** (central nervous system).
(3) **Pharyngeal clefts**.
(4) Segmented muscle blocks (**myotomes**).
(5) **Post-anal tail**.
(6) Circulation in which blood flows forward ventrally and backwards dorsally.
(The above features constitute the **chordate characters**).

Sub-phyla

UROCHORDATA

Notochord present only in larva where it is confined to the **tail**. Adult is sessile filter-feeder.

Ciona: solitary 'sea squirt'; marine, sessile, looks very unchordate-like.

HEMICHORDATA

Notochord restricted to **proboscis**.

Balanoglossus: marine 'acorn-worm'; burrows into sand by means of muscular proboscis.

Acorn worm

proboscis

collar

CEPHALOCHORDATA

Notochord extends into **head**; other chordate characters clearly shown.

Amphioxus: marine, fish-like 'lancelet'; filter-feeder.

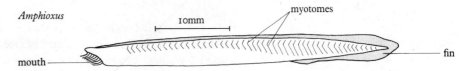

Amphioxus

VERTEBRATA (CRANIATA)

Notochord replaced in adult by **vertebral column** (backbone); brain enclosed in **cranium** ('brain-case').

The vertebrates are sub-divided into the following groups:

CYCLOSTOMATA

Jawless fishes; eel-like with round suctorial mouth and rasping tongue; no fins or scales; semi-ectoparasitic on fishes; numerous gills.

lamprey: attaches itself to body of fish, rasping and sucking at its tissues.

Lamprey

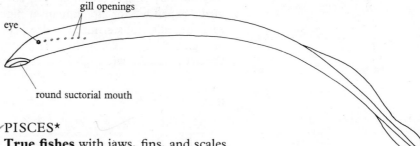

* In more detailed clasifications this term is not used, being replaced by several different groups of equal rank. However, for simplicity (and because the author likes it!) it is retained here.

PISCES*

True fishes with jaws, fins, and scales.

ELASMOBRANCHS

Cartilaginous fishes; spiracles; five pairs of gill slits; heterocercal tail fin; broad pectoral and pelvic fins; no air sac.

Dogfish: bottom-living scavenger.

Sharks: active carnivorous swimmers.

Also skates and rays.

TELEOSTS

Modern bony fishes; no spiracles; four pairs of gill slits covered by operculum; homocercal tail fin; fan-like pectoral and pelvic fins; air sac which usually serves hydrostatic function (swim bladder).

Carp, trout, cod, eel, plaice, sole.

DIPNOANS

'**Lung fish**' of South America, Africa and Australia; live in oxygen-deficient swamps; breathe air by means of lungs.

AMPHIBIA

Part aquatic, part terrestrial; simple sac-like lungs; **moist skin** used as supplementary respiratory surface; no scales; breed in water, fertilization external; aquatic larva (**tadpole**) undergoes metamorphosis into terrestrial adult; gills in larva, lungs in adult.

Newts and salamanders: tail present in adult as well as larva.

Frogs and toads: tail present in larva but lost in adult.

Shark

Ray

Cod

Lung fish

Newt

Frog

> Vertebrates from this point onwards all possess extra-embryonic membranes, including an amnion, during development.

Lizard

REPTILIA

Fully terrestrial, dry skin with **scales**; respiration exclusively by lungs; internal fertilization; eggs laid on land and enclosed in shell (**cleidoic**).
Lizards, snakes, crocodile and alligator, turtles and tortoise. Extinct forms include the dinosaurs.

Crocodile

Snake

Tortoise

AVES

Birds; similar to reptiles in many respects, but scales replaced by feathers, and forelimbs by wings; toothless jaws covered by horny beak; sternum of pectoral girdle expanded for attachment of flight muscles; well developed cleidoic egg.
Pigeon, gulls etc: sternum has deep keel associated with well developed powers of flight.
Ostrich, rhea, kiwi: flightless, running birds, large and heavy; sternal keel reduced.

Gull

Ostrich

MAMMALIA

Skin typically has **hair**; young nourished by milk secreted generally by mammary glands.

MONOTREMES

Primitive egg-laying mammals found in Australia; nourished by modified form of milk.
Duck-billed platypus (*Ornithorhynchus*) and spiny anteater (*Echidna*).

Duck-billed platypus

MARSUPIALS

'Pouch-mammals'; young born in miniature state migrate into pouch (marsupium) where they are fed on milk from mammary glands; simple, generally non-allantoic, placenta.

Opposums, Tasmanian wolf (*Thylacinus*), Koala bear, kangaroos.

Kangaroo

EUTHERIANS

Mammals possessing a true allantoic placenta; young born in a more mature state; marsupium absent.

Insectivores (e.g. shrew, mole), rodents (e.g. rats), carnivores (e.g. cats, dogs), ungulates (cattle, sheep, horses, tapirs), cetaceans (e.g. whales and porpoises), proboscideans (elephants), chiropterans (bats), and primates (e.g. lemurs, monkeys, apes, man).

Elephant shrew   **Man**

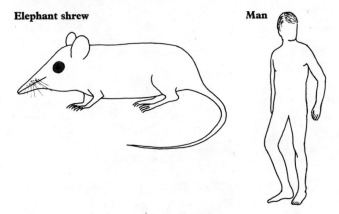

Appendix 2

Who's Who in the Plant Kingdom

Phylum **Thallophyta** ('thalloid' plants)

Main features
(1) Simple plants where the plant body is a **thallus**, i.e. not differentiated into root, stem, and leaf.
(2) Show wide diversity of form and size, ranging from microscopic unicellular organisms to large seaweeds.

Sub-groups

ALGAE
Photosynthetic thallophytes which themselves exhibit a wide range of form.

CHLOROPHYTA*
Green algae; chlorophyll as photosynthetic pigment.
Chlamydomonas: pond-dwelling unicell with two flagella and cup-shaped chloroplast.
Volvox: spherical colony of up to 20,000 *Chlamydomonas*-like cells.
Pleurococcus: very common non-motile unicell, found on damp tree trunks, walls etc.
Spirogyra: one of many filamentous algae; consists of a chain of identical cells, each of which contains one or more spiral chloroplasts.

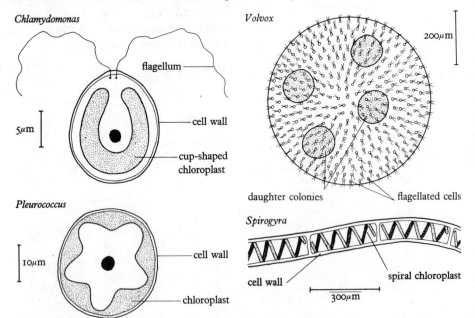

Chlamydomonas

flagellum

5μm

cell wall

cup-shaped chloroplast

Pleurococcus

10μm

cell wall

chloroplast

Volvox

200μm

daughter colonies

flagellated cells

Spirogyra

cell wall

spiral chloroplast

300μm

* This group includes unicellular forms which are also considered members of the sub-group Mastigophora of the Protozoa in the animal kingdom. This merely illustrates the artificiality of classification, and the difficulty of separating primitive unicells into either animals or plants. Some authorities consider that all unicellular organisms should be placed together in a separate kingdom called the **Protista**, a viewpoint that has much to commend it.

Fucus

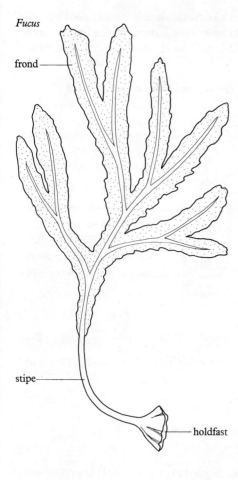

frond

stipe

holdfast

PHAEOPHYTA

Brown algae; photosynthetic pigment is chlorophyll plus a brown pigment (fucoxanthin); multicellular.

Fucus: common seaweed of the intertidal zone; thallus differentiated into holdfast for clinging to rocks, tough stipe, and flat slippery fronds.

Laminaria: similar to *Fucus* but with large elongated fronds.

RHODOPHYTA

Red algae; contain chlorophyll plus red pigment (phycoerythrin) and blue pigment (phycocyanin) of which the red predominates; microscopic, filamentous, or flattened.

CYANOPHYTA

Blue-green algae; contains chlorophyll plus phycoerythrin and phycocyanin, but in this case the blue pigment predominates; unicellular, colonial, or filamentous. (See note on p. 412 on systematic position of blue-green algae.)

CHRYSOPHYTA

Gold-brown algae; contains chlorophyll plus variable quantities of carotene (orange); unicellular, colonial, or filamentous; include diatoms, common in marine and fresh water.

FUNGI

Non-photosynthetic thallophytes; saprophytic and parasitic; plant body usually consists of a network (**mycelium**) of filamentous **hyphae**.

Mucor: 'pin mould', lives saprophytically on jam, bread, etc.

Neurospora and *Penicillium* are close relatives of *Mucor*.

Peronospora: parasite of various flowering plants.

Phytophthora, potato blight fungus, and *Pythium* which causes 'damping off' of seedlings are closely related to *Peronospora*.

MUSHROOM: mycelium saprophytic in soil; familiar 'mushroom' is the spore-producing reproductive body.

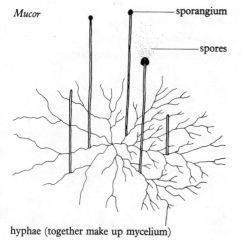

Mucor

sporangium

spores

hyphae (together make up mycelium)

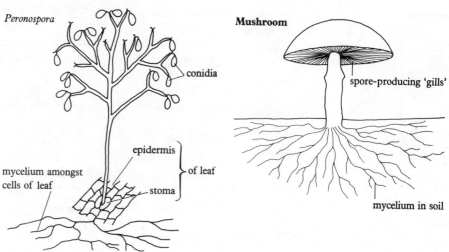

Peronospora

conidia

epidermis

of leaf

mycelium amongst cells of leaf

stoma

Mushroom

spore-producing 'gills'

mycelium in soil

YEAST: curious unicellular fungus which feeds saprophytically on sugars; can respire anaerobically and reproduces by budding.

Yeast (budding)

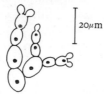

20μm

BACTERIA: controversial as to whether or not bacteria should be included with the fungi; have simplified cell structure and show wide variety of feeding methods of which saprophytism is but one; at the cell level they show a number of similarities to the blue-green algae.

Bacteria

Generalized bacterial cell in detail

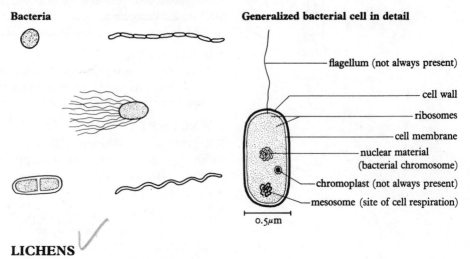

flagellum (not always present)

cell wall

ribosomes

cell membrane

nuclear material (bacterial chromosome)

chromoplast (not always present)

mesosome (site of cell respiration)

0.5μm

LICHENS

Plant body consists of alga and fungus living together to their mutual benefit.

Xanthoria: common orange-coloured lichen which grows on damp rocks and walls.

Phylum **Bryophyta**

Main features
(1) Generally more complex than the thallophytes, with simple leaves or leaf-like form.
(2) Show clear-cut **alternation of generations** between haploid gamete-producing gametophyte and diploid spore-producing sporophyte.
(3) Gametophyte is the more prominant phase in the life cycle.
(4) Gametophyte anchored to ground by filamentous **rhizoids**.
(5) Spore-producing sporophyte is attached to, and derives nourishment from, the gametophyte.
(6) The most obvious part of the sporophyte is the **spore-capsule** which is borne at the end of a slender stalk above the gametophyte.

Sub-groups
HEPATICAE
Liverworts; aquatic or semi-terrestrial gametophyte generally simple with thalloid body; rhizoids for anchorage; no stem.

Pellia: common liverwort growing on banks of streams etc.

MUSCI
Mosses: similar to liverworts but gametophyte more complex with body differentiated into simple stem and leaves; occur mainly in damp places.

Funaria: common moss growing on rock, waste ground, walls, etc.
Mnium: woodland moss similar to *Funaria*.
Polytrichum: large moss inhabiting moorlands.
Sphagnum: bog moss which thrives in areas where the soil is on the acidic side.

Liverwort

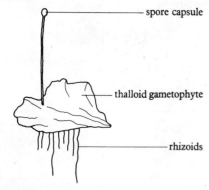

spore capsule

thalloid gametophyte

rhizoids

Moss

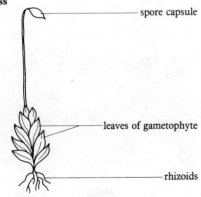

spore capsule

leaves of gametophyte

rhizoids

> Plants from this point onwards have vascular tissues for transport of water, mineral salts and soluble food substances. They are sometimes grouped together as the **Tracheophyta**.

Hart's tongue fern

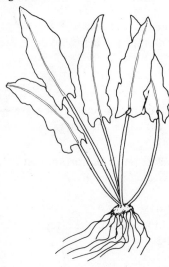

Phylum **Pteridophyta**

Main features
(1) Like bryophytes, show clear-cut **alternation of generations**, but in this case the **sporophyte** is the most prominent phase in the life cycle.
(2) Gametophyte reduced to small simple **prothallus**.
(3) Sporophyte has roots, stems, and leaves with **vascular tissues**.

Sub-groups
FILICALES
True ferns; large prominent leaves bearing sporangia on undersides.

Dryopteris: common fern of woods and hillsides with large divided leaves (fronds) and underground rhizome (stem) (*see* Fig. 25.4 p. 277). Compare hart's tongue fern with undivided leaves.

EQUISETALES
Horsetails; upright stem bears whorls of small leaves at nodes and spore-producing 'cone' at apex.

Equisetum: the only living genus; occurs in wet places.

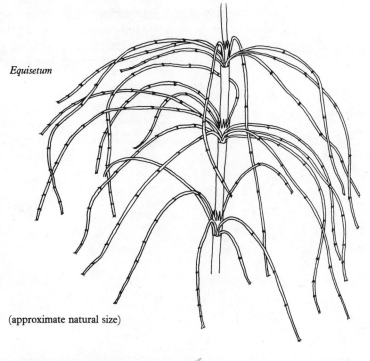

Equisetum

(approximate natural size)

Lycopodium

(approximate natural size)

Selaginella

(approximately ½ natural size)

LYCOPODIALES
Club mosses; not mosses at all, but small-leaved pteridophytes. Leaves generally densely arranged on branching stem.

Lycopodium: common club moss of moorland and hills.
Selaginella: produces two types of spore: microspore and megaspore which develop into male and female prothalli respectively. Prothalli, particularly male, much reduced.

Phylum **Spermatophyta** ('seed plants')

Main features
(1) Separate male and female spores (**pollen grains** and **embryosac** respectively).
(2) Embryosac enclosed in **ovule** which, after fertilization, develops into **seed** (hence name of phylum).
(3) Complex **vascular tissues** in roots, stems, and leaves.
(4) Main plant is the **sporophyte**, the gametophyte being severely reduced.

Sub-groups

GYMNOSPERMS
'Naked seeds', meaning that the ovules are unprotected; they are generally carried on spore-bearing leaves massed together into cones. No vessels in xylem, only tracheids.

Pinus: the familiar pine tree with needle-like leaves and prominent cones.
Also: fir, larch, cypress, cedar, redwood (*Sequoia*).

ANGIOSPERMS
Flowering plants; ovules protected within ovary; after fertilization ovary develops into fruit; xylem contains vessels.

DICOTYLEDONS
Embryo has two seed leaves (cotyledons); leaves generally have network of veins; stem contains ring of vascular bundles; secondary growth occurs.
Ranunculaceae (e.g. buttercup); Rosaceae (e.g. rose); Cruciferae (e.g. wallflower); Leguminosae (e.g. pea); Labiatae (e.g. deadnettle); Compositae (e.g. daisy); Hypericaceae (e.g. St John's wort); Euphorbiaceae (e.g. Para rubber tree); Cactaceae (e.g. prickly pear cactus). Dicotyledons also include familiar trees e.g. oak, beech, birch, lime, elm, and chestnut.

Pine

St John's wort

Prickly pear cactus

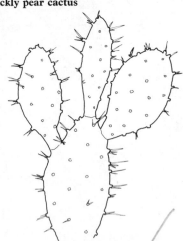

Horse chestnut

Iris

MONOCOTYLEDONS
Embryo has one seed leaf (cotyledon); leaves generally have parallel veins; stem contains scattered vascular bundles; secondary growth does not occur so, with the exception of palms, monocotyledons do not achieve great size.
Liliaceae (e.g. lily); Juncaceae (e.g. rush); Iridaceae (e.g. iris); Orchidaceae (e.g. orchid); Graminae (e.g. wheat and other grasses); Palmaceae (e.g. coconut palm).

Appendix 3   Composition of Foods

The composition of foods listed below are given per 100 g. Unless otherwise stated the figures are for the raw edible portion of the food. The data, based chiefly on McCance and Widdowson, are reproduced from the Ministry of Agriculture, Fisheries, and Food *Manual of Nutrition* (HMSO; 7th edition 1970).

Table 3.1   Composition per 100 g of raw edible portion of food

No.	Food	Inedible waste %	Energy value kcal	Energy value kJ	Protein g	Fat g	Carbohydrate (as monosaccharide) g	Calcium mg	Iron mg	Vitamin A (retinol equivalents) µg	Vitamin D µg	Thiamine mg	Riboflavine mg	Nicotinic acid Total mg	Nicotinic acid Equivalents mg	Vitamin C mg
Milk																
1	Cream, double	0	449	1,881	1.8	48.0	2.6	65	0	420	0.28	0.02	0.08	0	0.4	0
2	Cream, single	0	189	792	2.8	18.0	4.2	100	0.1	155	0.10	0.03	0.13	0.1	0.8	1
3	Milk, liquid, whole	0	65	272	3.3	3.8	4.8	120	0.1	44 (a) / 37 (b)	0.05 (a) / 0.01 (b)	0.04	0.15	0.1	0.9	1 (e)
4	Milk, condensed, whole, sweetened	0	322	1,349	8.2	9.2	55.1	290	0.2	112	0.12	0.10	0.40	0.2	2.0	3
5	Milk, whole, evaporated	0	166	696	8.5	9.2	12.8	290	0.2	112	0.12	0.06	0.37	0.2	2.0	2
6	Milk, dried, whole	0	492	2,061	26.6	27.7	37.6	813	0.7	246	0.30 (c) / 8.82 (d)	0.31	1.10	0.7	6.9	11
7	Milk, dried, half-cream	0	425	1,781	31.5	15.0	43.8	940	0.8	143	0.18 (c) / 8.82 (d)	0.36	1.35	1.0	8.2	10
8	Milk, dried, skimmed	0	329	1,379	37.2	0.5	46.9	1,277	1.1	4	0	0.30	1.73	1.1	9.7	10
9	Yoghurt, natural	0	57	239	3.6	2.6	5.2	140	0.1	39	0.02	0.05	0.19	0.1	0.9	0
10	Yoghurt, fruit	0	79	331	3.6	1.8	13.0	140	0.1	22	0.02	0.05	0.19	0.1	0.9	0
Cheese																
11	Cheese, Cheddar	0	412	1,726	25.4	34.5	0	810	0.6	420	0.35	0.04	0.50	0.1	5.2	0
12	Cheese, cottage	0	115	482	15.3	4.0	4.5	80	0.4	27	0.02	0.03	0.27	0.1	3.2	0

(a) Summer value   (b) Winter value   (c) natural value   (d) with fortification   (e) less than 1 mg

No.	Food	Inedible waste (%)	Energy value (kcal)	Energy value (kJ)	Protein (g)	Fat (g)	Carbo-hydrate (as mono-saccharide) (g)	Calcium (mg)	Iron (mg)	Vitamin A (retinol equiva-lents) (µg)	Vitamin D (µg)	Vitamin Thia-mine (mg)	Ribo-flavine (mg)	Nicotinic acid Total (mg)	Nicotinic acid Equiva-lents (mg)	Vitamin C (mg)
Meat																
13	Bacon, average	13	476	1,994	11.0	48.0	0	10	1.0	0	0	0.40	0.15	1.5	4.0	0
14	Beef, average	17	313	1,311	14.8	28.2	0	10	4.0	0	0	0.07	0.20	5.0	7.8	0
15	Beef, corned	0	224	939	22.3	15.0	0	13	9.8	0	0	0	0.20	3.5	7.7	0
16	Beef, stewing steak, raw	0	212	888	17.0	16.0	0	10	4.0	0	0	0.07	0.20	5.0	8.2	0
17	Beef, stewing steak, cooked	0	242	1,014	29.0	14.0	0	8	5.0	0	0	0.05	0.22	5.0	10.4	0
18	Chicken, raw	31	144	603	20.8	6.7	0	11	1.5	0	0	0.04	0.17	5.9	9.5	0
19	Chicken, roast	—	184	771	29.6	7.3	0	15	2.6	0	0	0.04	0.14	4.9	10.0	0
20	Ham, cooked	0	422	1,768	16.3	39.6	0	13	2.5	0	0	0.50	0.20	3.5	7.2	0
21	Kidney, average	0	105	440	16.9	4.2	0	14	13.4	300	0	0.30	2.00	7.0	11.1	12
22	Lamb, average, raw	17	331	1,387	13.0	31.0	0	10	2.0	0	0	0.15	0.25	5.0	7.7	0
23	Lamb, roast	—	284	1,190	25.0	20.4	0	4	4.3	0	0	0.10	0.25	4.5	9.8	0
24	Liver, average, raw	0	139	582	16.5	8.1	0	8	13.9	6,000	0.75	0.30	3.00	13.0	17.1	30
25	Liver, fried	0	276	1,156	29.5	15.9	4.0	9	20.7	6,000	0.75	0.30	3.50	15.0	22.4	20
26	Luncheon meat	0	325	1,362	11.4	29.0	5.0	18	1.1	0	0	0.40	0.20	3.5	6.1	0
27	Pork, average	15	408	1,710	12.0	40.0	0	10	1.0	0	0	1.00	0.20	5.0	7.7	0
28	Pork chop, grilled	40	527	2,208	18.6	50.3	0	8	2.4	0	0	0.80	0.20	5.0	9.2	0
29	Sausage, pork	0	369	1,546	10.4	30.9	13.3	15	2.5	0	0	0.17	0.07	1.6	3.9	0
30	Steak and kidney pie, cooked	0	304	1,274	13.3	21.1	16.2	37	5.1	126	0.55	0.11	0.47	4.1	6.0	0
31	Tripe	0	60	251	11.6	1.0	0	70	0.7	10	0	0.18	0.10	3.5	5.7	0
Fish																
32	Cod, haddock, white fish	40	69	289	16.0	0.5	0	25	1.0	0	0	0.06	0.10	3.0	6.0	0
33	Cod, fried in batter	0	199	834	19.6	10.3	7.5	80	0.5	0	0	0.04	0.10	3.0	6.7	0
34	Fish fingers	0	198	804	13.4	6.8	20.7	50	1.4	0	0	0.12	0.16	1.8	3.9	0
35	Herring	37	190	796	16.0	14.1	0	100	1.5	45	22.25	0.03	0.30	3.5	6.4	0
36	Kipper	40	220	922	19.0	16.0	0	120	2.0	45	22.25	0	0.30	3.5	6.9	0
37	Salmon, canned	2	133	557	19.7	6.0	0	66	1.3	90	12.50	0.03	0.10	7.0	10.6	0
38	Sardines, canned in oil	0	285	1,194	20.4	22.6	0	409	4.0	30	7.50	0	0.20	5.0	8.6	0
Eggs																
39	Eggs, fresh	12	158	662	11.9	12.3	0	56	2.5	300	1.50	0.10	0.35	0.1	3.0	0
Fats																
40	Butter	0	745	3,122	0.5	82.5	0	15	0.2	995	1.25	0	0	0	0.1	0
41	Lard, cooking fat															

Preserves etc.

No.	Food	1	2	3	4	5	6	7	8	9	10	11	12	13	14
44	Chocolate, milk	0	578	2,422	8.7	37.6	54.5	246	1.7	6.6	0.03	0.35	1.0	2.5	0
45	Honey	0	288	1,207	0.4	0	76.4	5	0.4	0	0	0.05	0.2	0.2	0
46	Jam	0	262	1,098	0.5	0	69.2	18	1.2	2	0	0	0	0	10
47	Ice cream, vanilla	0	192	805	4.1	11.3	19.8	137	0.3	1	0.05	0.20	0.1	1.1	1
48	Marmalade	0	261	1,094	0.1	0	69.5	35	0.6	8	0	0	0	0	10
49	Sugar, white	0	394	1,651	0	0	105.0	1	0	0	·	0	0	0	0
50	Syrup	0	297	1,244	0.3	0	79.0	26	1.4	0	0	0	0	0	0

Vegetables

No.	Food	1	2	3	4	5	6	7	8	9	10	11	12	13	14
51	Beans canned in tomato sauce	0	92	385	6.0	0.4	17.3	62	2.1	50	0.06	0.04	0.5	1.5	3
52	Beans, broad	75	69	289	7.2	0.5	9.5	30	1.1	22	0.28	0.05	4.0	5.0	30
53	Beans, haricot	0	256	1,073	21.4	0	45.5	180	6.7	0	0.45	0.13	2.5	6.1	0
54	Beans, runner	14	15	63	1.1	0	2.9	33	0.7	50	0.05	0.10	0.9	1.2	20
55	Beetroot, boiled	20	44	184	1.8	0	9.9	30	0.7	0	0.02	0.04	0.1	0.4	5
56	Brussels sprouts, raw	25	32	134	3.6	0	4.6	29	0.7	67	0.10	0.16	0.7	1.4	100
57	Brussels sprouts, boiled	0	16	67	2.4	0	1.7	27	0.6	67	0.06	0.10	0.4	0.9	35
58	Cabbage, raw	30	28	117	1.5	0	5.8	65	1.0	50	0.06	0.05	0.2	0.5	60
59	Cabbage, boiled	0	8	34	0.8	0	1.3	58	0.5	50	0.03	0.03	0.2	0.3	20
60	Carrots, old	4	23	96	0.7	0	5.4	48	0.6	2,000	0.06	0.05	0.6	0.7	6
61	Cauliflower	30	24	101	3.4	0	2.8	18	0.6	5	0.10	0.10	0.6	1.4	70
62	Celery	27	8	34	0.9	0	1.3	52	0.6	0	0.03	0.03	0.3	0.5	7
63	Lentils, dry	0	295	1,236	23.8	0	53.2	39	7.6	6	0.50	0.25	2.5	6.3	0
64	Lettuce	20	11	46	1.1	0	1.8	26	0.7	167	0.07	0.08	0.3	0.4	15
65	Mushrooms	25	7	29	1.8	0	0	3	1.0	0	0.10	0.40	4.0	4.5	3
66	Onions	3	23	96	0.9	0	5.2	31	0.3	0	0.03	0.05	0.2	0.4	10
67	Parsnips	26	49	205	1.7	0	11.3	55	0.6	0	0.10	0.09	1.0	1.3	15
68	Peas, fresh raw or quick frozen	63	63	264	5.8	0	10.6	15	1.9	50	0.32	0.15	2.5	3.5	25
69	Peas, fresh, boiled, or quick frozen boiled	0	49	205	5.0	0	7.7	13	1.2	50	0.25	0.11	1.5	2.3	15
70	Peas, canned, processed	0	96	402	7.2	0	18.0	29	1.1	67	0.06	0.04	0.5	1.6	2
71	Peppers, green	18	21	88	1.2	0.2	3.7	9	0.7	42	0.08	0.08	0.5	0.7	128
72	Potatoes, raw	27 (g) / 14 (h)	76	318	2.1	0	18.0	8	0.7	0	0.11	0.04	1.2	1.8	8–30 (i)
73	Potatoes, boiled	0	79	331	1.4	0	19.7	4	0.5	0	0.08	0.03	0.8	1.2	4–15 (i)
74	Potato chips, fried	0	236	989	3.8	9.0	37.3	14	1.4	0	0.10	0.04	1.2	2.2	6–20 (i)
75	Potatoes, roast	0	123	515	2.8	1.0	27.3	10	1.0	0	0.10	0.04	1.2	2.0	6–23 (i)
76	Spinach	25	21	88	2.7	0	2.8	70	3.2	1,000	0.12	0.20	0.6	1.3	60
77	Sweet corn, canned	0	95	398	2.6	0.8	20.5	5	0.5	35	0.03	0.05	0.9	0.3	4
78	Tomatoes, fresh	0	14	59	0.9	0	2.8	13	0.4	117	0.06	0.04	0.6	0.7	20
79	Turnips	16	17	71	0.8	0	3.8	59	0.4	0	0.04	0.05	0.6	0.8	25
80	Watercress	23	14	59	2.9	0	0.7	222	1.6	500	0.10	0.16	0.6	2.0	60

(f) some margarines contain carotene (g) refers to old potatoes (h) refers to new potatoes (i) vitamin C falls during storage

No.	Food	Inedible waste (%)	Energy value (kcal)	Energy value (kJ)	Protein (g)	Fat (g)	Carbohydrate (as monosaccharide) (g)	Calcium (mg)	Iron (mg)	Vitamin A (retinol equivalents) (µg)	Vitamin D (µg)	Thiamine (mg)	Riboflavine (mg)	Nicotinic acid Total (mg)	Nicotinic acid Equivalents (mg)	Vitamin C (mg)
	Fruit															
81	Apple	21	46	193	0.3	0	12.0	4	0.3	5	0	0.04	0.02	0.1	0.1	5
82	Apricots, canned	0	106	444	0.5	0	27.7	12	0.7	166	0	0.02	0.01	0.3	0.3	5
83	Apricots, dried	0	182	763	4.8	0	43.4	92	4.1	600	0	0	0.20	3.0	3.4	0
84	Bananas	40	76	318	1.1	0	19.2	7	0.4	33	0	0.04	0.07	0.6	0.8	10
85	Black currants	2	28	117	0.9	0	6.6	60	1.3	33	0	0.03	0.06	0.3	0.3	200
86	Cherries	15	46	193	0.6	0	11.8	18	0.4	20	0	0.05	0.06	0.3	0.4	5
87	Dates	14	248	1,039	2.0	0	63.9	68	1.6	10	0	0.07	0.04	2.0	2.3	0
88	Figs, dried	0	213	892	3.6	0	52.9	284	4.2	8	0	0.10	0.13	1.7	2.2	0
89	Gooseberries	1	27	113	0.9	0	6.3	22	0.4	30	0	0.04	0.03	0.3	0.4	40
90	Grapefruit	50	22	92	0.6	0	5.3	17	0.3	0	0	0.05	0.02	0.2	0.3	40
91	Lemons	64	7	29	0.3	0	1.6	8	0.1	0	0	0.02	0	0.1	0.1	50
92	Melon	41	23	96	0.8	0	5.2	16	0.4	160	0	0.05	0.03	0.5	0.5	25
93	Oranges	25	35	147	0.8	0	8.5	41	0.3	8	0	0.10	0.03	0.2	0.3	50
94	Orange juice, canned unconcentrated	0	47	197	0.8	0	11.7	10	0.4	8	0	0.07	0.02	0.2	0.2	40
95	Peaches, fresh	13	37	155	0.6	0	9.1	5	0.4	83	0	0.02	0.05	1.0	1.1	8
96	Peaches, canned	0	88	369	0.4	0	22.9	3.5	1.9	41	0	0.01	0.02	0.6	0.6	4
97	Pears, fresh	25	41	172	0.3	0	10.6	8	0.2	2	0	0.03	0.03	0.2	0.3	3
98	Pineapple, canned	0	76	318	0.3	0	20.0	13	1.7	7	0	0.05	0.02	0.2	0.3	8
99	Plums	8	32	134	0.6	0	7.9	12	0.3	37	0	0.05	0.03	0.5	0.6	3
100	Prunes	17	161	675	2.4	0	40.3	38	2.9	160	0	0.10	0.20	1.5	1.7	0
101	Raspberries	0	25	105	0.9	0	5.6	41	1.2	13	0	0.02	0.03	0.4	0.5	25
102	Rhubarb	33	6	25	0.6	0	1.0	103	0.4	10	0	0.01	0.07	0.3	0.3	10
103	Strawberries	3	26	109	0.6	0	6.2	22	0.7	5	0	0.02	0.03	0.4	0.5	60
104	Sultanas	0	249	1,043	1.7	0	64.7	52	1.8	0	0	0.10	0.30	0.5	0.6	0
	Nuts															
105	Almonds	63	580	2,430	20.5	53.5	4.3	247	4.2	0	0	0.32	0.25	2.0	4.9	0
106	Coconut, desiccated	0	608	2,548	6.6	62.0	6.4	22	3.6	0	0	0.06	0.04	0.6	1.8	0
107	Peanuts, roasted	0	586	2,455	28.1	49.0	8.6	61	2.0	0	0	0.23	0.10	16.0	20.8	0
	Cereals															
108	Barley, pearl, dry	0	360	1,508	7.7	1.7	83.6	10	0.7	0	0	0.12	0.08	2.5	2.2	0
109	Biscuits, chocolate	0	497	2,082	7.1	24.9	65.3	131	1.5	0	0	0.11	0.04	1.1	2.0	0
110	Biscuits, plain, semi-sweet	0	431	1,806	7.4	13.2	75.3	126	1.8	0	0	0.17	0.06	1.3	2.0	0
111	Biscuits, rich, sweet	0	496	2,078	5.6	22.3	72.7	92	1.3	0	0	0.12	0.04	1.0	1.5	0
112	Bread, brown	0	237	993	9.2	1.8	49.0	92	2.5	0	0	0.28	0.07	3.2	2.6	0

No.	Food														
116	Cornflakes	0	365	1,529	7.5	0.5	88.0	5	1.1	0	0	0.60	1.07	7.0	6.4	0
117	Custard powder, instant pudding, cornflour	0	353	1,479	0.5	0.7	92.0	15	1.4	0	0	0	0	0	0.1	0
118	Crispbread, Ryvita	0	318	1,332	10.0	2.1	69.0	86	3.3	0	0	0.37	0.24	1.4	1.3	0
119	Flour, white	0	348	1,458	10.0	0.9	80.0	145	1.9	0	0	0.28	0.04	1.8	2.8	0
120	Oatmeal	0	400	1,676	12.1	8.7	72.8	55	4.1	0	0	0.50	0.10	1.0	2.8	0
121	Rice	0	359	1,504	6.2	1.0	86.8	4	0.4	0	0	0.08	0.03	1.5	1.5	0
122	Spaghetti	0	364	1,525	9.9	1.0	84.0	23	1.2	0	0	0.09	0.06	1.7	1.8	0
	Beverages															
123	Black currant juice	0	229	960	0.2	0	60.9	14	0.5	0	0	0.01	0.02	0.1	0.1	206
124	Chocolate, drinking	0	410	1,718	5.6	6.8	87.0	25	12.0	2	0	0.03	0.09	0.5	1.4	0
125	Cocoa powder	0	446	1,869	18.8	22.5	45.0	52	15.0	7	0	0.08	0.30 (j)	1.7	4.8 (j)	0
126	Coffee, ground	0	0	0	0	0	0	0	0	0	0	0	0.20 (j)	10.0 (i)	10.0 (i)	0
127	Coffee, instant	0	156	654	4.0	0	35.5	140	4.0	0	0	0	0.10 (j)	45.0	45.7	0
128	Tea, dry	0	0	0	0	0	0	0	0	0	0	0	0.90 (j)	6.0 (i)	6.0 (j)	0
	Alcoholic Beverages															
129	Beer, mild, draught	0	25	105	0.2	0	1.6	10	0	0	0	0	0.05	0.7	0.7	0
130	Spirits, 70% proof	0	222	930	0	0	0	0	0	0	0	0	0	0	0	0
131	Wine, red	0	67	281	0.2	0	0.3	6	0.8	0	0	0.01	0.02	0.2	0.2	0
	Puddings, Cakes etc.															
132	Apple pie	0	294	1,231	3.2	14.4	40.4	42	0.8	2	0	0.08	0.02	0.6	0.9	2
133	Buns, currant	0	328	1,374	7.8	8.5	58.6	88	1.8	24	0.27	0.14	0.10	1.8	2.1	0
134	Fruit cake, rich	0	368	1,542	4.6	15.9	55.0	71	1.8	56	0.80	0.70	0.70	0.3	1.2	0
135	Jam tarts	0	391	1,638	3.2	13.8	67.7	50	1.3	0	0	0.06	0.01	0.6	0.8	0
136	Plain cake, Madeira	0	430	1,802	7.1	24.0	49.7	67	1.4	82	1.20	0.70	0.10	0.7	1.8	0
137	Rice pudding	0	142	595	3.6	7.6	15.7	116	0.1	96	0.08	0.05	0.14	0.2	1.0	1
138	Soup, tomato, canned	0	67	281	0.9	3.1	9.4	18	0.3	46	0	0.03	0.02	0.5	0.2	6
139	Trifle	0	162	679	3.1	5.6	26.5	75	0.6	73	0.30	0.04	0.10	0.3	1.1	2

(i) 90 to 100 per cent is extracted into an infusion

Appendix 4

Keeping a Practical Notebook

It goes without saying that the student should keep a practical notebook as a personal record of his or her investigations and observations. For this purpose a hard-backed book with alternating plain and lined paper (A4 size) is recommended.

You will be constantly faced with the question of what to enter in your notebook. The following notes are intended as a guide.

Dissection

(1) Write a *brief* summary of your technique, stressing any special points of procedure which you discovered for yourself.
(2) Draw the relevant parts of the completed dissection and label the appropriate structures. (Advice on drawing is given on p. 430.)
(3) Make sure your drawing has a heading, and indicate what view it represents, e.g. dorsal, ventral etc.

Microscopic work

(1) If you have made your own preparation (as opposed to using a prepared slide), give a brief account of your method.
(2) Draw the relevant parts of the object and label the appropriate structures.
(3) Be sure you state precisely what the object represents, e.g. whole mount, transverse section, etc.
(4) Give an indication of the scale.

Live specimens

Record your observations in the form of sketches and/or short notes, as appropriate. Get into the habit of recording your observations quickly and neatly, *while* you are observing the specimen. With experience you will learn to judge what is important and worth recording.

Experiments

Most students shudder at the thought of having to 'write up an experiment'. However, the labour is alleviated if you do as much of the writing-up as you can *during* the experiment. In biological experiments there are often odd moments when this can be done.

The format and presentation which you adopt in your notebook depends on the particular experiment. In general you should give an account of your **method**, a summary of your **results**, and a statement of your **conclusions**.

Note: The anatomical drawings in this manual are intended as a *guide* to help you identify structures which you yourself observe. You will gain very little by copying them direct into your notebook!

Appendix 5   Dissection

The object of dissection is to reveal the anatomy, not to destroy it. In achieving this, there are certain rules which every biologist should observe:

(1) Instruments should be kept in good condition: always clean and grease them after use; cutting tools should always be sharp.

(2) Except in special circumstances, pin the animal to a board or to the bottom of a dissecting dish so that the body wall is stretched.

(3) Where appropriate dissect the animal under water; this supports its organs and facilitates the separation of its tissues.

(4) If you are not dissecting in water, keep your dissection moist at all times. If you wish to leave it for any length of time, cover it with a damp cloth.

(5) Consider before each cut what organ is being sought and where it is likely to be; never cut or remove anything without knowing what it is.

(6) In dissecting such structures as nerves and blood vessels, work along, not across, their course.

(7) When following a nerve or blood vessel cut upwards towards you, not downwards towards the object.

(8) Remove only those structures which, if left in position, would obscure the structures you want to expose.

(9) In the final stage of your dissection make sure all the structures that you wish to show are clearly displayed.

(10) In displaying your dissection make judicious use of pins to separate structures from each other.

Appendix 6   Viewing Small Objects

Objects which are too small to be seen satisfactorily with the naked eye but too large for the ordinary microscope, may be observed with the aid of a hand lens or binocular stereo-microscope.

If you are using a hand lens, best results are obtained by having your eye as close to the lens as possible. Make sure the object is adequately illuminated against a contrasting background.

The binocular microscope operates on the same basis as a normal (monocular) microscope except that its magnification is less and, as it has two eye-pieces and two objectives, the image it gives is stereoscopic (three-dimensional). The distance between the two eye-pieces can be adjusted to suit the viewer.

Since it operates at comparatively low magnifications, the distance between the objectives and the specimen is sufficiently great for comparatively large objects to be placed under the binocular microscope. It is possible, for example, to place a dissection under the microscope, and indeed to carry out the dissection while looking through the microscope.

For gaining best results with a binocular microscope, it is important that the specimen should be correctly illuminated, preferably from a point-source of light, and that it is viewed against a contrasting background.

Appendix 7   The Compound Microscope

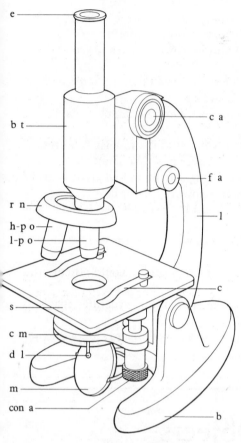

Fig. A1 A typical compound microscope. e, eye piece; b t, body tube; r n, rotating nosepiece; h-p o, high power objective; l-p o, low power objective; s, stage; c m, condenser mount; d l, diaphragm lever; m, mirror; con a, condenser adjustment; b, base; c, clip; l, limb; f a, fine adjustment; c a, coarse adjustment.

Structures which are too small to be observed by other means are studied with the aid of the compound microscope. In this instrument light rays which have passed through the specimen are transmitted through two lens systems, the objective and eyepiece. A typical student microscope is illustrated in Fig. A1.

The microscope is an expensive precision instrument and should be treated as such. When setting it up, adopt the following procedure:

Low power

Adjustment of lenses

(1) Place the microscope firmly on the table, not on books or papers. Set the microscope squarely opposite the source of illumination.

(2) The lenses must be quite clean. To test, hold them so that the light is reflected from their surfaces. Dirt or moisture should be removed by gentle wiping with a clean cloth or lens paper. Vigorous rubbing when grit is present scratches the lenses and makes them useless.

(3) Rack up the coarse adjustment until the objective lenses are about 20 mm above the stage. Turn the nosepiece so that the low-power objective is in use. Make certain that the objective has clicked exactly into line with the microscope tube.

(4) Note that the fine adjustment is a right-handed screw. Screwing in a clockwise direction *lowers* the objective.

(5) Place the slide to be examined on the microscope stage.

(6) Looking at the microscope from the side, rack down the coarse adjustment till the low-power objective is about 5 mm above the slide. Then, looking through the microscope, rack up the coarse adjustment till the object is in focus. (When looking through the microscope NEVER rack downwards to focus an object unless you know *with certainty* that by focusing downwards a very small distance the image will come into view).

(7) Take great care to focus accurately in order to avoid eyestrain. Keep both eyes open. Get accustomed to using either eye.

Adjustment of illumination

(1) The whole field of the microscope should be evenly illuminated. The best source of illumination is natural daylight or a diffuse bulb. If a filament lamp is used, interpose a thin sheet of paper between the bulb and the microscope. A point source of illumination is suitable only for work with very high powers.

(2) Adjust the flat mirror until light from the source is thrown up the microscope.

(3) Focus the condenser. To do this, adjust it until an object placed just in front of the source of light, reflected by the mirror, is seen in focus at the same time as the object on the microscope slide.

Always use the condenser focused; *never* use the condenser with the concave mirror. The lenses of the condenser are adjusted to give optimum illumination only when focused with the *flat* mirror.

(4) If the condenser is not in use, illuminate with the aid of the concave mirror. This method of illumination is, for most purposes, inferior to that using the condenser.

(5) The purpose of the condenser is to increase the illumination and to bring rays of light from a wide angle to bear on the object. Have this in mind when you are adjusting it.

(6) Open or close the diaphragm to the required extent. The condenser should be used with the diaphragm as wide open as possible, without admitting too

great an intensity of light. The *definition* of the image will then be at its best. If it is not possible to open the diaphragm widely without admitting too great a light intensity, place a sheet of paper between the microscope and the lamp.

N.B. A common cause of poor definition is that the object is over-illuminated. Best definition is often obtained by cutting down the light, not increasing it.

High power

Adjustment of lenses

(1) After the object is well defined under the low power, move the slide so that the part which is to be observed in greater detail is exactly in the centre of the field.

(2) Turn the nose-piece until the high-power objective clicks into place. The object should automatically come *approximately* into focus. If it does not do so, observe the microscope from the side, and rack the tube down until the lens is about one millimetre from the slide; then focus by racking *up*.

Adjustment of illumination

The diaphragm, etc., should be adjusted until the optimum intensity of illumination is obtained.

Magnification

(1) Do not use a higher power than is necessary. Far more can be made out with the low power in good illumination conditions, than under the high power with bad conditions. Also the larger the region of the object viewed in the field at the same time, the easier it is to interpret what is seen.

(2) Carry in your mind the degree of magnification. The following table shows the magnification given by typical lens combinations:

	Eyepieces	
	×6 (No. 2)	×10 (No. 4)
Objective 16 mm (×10)	60	100
Objective 4 mm (×40)	240	400

} Magnification

(3) Always enter a *rough scale* with any drawings that you make, based on the degree of magnification and the apparent size of the object.

Oil immersion

If particularly high magnifications are required an oil-immersion objective lens may be used. This is a special optical system in which a fluid of the same refractive index as the lens itself, is placed between the objective lens and the specimen. The fluid permits a larger cone of rays to enter the objective from the object than is otherwise possible, and this increases the resolving power obtainable. The fluid used is generally cedar-wood oil. A drop of the oil is placed on the coverslip above the specimen and the objective is lowered until the lens comes into contact with the oil. The object is then viewed with appropriate illumination in the normal way.

Dark-ground illumination

For small transparent objects it is often best to view the specimen as a bright object against a dark background. This involves the use of dark-ground illumination. In this technique the direct light illuminating the object must not enter the objective: the only light rays entering the microscope must be those which have been reflected or scattered by the object itself.

For low-power work this is achieved by interposing an opaque stop in the centre of the condenser. For high-power work a special dark-ground illuminator must be used.

Observations on living organisms under the microscope

Living organisms can be viewed under the microscope either on a slide or in a watchglass. The following points should be noted:

(1) Take care not to spill water on the stage of the microscope, and especially on the condenser.

(2) Take care not to wet the objective.

(3) When searching for organisms in a watchglass, place the latter on a slide. You can then move the slide easily on the stage and avoid the possibility of harming the condenser.

(4) *Never* tilt the microscope when there is living material on the stage.

Cleaning the microscope

Like all scientific apparatus the microscope should be kept clean. Above all, it must be kept free from traces of water (particularly sea water), fixatives, stains and reagents which readily corrode the instrument. *See that the microscope is clean and dry and in good order when you put it away.*

Tracing faults

If good definition is not obtained:

(1) Is the slide clean?

(2) Is the objective centred?

(3) Are the lenses free from dirt and moisture?

(4) Is the condenser adjusted and focused?

(5) Is the diaphragm adjusted?

(6) Is the microscope squarely placed in front of a suitable source of illumination?

(7) Is the lens itself faulty? If so, report the fact.

Appendix 8   Phase-contrast Microscopy

In normal microscopy structures in a specimen that would otherwise be transparent can be made visible by staining. However, these structures, by virtue of slight differences in density or refractive index, also produce invisible changes of phase in the light that passes through them. In phase-contrast microscopy these changes in phase are converted into corresponding changes of amplitude, resulting in a high-contrast image in which the distribution of light rays is related to the changes in phase. Provided there are variations in density or refractive index, any transparent object may be viewed this way. As fixation and staining are unnecessary, the technique can be used for examining living material that would otherwise be difficult or impossible to see.

Examining an object with phase-contrast involves placing a special annular disc beneath the condenser and using an objective fitted with a phase plate. Setting up the phase-contrast microscope can be a tricky business, particularly as the annuli and phase plates must be exactly aligned. However, several manufacturers are now producing student microscopes with built-in phase contrast equipment which is easy to handle and produces satisfactory results.

Appendix 9   Measuring the Size of an Object under the Microscope

This can be done using an **eyepiece micrometer scale**. This is a glass scale mounted in the focal plane of the eyepiece so it can be seen in the field of view at the same time as an object is being examined under the microscope.

Obviously to be of any use the eyepiece micrometer scale needs to be calibrated. This can be done by placing a **stage micrometer** under the microscope. This is a glass slide on which is etched a series of vertical lines separated by distances of 1.0 mm, 0.1 mm and 0.01 mm. By superimposing the images of the eyepiece micrometer and stage micrometer scales, the former can be calibrated so the size of a given object viewed under the microscope can be accurately estimated.

When calibrating, adopt the following procedure. Put the stage micrometer on the stage of the microscope, and bring its lines into focus. Move the stage micrometer until one of its lines coincides exactly with one of the numbered lines on the eyepiece scale. Count the number of lines on the latter which fills the space between the line that you have selected on the stage micrometer and the next one.

If the distance between the two lines on the stage micrometer is 100 μm and it is found that x eyepiece divisions exactly fill this space, then the value of one eyepiece division is $\frac{100}{x}$.

It is now possible to purchase a 100 μm scale, printed in transparent film, that can be used as an eyepiece graticule and/or stage micrometer.

Appendix 10   Preparing Material for Viewing under the Microscope

Temporary preparations

Observations on living material under the microscope are often very valuable. The material should be mounted in a drop of water, saline solution or glycerine on a slide and a coverslip applied. Anaesthetic fluids, fixatives, and stains may be introduced by a method called **irrigation**. A drop of the reagent is placed on the slide so that it just touches the edge of the coverslip. Fluid is then withdrawn from the opposite side of the coverslip by means of a piece of filter paper or blotting paper, and the reagent flows in to replace the fluid taken out. Care should be taken that there is always some fluid touching the coverslip to replace that removed. With delicate specimens be careful the organisms are not swept away by too rapid a rush of fluid.

Permanent preparations

The making of permanent preparations is no substitute for examining living material. However, many structures are difficult to see in the living material. In such cases the processes of fixation, staining, and mounting can make the study of details much easier.

The making of a permanent preparation involves the following processes:

Fixation. The purpose of this is to kill the living tissues with the minimum distortion, so as to permit subsequent staining and mounting of the preparations. Suitable fixatives include 70 per cent ethanol, formaldehyde (formalin), and Bouin's fluid.

Staining. The purpose of this is to colour structures which would otherwise be difficult, if not impossible, to see under the microscope. Staining is normally carried out during the dehydration process at the appropriate ethanol concentration. This will vary with the stain used: aqueous stains should be used before dehydration; stains in 50 per cent ethanol after dehydration in 50 per cent ethanol, and so on.

Differentiation. The purpose of this is to sharpen the contrast between, e.g. nuclei and cytoplasm. It is advisable to examine the specimen under the microscope during the differentiation process.

Dehydration. The purpose of this is to remove all traces of water from the stained material. This is carried out by passing it through a series of ethanols of gradually increasing strength. The appropriate staining technique is interpolated into this series at the correct point.

Clearing. The purpose of this is to remove the ethanol and render the material transparent. Suitable clearing agents are xylene and clove oil; when completely cleared the material will sink to the bottom. If dehydration has not been complete a milky precipitate may be formed in the clearing agent, or the specimen may remain floating. In either case it must be returned to absolute ethanol until dehydration is complete. The specimen must not be placed in balsam until completely cleared.

Mounting. The purpose of this is to embed the material in a suitable medium for observation under the microscope. When the material is in the clearing agent a small drop containing it is placed on a slide, excess fluid is drained off with blotting paper, a drop of Canada balsam is added and a coverslip applied. The slide should be left in a warm place: the balsam is dissolved in xylene and, as this solvent evaporates, sets hard so that the coverslip remains fixed.

The above procedures apply in general to the making of any permanent preparation. However, the details vary according to the material and the stain which is to be used.

The manipulations may be carried out either with the specimen on a slide (as in the case of smears or sections), or with the specimen in a watchglass (as in the case of complete organisms or pieces of tissue). In the former case the slide, with the specimen attached to it, is immersed for the appropriate time in a series of dipping jars. In the latter case the specimen is transferred from one reagent to another in a series of watchglasses.

When viewing the specimen under the microscope during the differentiation process, make sure the underside of the slide or watchglass is dry.

Several useful staining techniques are given below. For other techniques the reader should consult a handbook of microscopic technique:

Borax carmine

Specially suitable for whole mounts of animal material (i.e. solid pieces of tissue), this technique involves the use of a single stain.

Short method

(i) Transfer the specimen to 50 per cent ethanol, if not already in it.

(ii) Stain in **borax carmine** until the specimen is just thoroughly penetrated (about ten minutes).

(iii) Differentiate in acid ethanol until the material is pale pink. While differentiating examine under the low power of the microscope. Nuclei should be pink against a light background. If understained return the specimen to borax carmine; if overstained leave in acid ethanol.

(iv) Dehydrate in 90 per cent ethanol (ten minutes), and two successive lots of

absolute ethanol (5–30 minutes each, depending on thickness of specimen).
(v) Clear in xylene.
(vi) Mount in Canada balsam, supporting the coverslip with strips of paper or celluloid if necessary. Leave on hotplate until balsam hardens[1].

Long method
Stain the specimen in borax carmine for up to 24 hours. Differentiate in acid ethanol for between several days and six weeks. Examine the specimen under the microscope at intervals to determine the progress of differentiation. When the material is sufficiently differentiated, proceed as for the short method.

Borax carmine stains nuclei and cytoplasm pink, but since the acid ethanol removes the stain more completely from the cytoplasm than from the nucleus, there will be a difference of colour.

Haematoxylin and eosin
This is a double staining technique in which the material is treated with two stains in succession. It is specially suitable for sections of animal material, and for smears.
(1) Bring the material to be stained into 50 per cent ethanol, if not already in it.
(2) Stain in **haematoxylin** until the specimen is dark blue (two to five minutes).
(3) Blue in tap water.

While blueing, examine under low power. Nuclei should be blue; cytoplasm light or colourless. If understained return the specimen to haematoxylin; if overstained differentiate in acid ethanol.
(4) Dehydrate in 70 per cent, then 90 per cent ethanol (about three minutes each).
(5) Counterstain in **eosin** for two to five minutes.
(6) Replace in 90 per cent ethanol.

The specimen may be examined quickly under low power. Again make certain the underside of the slide is dry. If understained, return the specimen to eosin; if overstained leave it in the ethanol.
(7) Complete dehydration in absolute ethanol for about five minutes.
(8) Clear in xylene and mount in Canada balsam. Leave on hotplate until balsam hardens.

Safranin and light (or fast) green
This is a double staining technique which is suitable for botanical tissues, including sections. Fast green may be substituted for the light green: this has the advantage of fading less rapidly.
(1) Stain in **safranin** (ten minutes).
(2) Dehydrate in 50, 70, and 90 per cent ethanol, spending one minute in each.
(3) Complete dehydration in two successive lots of absolute ethanol (three to five minutes each).
(4) Counterstain in **light green** in clove oil (one minute).
(5) Clear and wash in clove oil (five minutes).

Examine under the microscope and if the material is understained with safranin, or overstained with light green, pass down through the ethanols and re-stain with safranin.
(6) Mount in Canada balsam and leave on hotplate.

Safranin and light green stain cytoplasm and cellulose green; lignified tissues and nuclei red; and chloroplasts pink.

Safranin and light green in cellosolve[2]
It is possible to carry out the dehydration and double staining together in a single solution. This consists of **safranin and light green in cellosolve** and is available, made up ready for use, from certain suppliers.
(1) Stain and dehydrate in safranin and light green in cellosolve (five to ten minutes).

[1] Canada balsam, ready for use, is available in tubes from which it can be made to flow by gentle squeezing. Other mounting media, e.g. euparal, can be used instead of balsam. Smears and their sections can be sprayed with tryolac spray which is, in effect, a liquid cover glass.
[2] 2-ethoxyethanol.

(2) Wash in cellosolve.

The cellosolve slowly removes the stain: watch under the microscope until the required intensity of staining is achieved.

(3) Transfer material into a mixture of equal volumes of cellosolve and xylene, then into pure xylene (two changes) for clearing.

(4) Mount in Canada balsam and leave on hotplate.

Appendix 11 Drawing

Drawings are an aid to precise observations and for this reason they are an important part of laboratory work. The biologist is not expected to be an artist, but to become, in some degree, a draughtsman. Clear and accurate line drawings are preferable to rough sketches on the one hand, or to elaborately shaded pictures on the other.

First consider what you want to show. Then plan your drawing so the various parts are in proportion and fit on the page. Small marks indicating the length and breadth of the drawing are a great help in planning, and a faint outline can be rapidly drawn to show the relative positions of the parts. The final outline should be drawn with clean firm lines and details should be put in clearly with a sharp pencil. If important details are too small to be shown in proportion, they can be shown in an enlarged diagram at the side. Do not shade unless you are a competent artist, and only in special circumstances should it be necessary to use coloured crayons. It should be possible to make the drawing perfectly clear by the judicious use of thick and thin lines and careful cross-hatching. Get into the habit of making your drawings large and clear.

As important as the drawing is the **labelling**. This should be done neatly in pencil. Each label should be connected to the appropriate part of the drawing by a clear guideline or pointer. Do not label too close to the drawings, and never write on the drawing itself. Always make sure that each drawing is fully labelled before you leave it.

It is sometimes appropriate, particularly when drawing live specimens, to make short succinct notes close to the labels. Such **annotated drawings** are particularly valuable as they combine a record of structure with functional observations.

Appendix 12 Breeding *Xenopus laevis*

Use mature specimens, at least two years old. Keep them at room temperature in a tank with perforated cover. Feed them twice a week on mealworms, earthworms or liver and change the water whenever necessary.

Identification of sexes

Mature females are much larger than males and they have labia ('flaps') surrounding the cloaca.

Bringing them into season

Mature toads can be brought into season at any time of the year by injecting them with pregnyl (chorionic gonadotrophin) which is available from suppliers in ampoules of stated capacity.

The toads have to be treated with specific amounts of pregnyl. This should be injected with a hypodermic syringe into one of the dorsal lymph sacs. These are located in front of the row of 'stitch marks' in the lumbar region of the body. The toad should be held firmly and the needle inserted as shown in Fig. A2. Keep the needle close to the skin so it does not penetrate the tissues beneath.

To allow for failure, inject at least three pairs of toads. Keep them in clean water in separate tanks at room temperature for 24 hours beforehand. For injection make up 500 international units (IU) of pregnyl in 1.0 cm^3 of distilled water or amphibian saline. Proceed as follows:

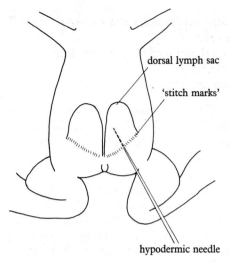

dorsal lymph sac

'stitch marks'

hypodermic needle

Fig. A2 Injecting the African clawed toad *Xenopus laevis*.

Primer injection

Two to four days before the eggs are required give the toads a primer injection of pregnyl: male–50 IU; female–100 to 500 IU.

Giving the female more than 100 IU may result in her spawning prematurely. Continue to keep the toads in separate tanks at room temperature.

Final injection

Twelve hours before the eggs are required give the toads a final injection of pregnyl: male–100 IU; female–300 IU.

Put the male and female toad together in the same tank and, using a thermostatically controlled heater, maintain at a minimum of 22°C. The tank should measure at least 30 × 20 × 20 cm. Reluctant females can sometimes be encouraged to spawn by raising the temperature to (not more than) 28°C.

It may be found helpful to spread a sheet of muslin over the bottom of the tank and cover this with a single layer of smooth rounded stones. Eggs will fall between the stones where they will be protected, and can be lifted from the tank on the muslin sheet. Otherwise they must be lifted out individually with a wide pipette.

Keep the eggs in bowls at about 23°C.

Allow at least six weeks to elapse before injecting the same toads again. One or two preliminary injections two to three weeks before the eggs are needed may sometimes raise the chances of success.

Maintaining the tadpoles

The tadpoles should be kept in a tank at about 25°C. The water should be changed and the tank cleaned once a week. The tadpoles are herbivorous and can be fed on powdered nettle leaves (available from suppliers). Do not overfeed: the food should disappear within 24 hours.

At about the time of metamorphosis, when the tail begins to be resorbed, they become carnivorous and should be fed on water fleas, small worms, etc., and later shredded heart tissue.

During metamorphosis it is advisable to oxygenate the tank and provide some means by which the young toads can climb out of the water.

Appendix 13   Rearing Chick Embryos

Obtain newly laid fertile eggs from a supplier, not longer than ten days before they are required. To allow for infertility (which is generally around 15 per cent), more eggs should be obtained than are needed. Arrest development by keeping them in a refrigerator at 10–12°C.

If three-day embryos are required, move the eggs to an incubator 72 hours before the embryos are needed. If older embryos are needed, move the eggs to the incubator at the appropriate length of time beforehand.

The incubator should be maintained at a steady 38°C. It must have some means of ventilation and the atmosphere inside it should be kept moist by means of an open container of water.

The eggs should be placed sideways on a tray. They should be rotated laterally through 180° each day to prevent the embryo sticking to the shell. When the egg is turned the embryo and yolk sac will automatically roll round so the embryo lies towards the upper side of the egg.

Appendix 14   Experiments with Live Animals

Experiments on live animals should be undertaken with extreme discretion and with due regard to the law. This may be summarized as follows as it applies to the United Kingdom.

Invertebrates: No legislation, but you are urged to anaesthetize if in doubt. As a general anaesthetic for cold-blooded animals MS-222 Sandoz is recommended.

Vertebrates: The 1876 Cruelty to Animals Act states that experiments may be performed on vertebrates only if nothing is 'liable to cause pain, stress, interference with or departure from, the animal's normal condition or well-being'.

This means:

(1) *No* dietary deficiency experiments.
(2) *No* electric shocks.
(3) *No* poisons or drugs.
(4) *No* inoculation with parasites.
(5) *No* decerebration, but pithing (destruction of the entire brain) is permitted.
(6) *No* injection of injurious substances, but injecting *Xenopus* with gonadotrophin is permitted.
(7) Experimental surgery on amphibian larvae and unhatched chick embryos is permitted.

Note: Pithing should be carried out only by a qualified teacher or technician, *not* by students.

Reference:
J. J. Bryant, *Biology Teaching in Schools Involving Experiment or Demonstration with Animals or with Pupils* (Association for Science Education, 1967).
Schools Council, *Recommended Practice for Schools Relating to the Use of Living Organisms and Material of Living Origin* (English University Press, 1974).

Appendix 15   Sterilization

It is necessary for instruments, culture media, and glassware to be sterilized for experiments involving micro-organisms, live embryos, tissue culture, etc.

Sterilization is most efficiently and conveniently carried out in an **autoclave** or, failing that, a **pressure cooker**. For most purposes it is sufficient to sterilize at 6.7 kg (15 lbs) pressure for 20 minutes.

Instruments, pipettes, syringes, etc., should be wrapped in paper or metal foil before being autoclaved and left wrapped afterwards until required. Test tubes should be plugged with cotton wool beforehand and the caps of screw-topped bottles should be loose.

After unsealing, instruments can be quickly re-sterilized by heating in the flame of a bunsen burner or spirit lamp, or by dipping in 50–70 per cent ethanol.

Bench tops should be washed with an antiseptic, e.g. three per cent solution of lysol, before experiments are started.

For sterilization of skin, swab the skin with cotton wool soaked in, e.g. 70 per cent ethanol. Alternatively use a pre-packaged medical swab obtainable from supplier.

Appendix 16   Kymograph

This apparatus usually consists of a vertical cylindrical **drum**, which can be made to rotate at a variety of speeds by either a clockwork or electric motor. It is used for recording movements, contraction of muscle, and so on. The speed can be adjusted according to requirements.

Movements can be recorded by a pen writing on white paper. Recording pens usually consist of a small well for holding the ink, with a narrow outlet to serve as the writing point. It is essential to use ink that does not clog: eosin is very satisfactory.

If kymograph recordings are to be of any value it is important to provide a **time scale**. With some kymographs the speeds at which the drum rotates are supplied, and a time scale can be worked out from this information. In cases where the speeds are not known, a time scale can be recorded on the revolving drum by means of a tuning fork vibrating at a known frequency (e.g. 100/s): one of its arms is fitted with a point that will write on the revolving drum. Alternatively, and more expensively, an electrical time marker can be used: driven from the mains, the writing point of the marker will blip at a given frequency.

Now available on the market are chart recorders incorporating a kymograph, electronic stimulator, time marker, and various other accessories. The stimulator will give single pulses, or repetitive pulses ranging from 1 per ten seconds to 100 per second.

Appendix 17

Apparatus for Delivering Electric Shocks

If an electronic stimulator is not available, the apparatus shown below may be used for delivering single or repetitive shocks.

Two electrical circuits are involved. The primary circuit incorporates a battery, key, and the primary coil of an inductorium.

The secondary circuit includes the secondary coil of the inductorium, the electrodes, and a short-circuit key which should be opened during an experiment and kept closed between experiments. A stimulus marker can be included in the primary circuit as shown in the diagram below.

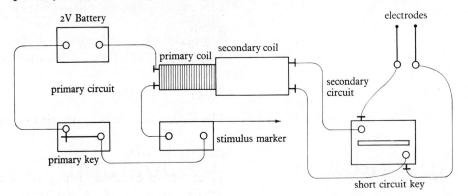

Principle

When the key in the primary circuit is closed an electromagnetic field is built up round the primary coil and a brief current is induced in the secondary circuit. This is called the *make* induction shock. If now the key in the primary circuit is opened the electromagnetic field falls off rapidly and a second shock, the *break* induction shock, passes through the secondary circuit.

It follows that the intensity of the induction shock depends on the relative proximity of the primary and secondary coils of the inductorium. If they are pushed close together the whole of the secondary coil will be influenced by the primary coil's electro-magnetic field and the current resulting in the secondary circuit will be large. If, on the other hand, the two coils are pulled far apart a relatively small portion of the secondary coil will come under the influence of the primary's electromagnetic field and the current will be correspondingly weaker. Even weaker shocks can be achieved by turning the secondary coil sideways. Owing to the more rapid change of primary current when the circuit is broken, a larger secondary current is induced at break than at make.

The contact at the end of the primary coil can be made to vibrate at a high frequency thus providing repetitive shocks: if this is required terminals **C** and **R** should be used. For single shocks use terminals **C** and **S**.

The system can be tested by connecting a pair of headphones to the output terminals of the secondary coil, or to the electrodes.

Appendix 18   Recipes for Biological Reagents

The reagents listed below are the more common ones that are constantly required for biological work. Details of other more specialized reagents are given at the ends of the particular laboratory investigations in which they feature.

Acetic orcein

Dissolve 3.3 g of orcein in 100 cm^3 of glacial acetic acid by gently boiling under reflux for about six hours. Filter. This gives a stock solution which can be stored.

When required for staining, dilute 10 cm^3 of stock solution with 12 cm^3 of distilled water. The diluted stain deteriorates quickly.

Acetic ethanol

Three parts of ethanol to one part of glacial acetic acid.

Acetocarmine

To 45 cm^3 of glacial acetic acid add 1.0 g of carmine. Mix and add 55 cm^3 of distilled water. Boil. Cool and filter.

Benedict's reagent

Dissolve 173 g of hydrated sodium citrate and 100 g of hydrated sodium carbonate in approximately 800 cm^3 of warm distilled water. Filter and make the filtrate up to 850 cm^3. This is solution **A**.

Dissolve 17.3 g of hydrated copper sulphate in approximately 100 cm^3 of cold distilled water. This is solution **B**.

Add **B** to **A**, stirring as you do so. Make up to 1 dm^3 with distilled water.

Borax carmine

Dissolve 4 g of borax in distilled water. Add 3 g of carmine and boil for 30 minutes. Add 100 cm^3 of 70 per cent ethanol, allow to stand for two days, then filter.

Bouin's fluid

Mix 5 cm^3 of glacial acetic acid and 25 cm^3 of formalin (40 per cent formaldehyde) with 75 cm^3 of a saturated aqueous solution of picric acid. The resulting solution will keep indefinitely.

Chloral hydrate solution

Dissolve 128 g of chloral hydrate in 80 cm^3 of distilled water.

Cobalt chloride paper

Dip a strip of filter paper, measuring approximately 2 × 5 cm, into a five per cent solution of cobalt chloride. Dry and keep in a dessicator.

Cobalt thiocyanate paper

Drip a strip of filter paper, measuring approximately 2×5 cm, into a 25 per cent solution of cobalt thiocyanate. Dry and keep in dessicator.

Cotton blue in lactophenol

Dissolve 1 g of cotton blue in 100 cm^3 of lactophenol.

Eosin (alcoholic)

Dissolve 1 g of alcohol-soluble eosin in 100 cm^3 of 90 per cent ethanol.

Fabil

The following solutions are required:
Lactophenol: phenol (crystals), glycerol, lactic acid, and distilled water in equal parts by weight.
A Aniline blue: 0.5 per cent in lactophenol.
B Basic fuchsin: 0.5 per cent in lactophenol.
C Iodine, 3 g; potassium iodide, 6 g; lactophenol 1 dm^3.
To make up the stain mix the stock solutions in the proportions of **A**, 4 : **B**, 1 : **C**, 5. Allow to stand for 12 hours, then filter. The stain will keep indefinitely.
(Noel, A. R. A., *School Science Review*, Vol. XLVII, No. 164, pp. 156–157).

Formalin

Commercial formalin contains approximately 40 per cent formaldehyde. For use as a fixative, add 90 cm^3 to 10 cm^3 of commercial formalin. The resulting 10 per cent formalin is suitable as a general fixative.

Glycerine

For mounting botanical sections use a concentration of five per cent. Prepare by making up 5 g of pure glycerine to 100 cm^3 with distilled water.

Haematoxylin (Ehrlich's)

Dissolve 2 g of haematoxylin in 100 cm^3 of absolute ethanol. This is solution **A**.
Add 3 g of aluminium potassium sulphate to 100 cm^3 of glycerol, 100 cm^3 of distilled water, and 10 cm^3 of glacial acetic acid. This is solution **B**.
Mix **A** and **B** and allow to ripen in an unstoppered bottle in bright daylight for two to three weeks.

Iodine

As a general stain for plant material, dissolve 1.0 g of iodine crystals and 2.0 g of potassium iodide in 300 cm^3 of distilled water.

Lactophenol

Dissolve 100 g of phenol in 100 cm^3 of distilled water without heating. Then add 100 cm^3 of lactic acid and 200 cm^3 of glycerine. Store in a brown glass bottle.

Light green

Dissolve 1 g of light green in a mixture of 25 cm^3 of absolute ethanol and 75 cm^3 of clove oil. (Fast green, which fades less rapidly, may be used as an alternative to light green).

Methylene blue

Mix 1 g methylene blue, 0.6 g sodium chloride and 100 cm^3 distilled water.

Millon's reagent

In a fume cupboard dissolve 1 cm^3 of mercury in 9 cm^3 of concentrated nitric acid. Dilute with 10 cm^3 of distilled water.

Methyl green (acetic)

Make up a 1 per cent solution of acetic acid in distilled water. To 100 cm^3 of this add 1 g of methyl green.

Noland's solution

Dissolve 20 mg of gentian violet in a mixture of 80 cm^3 of phenol (saturated solution in distilled water), 20 cm^3 of formalin (40 per cent formaldehyde) and 4 cm^3 of glycerol.

Phloroglucin

Dissolve 2–5 g of phloroglucin in 100 cm^3 of 95 per cent ethanol. (The solution should be acidified with concentrated hydrochloric acid before use.)

Ringer's solution (frog)

To 1.0 dm^3 of distilled water add 6.5 g of sodium chloride, 0.3 g of calcium chloride, 0.2 g of sodium bicarbonate, and 0.2 g of potassium chloride. Dissolve.

Ringer's solution (earthworm)

Dilute frog Ringer's solution with distilled water ($\times 6$).

Safranin

Dissolve 1 g of safranin in 100 cm^3 of 50 per cent ethanol.

Schultz' solution (chlor-zinc iodide)

Dissolve 110 g of zinc in 300 cm^3 of hydrochloric acid and evaporate to half the volume. In the course of the evaporation add a little more zinc to ensure complete neutralization of the acid. Dissolve 10 g of potassium iodide in the least possible quantity of water and add 0.15 g of iodine crystals. Mix thoroughly and, if necessary, filter through glass wool. Keep in tightly stoppered bottle in the dark.

Appendix 19   Statistical Tests

Chi-squared test

The **chi-squared test** enables us to assess the significance of differences between observed and expected results, and is particularly useful in analysing the results of genetic experiments.

The chi-squared value (χ^2) is a measure of the degree of deviation between an expected and observed result.

In mathematical terms:

$$\chi^2 = \Sigma \ \frac{d^2}{x}$$

where Σ means sum of,

d = deviation, i.e. difference between observed and expected results,

x = expected result.

From the X^2 value it is possible to calculate the probability that chance alone could be responsible for the deviation. The mathematics is complex, but the answer may be conveniently obtained from a table (Table A19.1).

It is generally accepted that when the probability of a deviation being due to chance is greater than 0.05 (5%), the deviation is *not significant*, i.e. there is no reason to consider that it is due to factors other than chance.

However, if the probability is 0.05 or less, then the deviation is regarded as *significant*, i.e. it is due, at least in part, to factors other than chance.

The chi-squared test is of particular value in establishing the significance, or otherwise, of deviations from expected ratios in genetic experiments. This is illustrated in the following very simple examples:

Example involving two classes of results (one degree of freedom)

Suppose that in a genetic experiment where the expected ratio is 3:1, we obtain a total of 40 plants, 32 with red flowers, and 8 with white flowers.

	red	white
Expected numbers:	30	10
Actual numbers:	32	8

$$\chi^2 = \left[\frac{(32-30)^2}{30}\right] + \left[\frac{(10-8)^2}{10}\right]$$

$$= \tfrac{2}{15} + \tfrac{2}{5} = 0.53$$

A χ^2 value of 0.53 corresponds to a probability of between 0.30 and 0.50, i.e. between 30 and 50 per cent.

Therefore, the deviation is not significant and can be regarded as the result of chance.

Example involving three classes of results (two degrees of freedom)

Suppose that in a genetic experiment where the expected ratio is 1:2:1, we obtain a total of 80 plants, 22 with red flowers, 42 with pink, and 16 with white. (What circumstances might produce such results?)

	red	pink	white
Expected numbers:	20	40	20
Actual numbers:	22	42	16

$$\chi^2 = \left[\frac{(22-20)^2}{20}\right] + \left[\frac{(42-40)^2}{40}\right] + \left[\frac{(20-16)^2}{20}\right]$$

$$= \tfrac{1}{5} + \tfrac{1}{10} + \tfrac{4}{5} = 1.10$$

A χ^2 value of 1.10 corresponds to a probability of between 0.50 and 0.70, i.e. 50–70 per cent.

Therefore, the deviation is not significant and can, again, be regarded as the result of chance.

The chi-squared test can also be applied to situations where there are four, five, or more classes of results (i.e. three, four, or more degrees of freedom).

Degrees of freedom	Number of classes	χ^2									
1	2	0.016	0.064	0.15	0.46	1.07	1.64	2.71	3.84	5.41	6.64
2	3	0.21	0.45	0.71	1.39	2.41	3.22	4.61	5.99	7.82	9.21
3	4	0.58	1.01	1.42	2.37	3.67	4.64	6.25	7.82	9.84	11.34
4	5	1.61	2.34	3.00	4.35	6.06	7.29	9.24	11.07	13.39	15.09
Probability that chance alone could produce the deviation		0.90 (90%)	0.80 (80%)	0.70 (70%)	0.50 (50%)	0.30 (30%)	0.20 (20%)	0.10 (10%)	0.05 (5%)	0.02 (2%)	0.01 (1%)

Table A19.1   Table of χ^2 values (based on Fisher)

Standard deviation

Imagine you are interested in an animal's body temperature and its weight. You measure each at regular intervals over a period of time and calculate the average or arithmetic mean.

How useful is the arithmetic mean? The fact is that on its own it is of little use for, although it tells us the animal's mean weight and body temperature, it gives us no information on the extent to which these two values have *fluctuated* during the period of time. This is obviously important for, in a warm-blooded animal, the body temperature will probably only fluctuate by a degree or two, whereas the body weight may fluctuate much more widely.

The extent to which a series of figures deviate from the arithmetic mean is expressed by the **standard deviation** (*SD*):

$$SD = \sqrt{\frac{\Sigma d^2}{n}}$$

where Σ means sum of,

d = difference between each number and the arithmetic mean,

n = the total number of numbers.

Standard error

It is sometimes necessary to assess the significance of differences between two or more sets of data.

Imagine you have two sets of data, each consisting of a series of numbers. In comparing the two sets of data you obviously need to know the extent to which differences between them are due to **chance** or sampling error, as opposed to other factors.

To establish this you require two pieces of information:

(1) The average of each set of data, i.e. the **arithmetic mean**.

(2) The extent to which the numbers within each set of data deviate from the arithmetical mean, i.e. the **standard deviation**.

The first step, then, is to calculate the arithmetic mean and standard deviation for each set of data.

The next step is to determine whether the difference between the standard deviation of the two sets of results is due to chance or is statistically **significant**, i.e. the result of factors other than chance. This is done by calculating the **standard error** (SE):

$$SE = \sqrt{\frac{SD_1{}^2}{n_1} + \frac{SD_2{}^2}{n_2}}$$

where SD_1 and SD_2 = standard deviation of each set of results,

n = the number of numbers in each set of results.

By general agreement, if the difference between the arithmetic mean of the two sets of results is more than twice the standard error, it is regarded as significant, i.e. due to factors other than chance. A difference of twice the standard error means that there is a 5 per cent probability that the difference is due to chance.

Example

Suppose we are interested in comparing the population of a certain organism in two different localities. In each locality we count the numbers of individuals in four different areas, each of which represents a sample.

The results are as follows:

Locality A

Sample no.	No. of individuals in each sample	Deviation from mean (d)	d^2
1	151	3	9
2	154	6	36
3	148	0	0
4	139	9	81
	Mean: 148		Total: 126

Locality B

Sample no.	No. of individuals in each sample	Deviation from mean (d)	d^2
1	130	2	4
2	124	8	64
3	112	20	400
4	162	30	900
	Mean: 132		Total: 1368

Applying the formula:
$$SD = \sqrt{\frac{\Sigma d^2}{n}}$$

Locality A
$$SD = \sqrt{\frac{126}{4}}$$

$$= 5.6$$

Locality B

$$SD = \sqrt{\frac{1368}{4}}$$

$$= 18.5$$

Now applying the formula: $SE = \sqrt{\frac{SD_1^2}{n_1} + \frac{SD_2^2}{n_2}}$

$$SE = \sqrt{\frac{5.6^2}{4} + \frac{18.5^2}{4}}$$

$$= 9.7$$

Difference between arithmetic mean of the results for localities **A** and **B**

$$= 148 - 132$$

$$= 16.$$

It is clear that this is considerably less than twice the standard error, so the difference is *not significant*, i.e. it can be ascribed to chance.

Application of standard deviation and standard error to frequency data
Suppose we want to investigate the occurrence of **variations** in a particular feature in two separate populations to see if there is a significant difference between them. In this case the procedure is a little more complex than in a simple population count because there are more data to take into account.

First we measure the variation under consideration: this may be the height of individuals in a population, the number of petals in a species of flowering plant, the number of bristles on the leg of a given species of insect, and so on. In some cases it is convenient to group the variations together into **classes**.

Next we count the number of individuals sharing each variation, or belonging to each class. These numbers are termed the **frequencies**.

The results can be presented in the form of a **frequency distribution histogram**, an example of which is shown in Fig. A4 (*see* also p. 443). The largest class, i.e. the most 'popular' variation, is known as the **mode** or **modal class**.

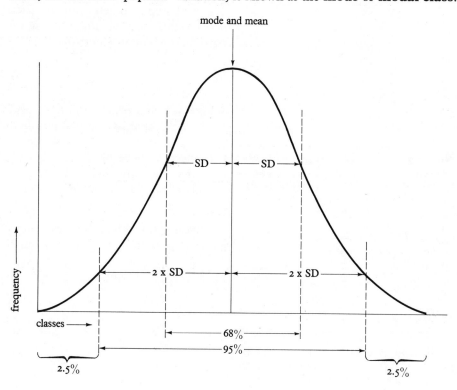

Fig. A3 Diagram of normal distribution curve.

The **mean** is the average variation for the population as a whole and is obtained from the formula:

$$\text{mean} = \frac{\Sigma xf}{\Sigma f}$$

where Σ means sum of,

x = measurements,

f = frequency.

In the example shown in Fig. A4 it so happens that the arithmetic mean is also the mode, but this need not always be the case.

The curve obtained by joining the tops of the vertical bars of the histogram is generally symmetrical on either side of the mode, and approximates to a **normal distribution curve**.

The area enclosed by the curve represents the total number of organisms in the population sample.

The **standard deviation** is the distance from the arithmetic mean to the steepest part of the curve (Fig. A3). In normal circumstances about 70 per cent of the population falls in the curve within one standard deviation on either side of the mean, and approximately 95 per cent falls within two standard deviations. In other words in normal circumstances *not more than approximately 5 per cent of the population depart from the mean by more than twice the standard deviation.*

Alternatively, and more accurately, the standard deviation can be calculated from the following formula:

$$SD = \sqrt{\frac{\Sigma fd^2}{\Sigma f} - \left(\frac{\Sigma fd}{\Sigma f}\right)^2}$$

Where Σ means sum of,

f = frequency,

d = deviation from mean.

The next step is to determine if differences in the occurrence of the variations between the two populations are significant. This can be done by calculating the **standard error of the difference** (SE):

$$SE = \sqrt{\frac{SD_2{}^2}{\Sigma f_1} + \frac{SD_2{}^2}{\Sigma f_2}}$$

Where SD_1 and SD_2 = standard deviations of the two population samples,
Σf_1 and Σf_2 = total number of individuals in each of the two population samples.

If the difference between the arithmetic means of the two populations is more than twice the standard error, then the difference is considered significant, i.e. due to factors other than chance.

EXAMPLE

An investigation is made into variations in the number of bristles projecting from a certain part of the body in a population of insects. The results* are at left:

A frequency distribution histogram corresponding to this data is shown in Fig. A4.

No. of bristles (classes)	No. of individuals belonging to each class (frequency)
2	I
3	3
4	6
5	9
6	10
7	9
8	5
9	2
10	I

* In reality the number of individuals investigated in this example would be insufficient, but the data serves to illustrate the calculations involved.

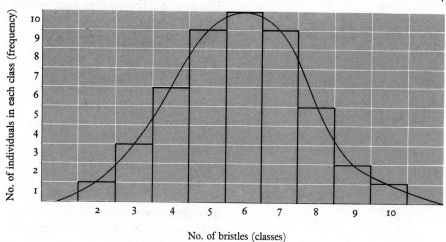

Fig A4 Frequency distribution histogram.

First we must calculate the arithmetic mean:

$$\text{mean} = \frac{\Sigma xf}{\Sigma f}$$

$$= \frac{271}{46}$$

$$= 5.9$$

Next we must calculate the standard deviation. If we did this by the method described on p. 439, the arithmetic would be tiresome, so we will employ the formula given on p. 442 in which the mean need not be exact. In this example the mean may be taken as 6 without in any way affecting the result.

First we calculate the deviations from the mean as shown in the table below:

No. of bristles (classes)	No. of individuals belonging to each class (frequency, f)	Deviations from mean (d)	fd	d^2	fd^2
2	1	-4	-4	16	16
3	3	-3	-9	9	27
4	6	-2	-12	4	24
5	9	-1	-9	1	9
6	10	0	0	0	0
7	9	$+1$	$+9$	1	9
8	5	$+2$	$+10$	4	20
9	2	$+3$	$+6$	9	18
10	1	$+4$	$+4$	16	16
Mean 6	Total (Σf) 46	Total (Σfd) -5			Total (Σfd^2) 139

Applying the formula
$$SD = \sqrt{\frac{\Sigma fd^2}{\Sigma f} - \left(\frac{\Sigma fd}{\Sigma f}\right)^2}$$

$$SD = \sqrt{\frac{139}{46} - \left(\frac{-5}{46}\right)^2}$$

$$= 1.7$$

In this example seven individuals out of 46 (i.e. over 15 per cent) depart from the mean by more than twice the standard deviation.

Suppose that in a second population of the same size the arithmetical mean is 4 and the standard deviation 1.8. Is the difference between the two population samples significant? To answer this we must calculate the standard error of the difference.

Applying the formula
$$SE = \sqrt{\frac{SD_1{}^2}{\Sigma f_1} + \frac{SD_2{}^2}{\Sigma f_2}}$$

$$SE = \sqrt{\frac{1.4^2}{46} + \frac{1.8^2}{46}}$$

$$= 0.37$$

The difference between the arithmetic means of the two populations = 2. This is considerably greater than twice the standard error. It is, therefore, probable that in this case the difference between the two populations is significant, i.e. due to factors other than chance.

Sources of Further Information

This is not an exhaustive bibliography. It is a selection of references which I, personally, have found useful for laboratory work, etc. and which I would wish, therefore, to recommend to others. Within each category elementary books are mentioned first, more advanced or specialized ones later.
A general reading list is given in *Biology: A Functional Approach*.

DISSECTION

H. G. Q. Rowett, *Dissection Guides* (UK: John Murray, 1950-53; U.S.A.: Holt, Rinehart and Winston)
Clear line diagrams, accompanied by instructions, on dissection of the dogfish, frog, rat, and selected invertebrates; available separately or combined in a single volume.

C. Gans and T. S. Parsons, *A Photographic Atlas of Shark Anatomy* (Academic Press, 1964)
A dissection guide consisting of 40 labelled photographs of the spiny dogfish *Squalus acanthias*, with brief explanatory notes. Very useful.

J. T. Saunders and S. M. Manton, *Practical Vertebrate Morphology* (Oxford University Press, 4th edition 1969)
This advanced manual includes instructions on dissections of lamprey, skate, bony fish, salamander, lizard, grass snake, and pigeon.

W. S. Bullough, *Practical Invertebrate Anatomy* (Macmillan, 2nd edition 1958)
Contains instructions with diagrams on over 100 invertebrate species including microscopic ones. Instructions on dissection are given where appropriate.

ANIMAL HISTOLOGY AND EMBRYOLOGY

H. G. Q. Rowett, *Histology and Embryology* (John Murray, 3rd edition 1966)
Like her dissection guides, this book contains clear diagrams and a minimum of text. The book includes mammalian histology, the embryology of the frog and chick, and invertebrate histology.

W. H. Freeman and Brian Bracegirdle, *An Atlas of Histology* (Heinemann Educational Books, 1966)
A collection of photomicrographs, accompanied by labelled diagrams, of mammalian tissues and organs. Includes some useful notes and schematic diagrams. Eminently suitable for advanced classes in schools and first degree students.

W. H. Freeman and Brian Bracegirdle, *An Atlas of Embryology* (Heinemann Educational Books, 2nd edition 1967)
Similar format to the previous book. Covers the development of *Amphioxus*, frog, and chick. Apart from some introductory notes on basic embryology, there is no text.

A. W. Ham, *Histology* (U.S.A.: Lippincott; U.K.: Pitman; 5th edition 1965)
Standard work of reference on mammalian histology in which microscopic structure is related to function. Full of useful information.

W. Bloom and D. W. Fawcett, *A Textbook of Histology* (Saunders, 8th edition 1962)
In this updated version of Maximow and Bloom's famous histology text, the fine structure of mammalian cells is integrated with traditional histology. Excellent for reference.

PLANT HISTOLOGY

A. C. Shaw, S. K. Lazell and G. N. Foster, *Photomicrographs of the Flowering Plant* (Longmans Green 1965)
Covering the stem, root, leaf, and flower, each photomicrograph is accompanied by a labelled line diagram. There is no text.

Brian Bracegirdle and Patricia H. Miles, *An Atlas of Plant Structure* (Heinemann, 2 volumes, 1971–1973)
A collection of photomicrographs, each accompanied by a labelled diagram, illustrating a wide range of plant structures from algae to angiosperms. Vol. 2 contains information on the gross anatomy of plants.

Mary-Anne Burns, *The Arlington Practical Botany, Book I Plant Anatomy* (Arlington Books, 1964)
A book of excellent drawings and notes in which emphasis is laid, not on topographical anatomy, but on the recognition of plant tissues.

EXPERIMENTAL WORK: GENERAL

Biological Sciences Curriculum Study (BSCS) publications
Produced by the American Institute of Biological Sciences, the BSCS programme consists of three alternative texts and ancillary publications. The main texts are *Biological Science: an Inquiry into Life* (BSCS Yellow Version), *Biological Science: Molecules to Man* (BSCS Blue Version), and *High School Biology* (BSCS Green Version). Each is accompanied by appropriate laboratory blocks, teachers' guides, etc.
   The BSCS programme provides a useful source of laboratory investigations for use in colleges and schools.

Nuffield Foundation Science Teaching Project: Biology (Longman and Penguin)
The Nuffield O level programme consists of five texts, each accompanied by a teachers' guide. Some of the laboratory investigations, particularly those in Text V, are more appropriate for A level than O level.
   The Nuffield A level publications include four Laboratory Guides, each accompanied by a teachers' guide. Although some of the investigations use difficult techniques or take a lot of preparing, the texts collectively provide a magnificent source of material and ideas for those who like to try new experiments.

G. D. Brown and J. Creedy, *Experimental Biology Manual* (Heinemann Educational Books, 1970)
This manual includes a wide range of experiments in microbiology, genetics, biochemistry, physiology, and behaviour. The instructions are clear and succinct.

P. Abramoff and R. G. Thompson, *An Experimental Approach to Biology* (W. H. Freeman, 1967)
Thirty-five laboratory exercises for first-year American college students. Some of the exercises, particularly those relating to cell biology, are useful.

G. Wald *et al*, *Twenty-six Afternoons of Biology* (Addison-Wesley, 2nd edition 1967)
This account of experiments featuring in the introductory biology course at Harvard University is full of useful ideas even though many of the experiments are demanding of apparatus.

EXPERIMENTAL WORK: GENETICS AND ANIMAL PHYSIOLOGY

J. J. Head and N. R. Dennis, *Genetics for O level* (Oliver and Boyd 1968)
Laboratory investigations, some too sophisticated for O level, run through this book (in green print). There are useful photomicrographs of mitotic and meiotic figures.

Gordon Haskell, *Practical Heredity with Drosophila* (Oliver and Boyd 1961)
In this useful little book the author explains how *Drosophila* should be handled and he suggests a series of suitable breeding experiments.

W. D. Zoethout, *Laboratory Experiments in Physiology* (U.S.A.: Mosby; U.K.: Henry Kimpton; 6th edition 1963)
Over 200 experiments in animal physiology (mainly mammalian), many of which are not unduly complicated.

R. B. Clark, *A Practical Course in Experimental Zoology* (Wiley, 1966)
Based on the author's own course, the experiments are designed for first-year university students in the U.K. Some of them can be adapted for more elementary use.

EXPERIMENTAL WORK: PLANTS

L. J. F. Brimble, *Intermediate Botany* (Macmillan, 1952)
This well-known text has suggestions for laboratory work at the end of each chapter, and a very useful classification of angiosperms at the end of the book.

W. O. James, *An Introduction to Plant Physiology* (Oxford University Press, 6th edition 1963)
In this classic textbook there is a section on laboratory work, consisting of a list of suggested experiments, at the end of each chapter.

W. M. M. Baron, *Organization in Plants* (Arnold, 2nd edition 1967)
There is an extensive series of experimental procedures in the appendix. In designing these the author spares us from unnecessary expense. Very useful.

EXPERIMENTAL WORK: ECOLOGY

D. P. Bennett and D. A. Humphries, *Introduction to Field Biology* (Arnold, 1965)
A general introduction to the practical aspects of ecology. Plenty of information on techniques and a useful section on statistical methods.

W. H. Dowdeswell, *Practical Animal Ecology* (Methuen, 1959)
A useful book dealing exclusively with animal work. There are separate sections on terrestrial, freshwater and marine habitats.

Maurice Ashby, *Introduction to Plant Ecology* (Macmillan, 1963)
Less practical than the other two, but a useful introduction to plant ecology with a helpful section on the nature of plant communities.

T. Lewis and L. R. Taylor, *Introduction to Experimental Ecology* (Academic Press, 1967)
More advanced than the previous books but very useful to the teacher on both the practical and theoretical aspects of this subject. Full of ecological exercises and experimental data. There is a useful section on techniques, and keys to common land invertebrates.

In addition to the books listed above: *The School Science Review* (published for the Association for Science Education by John Murray) and *The Journal of Biological Education* (published for the Institute of Biology by Academic Press) contain much useful information on experimental work suitable for advanced work in schools and equivalent levels.

Some of the teaching notes from the last twenty years' issues of *The School Science Review* have been brought together in book form. Published by John Murray, several titles are already available and others are scheduled for the future.

Also recommended is the *Investigations in Biology Series* (Heinemann Educational Books). These moderately priced booklets contain much helpful information on laboratory work.

ANIMAL AND PLANT TYPES

R. Freeman (Introducer), *Classification of the Animal Kingdom, an Illustrated Guide* (English Universities Press, 1972)
A reasonably detailed and very well illustrated classification of animals. Its attractive layout makes it possible to see the wood for the trees.

M. A. Robinson and J. R. Williams, *Animal Types* (Hutchinson, 2 volumes, 1970–1971)
With modern biology courses paying so little attention to animal types, there is a need for a book which surveys the animal kingdom in a simple and attractive way. These two slim volumes fill the bill admirably.

H. Godwin, *Plant Biology* (Cambridge University Press, 4th edition 1945)
The middle chapters of this little textbook deal with selected plant types. The drawings are based on actual specimens and are, therefore, useful in the laboratory.

ANIMAL AND PLANT IDENTIFICATION

J. L. Cloudsley-Thompson and John Sankey, *Land Invertebrates* (Methuen, 1961)
An admirable little book, well illustrated with line diagrams and containing simple identification keys, eminently suitable for schools.

T. T. Macan, *A Guide to Freshwater Invertebrate Animals* (Longmans Green, 1959)
Similar to the previous book in format and depth but covering freshwater invertebrates. Good diagrams and useful identification keys.

Helen Melanby, *Animal Life in Fresh Water* (Methuen, 6th edition 1963)
No keys, but very clear diagrams and excellent background information on the various animals surveyed.

John Barratt and C. M. Yonge, *Pocket Guide to the Sea Shore* (Collins, 1958)
A well illustrated survey of the organisms that live in rock pools, etc., on the sea shore.

W. Keble Martin, *The Concise Flora in Colour* (Ebury Press and Michael Joseph, 2nd edition 1965)
This best-selling survey of the families of flowering plants is beautifully illustrated with the author's own paintings.

F. K. Makins, *Concise Flora of Britain* (Oxford University Press, 2nd edition 1957)
A reliable flora suitable for use in elementary biology courses. Will fit into the average pocket.

Also useful for identification purposes are *The Observer's Books* (Warne), *Wayside and Woodland Series* (Warne), and Collins' Pocket Guide Series.

QUESTIONS AND PROBLEMS

Joseph J. Schwab (Supervisor), *BSCS Biology Teachers' Handbook* (Wiley, 1963)
This book contains 44 *Invitations to Enquiry*, involving the formulation of hypotheses and interpretation of data in a wide range of investigations.

Garrett Hardin, *Biology: Its Principles and Implications* (Freeman, 2nd edition 1966)
Some of the questions and problems at the ends of the chapters are stimulating and thought-provoking.

S. W. Hurry and D. G. Mackean, *Enquiries in Biology* (John Murray, 1968)
Seven investigations involving interpretation of experimental evidence and data. The investigations include muscle action, breathing, photosynthesis, and moulting in insects. There is a teachers' guide.

P. J. Kelly (Editor), *Study Guide: Evidence and Deduction in Biological Science* (Penguin, 1970)
This is the largest of the Nuffield Advanced Science biology publications. Providing ample opportunities for the handling of biological data, it is divided into four sections covering maintenance of the organism, populations, development, and co-ordination.

J. M. Eggleston, *Problems in Quantitative Biology* (English Universities Press, 1968)
If you like grappling with quantitative data and numerical problems, you will enjoy this book. The problems cover a wide range of topics, including physiology and genetics.

Margaret K. Sands, *Problems in Plant Physiology* (John Murray, 1971)
Problems involving analysis of data in photosynthesis, germination and growth, water relations, translocation, respiration, and plant hormones. There is a teachers' edition with answers.

Students are also advised to examine past examination questions. Past papers are generally obtainable from the offices of the various examining boards or from certain booksellers.

GENERAL LABORATORY TECHNIQUES

Peter Fry (Editor), *Biological Science—Laboratory Book: a Technical Guide* (Penguin, 1971)
One of the Nuffield Advanced Science publications. It contains instructions for teachers and technicians on culture methods, chemical recipes, and apparatus.

James J. Needham *et al*, *Culture Methods for Invertebrate Animals* (Dover Publications, 1937)
Compiled by the American Association for the Advancement of Science, this is an old book, but it contains a wealth of information on the rearing and maintenance of a wide range of invertebrates.

P. Hunter-Jones, *Rearing and Breeding Locusts in the Laboratory* (Anti-Locust Research Centre, 1966)
For anyone who wishes to keep locusts in the laboratory this booklet (only 12 pages) is indispensable.

The UFAW Handbook on the Care and Management of Laboratory Animals (Livingstone, 1967)
One thousand pages on every conceivable aspect of keeping and handling animals in the laboratory. Edited by the staff of the Universities Federation for Animal Welfare. Useful to have around!

Schools Council, *Recommended Practice for Schools Relating to the Use of Living Organisms and Material of Living Origin* (English University Press, 1974).
Drawn up by the Schools Council working party, this short book deals with the choosing, maintenance and humane usage of living organisms in schools.

USE OF THE MICROSCOPE

C. A. Hall and E. F. Linssen, *How to use the Microscope* (A. & C. Black, 6th edition 1968)
Sub-titled *A Guide for the Novice*, this is a practical introduction to microscopy with no physical background. There is a useful section on the use of the hand lens and simple (single lens) microscope.

L. C. Martin and B. K. Johnson, *Practical Microscopy* (Blackie, 3rd edition 1958)
Concise account with explanation of the physics of different types of microscope, including the phase contrast, polarizing, and electron microscopes.

J. D. Cassartelli, *Microscopy for Students* (McGraw-Hill, 1969)
A very practical book by an experienced microscopist who, as technical representative of a well-known manufacturer, is familiar with the kinds of questions asked by biologists at all levels. Includes sections on the phase contrast and stereomicroscopes.

HISTOLOGICAL METHODS

R. R. Fowell, *Biology Staining Schedules for First Year Students* (H. K. Lewis, 1964)
A concise summary of various zoological and botanical staining procedures. Very useful.

Biological Stains and Staining Methods (British Drug Houses Ltd.)
Short booklet (50 pages) full of information on stains and histological procedures, including methods suitable for bacteria.

C. F. A. Pantin, *Notes on Microscopical Technique for Zoologists* (Cambridge University Press, 1946)
For anyone who wishes to prepare wax sections, this little book is indispensable. In addition to staining techniques, there are clear instructions on narcotization, fixation, dehydration, impregnation, and embedding.

Ann Preece, *A Manual for Histologic Technicians* (U.S.A. Little, Brown & Co.; U.K. Churchill-Livingstone, 3rd edition 1972)
Emphasizing the mastering of histological skills, this book is particularly aimed at the technical student. A range of staining procedures are summarized in an extensive appendix.

Edward Gurr, *A Practical Manual for Medical and Biological Staining Techniques* (Leonard Hill, 2nd edition 1956)
This book has become a classic since it was first published in 1952. An excellent source of information on all manner of staining techniques.

J. B. Gatenby and H. W. Beams, *The Microtomist's Vade-Mecum* (Churchill, 11th edition 1950)
A mine of information on zoological and botanical staining procedures, covering just about all aspects of the subject. Recommends the best methods to adopt for specific tissues.

Slide Sets

A series of slide sets is now available for use with *Biology, A Functional Approach* and with the *Students' Manual*. A complete list of the slides is given below. The slide sets correspond to the chapters of the textbook and the manual as stated. Each set is accompanied by an explanatory booklet.

Notes and Abbreviations

Electron micrographs are indicated by the abbreviation 'EM'. All other slides are photomicrographs. The following abbreviations have been used throughout:

E	Entire
VS	Vertical section
TS	Transverse section
LS	Longitudinal section
HS	Horizontal section
VLS	Vertical longitudinal section
RLS	Radial longitudinal section
TLS	Tangential longitudinal section
Phase	Phase contrast microscope
Pol	Polarizing microscope
LP	Low power
MP	Medium power
HP	High power
VLP	Very low power
EM	Electron micrograph
SEM	Scanning electron micrograph

Set N1 Cells
For use with chapters 2–6
1 Pavement epithelial cells (cheek), phase, HP
2 Pancreas exocrine cells, HP
3 EM pancreas exocrine cell
4 EM pores in nuclear membrane
5 EM Golgi apparatus
6 EM mitochondrion
7 EM cell membrane
8 EM flagella, TS
9 EM microtubules, TS
10 Plant cells from TS pine leaf
11 EM pit in plant cell wall
12 EM plant cell wall and cytoplasm
13 *Amoeba proteus*, E
14 *Chlamydomonas*, E, phase, HP
15 *Euglena*, E, phase, HP
16 *Paramecium*, E, phase, HP
17 *Opalina*, E, phase, HP
18 *Spirogyra* cell, phase, HP
19 Pigment cells (frog's skin), HP
20 Nerve cell, HP
21 Sperm cells (mouse), phase, HP
22 EM sperm (bat), LS
23 Onion epidermal cells, normal, interference microscope
24 Onion epidermal cells, plasmolysed, interference microscope

Set N2 Tissues and Organization
For use with chapters 2–6
1 Pavement epithelium (mesothelium), phase
2 Cuboidal epithelium (kidney), bright field
3 Cuboidal epithelium (thyroid), phase
4 Simple columnar epithelium (small intestine)
5 Ciliated epithelium (roof of mouth of frog), phase
6 Ciliated pseudostratified epithelium (trachea), phase
7 Glandular epithelium (goblet cells, small intestine), phase
8 Glandular epithelium (goblet cells, large intestine)
9 Stratified epithelium (frog's skin), phase
10 Areolar connective tissue
11 Collagen connective tissue (tendon LS)
12 Adipose tissue
13 Adipose tissue stained for fat
14 Hyaline cartilage
15 Compact bone, TS, HP
16 Compact bone, TS, HP
17 Smooth muscle, LS, HP, phase
18 Photosynthetic tissue (*Elodea* leaf), phase
19 Plant parenchyma tissue, HP
20 Starch storage tissue (potato), HP
21 Collenchyma tissue, TS, HP, phase
22 Sclerenchyma tissue, TS, HP
23 *Hydra*, TS, LP
24 *Hydra* body wall, HP

Set N3 Gaseous Exchange and Nutrition in Animals
For use with chapters 7–12
1 Lung, TS, LP
2 Lung, TS, HP
3 Lung injected, LP
4 Trachea, TS, LP
5 Trachea, TS (mucous glands), HP
6 Dogfish branchial region, TS, LP
7 Dogfish gill, VS, MP
8 Dogfish gill, HS, LP
9 Dogfish gill, HS, HP
10 Trachea of insect
11 Tooth, VS, Pol
12 Stomach, TS, LP
13 Gastric glands in stomach
14 Ileum, LS, LP
15 Ileum, LS, HP, villus
16 Ileum, LS, HP, Crypts of Lieberkuhn
17 Small intestine injected
18 Radula of snail
19 Cockroach mouthparts, LP
20 Mosquito mouthparts, LP
21 Cabbage white butterfly mouthparts, LP
22 Honey bee mouthparts, LP
23 Blowfly proboscis, LP
24 Gill of *Anodonta*, HS

Set N4 Nutrition and Transport
For use with chapters 7–12
1 EM chloroplast
2 Holly leaf, TS midrib
3 Holly leaf, TS lamina
4 Blood film (human), LP
5 Blood film (human), HP
6 Blood film, platelets, HP

7 SEM red blood corpuscles
8 Artery and vein, TS, LP
9 Cardiac muscle, LS, HP
10 EM cardiac muscle
11 Stomata (box leaf), surface, HP
12 Stomata (*Prunus* leaf), TS, HP
13 EM guard cells
14 Buttercup root, TS, LP
15 Root hairs (bean), HP
16 Buttercup root, TS, HP
17 Sunflower stem, TS, LP
18 Maize stem, TS, LP
19 Vascular bundle (Sunflower stem), TS, HP
20 Xylem vessels (*Cucurbita* stem), LS, HP
21 Junction of xylem vessels (*Cucurbita* stem), LS, HP
22 EM bordered pit with torus, spruce
23 Sieve tubes (*Cucurbita* stem)
24 EM sieve tube and companion cell

Set N5 Adjustment and Control
For use with chapters 13–17
1 Pancreas, LP
2 Pancreas, Islet of Langerhans, HP
3 Liver (pig), LP
4 Liver (pig), edge of lobule, HP
5 Liver (pig), centre of lobule, HP
6 Liver, injected
7 Liver, stained for glycogen
8 Kidney, HS, VLP
9 Kidney, injected
10 Kidney, cortex, TS
11 Kidney, glomerulus, HP
12 Kidney, glomerulus, injected, HP
13 Kidney, collecting ducts, LS
14 EM glomerular barrier
15 EM lining cell of Malpighian tubule
16 EM microvilli of Malpighian tubule cell
17 EM basal infoldings of Malpighian tubule cells
18 Marram grass leaf, TS
19 Sunken stoma (*Hakea* leaf), TS
20 Skin, VS (human)
21 Skin, VS (dog), erector-pili muscles
22 Skin, VS, sweat glands
23 White blood corpuscle engulfing bacteria
24 EM phagocyte

Set N6 Response and Co-ordination
For use with chapters 18–22
1 Motor nerve cells, MP
2 Motor nerve cells, HP
3 Nerve trunk LS, HP, Nodes of Ranvier
4 Earthworm, body wall and nerve cord, TS
5 Earthworm, nerve cord showing giant axons, TS
6 Synaptic knobs on surface of nerve cell (cat)
7 EM synaptic knob (fish)
8 Spinal cord (cat), TS, LP
9 Pyramidal cells, HP
10 Thyroid gland, HP
11 Adrenal medulla, HP
12 Pituitary, VLS, LP
13 SEM olfactory epithelium
14 Whole eye, VS, VLP
15 Retina, peripheral, MP
16 Retina, central fovea, MP
17 Retina, rods and cones, HP
18 Compound and simple eye of insect
19 Mammalian cochlea, LS, LP
20 Organ of Corti, HP
21 Striated muscle, LS, HP, bright field
22 Striated muscle, LS, phase, HP
23 EM myofibril, stretched and contracted
24 Knee joint (guinea pig), LS, LP

Set N7 Cell Division and the Life Cycle
For use with chapters 23–25
1–6 Stages in mitosis (peony)
7–14 Stages in meiosis (grasshopper)
15 Giant chromosome (*Drosophila*), phase
16 Yeast budding, HP, phase
17 *Hydra* budding
18 Moss, archegonia, VS
19 Moss, antheridia, VS
20 Moss, sporogonium, LS
21 Fern, VS prothallus showing archegonia, bright field
22 Fern, VS prothallus showing antheridia, phase
23 Fern, sorus, VS, LP
24 Fern prothallus, showing antheridia and archegonia (surface), LP

Set N8 Reproduction of Mammals and Flowering Plants
For use with chapters 24–27
1 Mammalian testis, LP
2 Mammalian testis, HP, tubule
3 Mammalian testis, showing sperm tails, HP
4 Human sperm, HP
5 Mammalian ovary, LP
6–9 Mammalian ovary, stages in development of Graafian follicle, HP
10 Mammalian ovary, corpus luteum, HP
11 Oviduct, TS
12 Uterus (rat), TS, LP
13 Uterus, TS, HP (uterine glands)
14 Umbilical cord, TS, LP
15 Placenta
16 Buttercup flower, LS, LP
17 Lily flower, TS, LP
18 Lily anther, TS
19 Mature pollen grains (lily)
20 Mixed pollen, phase
21 Stigma of primrose, LS, MP
22 Ovule megaspore mother cell (lily), TS
23 Ovule, embryo sac (fritillary), TS
24 Mature embryos (shepherd's purse), LS

Set N9 Animal and Plant Development
For use with chapters 26, 27
1 Frog embryo, blastula, VS
2 Frog embryo, gastrula, VS
3 Frog embryo, neural fold, TS
4 Frog embryo, neural tube, TS
5 Frog tadpole (stage of hatching), VLS
6 Frog tadpole, trunk region, TS
7 Chick embryo, TS (neural tube)
8 Chick embryo, TS (amniotic folds)
9 Chick embryo, TS (amnion)
10 Chick embryo, E (2 day)
11 Chick embryo, E (3 day)
12 Chick embryo, E (4 day showing allantois)
13 Onion root, LS, LP
14 Onion root, LS, zone of cell division, MP
15 Onion root, LS, zone of cell division, HP
16 Onion root, LS, zone of elongation, MP
17 Stem apex of lilac, LS, LP
18 Lime stem (young), TS, LP
19 Lime stem (old), TS, LP
20 Lime stem (old), TS, HP
21 Ash stem, TS
22 Ash stem, RLS
23 Ash stem, TLS
24 Lenticel, TS (*Syringa* stem), HP

Set N10 Associations and Evolution
For use with chapters 32–36
1 EM bacteriophage
2 EM disrupted bacteriophages
3 EM bacteriophages after release
4 *Peronospora* mycelium, HP
5 *Phytophthora*, asexual reproduction
6 Hookworm head
7 Trypanosomes, LP in blood film
8 Trypanosomes, HP
9 EM trypanosome
10 *Taenia*, scolex, E
11 Malarial parasite, pre-erythrocytic schizont
12 Malarial parasite, late trophozoite
13 Malarial parasite, developing schizont
14 Malarial parasite, mature schizont
15 Malarial parasite, oocysts in stomach wall of mosquito
16 Liver fluke, adult, E
17 Liver fluke, miracidia, E
18 Liver fluke, redia, E
19 Liver fluke, cercaria, E
20 Lichen (*Xanthoria*), TS
21 *Amphioxus*, head region, E
22 *Amphioxus*, pharynx region, E
23 *Amphioxus*, pharynx region, TS, LP
24 Larva of sea squirt (ascidian)

Index